BRITISH AIRCRAFT BEFORE THE GREAT WAR

...when set free in the air then will come the real test and the real danger; but who that considers man's pluck, luck and perseverance can doubt that success will ultimately be achieved?

F.W. Breavey, about 1880

The flying machine has come, and come to stay, and whether we like it or not, it is a problem that must be taken into serious consideration. If we are laggards we shall, unquestionably, be left behind, with the strong probability that before many years have passed over our heads, we shall have to change the colouring of our school maps.

Sir Hiram Maxim, 1908

BRITISH AIRCRAFT BEFORE THE GREAT WAR

Michael H. Goodall
and
Albert E. Tagg

Schiffer Military History
Atglen, PA

Book Design by Ian Robertson.

Library of Congress Catalog Number: 2001089019

Printed in China.
ISBN: 0-7643-1207-3

Published by Schiffer Publishing Ltd.
4880 Lower Valley Road
Atglen, PA 19310
Phone: (610) 593-1777
FAX: (610) 593-2002
E-mail: Schifferbk@aol.com.
Visit our web site at: www.schifferbooks.com
Please write for a free catalog.
This book may be purchased from the publisher.
Please include $3.95 postage.
Try your bookstore first.

In Europe, Schiffer books are distributed by:
Bushwood Books
6 Marksbury Avenue
Kew Gardens
Surrey TW9 4JF
England
Phone: 44 (0) 20 8392-8585
FAX: 44 (0) 20 8392-9876
E-mail: Bushwd@aol.com.
Free postage in the UK. Europe: air mail at cost.
Try your bookstore first.

Contents

Foreword

by Sir Peter Masefield

All devotees of the galaxy of historic aircraft which were designed, assembled and flown at Brooklands, Hendon, Eastchurch, Larkhill and numerous smaller sites during the years which led up to the First World War, will join me in welcoming this fine and meticulously researched book by Michael Goodall and the late Bert Tagg.

For the first time the leading particulars of most of these aircraft types have been gathered together, more than 900 of them well illustrated and presented in one volume. It constitutes a most valuable contribution to the detailed information about a remarkable period in aviation history and is a memorial to the bravery and inventiveness of the intrepid pioneers of that faroff era.

Importantly, among them are details of the first thirty Avro aircraft from the original Roe No.1 biplane which made the first and now authenticated, though very brief take off under its own (Antoinette) power on the Finishing Straight of the Brooklands Motor Course in June of 1908.

With them are the thirty-three Sopwith aircraft flown between July 1912 and the end of 1914. Between them the Avro and Sopwith aircraft, together with those of Shorts and Bristol, make up a significant proportion of all the British aircraft which took the air before the First World War. They were joined also at Brooklands by the nine adventurous Martin-Handasyde monoplanes and two Howard Wright monoplanes and one biplane on which Tom Sopwith made his first flight without the benefit of any dual instruction. In fact, on the Howard Wright biplane, Sopwith made a gallant attempt upon the British Empire Michelin Trophy by flying one hundred and fifty miles into Europe on the last day of 1910 before taking it on a successful tour of the United States.

Michael Goodall and the late, lamented Bert Tagg, both with such long and close associations with Brooklands, are to be congratulated on the production of such a valuable addition to the early history of British aircraft. The Wright Brothers had been an inspiration to the British pioneers and it is perhaps appropriate that this excellent volume has been published in the United States. A memorable photograph shows Wilbur and Orville, together with several of the earliest British pioneers outside Mussel Manor, Leysdown, Isle of Sheppey, Kent the clubhouse of the Royal Aero Club. The building still survives and is now a tavern.

It is, in addition to the details of the aircraft, a valuable reminder not only of the beginnings of British aircraft construction but also the fact that Brooklands grew from its small origins, on the World's first banked Motor Course, to become during two World Wars, the chief center of the British aircraft industry. In all for some eighty years Brooklands remained not only a significant reminder of early work but also, with both the Hawker and Vickers companies, a leader into the jet era.

Today, Brooklands lives on as "The Brooklands Museum Trust" where later generations can relish reminders of "The Great Days That Were" with an assembly of some of the most illustrious aircraft as well as motor cars of the historic past, starting in June of 1908.

I most warmly commend this book to all students of aeronautical history.

Introduction

It is hard to believe that less than a lifetime separates the early pioneers, with their flimsy creations of bamboo, piano wire and cotton bedsheeting, from today's airborne marvels machined from duralumin, stainless steel, titanium and carbon fiber. The first primitive machines struggled to get off the ground with just one person aboard – today's mighty airliners carry 500 passengers with ease at speeds that would have been unimaginable to the pioneers. But without those pioneers, with their boundless courage, enthusiasm and inventive genius, today's progress would not have been possible. Not all the pioneers were successful however and there were many deadends, but gradually the accumulated store of knowledge led to quickening progress up to the beginning of World War I.

It must not be forgotten that, after these novel machines were completed, some brave (or should one say foolish) fellow had to be found to coax it into the air. At first the pioneer aircraft constructors were often their own pilots, self taught, often alone in the pearly light of early dawn, shivering on the edge of some usually unsuitable grassland, waiting for an unreliable engine to drag them into the air.

Gradually flying training became more organized with the opening of flying schools at Brooklands, Hendon and Larkhill and other, smaller centers. Up to the outbreak of War on 4 August 1914, a total of 881 British subjects civil and military qualified for British Aviation Certificates from the Royal Aero Club of the United Kingdom. The first went to J.T.C. Moore-Brabazon (later Lord Brabazon of Tara) when, on 8 March 1910, he flew briefly a Short biplane at Shellbeach, Leysdown. The final pre-War Certificate (No.863) was awarded to Sergeant A.F. May, RFC.

Out of these 881 Certificates, 492 (56%) were gained by members of the British fighting forces (RFC and RNAS). The other 389 were awarded to civilian pilots from 11 June 1910. Out of the total, forty-eight British pilots (5 1/2%) were killed in accidents between the above date (the first, the Honorable C.F. (Charlie) Rolls, at Bournemouth in a Short biplane, through structural failure) and the last, before civil flying was abandoned at the start of the First World War when Lieutenant L.C. Hordern, flying a Henry Farman of No.5 Squadron RFC was killed at Gosport through an inadvertent (and familiar) stall and spin, following an engine failure.

This book would not have been possible without the help of my two old friends, now both sadly departed. H.F. "Fitz" Cowley started acquiring photographs of early British aircraft in the late 1920s, when many of the pioneer airmen were still alive and when it was possible to find photographs and brochures of early aircraft in bookshops and antique shops and even from the pioneers themselves. Consequently he was able to amass one of the finest photograph collections in the country. I was fortunate to inherit his collection and add it to my own.

The other major contributor and my co-author was A.E. "Bert" Tagg. Bert joined Hawkers at Kingston in the late 1930s and spent all of his working life with them and their successor companies. He was a skilled aircraft engineer and was able to add the flesh to the bare bones of my raw research material. Tragically he died before knowing that the book would be accepted by our publisher, the first to have been approached. Andrea and John Garner, Bert's daughter and son-in-law, have most kindly given me great assistance in editing the text and making it in an acceptable format for our publishers

Many other friends have given unstinting help and have searched their archives on our behalf and many local newspapers published our appeals asking for information on long lost aviation pioneers. We must particularly thank Jack Bruce, Stuart Leslie, Phillip Jarrett, Oliver Thompson and Dennis Manning for their evergenerous assistance. We are also indebted to Sir Peter Masefield for his helpful suggestions and for writing the foreword.

The reader of the text of this lengthy work will soon realize that some of the aircraft mentioned have no photograph or drawing shown. It is much to the regret of the authors that this is the case, but in spite of many years' search, some machines have proved completely elusive. It is hard to believe that any aeroplane, in an era when everything and everybody was being photographed, would not have been 'snapped' by the proud 'parent'. Perhaps this book will bring some of these photos to light at last. Even more frustrating are the photos that exist, but unlabelled, and have proved unidentifiable.

The small advertisements in the aeronautical press have tantalizing mention of aircraft and engines for sale, which may relate to undiscovered pioneer machines, but as they have neither names nor addresses, it is rarely possible to equate them to a particular aircraft.

A number of designs which were described in patents and/ or built in detailed model form have been included although there is no proof that full-size versions were ever completed. An ever-present problem for the early pioneers was the lack of finance, particularly to cover the cost of the very expensive contemporary engines and consequently many machines were never completed.

The pages of the aeronautical and motoring journals contained many small advertisements from pioneers wishing to buy inexpensive secondhand engines or else to hire them. There were also optimistic advertisements inviting investors to take a financial share in projects. It is not known how many of these appeals were successful.

The authors have done their best to make this a comprehensive account of every flying machine built in Britain prior to World War I, but are well aware that there must be many others which, nearly ninety years on, may never now be recorded.

Aircraft Manufacturers

AERIAL MANUFACTURING COMPANY monoplane with subsidiary stabilizers. (Aerial Manufacturing Co. of Great Britain and Ireland, Upper Charles St., Finsbury. Designer W.J. Potter, Elysia, Woodham Ferrers, Essex.)

The machine illustrated was built in 1909 and was flown by Alec Ogilvie at Camber Sands, near Rye in Sussex, at the end of the year for distances up to 250 yards.

The special feature was the undercarriage that folded forward in stages and was designed to provide extra lift at takeoff and to provide a braking effect on landing. This was achieved by mounting two aerofoils on the undercarriage, which presented a change of incidence as the undercarriage was raised or lowered. A fixed aerofoil as wide as the fuselage was also fitted forward of the airscrew together with additional small aerofoils on top of the rudder and below the rear fuselage. Another feature was the fuel tank of aerofoil section.

Power: 50hp believed to be a four-cylinder vee NEC two-stroke as used by Ogilvie in his Short-Wright biplane, driving a four-bladed propeller.

Data	Span	44ft
	Length	44ft

In 1910 they were reported to be building a Farman type biplane for Edward Mines.

In January 1910 they were advertising aeroplanes with very imposing names *'The British King'*, *'The British Queen'*, *'The Flying Scotsman'* and *'The Tradescent'*. None of them appear to have been sold and nothing further was heard of them.

Potter registered the following patents in 1910: 19198 (monoplane), 20108 (biplane), 21189 and 22475.

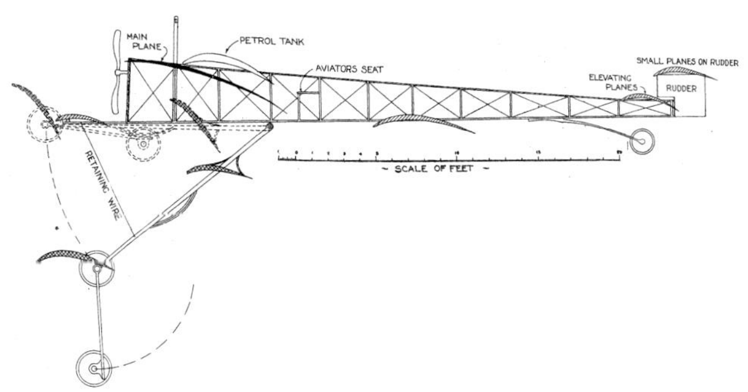

Aerial Manufacturing Company monoplane with folding undercarriage and subsidiary stabilizers covered by Patent No.21 189/1909.

AERIAL WHEEL monoplane. (Aerial Wheel Syndicate Ltd., Ralph Platts and George Sturgess, Mablethorpe, Lincolnshire.)

This most unorthodox monoplane arrived incomplete at Larkhill for the Military Trials in September 1912 but although entered as No.18 it took no part in the trials. The machine, which was

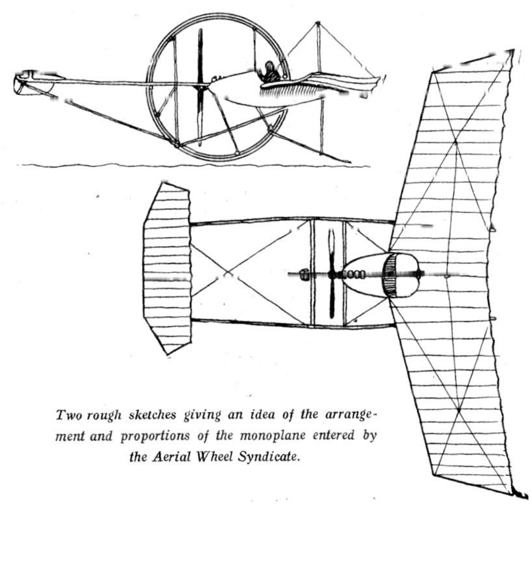

Two rough sketches giving an idea of the arrangement and proportions of the monoplane entered by the Aerial Wheel Syndicate.

Aerial Wheel monoplane. The design of this 1912 monoplane concentrated on the rough field performance for the Military Trials, but its flying ability was never tested.

built in Birmingham, was a tractor canard with swept wings and was powered by a 50hp NEC four-cylinder, water-cooled, two-stroke engine. A nacelle between the booms that supported the front elevator housed both engine and crew and was surrounded by a circular frame incorporating a revolving tread which, with skids under the wings, constituted the landing gear. Patent No.26 924/1908 was an early version by the Sturgess brothers.

Pilots were reluctant to test the Aerial Wheel and it is believed that the machine was abandoned unflown. It was still in existence in a hangar at the Midland Flying School at Billesley Common, King's Heath, Birmingham when it was wrecked by a gale in the autumn of 1915.

AEROPLANE BUILDING and FLYING SOCIETY gliders. (Sec. W. le Maitre, Workshop Kings Rd., Hammersmith. Shed at Kensal Rise Athletic Ground.)

J.D. North, later chief designer at the Grahame-White Company and Boulton and Paul Aircraft designed a biplane glider with front elevator and tail. This was built in 1910-1911 and donated to the society. Because of the lack of gradient at the ground, a launching apparatus consisting of a trolley on two ropes, giving a gradient of 1 in 6 was prepared. This seems to have been used successfully.

Later it was fitted with a 15hp JAP engine lent by one of the members. It had two propellers, one on each side of the tail. There is no evidence that the machine flew in this form.

A monoplane glider, with a span of 30 ft, designed by AN Myers and donated to the Society was in the shed at the ground in October 1910 and awaiting the fitting of an engine.

Aeroplane Building & Flying Society. One of many gliders built by enthusiasts, which led nowhere.

Data	Biplane glider	
	Span	27ft
	Length	30ft
	Chord	5ft
	Weight	130lb

AIRCO. (The Aircraft Manufacturing Co. formerly the Aircraft Co.; Head Office, St. Stephens House, Westminster; Aerodrome, Hendon; Works at Merton, Surrey; later at the Hyde and Edgware Rd., Hendon)

The early Farman biplane was copied by a number of firms, without an official license, the Bristol Boxkite being a notable example. The Airco firm was formed by George Holt Thomas in 1912 to produce Farman aircraft under license from Henry and Maurice Farman.

The first British built aircraft was a Maurice Farman Series 7, known as the Longhorn, which was flown at Hendon in November 1912 by Maurice Farman and Paul Verrier.

Thereafter various Farman types were built. These were used for operations by the RFC, before and during the early months of the war, but the Longhorn and Shorthorn were later used mainly for training purposes. Airco and other contractors built large quantities in Britain.

Flight operations were established at Hendon aerodrome and a works at Merton, which was later transferred to an associated company engaged on airship work, and this was replaced by larger premises at the Hyde, Hendon.

AIRCO: MAURICE FARMAN S.7 Longhorn

The MF S.7 Longhorn was a two-seater for general purposes and training, its main role being training although some served with the RFC in France and the Middle East well into 1915. Airco delivered the first aircraft of their own manufacture No.415 in September 1912; others were bought from France and in wartime from Brush Electrical Engineering Co. and other contractors.

Airco Maurice Farman Longhorn. This complicated design served as a trainer as late as 1918.

The Longhorn, with its front elevator and complex supporting structure, was a low performance, but safe aircraft and served as a trainer throughout the war.

Power: 80hp Renault eight-cylinder air-cooled vee.

Data	Span top	50ft 11in (15.52m)
	Span bottom	37ft 9 1/2 in (11.52m)
	Area	528sq ft (49sq m)
	Length	37ft 9 1/2 in (11.52m)
	Weight	1,320lb (599kg)
	Weight allup	1,925lb (871kg)
	Speed	65 mph
	Endurance	5hr
	Climb	15min to 3,280ft (1,000m)
	Climb	35min to 6,560ft (2,000m)

Airco. The MF 7bis seaplane version of the Longhorn had no front booms or foreplane.

Airco Maurice Farman Shorthorn. Another pre-war Maurice Farman type that trained pilots throughout the war.

AIRCO: MAURICE FARMAN S.11 Shorthorn

The MF S.11 Shorthorn was derived from the Longhorn by the removal of the front elevator and its structure. The biplane tail was discarded in favor of one of monoplane type, with twin fins and rudders of triangular shape, supported by booms tapering in side elevation.

The Shorthorn appeared late in 1913 and was introduced into the RFC in the following March. It is not certain when Airco produced their first aircraft, as some were also purchased from the Farman Company. The type was used in early operations but continued in use to the end of the war, mainly in the training role.

Production of the type continued at Airco into 1918, with the Whitehead Company also a contractor.

Power: 80hp Renault eight-cylinder air-cooled vee.

Data		
	Span top	51ft 9in (15.78m)
	Span bottom	38ft 8in. (11.76m)
	Area	561sq ft (52sq m)
	Length	30ft 6in (9.3m)
	Weight	1,441 lb (654kg)
	Weight allup	2,046 lb (928kg)
	Speed	72 mph (116kph)
	Endurance	3 3/4hr
	Climb	15min to 3,280ft (1,000m)

Airco Henry Farman F20. Served on operations into the early war years but by 1915 was used as a trainer. Also flown as a seaplane.

AIRCO: HENRY FARMAN F.20

The type was originally a 2/3-seater biplane but a large fuel tank usually replaced the third seat. Trials with various types of armament were carried out and the type was also flown as a seaplane. A typical pusher biplane of neat appearance, the top wing had long extensions and was fitted with ailerons. The tail unit comprised a single rudder with tailplane and elevators mounted on the top booms.

Power: 80hp Gnôme or Le Rhone seven-cylinder air-cooled rotary.

Data		
	Span top	45ft (13.5m) or 43ft 6in (13.25m)
	Span bottom	24ft 7in (7.5m)
	Chord top	6ft (1.9m)
	Chord bottom	4ft 10in (1.45m)
	Area	377 sq ft (35 sq m)
	Length	27ft 10in or 26ft 6in (8.06m)
	Height	10ft (3.15m)
	Weight	794 lb (360kg)
	Weight all up†	900 lb or 1,455 lb (660kg)
	Speed	62mph or 65mph
	Endurance	3hr
	Climb	8 min to 1,640ft (500m)

† (No passengers pilot and fuel only)

ALDERSHOT Aero Club glider

A Mr. Richardson, a club member, built this glider, of Farman type, in 1911.

ALDERSON monoplane (C.H.M.A. Alderson, Farnborough Aviation Works, Farnborough, Orpington, Kent)

Few details of this machine, reported built in 1910, can be traced. Alderson, who formed his company in 1909, had been actively interested in aviation since he experienced a balloon flight in France in 1880. In 1901 he assisted Thomas Moy with experiments with a model ornithopter and in 1905-1909, as a friend of Frederick J. Stringfellow, he assisted with the reclamation of the Stringfellow artifacts for donation to the Science Museum. A member of the Aeronautical Society and Aeronautical Institute and Club, he contributed to the building of the Candler Stella (*q.v.*) in 1909 at his works. In that year he appealed for funds to complete *his* aircraft, which may have referred to the Candler Stella, which was never completed.

However, in March 1910 *The Times* reported that Alderson had built a monoplane of 38ft wingspan with a 50hp four-cylinder engine. The weight without engine was 250 lb.

In August 1910 Alderson visited Cumberland, where he had been born in 1854 and while there flew a Cody kite. At this time he was reported to be building a biplane to be named *The Brae*, of which nothing further was heard. Alderson died in 1929.

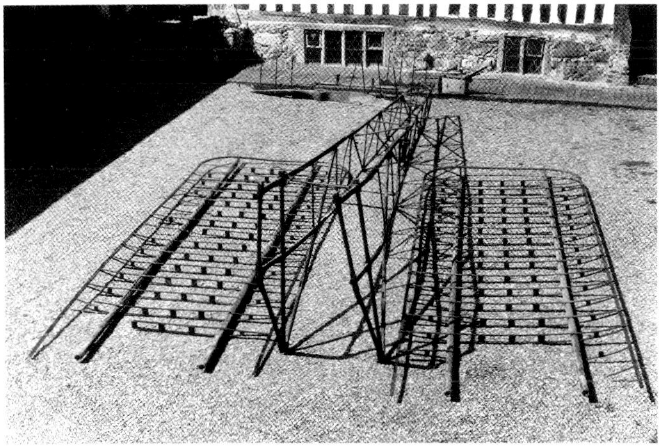

The remains of the Aldritt monoplane at Filching Manor. The bamboo main spars are noteworthy. (Courtesy Paul Foulkes Halbard).

'Bobby' Allen ornithopter.

ALDRITT monoplane
(Aldritt's Garage, Portlaiose, Ireland)

This Blériot type monoplane was built by the Aldritts in 1912, but does not seem to have been very successful. The three-cylinder engine was also designed and built by the Aldritts, but was rather heavy. The main wing spars were made of large diameter bamboo poles, possibly from a local market garden.

The monoplane took off and flew several hundred yards, but the excessive weight caused it to make a very heavy landing. The engine was unfortunately sent to a scrap yard by mistake but this very early monoplane still exists, minus engine, as an exhibit at Filching Manor Motor Museum, near Polegate, Sussex.

Data	Span	c.40ft
	Length	c.30ft

ALEXANDER monoplane (Henry and Alfred Alexander & Co., Edinburgh)

This machine was of Demoiselle type and was exhibited at the Scottish Motor Show at Waverley Market in January 1910. The engine, shown separately, was described as having air-cooled cylinder barrels with water-cooled heads, and overhead valves and was of the Otto cycle type, i.e. a four-stroke.

The fuselage of the aircraft was made of steel tubing, and the wings were braced to it by struts of the same material, thus eliminating the use of wire bracing. The wings were double surfaced with wooden spars and ribs, and could be warped for lateral control. The tail unit was hinged on a universal joint.

Power: 20hp Alexander four-cylinder inline air and water-cooled.

Data	Span	19ft
	Wing area	136sq ft
	Weight	140lb

ALLEN Flying Bicycle ornithopter (Richard 'Bobby' Allen, Bradford, Yorkshire)

Designed by a policeman, the 'Bobby' Allen Flying Bicycle was exhibited at Stoney Bridge and Peel Park Gala in June 1908. It was built as a bamboo structure on a bicycle and had a wingspan of 33ft.

The machine had three pairs of narrow chord, beating wings driven by a belt passing over the rims of the cycle wheels. Further pulleys at the top of the structure, with cranks converted the drive into reciprocating motion, which actuated the wings. A tall rectangular fin and rudder were mounted at the front, the latter controlled by the cycle handlebars.

The machine refused to fly or even to move along the ground.

ALVAREZ monoplane (C.G. Spencer and Co. of Highbury, tested at the Welsh Harp at Hendon)

This machine was taken aloft by a 23,000 cubic ft balloon. A larger machine was reported to be under construction in December 1906.

Power: 2hp

Data	Span	40ft
	Wing area	400 sq ft
	Weight	150 lb

Alvarez Monoplane. Made by the firm of C.G. Spencer in 1904 and dropped from a balloon.

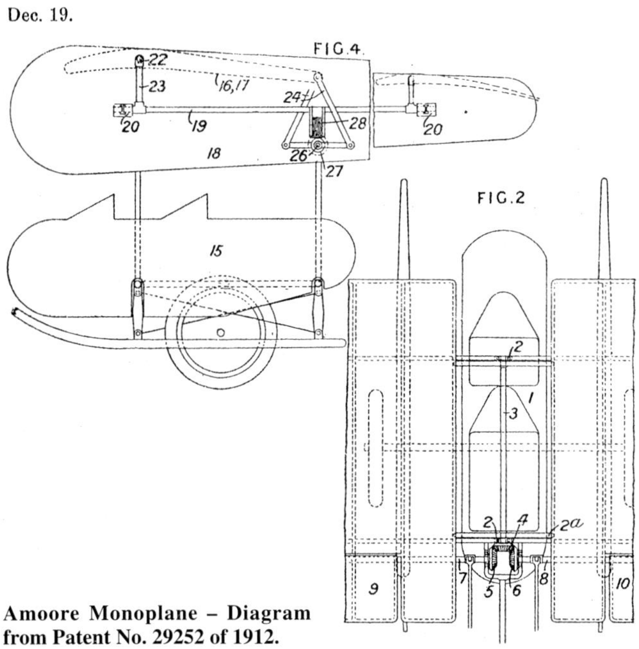

Amoore Monoplane – Diagram from Patent No. 29252 of 1912.

AMOORE monoplane (Frank Amoore, Dover, Kent and 4 City Rd., London EC)

This machine was under construction late in 1913 and early 1914, but appears not to have been completed. It incorporated a lateral pendulum arrangement, to maintain the occupant in a perpendicular attitude, by controlling the seat, which swung below the mainplane. This movement, relative to the airframe, automatically operated the ailerons to correct the attitude of the aircraft. The construction was mainly of light steel tubing. British patent No.29252 of 1912 refers.

The system was similar to that patented in France and Britain by Mons J.A. Moreau No.9445/1911, except that the Moreau Aérostable system operated in the longitudinal sense.

Power: 80hp Dutheil-Chalmers either four- or six-cylinder water-cooled.

Data

	Span	36ft
	Length	24ft 3in
	Weight	1,000 lb

ANDERSON and SINGER glider

This Chanute type biplane glider was built at Aberdeen in 1911 and was flown successfully. Glides up to 30 yards were reported after September 1911.

Data

	Span	24ft
	Wing area	288 sq ft

ARMSTRONG WHITWORTH FK.1 (Sir W.G. Armstrong Whitworth Co., Gosforth, Newcastle on Tyne)

This two-seater tractor scout biplane, designed by Dutchman Frederick Koolhoven, was the first type built by the Aircraft Dept of this large company. It was a typical biplane design of the period and flew for the first time in September 1914.

The fuselage was a wooden girder which tapered to a horizontal knife-edge. The single bay wings were not staggered and the top wings were connected at the aircraft centerline, where two inverted vee struts supported them. Parallel chord ailerons were replaced with tapered surfaces with a large rear overhang. No tailplane was originally provided but one was added during the early trials.

The machine was under-powered, having been planned for an 80hp engine, which was not available at the time. The performance was not as good as contemporary types and no production resulted.

Power: 50hp Gnôme seven-cylinder air-cooled rotary.

Data Speed range 30-75mph

Armstrong Whitworth FK1. Koolhoven's first design for this large armament company.

Anderson & Singer glider. A 1911 amateur-built Chanute-type glider. Flight 21 October 1911 (p.923).

Armstrong Whitworth FK1 with modified ailerons.

ARMSTRONG monoplane (Gordon Armstrong, Beverley, near Hull, Yorkshire)

No doubt inspired by Blériot's Channel flight, the monoplane built by Gordon Armstrong, resembled a Blériot XI in general layout and its use of the 35hp Anzani fan type radial. The levered type suspension and the tip elevators were similar features but the rest of the aircraft appeared to be of original conception.

Armstrong monoplane No.1 as originally built with tailwheel. The pilot is Gordon Armstrong. (Courtesy M.G. Simmons).

The fuselage was a triangular section open girder structure with a single bottom longeron, the three longerons curving to a point at the extreme rear. There was no fin and the rectangular shaped rudder was hinged on a vertical post, braced by a diagonal strut. The tail was supported by a flexible strip, clamped to the bottom longeron.

The monoplane was housed at the East Riding Garage and tested at Beverley Westwood on 26 August 1910 and 2 September 1910, damage occurring on both occasions. The machine complete with engine was offered for sale in February 1912.

From small beginnings at Beverley, Armstrong shock absorbers were developed for cars, together with other engineering products, forming the basis of a large engineering company with several factories in Britain and overseas.

ARUNDEL HOUSE SCHOOL CLUB glider. (Sec. R.P. Grimmer, Senior Pupil, R.F. Mann)

A biplane of Chanute type was built by Cyril Ridley (*q.v.*), pupil at the school in 1910. The two leading lights in the club were later partners building, in wartime, the Mann and Grimmer M1 biplane. A second glider was reported nearly complete in May 1910, built by pupil Owen Wilson-Jones of Surbiton.

ASL. The Aeronautical Syndicate Ltd. (Horatio C. Barber, Hendon Aerodrome, Sheds 5, 6 and 7. Testing at Larkhill discontinued in September 1910. Head Office at 30 Moorgate St., London EC.1)

ASL monoplane No.2

This second machine for Horatio Barber was built by Howard Wright at Battersea in January and February 1910 and was tested at Larkhill, when it flew in the hands of B. Woodrow on the 6 March. (For first machine see Barber Monoplane)

The ASL No.2 was a pusher monoplane of wooden construction with divided four wheel undercarriage and double nose wheels, together with wheels at the wing tips. The tapered wings had considerable camber and control by warping was aided by hinging the rear spar at the fuselage joint.

The Armstrong monoplane as modified after crashing on its first flight. The tailwheel has been replaced by a Blériot-type bamboo tail-skid. (Courtesy M.G. Simmons).

ASL monoplane No.2. This 1910 monoplane built by Howard Wright for Horatio Barber was not a success.

The monoplane flew for short distances on several occasions and sustained damage but flew successfully again after repair. The ASL No.2 was succeeded by the Valkyrie also of pusher type.

Power: 60hp Green four-cylinder inline water-cooled with 8ft 2in diameter propeller.

Data		
	Mainplane Span	42ft
	Foreplane Span	12ft
	Chord	10ft tapering to 6ft
	Foreplane Chord	3ft
	Area	310sq ft
	Foreplane Area	36sq ft inc. elevators
	Length	31ft
	Cruising speed	35mph
	Weight	802lb

ASL VALKYRIE monoplanes, Types A, B and C

The first Valkyrie Type A was flown during September 1910 at Larkhill on Salisbury Plain, but was moved soon after to Hendon, where new works had been established in three sheds. It is uncertain whether No.1 was built at Larkhill or by Howard Wright, but all later machines were certainly made there. *Flight* of 24 September 1910 (p.774) illustrated No.1 in the air and stated that it was the fifth of a series of experimental models tested on Salisbury Plain over a period of seventeen months. This suggestion of two additional types to those identified may refer to major variations of the Barber and ASL2 monoplanes.

Production of Valkyrie aircraft continued until it was announced in April 1912 that the company was to be closed down, by which time it was reported that nearly 30 machines had been built. Valkyrie No.11 was the highest number referred to in reports of the time. Most machines were used by ASL for pilot training, some experimental mail flights and the carriage of light freight. On 25 May 1911 pupil Benson was killed in the crash of Type A No.7.

Barber had been discouraged when refused permission to demonstrate any of his five available aircraft, when the Parliamentary Defense Committee visited Hendon on 12 May 1911. Despite this, he presented four machines to the Government in July, three of which had Green engines and one with a Gnôme. Two of these were supplied to the Army and two to the Navy at Eastchurch, later moved to Shorncliffe, but it seems that they were never flown.

ASL *Valkyrie* prototype.

Various versions of these pusher canard monoplanes were built and flown at Hendon, the original being a single-seater, but other types able to carry one or two passengers, including some referred to as "military" types followed. All were of similar basic type and construction with wooden open fuselages with fabric covered wings and flying surfaces, all extensively wire braced.

The pilot was seated in an exposed position on the level of the top longerons, just in front of the engine; on the three-seater the passengers were positioned on either side of the engine. The radiator on the Green-engined machines was positioned horizontally below the wing center section. A gravity fuel tank above the engine was supplied from a tank by a plunger pump added to the magneto drive of the Green. On the Gnôme engined Type B the combined fuel and oil tank was horseshoe shaped and carried on the front of the engine mounting.

No.1 flew originally with rudders close to the structure and no "blinker" fins at the front. Extensions to mount the rudders further aft and the "blinkers" were added to improve control and became standard for all subsequent machines.

ASL *Valkyrie* type A. Single-seater practice machine of 1910.

Valkyrie Type A monoplane

This was a single-seater for solo practice flying. First flight was on 13 September 1910.

Power: 35hp Green four-cylinder inline driving a 7ft 3in diameter propeller.

Data	Mainplane span	34ft Single surfaced (*The Aero* 31ft 6in)
	Foreplane span	14ft Single surfaced
	Mainplane chord	6ft
	Foreplane chord	3ft
	Mainplane area	190 sq ft
	Foreplane area	42 sq ft
	Elevators area	20 sq ft
	Length	22ft increased to 28ft
	Width between skids	8ft
	Weight	520 lb

ASL *Valkyrie* type B. A passenger could be carried. Used for training and racing.

Valkyrie Type B monoplane

A single-seater for racing or cross-country work. A passenger could be accommodated behind the pilot. First flight was on 12 November 1910.

Power: 50hp Gnôme seven-cylinder air-cooled rotary driving a 8ft diameter propeller.

Data	Mainplane span	31ft Single surfaced
	Foreplane span	12ft Single surfaced
	Mainplane chord	6ft
	Foreplane chord	3ft
	Mainplane area	168 sq ft
	Foreplane area	42 sq ft
	Elevators area	20 sq ft
	Length	26ft
	Width between skids	8ft 4in
	Weight	550 lb
	Max speed	70mph
	Price	£920

Valkyrie Type C monoplane

Three-seater with passengers seated either side of the engine. First flight was on 13 September 1913.

Power: 60hp Green four-cylinder inline water-cooled.

Data	Mainplane span	39ft double-surfaced
	Foreplane span	14ft
	Length	29 ft 9in
	Mainplane chord	7ft 6in
	Foreplane chord	3ft
	Mainplane area	302 sq ft
	Foreplane area	42 sq ft
	Width between skids	9 ft
	Weight	800 lb
	Max speed	70mph
	Price	£1000

ASL VIKING 1 biplane

This was the last design by Horatio Barber and was constructed at Hendon towards the end of 1911. It flew in January 1912 shortly before the closure of the company in April. The pilot and passenger were seated side-by-side behind the rotary engine, which drove the two tractor propellers through chains. Fuel was supplied from two external tanks alongside the cockpit and one in the rear fuselage. The rubber-sprung main undercarriage was supplemented by sprung nose, tail and wing tip skids. The ailerons were mounted on three rear interplane struts. The rudder was above, and the fin below the fuselage.

ASL *Viking*. Horatio Barber's last design before the company closed in 1912.

ASL *Valkyrie* type C. The most powerful of the Valkyries could lift three people.

ASL *Viking* after conversion to a single propeller seaplane by Chanter.

Astley's second machine of 1911 was probably only a reconstruction of No.1.

The first flight was at Hendon on 18 January 1912. The Viking was sold to the Chanter Aviation School at Shoreham, where it was converted into a seaplane. At the same time the twin propeller system was discarded and replaced by a single conventional tractor propeller.

Power: 50hp Gnôme driving two 8ft 6in diameter propellers through chains.

Data		
	Mainplane span	31 ft
	Tailplane span	9 ft
	Elevator span	9 ft
	Mainplane chord	4 ft 6 in
	Tailplane chord	2 ft 9 in
	Elevator chord	2 ft
	Mainplane area	310 sq ft
	Foreplane area	24 sq ft
	Elevator area	18 sq ft
	Length	29 ft 6 in
	Cruising speed	55 mph
	Endurance	6 hrs
	Weight	800 ib

ASTLEY monoplanes Nos.1 and 2 (H.J.D. Astley, Chequers Court, Ellesborough, near, Tring, Buckinghamshire)

Astley's first machine was at Brooklands in December 1909, where it was photographed with a damaged undercarriage, an occurrence which took place during trials.

The aircraft had an enclosed rectangular section fuselage mounted on a rather weak-looking Blériot-style undercarriage. The wing spars passed through the fuselage just below the top longerons, the pilot being seated behind the rear spar. The wing was braced by a tall cabane and to the undercarriage and carried small triangular ailerons, hinged at the trailing edge. The fixed fin and tail surfaces were triangular and extended well forward along the fuselage. The elevator was rectangular and in one piece, with small triangular rudders above and below. Separate wheels were provided for control of the moving surfaces, that for the elevator being mounted outside the fuselage on the top longeron.

Astley built a second monoplane in 1910, using the remains of the first, including the engine, in its construction. The undercarriage and front portion of the fuselage were much as before, but the pilot's seat was repositioned on a structure below the bottom longerons. The full length top longerons of No.1 were shortened and joined to the bottom longerons, midway along the fuselage, where a small tail wheel was fitted on a vertical post. The lower longerons were presumably increased in strength to deal with the loads from the tail, which was of triangular cruciform shape. No fabric was applied to the fuselage. The ailerons were inset into the trailing edges of the wings, which otherwise resembled those of No.1.

Astley monoplane No.1.

No.2 was probably only capable of taxiing and Astley soon discarded these machines and learnt to fly on a Sommer biplane of the Universal Aviation Co. operating from Shed No.17 at Brooklands, obtaining RAeC certificate No.48 on 24 January 1911. He subsequently carried out many exhibition flights and flew competitively in the UAC *Birdling* and Blériot aircraft, including a cross-Channel flight to France and into Germany. He was regarded as a pilot of considerable ability, but was killed in a Blériot, while performing aerobatics at low level at Belfast on 21 September 1912.

Power: 30-40hp NEC four-cylinder two-stroke inline water-cooled

Data Span 24ft
Length 24ft

AVRO (ROE) aircraft (Alliott Verdon Roe, later A.V. Roe and Co.)

The early machines built by A.V. Roe were initially identified as Roe types, later the name AVRO was adopted. Manufacture of his first machine was carried out at the home of his brother, Dr. H.V. Roe at 47 West Hill, Putney in 1907, with trials at Brooklands from December until 17 July 1908. The Roe 1 triplanes were also partly made at Putney, but erection and flight trials took place at the Lea Valley, Walthamstow, or at Blackpool.

A.V. Roe and Co. was formed in January 1910 and, from that date, manufacture was carried out at Brownsfield Mills, Great Ancoats St. Manchester with flying at Brooklands. The company became a limited company in January 1913, and moved the factory to Clifton St. Miles Platting, Manchester, in April 1913. The training operations, carried out at Brooklands, had been moved to Shoreham in the previous October, and although some use was made of Brooklands, this ceased by the end of the year. Flying continued from Shoreham and elsewhere until Hamble aerodrome became available much later in wartime.

AVRO aircraft have been extensively researched and are comprehensively dealt with in many publications, to which reference should be made for more complete descriptions and data.

Roe I Avroplane at Brooklands fitted with 9hp JAP engine.

Roe 1 Avroplane. AV Roe may have flown on this, his first machine, after fitting an Antoinette engine in 1908, but not officially.

The company became a major part of the British aircraft industry and elements of it continue in being to the present day.

ROE I biplane (Avroplane)

The machine was a pusher type canard biplane, mounted on an undercarriage consisting of two pairs of main wheels, spaced well apart, and with small wheels at the wing tips. There was covering on the lower surfaces of the wings and flying controls only, leaving the ribs exposed above.

There was insufficient power to fly with the JAP engine originally used, but towing by car along the Finishing Straight at Brooklands achieved short glides. A later claim to have flown on 8 June 1908 with the Antoinette engine was not accepted officially, through lack of evidence, although hops of a few feet in height were probably accomplished. The Roe I was dismantled in July 1908.

A modern replica of the biplane, and its shed, now exists at the Brooklands Museum.

Power: 6hp JAP vee twin air-cooled driving a four-bladed metal propeller. Replaced by a 24hp Antoinette eight-cylinder vee water-cooled driving a two-bladed metal propeller

Data Span top 36ft
Additional wing surfaces added to the center section on the fitting of the Antoinette, converting the ma-chine into a semi-triplane.
Length 23ft
Weight 650 lb loaded

ROE I Triplane (Bulls-Eye Avroplane)

Two of these triplanes were built; No.1 was made at Putney and erected and tested at Lea Valley, Walthamstow. Flight trials of No.2 were carried out at Wembley Park and Brooklands.

A.V. Roe had two triplanes of different types, under construction in the railway arch workshops in the Lea Valley in 1909. The larger one, for car dealer Friswell, was not completed and was auctioned incomplete. The smaller machine was assembled and flown on 23 July 1909, becoming the first all-

Roe I Triplane. The first all-British machine to fly in 1909 at Lea Marshes.

British machine to fly. It was then taken to Blackpool with second machine of similar type, which was not able to fly at the Meeting. The first triplane was discarded after the Blackpool Meeting and has been on display at the Science Museum since 1925.

The second triplane, with a tail skid instead of a wheel, was taken to the Old Deer Park, Richmond, but since facilities were unsuitable, was moved in November 1909 to Wembley Park, where it flew on 6 December 1909. Brooklands aerodrome was properly established in 1910, and Roe moved the triplane there, housing it in Shed No.14. It flew in various forms until dismantled in May 1910.

Roe I Triplane. This is the second variant at Wembley at the end of 1909.

Variations were made in the course of development which included a tapered fuselage to the second aircraft, repositioned fuel tank and a later type undercarriage on this machine known as the 'Two-and-a-Bit Plane'.

On both machines the tailplanes were fixed lifting surfaces and pitch control was by varying the incidence of the mainplanes, which could also be warped to provide roll control. Additional fin area between the outboard struts of the triplane tail was originally fitted to No.1. This system was inadequate for fully controlled flight.

Power: 6hp JAP vee twin air-cooled with four-bladed metal propeller; also a 9-10hp JAP vee twin was fitted to the first machine at a later stage. A 14-20hp JAP vee four-cylinder air-cooled with four-bladed metal propeller was fitted to the second machine.

Data		
	Mainplane span	20 ft
	Mainplane chord	3ft 6in
	Mainplane area	210 sq ft
	Tailplane span	10ft
	Tailplane chord	3ft 6in
	Tailplane area	105 sq ft
	Length	20 ft
	Weight loaded	450 lb

Two upper planes later extended and lower planes reduced on the second machine only, then nicknamed 'Two-and-a-Bit Plane'.

ROE II triplanes (No.1 *Mercury*)

Two of these machines were built at Brownsfield Mills and flown at Brooklands. They were built of better materials and had improved design features. Most of the fuselage was covered with light wooden sheeting, the rest with fabric. The wings and tail unit were now inter connected, to enable the incidence of both to be controlled.

A triangulated undercarriage, as fitted to the 'Two-and-a-Bit Plane', was used and although wing warping was used ini-

Roe 1 Triplane The version known as the 'Two-and-a-bit-plane'.

Roe II Triplane. Named *'Mercury'* two of these machines flew in 1910.

tially, this was superseded by ailerons fitted to the center wing from 27 April 1910, and a rudder the full height of the tail unit. A water-cooled Green engine with small vertical radiator behind, provided the power. A second cockpit in front of the pilot could accommodate a passenger.

No.1 was exhibited at the Manchester Aero Club Exhibition at White City, Manchester, on 4 March 1910, when it was named *Mercury*; then it appeared at the Olympia Aero Show in London between 11-19 March 1910. The first flight trials followed at Brooklands, with several minor incidents, before the machine was prepared for dispatch by rail to Blackpool for the flying meeting. Roe II No.1, with the successor Roe III, were accidentally burnt on the train on 27 July 1910.

No.2 was built for W.G. Windham, being ordered by him at Olympia. It flew for the first time on 26 May 1910. and remained at Brooklands, being used intermittently by the owner.

Power: 35hp Green four-cylinder inline water-cooled driving a 8ft diameter wooden propeller.

Data		
	Mainplane span	26 ft
	Mainplane chord	3ft 6in
	Mainplane area	280 sq ft
	Tailplane span	8ft 4in
	Tailplane chord	3ft
	Tailplane area.	75 sq ft
	Length	23 ft
	Weight allup	550 lb

ROE III triplanes

This was a larger machine capable of carrying a passenger in front of the pilot, facing either forward or aft. A major change from Roe II was to fix the wings at the fuselage and rely on the tail unit for control in pitch. Ailerons were fitted to the top wing on the first machine and on the center wing on the three subsequent machines. The substantial undercarriage was of Farman type with four wheels, although it is believed that on the fourth machine, only two were used. A JAP engine was fitted to the first machine and Green engines to the remainder.

Roe III Triplane. This machine was destroyed by fire on the train on the way to the Blackpool Meeting. (Green engine).

Roe III Triplane. Four of these were built, including one exported to the United States. (JAP engine).

This was the first Avro machine capable of fully controlled flight. 'AV' had flown "straights" on it for the first time on 4 July 1910. By 9 July 1910 he was making turns and figures of eight, being airborne for up to twenty-five minutes. He took his tests for the RAeC certificate on the 20 July 1910 and was issued with No.18 a week later.

Power: 35hp JAP eight-cylinder air-cooled vee.
35hp Green four-cylinder inline water-cooled.

Data		
	Top and center span	31ft
	Bottom span	20ft
	Mainplane chord	3ft 6in
	Mainplane area	287 sq ft
	Length	23ft
	Tailplane span	8ft 4in
	Tailplane chord	3ft
	Tailplane area	75 sq ft
	Weight allup	750 lb

No.1 35hp JAP first flown at Brooklands 24 June 1910. For sale May 1911.

No.2 35hp Green first flown at Brooklands 9 July 1910. Burnt on train on way to the Blackpool Flying Meeting 27 July 1910.

No.3 35hp Green Replacement built urgently after loss of No.2. Assembled at Blackpool. First flown at Blackpool 1 August 1910. Destroyed at Boston on 8 August 1910

No.4 35hp Green supplied to the Harvard Aeronautical Society less engine. First flown at Boston, Massachusetts in the United States using engine from No.3. Crashed but rebuilt.

ROE IV triplane

This single-seater, developed from the earlier machines, was the last of A.V. Roe's triplanes. It incorporated a single tailplane with elevators but reverted to warping wings with the bottom wing of narrow chord. The single machine built was retained at Brooklands for the Avro Flying School. It was crashed on several occasions, including a crash on 14 February 1911, when the opportunity was taken, during the subsequent repair, to in-

Roe IV Triplane. The last of AV Roe's triplanes first flew in September 1910.

Avro Farman-type biplane built for the Edwards brothers.

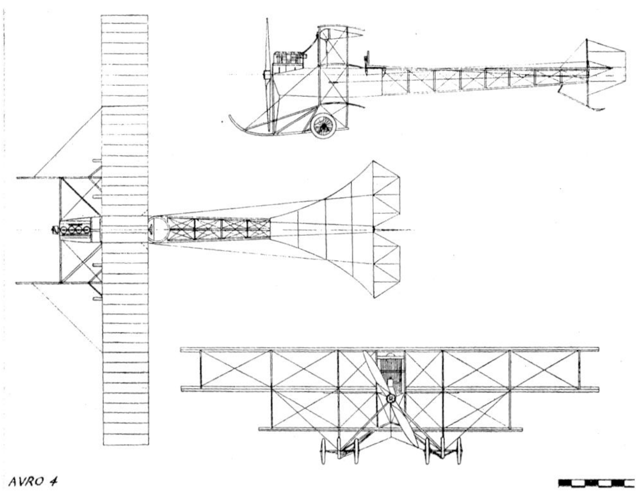

sert an extra 4ft length in the fuselage. The Roe IV first flew in September 1910 and was operated until August 1911.

Power: 35hp Green four-cylinder inline water-cooled.

Data		
	Top and center span	32 ft
	Top and center chord	3ft 6in
	Bottom span	20 ft
	Bottom chord	3 ft
	Area	284 sq ft
	Weight allup	650 lb
	Length	30 ft
		(Later extended to 34 ft)

AVRO Farman type biplane

In 1910 a machine was constructed for M.F. Edwards of Bolton with a 20hp two-cylinder two-stroke engine, which although identified as an Avro engine, was designed by M.F. and W. Edwards and made by H.W. Cowley and Co. of Daubhill, Bolton. No details of the machine are available, but it was owned by a Manchester group of enthusiasts, including four Edwards brothers, until advertised for sale at the end of 1912.

AVRO Curtiss type biplane

This aircraft, ordered by Capt. E.W. Wakefield of Kendal, was built in 1911 at Manchester and then transferred to Brooklands, where it flew on the 19 July 1911 as a land-plane. After completion of flight tests it was delivered to Lake Windermere and fitted with a single central float and cylindrical floats below the lower wing tips, flying for the first time in this form on 25 November 1911. The machine was the subject of criticism and legal action by Wakefield against A.V. Roe and Co., but operated successfully until the following year under the name of *Lakes Water Bird* (*q.v.*).

AVRO monoplane

A scheme for a monoplane was advertised in *'The Aviator's Storehouse'*, the AVRO catalogue, in 1910. The machine was not built. The price less engine was £200.

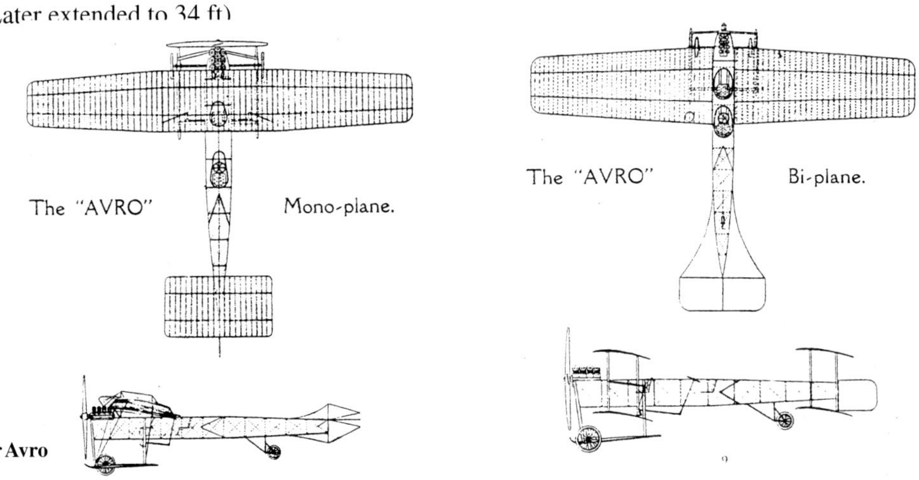

From the *Aviator's Storehouse*, the projects for Avro monoplane and biplane.

Power: 20 or 40hp Avro engine. These were two-stroke engines designed by M.F. and W. Edwards and made by H.W. Cowley of Bolton. The smaller engine was a horizontally opposed two-cylinder air-cooled type, the larger a four-cylinder. 20hp or 35hp JAP or NEC with two or four cylinders were also offered. The JAPs were air-cooled two or four-cylinder vee types. The NECs were water-cooled two-stroke inlines and vee-fours. 35hp Green four-cylinder inline water-cooled could also be used.

Data			
	Span	36ft	
	Area	220 sq ft	
	Price	20hp Avro	£325
	with engine	20hp JAP or NEC	£375

AVRO biplane

Also advertised in '*The Aviator's Storehouse*' and again, never built, was a biplane with the engine alternatives as for the monoplane. The price less engine was £225

Data			
	Span	29ft	
	Area	350 sq ft	
	Length	21ft	
	Height	7ft 6in	
	Weight	250 lb less engine	
	Prices	20hp Avro	£350
	with engine	20hp JAP or NEC	£400
		35hp JAP or NEC	£500
		35hp Green	£550
		40hp Avro	£450

AVRO Type D biplanes

The first of these biplanes arrived from Manchester in March and was flown for the first time on 1 April 1911 at Brooklands by Howard Pixton. The fuselage and tail surfaces were similar to those of the Roe IV, but with two bay warping biplane wings, plus a partial center section bay. A further six were built in about a year after the first had flown, and these varied in a number of respects. Several types of engine were used, radiator positions were changed, the tailplane shape became rectangular and two machines were converted to sesquiplanes for

The first Avro Type D at Barrow for seaplane trials.

competition work. Although the basic Type D was a two-seater, provision of an extra fuel tank and other changes converted the machine to a single-seater The prototype was sold for £700 to Cdr. Schwann RN, for seaplane experiments and was used with seven types of floats, both single and twin type, flown mainly by S.V. Sippé at Barrow-in-Furness.

Power: 35hp Green four-cylinder inline water-cooled.
45hp Green with drilled cylinder barrels.
60hp ENV type F eight-cylinder water-cooled vee.
35hp Viale five-cylinder air-cooled radial.
50hp Isaacson seven-cylinder air-cooled radial.

Data		
	Span top	31ft
	Bottom	31ft
	Chord	5ft
	Area	310 sq ft
	Length	28 ft (No.1 26 ft)
	Height	9 ft 2in
	Weight allup	500 lb

Sesquiplane

	Span top	33ft
	Bottom	23ft (No.2 28ft)
	Chord	5ft
	Area	279 sq ft (No.2 329 sq ft)
	Weight allup	550 lb
	Speed	45-50 mph
	Seaplane	40 mph

The first Avro Type D as originally built.

Avro Type D biplane with 60hp ENV type F in 1911.

Avro Type D. Third aircraft with sloping radiator at Shoreham. Pilot A.E. Geere.

The fourth Type D at Brooklands in October 1911.

No.1 35 and 45hp Green. First flown at Brooklands 1 April 1911. Converted to a seaplane at Barrow-in-Furness from August 1911. Last reported in use April 1912.

No.2 60hp ENV Intended for Circuit of Britain. First flown Brooklands 18 July 1911, but crashed 22 July 1911. Converted to sesquiplane with large triangular tailplane. Extensions fitted to lower wings were unsuccessful.

No.3 35hp Green with slanting radiator. First flown Brooklands 12 October 1911. Used by the Avro School at Brooklands and Shoreham until May 1914.

No.4 45hp Green with fore and aft radiator. Single-seater. First flown at Brooklands 12 October 1911. Abortive attempt at Michelin Prize but used mainly by the Avro School at Brooklands and Shoreham until May 1914.

No.5 35hp Green. Sesquiplane similar to No.2. First flown Brooklands October-November 1911. Reported for sale at Shed No.4 in May 1912 and scrapped in December at Brooklands.

Avro Type D. Seventh aircraft with 50hp Isaacson radial at Shoreham.

No.6 50hp Viale. School machine. First flown at Brooklands 20 November 1911 and crashed at Abingdon 29 January 1912. The engine was installed by Mons Ducrocq, agent for Viale, and Jack Alcock, later of Atlantic fame.

No.7 50hp Isaacson. School machine. Completed November-December 1911, used at Brooklands and Shoreham until May 1914.

AVRO Duigan biplane

This biplane was made to an order from John R. Duigan, an Australian, who had built his own Wright type glider and Farman-type powered aircraft in 1909-1911, with which he achieved a number of successful flights. He came to England to learn more of current practice and to receive flying instruction.

On arrival he ordered a biplane from Avro, which was built at Manchester and delivered to Huntingdon, when completed in December 1911. The machine was not a great success and Duigan took it back to Brooklands. After some limited straight flights in February, Duigan joined the Avro school. At Brooklands the original Alvaston engine was replaced by an ENV, and together with improvements, the machine was eventually capable of flying for one hour at up to 600ft with the pilot only; flights with a passenger were limited to "straights".

Duigan returned to Australia and his Avro was bought by the Lakes Flying Co. for £180 and was moved to Windermere on 4 June 1912. At Windermere it was largely rebuilt with a new engine and became the Lakes *Sea Bird* (*q.v.*).

This Avro biplane had a deep fuselage with radiators positioned either side of the front cockpit; wing warping was employed. The undercarriage had a leaf spring axle with a central skid. The skid became a feature of many later Avro aircraft. In its final form this machine was the basis of the next type, a much-improved machine, which eventually led to the outstanding Type 504.

Avro biplane for Australian John Duigan. As built originally with Alvaston engine in 1911.

Power:	40hp Alvaston two-cylinder horizontally opposed water-cooled 35hp ENV type D eight-cylinder water-cooled vee.	
Data	Span	34ft
	Chord	4ft 6in
	Length	29ft 6in
	Area	300 sq ft
	Speed	40mph

AVRO Type. E and Es (later known as Types 500 and 502)

This was a slightly larger version of the biplane built for John Duigan, more powerful and able to fly successfully as a two-seater. It was built to meet an official requirement and was often referred to as the 'Military Biplane'. The first flight took place at Brooklands on 14 March 1912 in the hands of Lt. Wilfred Parke RN. Testing proceeded, following which an entry was made for the Mortimer Singer Prize of £500, offered to an army or navy officer who flew the longest flight with a passenger, in Britain, before the end of March. Parke left

The Avro Duigan biplane with ENV engine in 1912.

Avro Type E (500) in original form with deep radiators.

Avro Type E (500) after change to radiators.

Avro Type E (500) with 60hp ENV and radiator on center section of top wing.

Brooklands with W.H. Sayers on 20 March 1912 for Hendon for his bid, but the aircraft suffered engine failure and crashed badly on takeoff. Sayers in the front cockpit was trapped by the radiators and was extricated with difficulty; the radiators were reduced in height in the course of the rebuilding.

Trials at Farnborough were carried out successfully in June 1912, and subsequently the engine was replaced by the new ABC. Testing recommenced in the hands of Raynham and later R.L. Charteris of the engine company until early in 1913, when the ENV was refitted. The machine was then handed over to the Avro school at Shoreham, where it flew until 29 June 1913, when it was crashed by a pupil and burned. The pupil was the first fatality in an Avro aircraft.

A second machine of the type, but with a 50hp Gnôme, was built and was flown by Parke at Brooklands on 8 May 1912. This proved to have a better performance and resulted in a early War Office order for three machines. The third of these was given its first flight from Eccles cricket ground near the Manchester factory. It was now also known as the Type 500. Further orders followed from the War Office, Admiralty, the Portuguese Government and civilian sources. It was during the time that these machines were in production that the company moved its Manchester premises in April 1913.

A number of changes were made to the aircraft during its operational life, including fitting ailerons to both top and bottom wings. The sprung rudder was replaced by a separate tail skid also various types of wing tip skids, some with wheels, were fitted. Later the rudder shape was changed to the more familiar comma shape of future Avro aircraft.

The War Office ordered a batch of five of the single-seater version, which were basically the same as the two-seaters and subject to the same modifications. These were identified as Type Es or Type 502.

Flight of 30 March 1912 and *The Aeroplane* of 11 April 1912 carried drawings and data for the prototype, with dimensions slightly at variance in some respects and neither agreed

Avro Type E (500) with Gnôme engine and wheels on outriggers at wingtips.

Avro Type E (500) with simple skids at wingtips.

Avro Type 502 ES with the RFC.

on the position of the radiators. The risk to the passenger was commented on by *The Aeroplane*. *Flight* showed auxiliary radiators just below the top wing, probably fitted later and the cause of Sayers' entrapment.

Type E prototype

Power: 60hp ENV type F eight-cylinder water-cooled vee.
60hp ABC eight-cylinder water-cooled vee.

Data *Flight*

Mainplane span	36 ft
Mainplane chord	4ft 10in *The Aeroplane* 4ft 9in
Mainplane gap	5 ft
Mainplane area	338 sq ft *The Aeroplane* 332 sq ft
Tailplane span	8 ft
Area inc. elevators	32 sq ft *The Aeroplane* 33 sq ft
Length	31ft *The Aeroplane* 30ft 6in
Height	9 ft 9in
Weight	1,200lb (Also quoted as 1,100lb)
Weight allup	1,650lb
Speed	60 mph

Type E (Avro Type 500)

Power: 50hp Gnôme seven-cylinder air-cooled rotary driving a 9ft diameter propeller.

Data

Span	36ft
Weight	900lb
Weight allup	1,300lb
Length	29ft
Area	335 sq ft
Max speed	62 mph

Production: Eighteen aircraft of various versions.

Type E 1 Prototype. First flight 14 March 1912 at Brooklands. Destroyed at Shoreham 29 June 1913.

Type E (Avro 500) seven machines delivered between May 1912 - April 1913 to the War Office. Serial Nos.404-406, 430, 432-433, 448 used at CFS Upavon.

Two machines delivered May 1913 and February 1914 to Admiralty contract, serial Nos.41 and 150 used at Hendon and Eastchurch.

One machine to Portuguese Government delivered Lisbon October 1912 named *Republica*.

One machine first flight July 1913 to private owner at Hendon later with RNAS serial No.939.

One machine to Hall School at Hendon January 1914, later with the RFC serial No.491.

Type Es (Avro 502) 5 machines delivered between April-June 1913. to War Office contract. Serial Nos.285, 288-291.

AVRO-BURGA monoplane

While the Type E prototype was being built at Manchester, a monoplane to the design of Lt. Burga of the Peruvian Navy was in hand, in which a unique system of lateral control by rudders, above and below the fuselage, was featured. These operated in opposed directions and served instead of ailerons or wing warping. Wings of alternative sections and set at various angles of incidence could be fitted. The narrow fuselage, tail unit and undercarriage owed much to the Avro Type E.

Construction took place throughout 1912 and first flight at Shoreham on 12 November 1912, piloted by H.R. Simms. After severe damage in January 1913 the monoplane was returned to Manchester but was not repaired.

Power: 50hp Gnôme seven-cylinder air-cooled rotary.

Data Length 29ft

The monoplane built by A.V. Roe for Lt. R. Burga of the Peruvian Navy.

AVRO Type F monoplane

One of the first aeroplanes where the pilot was seated inside the fuselage for the maximum protection. The pilot entered through a door in the roof of the cabin and was provided with a celluloid windscreen and various windows for vision. In addition a circular opening was provided on each side. The fuselage was built with a joint, at a point midway between the wing and the tail, for ease of dismantling and transport. The undercarriage and tail units were similar to those of the type E. The wing, of single spar construction, was attached at the fuselage centerline and was braced to the fuselage by a kingpost below and a pylon above.

The monoplane was flown at Brooklands on 1 May 1912 for the first time by W. Parke, who was involved at the end of May in a forced landing at Weybridge. After repairs at Brooklands it saw little use until 13 September, when being flown by R.H. Barnwell, it turned over on landing and was not repaired.

Power: 35hp Viale five-cylinder air-cooled radial previously fitted in the sixth Type D. Currently preserved in the Science Museum.

Data		
	Span	28ft
	Length	23ft
	Area	158 sq ft
	Weight	550lb
	Weight allup	800lb
	Max speed	65 mph

Avro Type F enclosed monoplane of 1912.

AVRO Type G biplane

The monoplane was followed by a two-seater cabin biplane, two of which were intended as entries Nos.6 and 7 in the Military Aeroplane Competition of August 1912; this was amongst the first enclosed type biplanes in the world.

The crew were seated one behind the other in the narrow fuselage which filled the gap between the wings. The warping wings, the tail unit and the undercarriage were basically components of the Type 500. The non-availability of the new 60hp ABC engine prevented completion of the second machine and, with limited time available, the first machine with a 60hp Green was delivered to Larkhill for the trials unflown. First flight by Wilfred Parke took place on 7 August 1912 in poor conditions, and the machine overturned on landing from a curtailed flight. The damage was sufficient for the aircraft to be returned to Manchester for repair, but it was back at Larkhill a week later. Parke participated further in the trials, making a notable recovery from a spin in what became known as 'Parkes Dive' and although attempting all the tests, the machine did not take a major award, mainly owing to its poor rate of climb. On 24 October F.P. Raynham, flying the Type G from Brooklands, established a British Duration Record of 7hr 31min, which was broken later that day by H.G. Hawker flying the Sopwith-Wright biplane.

Avro Type G enclosed biplane at Larkhill for the Military Trials.

The Type G was transferred to Shoreham, but was not greatly in evidence, although still in the hangar there in February 1913.

Power: 60hp Green four-cylinder inline water-cooled driving a 10ft diameter propeller.

Data		
	Span	35ft 3in
	Chord	4ft 9in
	Area	310 sq ft
	Tailplane span	8ft
	Tailplane area	20 sq ft
	Elevators area	14 sq ft
	Length	28ft 6in
	Height	9ft 9in
	Weight	1,191 lb
	Weight allup	1,792 lb
	Initial rate of climb	105 ft per min
	Max speed	62 mph
	Range	345 miles

The Avro Type 501.

Avro Type 501 in service with the RNAS.

AVRO Type 501

This machine, built and tested initially in amphibian form, was flown for the first time by H. Stanley Adams at Windermere in January 1913.

The large central float, designed by O.T. Gnosspelius, housed three wheels, two at the rear, one at the front, and was sprung. This undercarriage, together with small wingtip floats, contributed greatly to the weight and was replaced by twin floats, with a tail float fitted to the sprung rudder. In this form the aircraft was delivered to the Isle of Grain for trials, but proved to be unacceptable as a seaplane. The machine was converted to a land-plane and delivered to Eastchurch as serial No.16, against Contract No.77177/1913. It operated from Dover and Dunkirk, without much success, being finally deleted in February 1916.

The construction of the 501, which generally followed previous Avro practice, had two bay wings with strut braced top wing extensions and parallel chord ailerons, later inversely tapered ailerons were fitted to the top wings only.

Power: 100hp Gnôme fourteen-cylinder air-cooled rotary.

Data

Span top	47ft 6in
Bottom	39ft 6in
Area	478 sq ft
Height	12ft 6in
Weight	1,740 lb
Weight allup	2,700 lb
Max speed	55 mph seaplane, 65 mph (land-plane)

AVRO Type 503 (originally Type H)

This larger version of the Type 501 was first flown at Shoreham by FP Raynham on 28 May 1913, with a similar twin float arrangement to that used finally on the 501. After completion of trials the machine was sold to the German Navy and was identified as D12. in that service. Three more of the type Serial Nos.513 were built for the RNAS against Contract No.CP36208/1913 and were delivered to the Isle of Grain in September-October 1913 as seaplanes. All three were later converted to land-planes for use as trainers at various stations but were all deleted by January 1916.

Avro Type 503 seaplane at Shoreham in 1913.

Avro Type 503 landplane in service with the RNAS. No.52.

Power: 100hp Gnôme fourteen-cylinder or Gnôme Monosoupape nine-cylinder air-cooled rotaries driving a 8ft 9in diameter propeller. Replaced in Nos.523 by 60hp Le Rhone in service.

Data

Span top	50ft
Span bottom	47ft
Chord	6ft
Gap	6ft 9in
Area	567 sq ft
Length	33ft 6in
Height	12ft 8in
Weight allup	2,200 lb
Initial rate of climb	225 ft per min
Max speed	50 mph (seaplane)

AVRO Type 504

When A.V. Roe initiated work in November 1912 on a development of the Type 500, he could not have anticipated the success that this machine would have. It became mainly a training machine during the war period and with progressive improvements, continued in this role until 1932. Many other uses, both

Avro Type 504 prototype at Hendon in September 1913.

civil and military, were found for this outstanding aircraft, which was also built in large numbers by other contractors in wartime and in various countries around the world. The major changes to the aircraft or its role were identified by a suffix and ranged from 504A to 504S. These variants were started in 1915 and so are outside the period of this work.

Although Roe must rightly be given credit for the basic design, credit must also be given to his three designers, Chadwick, Taylor and Broadsmith, whose contributions were to make the overall design one which was outstanding.

The prototype 504 flew at Brooklands on 18 September 1913, having been partly built at Brownsfield Mills and completed at Clifton St. works and then delivered to Brooklands the day before. Just two days after the first flight it competed in the Aerial Derby, in which contest Raynham finished fourth at 66.5 mph. Following further trials the machine was returned to Manchester for various modifications, including replacing the bulky cowling with one of circular form. The tapered ailerons, which were fixed at the inboard end, and were warped separately from the wing structure, were replaced with hinged parallel units. A number of less obvious changes were also made. After further flight trials the machine was bought by the *Daily Mail* and, from April 1914, was flown as a twin float seaplane. The outbreak of war saw the prototype 504 seaplane commandeered at Shoreham by the Navy, but it crashed on land during the delivery flight in June 1914 and was not rebuilt.

Avro 504 in its earliest production form in August 1914.

Production for the War Office began in mid-1913, when a contract for twelve aircraft was issued and further aircraft for trials and for private owners, probably four, were put in hand. The Admiralty also placed orders for seven machines in early 1914 and a further forty-four were ordered by the War Office, before the end of the year.

The single 504 fitted with the ABC engine was one sold to Armstrong Whitworth Ltd. to test the first engine of the type made by that company. The engine was fitted to a machine which arrived at Brooklands from Manchester on 16 February 1913, but the engine was not a success and after only one flight in April the project was abandoned.

The prototype 504 converted to a seaplane with modified cowlings and ailerons. Summer 1914.

Power: 80hp Gnôme and Gnôme Monosoupape seven-cylinder air-cooled rotaries.
100hp ABC eight-cylinder water-cooled vee.

Data		
	Span	36ft
	Area	330 sq ft
	Length	29ft 5in
	Height	10ft 5in
	Weight land-plane	924 lb
	Weight allup	1,574 lb (Prototype 1,550 lb)
	Weight seaplane	1,070 lb
	Weight allup	1,717 lb
	Climb to 3,500 ft	prototype 9min 30sec Gnôme 7min Monosoupape
	Max speed	land-plane 82 mph seaplane 75 mph

Production.

Prototype. First flight at Brooklands 18 September 1913, destroyed. June 1914. S/No.889.

Four Aircraft for trials and private sales.

Twelve to the War Office. Serial Nos.376, 390, 397-398, 637-638, 652, 665, 685, 692, 715-716

Forty-four to the War Office. Serial Nos.750-793.

Seven to the Admiralty. Serial Nos.179, 873-878

AVRO Type 506 (Originally Type J)

This larger version of the Type 508 was to be a two-seat, twin float seaplane, but it was not built and remained only as a project in 1913.

Power: 160hp Gnôme fourteen-cylinder air-cooled rotary.

Data		
	Span	70ft
	Area	980 sq ft
	Length.	44ft
	Height	15ft 5in
	Weight allup	3,800 lb
	Endurance	4 hr

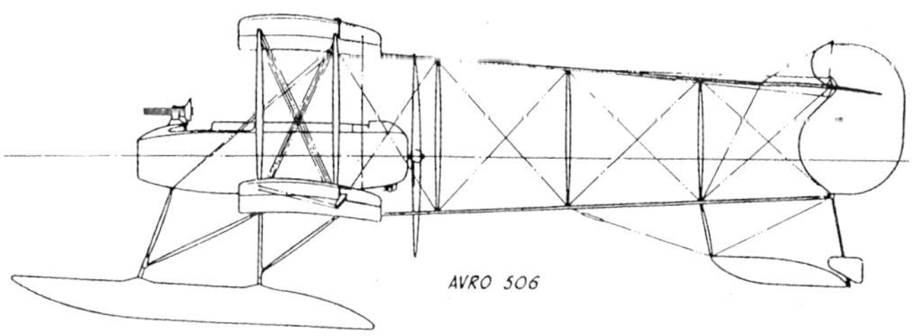

Avro 506 project of 1913.

AVRO Type 508

This, the first pusher of Avro design, was intended for reconnaissance purposes, with the observer well forward in the front cockpit. The tail booms were of steel tube but the nacelle and wings were of wooden construction, derived from earlier Avro types. The undercarriage and engine installation were based on the units of the Type 504.

The single prototype was exhibited incomplete in January 1914 at a show in Manchester and in March, as a finished aircraft, at the Olympia Aero Show. The machine is known to have been available for use at Brooklands in April 1915 and may possibly have been tested as a seaplane in the Southampton area, as reported in *The Aeroplane* of 9 September 1914. The machine was sold to the Hall School of Flying at Hendon, but seemingly was not used, being still there in a dismantled state in April 1916.

Avro 508 pusher biplane at Brooklands in 1914.

Power: 80hp Gnôme seven-cylinder air-cooled rotary.

Data		
	Span	44ft
	Area	468 sq ft
	Length	26ft 9in
	Height	10ft
	Weight	1,000 lb
	Weight allup	1,680 lb
	Max speed	65 mph
	Endurance	4hr

AVRO Type 509

A contract, valued at £3,384, was received from the Admiralty in January 1914 for a twin-float seaplane with two engines driving pusher propellers. The scheme for this 2/3 seater machine, fitted with radio equipment and a heavy gun in the nose, had been prepared in the previous November. The Navy allocated Serial No.94 to this aircraft, and it appeared on the lists of HM Naval Aircraft at least until October 1914, noted as being on order. Presumably some construction actually took place in the period since January, but the machine was not completed.

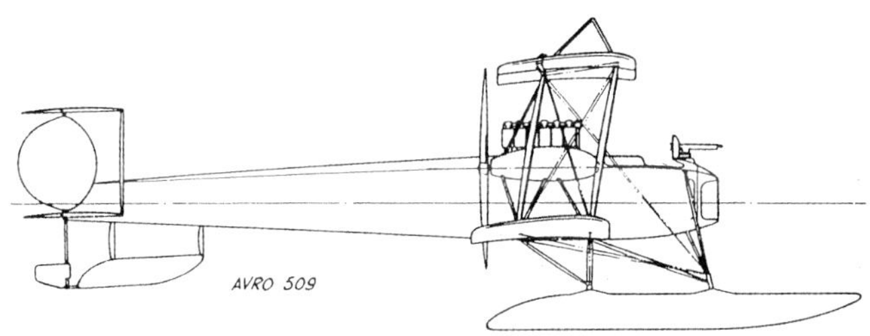

Avro 509 twin pusher biplane No.94, which did not reach service with the RNAS.

Power: Two 120hp Austro-Daimler six-cylinder inline water-cooled.

Data		
	Span top	80ft
	Span bottom	56ft
	Area	931sq ft
	Length	44ft 3in
	Weight	2,800 lb
	Weight allup	4,510 lb
	Max speed	70mph

AVRO Type 510

The first of these machines was intended to compete in the Seaplane Circuit of Britain to be held between 115 August 1914, starting from Southampton Water. The contest was canceled and the machine was immediately taken over by the Admiralty for £2,500 and five more of the same type were ordered. The

Avro 510 seaplane prototype at Calshot in August 1914.

prototype was delivered by rail to Calshot, where first flight took place in August. F.P. Raynham carried out the flying on behalf of the company. The production aircraft were delivered direct to the RNAS stations at Killingholme and Dundee between December 1914 and April 1915 and were the subject of criticism, which delayed their final acceptance.

The Type 510 had considerable top wing overhang braced to kingposts and was a seaplane with twin main floats, as well as tail and wing tip floats. The first machine sat low on the water and had smoothly shaped floats, mounted on a complex structure of steel tubes, and were sprung internally. Subsequent aircraft embodied fixed pontoon like floats on a revised chassis which raised the machine higher and was probably detrimental to the performance. A fin was not fitted initially to the first machine, but was later added to all aircraft.

The machine was normally flown from the rear cockpit, although dual controls were fitted. The Navy found the 510 to be a disappointing machine with poor performance, and it was mainly used as a single-seater.

Power: 150hp Sunbeam eight-cylinder water-cooled with front radiator. (Later named Crusader)

Data		
	Span top	63ft
	Span bottom	38ft
	Chord	5ft 9in
	Gap	6ft 6in
	Area	564 sq ft
	Length	37ft 6 in later 38ft
	Weight	2,080 lb
	Weight allup	2,800 lb
	Max speed	70 mph
	Climb to 1,000ft	4 1/2 min
	Climb to 3,000ft	20 min
	Endurance	4 1/2 hr

Production

No.881 'Circuit' machine to Calshot August 1914. Commandeered by the Admiralty at outbreak of war.
No.130 To Killingholme. June - December 1914.
No.131 To Killingholme 5 February 1915. First flight 15 April 1915. by Raynham.
No.132 To Killingholme 17 February 1915. First flight 15 April 1915. by Raynham.
No.133 To Dundee. First flight 31 March 1915.
No.134 To Dundee 4 April 1915. First flight 12 April 1915.

AVRO Type 511 *'Arrowscout'*

This single-seat scout biplane made its first appearance at the Olympia Aero Show of March 1914. It flew for the first time at Brooklands shortly before the Aerial Derby of May 23, for which it carried racing number 14. The contest was postponed due to poor weather, but Raynham flew to Hendon from Brooklands and demonstrated the aircraft. The race was flown on 6 June, but the *'Arrowscout'* was by this time modified to become the Type 514, with alternative wings and other changes to improve its disappointing performance.

Avro Type 511 *Arrowscout*. Later modified with straight wings to become Type 514.

The swept wings, a new departure, introduced design problems and justified the manufacture of the alternative straight wings as a safeguard. Another new feature was the fitting of air brakes, which formed the inboard sections of the lower wings, to reduce the landing speed.

Power: 80hp Gnôme Monosoupape seven-cylinder air-cooled rotary.

Data		
	Span	26ft
	Chord	6ft 6in
	Gap	5ft 3in
	Area	236 sq ft
	Length	22ft 9in
	Weight	675 lb
	Weight allup	1,165 lb
	Max speed	95-100 mph

AVRO Type 512

A projected land-plane with one 65hp Austro-Daimler engine. Span 26ft.

AVRO Type. 513

A design study made in March 1914 for a two-seat, twin float bomber seaplane powered by two 80hp. Gnôme engines. The wings were designed to fold, and floats and flying surfaces were interchangeable with those of the Avro 510.

Data		
	Span top	72ft
	Span bottom	47ft.
	Length	36ft 6in.

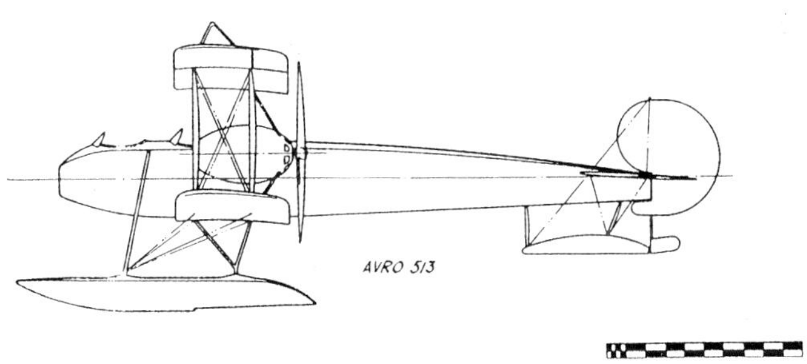

Avro 513 project of 1914.

AVRO Type 514

The main changes to the Type 511, to improve its performance, were to replace the wings with unswept units with conventional interplane struts and to redesign the undercarriage, which was an unsprung vee type and without the central skid. The aircraft was entered as No.20 in the postponed Aerial Derby, but its undercarriage failed on takeoff at Brooklands, before the event.

Although the 514 was repaired at Manchester and flown by Raynham at Southport in July, no further development took place.

AVRO Type 515

This was a September 1914 design for a biplane with a 150hp Sunbeam engine.

B

BADEN-POWELL gliders (Major B. Fletcher Smythe Baden-Powell, 32 Prince's Gate, London)

Major Baden-Powell was the brother of the Chief Scout He had experience of military ballooning and later carried out experiments with manlifting kites. His 36ft kite of 500 sq ft area was the first to lift a man, on 27 June 1894. In 1897 he made a small ornithopter glider which was not successful. In 1904 he experimented with gliders from a chute over water at the Crystal Palace, assisted by J.T.C. Moore-Brabazon. The gliders were tried in both biplane and monoplane form, achieving short glides

Baden-Powell gliding experiments at Crystal Palace in 1904. The monoplane glider.

Baden-Powell ornithopter gliding machine of 1898.

Baden-Powell gliding experiments at Crystal Palace in 1904. The biplane glider.

on a number of occasions. The crude structure illustrated may have been involved in these trials, but could have been a mockup of the later Quadruplane.

BADEN-POWELL quadruplane

Designed by Major Baden-Powell, and built in 1909, this machine appeared at the Dagenham Flying Ground in that year. It was basically a pusher biplane with large extra surfaces fore and aft. The pilot sat on the lower center section with the engine directly behind. The fuselage consisted of four bowed metal tubes, meeting together at both front and rear, and braced by spacing members. The lower portion of the forward fuselage was covered in, perhaps for alighting on water. The front surface was the elevator, the larger rear surface was fixed. The wings were rotatable around the main spar for steering. The machine did not fly. The structure of this and his other machines was covered by patent number 6443 of 1906.

Power: 10-12hp Buchet three-cylinder air-cooled fan type semi-radial.

Data		
	Span.	22ft
	Length	24ft

BADEN-POWELL SCOUT monoplane

This single-seat pusher monoplane was designed by Baden-Powell and built by Handley Page at Barking in 1909. It was exhibited at the Stanley Show in November 1910. The machine, also known as *The Midge*, had a notice 'Scout Aero Club' alongside.

Like his earlier machine, the Scout was constructed with four curved longerons meeting at the nose and tail, these were made of bamboo and were spaced at the center by a substantial X-shaped wooden member, which also formed the engine mounting and the rear support for the pilot's seat. The machine was controlled by the front elevator, this surface being divided to provide directional control. The tailplane and fin were apparently fixed surfaces. There is no record that the aircraft flew.

Power: 10-12hp Buchet three-cylinder air-cooled semi-radial driving a 5ft 6in diameter propeller.

Data		
	Span	22ft
	Weight	140 lb
	Wing area	90 sq ft
	Total area	150 sq ft
	Length	20ft

Baden-Powell. Mockup predecessor of the quadruplane.

Baden-Powell quadruplane of 1909 at Dagenham.

Baden-Powell Scout monoplane at the Stanley Show in 1909.

Baden-Powell Scout (*Midge*) monoplane under construction.

Baird monoplane built at Rothesay in 1909-1910.

BADEN-POWELL quadruplane 1910-1911 version
The construction of another machine appears to have begun in 1910, which was to be powered by a 50hp Antoinette engine, driving twin pusher propellers. It seems that it was not completed, and the only information is that conveyed by the adjacent illustration. Patent No.6443/1906 has some relevance to this and to the earlier machine.

Baden-Powell quadruplane of 1910-1911 partly built.

BAIRD monoplane (Andrew B. Baird, 115 High St., Rothesay, Isle of Bute, Scotland)
Constructed by a master blacksmith in 1909-1910, this was a single-seat tractor monoplane, which was completed and tested at Ettrick Bay in September 1910. Although the machine rose into the air briefly, it was not capable of sustained flight, and was damaged and not repaired. The machine was said to embody features of the Blériot and Antoinette with a Demoiselle type tail.

The fuselage was of triangular section with bamboo longerons and steel tube cross members braced by wires. The wings were heavily cambered, of parallel chord and covered with a single layer of fabric.

The engine is believed to have been made by Alexander Bros. of Edinburgh, as shown at the Scottish Motor Show in January 1910.

Power: 20hp Alexander four-cylinder inline air and water-cooled

Data	Span	29ft
	Wing area	180 sq ft
	Weight	380 lb

BAKER gyropter (F.W. Baker & Co., New Rd., Stourbridge, Worcestershire)
Baker is reported to have built a twin rotor machine with a 40hp engine, for which there are no other details. He is known to have built machines for other people at his motor works, such as the Foley biplane of 1910.

BALSTON ornithopters (R.M. Balston, Mereworth, Kent)
Little is known of Balston's machines, the earliest being reported built at Cadland, near Southampton in 1895-1900, referred to as an 'orthornicopter'. He built a large model to compete in the *Daily Mail* contest at Alexandra Palace in 1907, when A.V. Roe won the main prize. The man carrying machine followed in 1908 and was built mainly of bamboo. His patent No.23235 of 1905 relates to folding wings for aircraft to facilitate transport.

Power: 1 1/2hp driving two-bladed propeller

Data Span 17ft

Balston's ornithopter model of 1907.

Balston's full-size version of 1907-1908.

Barnes monoplane of 1909.

BARBER monoplane (Horatio C. Barber)
This machine was constructed by Howard Wright and designer W.O. Manning at Battersea during 1908-1909 and was tested at Larkhill from June 1909.

It was a two-seater with passenger in front and incorporated Barber's patented system of automatic stability (Patent No.1999 of January 1909). This system required the wings to be hinged at the fuselage so that the dihedral angle could be varied by the control of the bracing wires, which passed over kingposts. Wing warping appears to have replaced the original scheme for wing tip ailerons. During the course of the trials various features were modified, including the shape of the tail surfaces, which were changed to triangular units and the wing tip wheels which were discarded.

The machine did not fly and the trials were abandoned. Barber then went on to form in April 1909, the Aeronautical Syndicate Ltd. which produced machines mainly of the Valkyrie type.

Power: 50hp Antoinette eight-cylinder water-cooled vee driving contra-rotating propellers.

Data		
	Span	32ft
	Length	27ft
	Wing area	200 sq ft
	Weight allup	1,000 lb

Barber monoplane, built at Battersea by Howard Wright in 1908-1909, seen at Larkhill. (Courtesy Norman Parker).

BARNES monoplane (G.A. Barnes, Abbey Wood, Kent)
This machine, described as 'similar to an Antoinette', was shown at the Stanley Show in the Agricultural Hall, Islington, in November 1909 in skeleton form.

A single-seat tractor monoplane, it was designed and built by G.A. Barnes, a well known racing motorcyclist, during 1909. The power was provided by a 20hp JAP engine and it was reported to have flown for about a mile and a half at low altitude when tested in October 1909 at Abbey Wood, but was damaged on landing.

The fuselage was of vee-section and the wing tapered in plan.

BARNWELL Bros. aircraft (Frank Sowter Barnwell and Richard Harold Barnwell. Grampian Motor & Engineering Co., Causewayhead, Stirling, Scotland)
The brothers were members of the Scottish Aeronautical Society, but ran the Grampian Company after training in shipbuilding. Between them they were responsible for the construction of six machines, before they moved on to appointments with other aviation concerns. Their first types were biplane gliders, of which little is recorded. It is believed that two were built from 1905 at Balfron, their parent's home.

BARNWELL Bros. biplane
This lightly built biplane was a single seater pusher powered by a 7hp Peugeot engine. It did not fly when tested in 1908 at Cornton Farm, Causewayhead.

The Barnwell brothers built this unsuccessful biplane in 1908.

BARNWELL Bros. monoplane

Frank went to America in 1907 and met the Wright brothers and, upon returning, started the construction of a monoplane. Harold designed the engine The wing spars were made of bamboo and the wing ribs were in pockets in the covering of light sail canvas. The undercarriage had four main wheels and a tailskid. The machine was completed in December 1908 and reached 25 mph on the ground but would not lift off.

Power: 40hp Grampian two-cylinder horizontally opposed air-cooled vee.

Barnwell 1908 monoplane which did not fly.

BARNWELL Bros. biplane

The next effort by the Barnwell brothers was constructed by their Grampian company and was a large single-seat canard biplane. Harold Barnwell succeeded in flying for eighty yards on 8 July 1909, but the machine was damaged on landing. After repairs and with the wingspan reduced to 45ft further trials were carried out on 8 September 1909 using a starting rail. A height of 25ft was reached before the machine was damaged beyond repair on 10 September 1909.

Power: 40hp Humber TT car engine four-cylinder inline water-cooled driving two 10ft diameter pusher propellers by chain.

Data		
	Span	48ft
	Chord	8ft
	Gap	7ft 6in
	Weight	1,568 lb

Barnwell biplane of 1909 flown briefly by Harold Barnwell.

Barnwell monoplane of 1911. Designed by Harold and flown successfully.

BARNWELL monoplane

This was a design by Harold, the elder of the two brothers and was built by the Grampian company including the engine. It is recorded as the sixth machine built by the Barnwells. It was a single-seat tractor low wing monoplane of conventional layout. The sloping radiator with an air intake above the engine crankcase provided a windscreen for the pilot. The axle of the substantial undercarriage with central skid was wire braced to the body.

On 14 January 1911 at Causewayhead near Stirling, Harold made the longest flight of any Scottish aircraft or pilot to that date, and followed this on 30 January 1911 with a flight reaching a height of 200ft, but damaged the machine on landing. For these flights the J.R.K. Law Prize of £50 was awarded by the Scottish Aeronautical Society. (14 January 1911. 600 yards at a height of 50ft: 30 January 1911 distance 1 mile, endurance 1 min 2 2/3 sec)

The damaged machine was repaired and further flights were carried out at Cambussdrennie Farm, Blair Drummond on 16 August 1911 and 13 October 1911, but these were the last to be reported.

Power: 40hp Grampian two-cylinder horizontally opposed water-cooled.

BARTELT ornithopter (F.T. Bartelt J.P., Corston Lodge, Brislington, Bristol)

This biplane-like machine with beating wings was designed by the chairman of the Polysuphin Co. Ltd. of Bristol and was exhibited at the Olympia Aero Show in March 1911.

The rectangular center structure was made of steel tubing and mounted the engine, radiators and chain and crank drive mechanism, which actuated the top and bottom wings alternatively

Bartelt ornithopter at Olympia Aero Show in 1911.

to provide a beating motion in both forward and reverse directions.

The wings, which were tapered, were also made of steel tubing and braced by wires. The fabric covering was loosely applied with the object of this billowing out on the down-stroke to hold the maximum of air.

No further reports of progress appeared and it is certain that this machine was not successful.

Power: 40hp Star four-cylinder inline water-cooled (113 m/m bore 126 m/m stroke)

Data	Span	30ft 6in
	Taper	12ft at root to 7ft at tip
	Length	12ft.

Barton-Rawson multiplane of 1905 at St Helens Isle of Wight.

BARTON-RAWSON multi-plane (Dr. F.A. Barton and F.L. Rawson, St. Helens, Isle of Wight)

This twin-floatplane of bamboo construction was built at St. Helens in 1905 by Dr. Barton, his son Dudley and Rawson with the intention of being powered. Owing to the lack of a light-weight engine, it was tested by towing behind a launch, but was wrecked in the course of these trials on 26 September 1905.

Power: (proposed) 35hp engine driving a 7ft diameter propeller.

Data	Span	34ft
	Area	1,200 sq ft
	Length	36ft
	Height	9ft
	Twin floats	20ft long 20in wide 4in deep
	Weight unpowered	240 lb

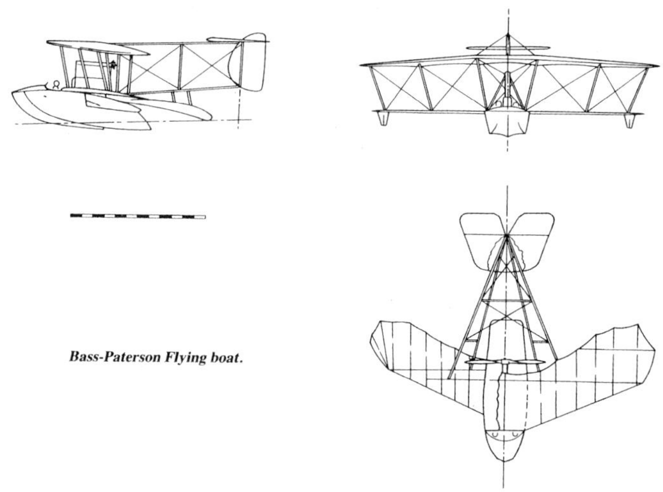

Bass-Paterson flying boat built by S.E. Saunders & Co. in 1914.

BASS-PATERSON flying boat (Capt. Ernest Bass and Compton C. Paterson)

This machine, also referred to as a 'Bat-Boat,' was built to the ideas of Paterson by S.E. Saunders Ltd. of Cowes, Isle of Wight. It was nearing completion in December 1914, but was destroyed by fire when on test in early 1915 before it could be flown.

The hull of Saunders 'Consuta' construction contained the engine, which drove a pusher propeller by chain. The two bay wings were swept aft at about 30 degrees and provided lateral control by warping. Wing tip floats were mounted flush to the undersides of the wings.

The open tailbooms carried a high mounted tailplane with elevators and large rudder and fin, together with a supporting float for the tail.

Power: 100hp Green six-cylinder inline water-cooled.

Data Hull length 13ft 9in

BASTIN ornithopter (T. Hugh Bastin, Clapham, London, SW)

Little is known of this machine, built between 1900 and 1902, except that it had four wings of 25ft span and was 50ft long.

Bastin ornithopter. A sketch which gives an idea of the appearance of this machine.

Batchelor monoplane Eastchurch 1910.

BATCHELOR monoplane (Albert Batchelor, Strood, near Rochester, Kent)

Batchelor built his aircraft in his spare time at the cement works, which he owned, starting work in September 1909. It was taken to Eastchurch when ready in July 1910. The machine had not flown by October and was abandoned.

The single-seat monoplane was fitted with a Blériot type castering undercarriage with skids and *Demoiselle* seating position; lateral control was by warping.

Power: 35hp Alvaston two-cylinder horizontally-opposed air-cooled driving a Cochrane propeller.

Data	Span	28ft
	Length	26ft
	Weight	480 lb

BATH monoplane.

Reported in August 1909, but no details available.

BAUMANN ornithopter (G. Baumann, Rutli Villa, Heaton Mersey, Manchester and at 4 Corporation St., Manchester.)

Patent No.23713/1909 was for an ornithopter with parachute, powered by a 10hp Universal engine. The machine, quaintly described as a Hornithopter, was for sale in February 1910.

BAZELEY glider (W.E. Bazeley, Scarning Road, East Dereham, Norfolk)

A brief report of a biplane glider with wing area of 140 sq.ft. under construction in January 1910.

BEARDMORE-DFW Tractor Biplane (Wm. Beardmore & Co. Dalmuir, Dumbartonshire)

A seaplane to compete in the 'Seaplane Circuit of Britain' was made by Beardmore under license from the Deutsche Flugzeug-Werke and was described in August 1914, at the time when the contest was abandoned.

The machine was a two-seater with the pilot occupying the rear seat. The fuselage was constructed of welded steel tube. The engine bay and the top decking to aft of the pilot's cockpit were covered in aluminum panels providing a well shaped entry; the remainder of the fuselage, which was basically of rectangular section, was covered with fabric. A high aspect ratio fin and variable incidence tailplane surfaces of triangular shape carried a rounded rudder and separate elevators.

The two bay wings were tapered and swept back, with the leading edges curving towards the tips. The top wings met at the aircraft centerline where they connected to the cabane structure of welded streamlined steel tubes above the engine. The bottom wings attached to the lower longerons and were braced to the top wings by pairs of steel struts and cables, being positioned to provide considerable stagger. The ailerons were hinged at right angles to the aircraft centerline giving them an elongated triangular shape.

The machine was equipped as a twin float seaplane with tail float for the contest, but was suitable for conversion to a land-plane. The coolant radiator was mounted under the fuselage between the mountings of the chassis.

Nothing is known of the use of the machine, which does not seem to have been taken over by the RNAS or the RFC at the outbreak of war. Beardmore was developing engines based on the Austro-Daimler and may well have retained the aircraft for engine testing. The company also undertook the design of aircraft, which were not adopted, but produced other types in quantity during the war. A height record of 25,450ft was established on 14 July 1914 by a similar German-built DFW at Leipzig.

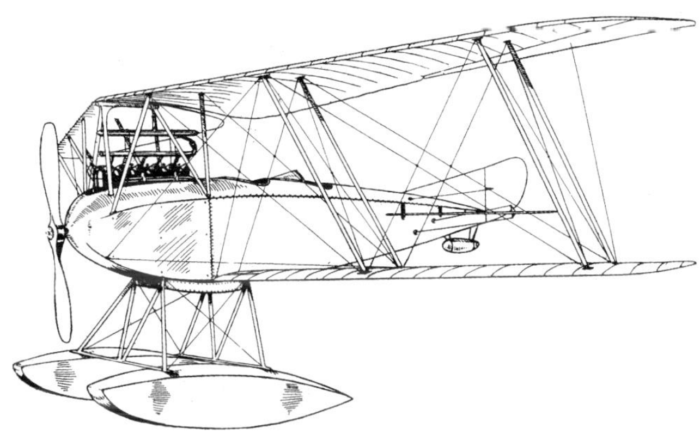

Beardmore-DFW built for the abandoned 1914 Seaplane Circuit of Britain. (Sketches by Geoffrey Watson).

Power: 120hp Beardmore Austro-Daimler six-cylinder inline water-cooled

Data	Span top	44ft
	Span bottom	40ft
	Area	450 sq ft
	Main floats	14ft long at 9 ft centers
	Length	25ft
	Weight	1,500 lb
	Weight allup	2,300 lb
	Speed range	45-80 mph
	Rate of climb	300ft per min
	Endurance	6 hr

Beatty-Wright biplane with 50hp Gyro engine.

BEATTY-WRIGHT biplanes (Beatty School of Flying, Hendon Aerodrome, London.)
The American aviator George W. Beatty had operated a flying school in America and, in the summer of 1913, he came to Britain bringing his Wright biplane with him. His initial aim was to demonstrate the Gyro engine and hopefully to obtain orders for it. However, he soon became involved with the Handley Page Flying School at Hendon as their chief (and only) instructor. By February 1914 Beatty had set up his own flying school using three Wright biplanes of quite primitive appearance, powered by Gyro or Wright engines. After the outbreak of war in August 1914, the School expanded and more aircraft were required for the initial training of service personnel. Several variants of the well tried Wright design were built by Beatty and his staff. The School remained in operation until at least the end of 1917.

Power: Two-seater 50hp Gyro, 50hp Gnôme or 50hp Beatty. Single seat pusher: 50hp Gnôme

Beatty-Wright. One of the Wright-type biplanes built by Beatty at Hendon (with Wright engine).

BEAUMONT biplane. (W.R.L. Beaumont, Birmingham Aero Club)
Beaumont abandoned this glider after it was smashed at its initial trials on 8 February 1914. There was mention of a monoplane under construction and due for completion in the autumn of 1914, on which work was presumably discontinued.

BEEDLE and SPIERS hydroplane (William Beedle and Wallace Spiers, The Lea, Garston, near Watford, Hertfordshire)
Nothing more was heard of this hydroplane, testing of which was due to start 'shortly' at Spiers home after the report of October 1913. It was to be fitted with a Beedle propeller.

BEER glider (H. Beer, Cardiff, South Wales)
This biplane glider of Wright type with front elevator and rear rudder had a span of 33ft and was constructed in 1912.

Beer glider. Built at Cardiff in 1912.

BELBIN cycleplane (H.W.G. Belbin, 1A William St., New Road, Battersea, London SW)
Belbin was a blacksmith with an inventive turn of mind and decided to enter for the Peugeot Prize of £400 for the first man to clear 20ft on a man powered cycle aeroplane. Belbin achieved

Belbin cycleplane built at Battersea.

6ft 11in, which so impressed Grahame-White that he gave him free unlimited access to Hendon aerodrome so that he could continue his experiments there.

The Belbin machine was of quite sophisticated appearance, with ailerons, elevator and rudder, with an auxiliary aerofoil surface mounted above the aviator's head.

Belbin went on to design, build and test amphibious bicycles.

BELL glider (Capt. H.S. Bell DSO, Boldon Racecourse, near Newcastle-on-Tyne)

A glider made by a member of the Northumberland Aero Club in 1910.

BELLAMY biplane (Monsieur Bellamy, Brooklands Track, Weybridge, Surrey)

Bellamy was a flamboyant Frenchman who had experimented in France and Italy in 1906 and claimed a flight of 500m at Modane in France on the biplane which he showed at a Milan exhibition, where it was suspended below a balloon. He arrived in England in December 1906 with a dismantled 'aeroplane', which he stated had been damaged in transit and established himself at Brooklands on the uncompleted site of the Railway Straight. His original experiments in France and Italy are believed to have made use of the Voisin-Archdeacon glider on floats much modified by Bellamy. The machine was described at the time in the *Auto* and ten years later in *Flight* and was quite different from that which Bellamy erected at Brooklands.

His arrival at Brooklands followed the offer of a prize by the Brooklands Automobile Racing Club of £2500 for the first aviator to fly a circuit of the track. Preparatory to attempting flight Bellamy carried out experiments with propellers on a catamaran on the lake in Mr. Locke-King's grounds.

However, at Brooklands, the Bellamy aircraft was constructed on the site, largely from bamboo, and was fitted with a Panhard engine driving a pusher propeller direct on the crankshaft and a tractor propeller on an extension shaft. The machine had a front mounted cruciform fin and horizontal plane, both of triangular shape, and a large flexible tailplane, serving as an elevator control, mounted midway between the top and bottom longerons of the rear structure. The whole machine looked extremely flimsy and unlikely to achieve sustained flight, which in the event it did not.

Bellamy biplane. *'The Great White Flying Bird'* which never flew. At Brooklands 1907.

Power: 35hp Panhard four-cylinder horizontally opposed water-cooled.

Data		
	Span	48ft
	Span tailplane	12ft
	Length behind wings	30ft

BELLAMY monoplane (Mons Bellamy, Petersham Meadows, Richmond, Surrey.)

Monsieur Bellamy's second known attempt at flight was with a single-seat tailless tractor monoplane on 18 August 1908. This is almost certainly the machine which was reported to be under construction at Old Oak Farm, Shepherds Bush, London in February 1908. The trials were carried out on Petersham Meadows, below the 'Star and Garter Hotel', Richmond, and although the machine taxied well, it failed to take off. The wing span was reported as 14ft which seems unlikely. Weight empty was 700lb and the engine was 30hp of unknown make.

In the summer of 1908 Bellamy was involved in a scheme to advertise a newspaper by a balloon flight across the Channel to France which did not transpire. In March 1909 he was again experimenting with a catamaran to test improved propellers, this time on the Thames at Hammersmith.

Bellamy monoplane 1908 at Petersham meadows near Richmond, Surrey.

BELMONT glider (Belmont Aeroplane Co., Ladywood, Birmingham)

In 1911 this company built a 40ft span glider to the design of F. Hill with a tapered wing. A low-powered engine was also fitted. (See also Birmingham Aero Club.)

Benton biplane B.II at Chalvey near Slough 1911.

BENTON B.I & B.II. biplanes (John Frederick Benton, Manor Farm, Chalvey, near Slough, Buckinghamshire)

The B.I and its development, the B.II, were built with the help of John Allen of Bray Rd., Maidenhead. Both were single-seat, two bay biplanes with equal span wings and open frame fuselages. Power was provided by a 35-50hp NEC vee four-cylinder, water-cooled two-stroke engine driving twin pusher propellers.

Lateral control was by single acting ailerons and control in pitch was by variable incidence wings covered by Patent No.24945/1910. A sprung undercarriage, to counter lateral drift, was also covered by Patent No.12375/1911. The Benton is believed not to have flown, and the B.II, although entered as No.31 in the 1911 Circuit of Britain Contest, was no more successful.

BENTON B.III, IV, V, VI and VII biplanes

The Benton B.III was a conventional tractor biplane with warping wings, powered by the same NEC engine as used on the earlier types. The machine was modified progressively to become the B.IV to B.VII, the most radical change being to introduce hinged trailing edges to the wings, operated by the pilot, to provide camber-changing flaps.

The B.III and its successors were flown successfully between 1912 and 1914 for distances of about a mile from the field at Chalvey. The final version survived in the hangar there until it was scrapped in 1919.

BERGAN monoplane (Harry Bergan, Lang's Arms Hotel, Grant St., Accrington, Lancashire)

Bergan was a member of the Manchester Aero Club who proposed to build a Blériot-like monoplane, with a 12hp engine, in the garage of his hotel in the autumn of 1909. This was in response to the publicity given to the acquisition of a Howard Wright monoplane by W.E. Cooke of Burnley, who challenged Bergan in the Press to meet him for a race at Hopton Common. Although a model of the aeroplane was flown, the full size machine was not completed.

BERRISFORD aeroplane (Ernest James Berrisford, Newcastle-under-Lyme, Staffordshire)

This machine was a motorcycle fitted with wings which was tested from a slope near Clayton Road, Newcastle in about 1912-1913. This was probably intended to be used as an act in public displays, for Berrisford was involved with the theatre, using the stage names of 'D'Artigan' and 'The Great Diavolo'. As a helper to Capt. Clayton of Mansfield, he flew in balloons and carried out parachute jumps. He also handled Clayton's Blériot and was involved in a taxiing accident at Worcester on 9 June 1910 in which a bystander was killed.

BETT aeroplane (James M. Bett & Co., Beverley Aeroplane Works, Barnes, Surrey)

A design with two 20hp coupled engines was proposed in 1909-1910 and was covered by Patents 27082/1909, 7333/1910 and 22001/1910 in conjunction with T.C. Murphy. The machine did not materialize.

BIGNOLD monoplane (Bignold, 51 Knollys Road, Streatham, London SW 16)

This Blériot type monoplane was advertised for sale in *Flight* in October 1910 for £30. It appears to have been in a crash as

Benton B.III and later versions were conventional tractor biplanes. The illustration is of the B.VII.

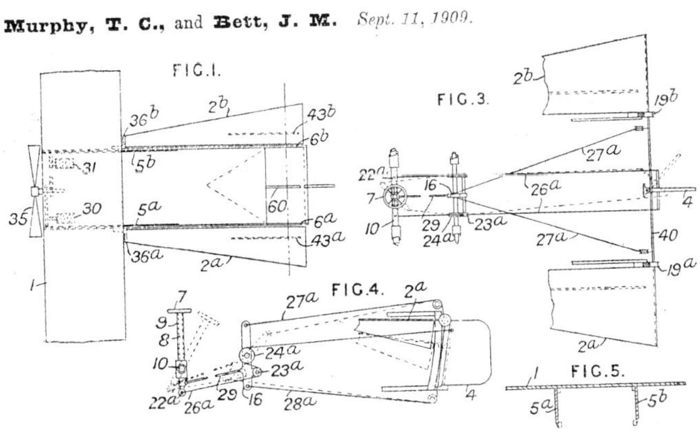

Bett monoplane. Sketches accompanying Patent No.22001/1910.

the wings were badly broken and the tail was slightly damaged.

Bignold had advertised in June 1910 for a 15/25hp engine. The monoplane was again advertised in December 1910 together with a small span monoplane for £10 and a Weiss pattern propeller for £6.25.

BILLING biplane (Eardley Billing, Aero Construction Co., Shed No.2 Brooklands)

This single-seat tractor biplane was constructed at Brooklands using the wings of the Voisin pusher of C.A. Moreing. The engine was a 40hp ENV Type D. It was originally flown with an uncovered fuselage but fabric was added later.

The machine was in use from May 1911 to the end of the season, becoming nicknamed the '*Oozley Bird*'. Eardley, who previously managed the Lane Gliding School, was the brother of Noel Pemberton Billing, and his wife ran the Bluebird restaurant at Brooklands until its closure at the outbreak of war.

Billing had previously made a ground trainer at Brooklands which was exhibited at the Stanley Show in November 1910. Two replica Billing biplanes were made in 1964 for the film '*Those Magnificent Men in their Flying Machines*'.

The Billing biplane was crashed on 4 October 1911 by N.S. Percival, who rebuilt it as the Percival Parseval I at the end of 1911. (*q.v.*)

BINGHAM monoplane (A. Bingham, East Grinstead, Sussex)

No information, other than dimensions, can be traced on this machine, reported in December 1910.

Data		
	Span	22ft
	Chord	4ft 6in
	Area	114 sq ft
	Length	22ft

BIRDLING monoplane & biplane. See Universal Aviation Co.

BIRMINGHAM AERO Co. (67 Moseley St., Birmingham)

This firm advertised in 1909 that it would manufacture aircraft to customers' designs or their own specially designed *Albatross* monoplane at prices between £250 and £600. There is no evidence that any were built.

BIRMINGHAM AERO CLUB gliders

This club was active from 1911 to 1914 and several members built full size and halfsize models. The secretary was G. Haddon Wood, who built two gliders and sent drawings and details of these to *Flight* and *The Aero*. The identified gliders built by club members are as follows:

J.H. Else-Mynard

Small monoplane built November 1909 and tried at Sutton Park. Chanute type built early in 1910.

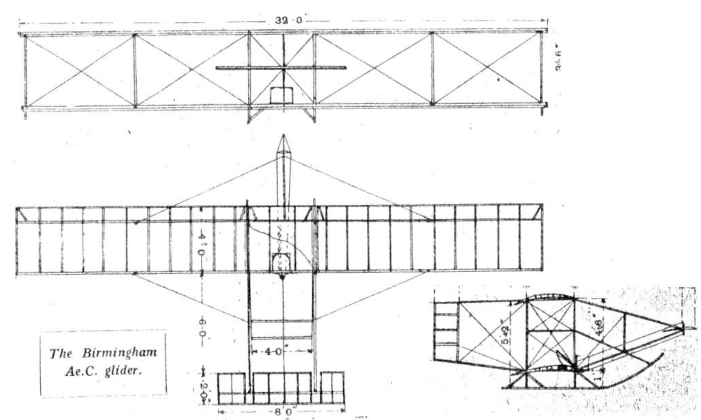

Birmingham Aero Club glider.

Billing biplane of 1911. Eardley Billing used the wings of a Voisin in its construction.

Else monoplane glider.

Haddon 1 glider.

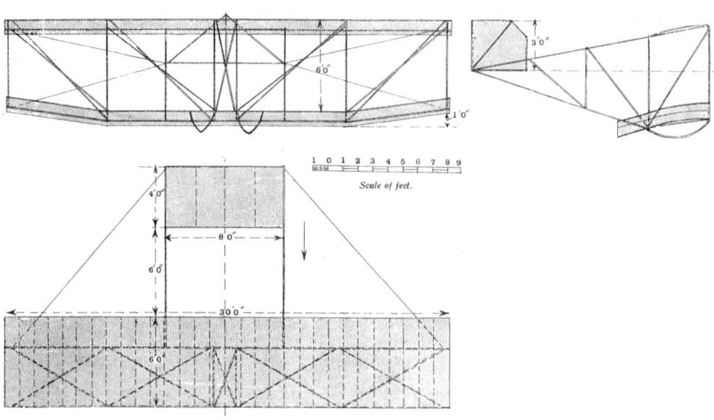

Plans of the second glider to be built by Mr. G. Haddon Wood, of the Birmingham Aero Club.

Haddon 2 glider.

Trykle glider.

Ernest Noble
Half-size monoplane built early in 1910 and tried in May.

G. Haddon Wood
Haddon 1. Built early 1911 and wrecked 26 August 1911. Wright type with front elevator and double-surfaced wings. Constructed of red deal.
Span 32ft Chord 4ft Area 292 sq ft Weight 130lb
Haddon 2. Completed January 1912; wrecked in a gale November 1912. Rear-mounted control surfaces and single-surfaced wings. Built of bamboo.
Span 32ft Chord 6ft Area 416 sq ft Weight 120 lb

Frank Warren
Built a 20ft span monoplane. Wrecked in a gale, unflown in October 1911.

R. Platts
Built a canard monoplane in March 1911 which was wrecked in a gale in October 1911. Subsequently rebuilt as a powered monoplane. Area 222 sq ft Weight 70 lb

Trykle
A biplane similar to *Haddon 1,* built of bamboo in early 1911. The top wing had upturned tips. Span of lower wing 20ft.

E. Prosser & A.M. Bonehill
These partners built a Chanute type glider in August 1910, which was wrecked at the same time as *Haddon 1* on 26 August 1911.

F. Hill
Designed a monoplane glider which was built by the Belmont Aeroplane Co. in 1911. Later converted to a hydroplane with a small engine.

N. Stamps
Completed a Blériot type monoplane glider in December 1913. Span 30ft Chord 7ft Area 210 sq ft Length 20ft

Club glider
Built from the remains of *Haddon 2* and *Trykle* gliders after they were wrecked in November 1912.

BLACK helicopter (R. Black, Mollinsburn and Cumbernauld, near Glasgow, Lanarkshire)
A report in 1909 described this machine as being fitted with two FN motorcycle engines driving a 'multi-blade fan and a two-bladed propeller above the head'. At the back were wings with a crossbar to control incidence. The engine was also reported as a seven-cylinder rotary
Black had previously made an unsuccessful biplane using the same powerplant.

BLACKBURN AIRCRAFT (Manufacture at Benson St., Balm Rd. & later at Olympia Works, Roundhay Rd., Leeds. Flight trials at Marske, Filey, Loft house Park & Roundhay Park, Leeds. Seaplane testing at Scalby Mills, Scarborough)
Robert Blackburn became interested in aviation while working in France, having seen Wilbur Wright and the French pioneers flying. He returned to Leeds in 1908 and with the help of Harry Goodyear built the first two Blackburn aircraft at Benson

St., following which, in 1911 the Blackburn Aeroplane Co. was formed, operating at Balm Rd. From June 1914 the company became the Blackburn Aeroplane and Motor Co. and took over a roller skating rink which became the Olympia Works.

BLACKBURN monoplane (Later referred to as the Heavy Type Monoplane)

This high wing monoplane carried the pilot and engine on a platform beneath the wing. The propeller was driven by chain and was carried on a shaft mounted just below the wing, with radiators on either side. The braced girder fuselage structure extended aft to carry a fixed tailplane and a cruciform tail with triangular surfaces, mounted on a universal joint for control in two directions; lateral control was by warping the wing.

Black helicopter of 1909.

Blackburn's first monoplane.

Blackburn's first monoplane was tested on the sands at Marske.

The machine was completed in September 1909 but was damaged beyond repair at Saltburn Sands on 24 May 1909. It had made short hops only and was dismantled later that year at Balm Rd.

Power: 35hp Green four-cylinder inline water-cooled driving a 8ft 6in diameter propeller at half engine speed by chain.

Data		
	Span	24ft
	Chord	6ft 5in
	Area	170 sq ft
	Length	23ft
	Weight allup	800 lb
	Speed	estimated 60 mph

A proposed military version with span increased to 30ft and length to 26ft was not built.

BLACKBURN monoplane (Second type, also referred to as Light Type Monoplane)

This single-seat machine was built in 1910 at Benson St. and was taken incomplete to the Blackpool Flying Meeting in July-August 1910, but was first flown at Filey on 8 March 1911. It later flew well and continued in use for some time.

The fuselage design was rather like an Antoinette, but with parallel chord warping wings and a triangulated undercarriage with twin skids.

Power: 40hp Isaacson seven-cylinder air-cooled radial with built in twenty-one reduction gearing

Data		
	Span	30ft
	Length	32ft

BLACKBURN *MERCURY* monoplanes

The third Blackburn design was basically a two-seat development of its predecessor and nine of these were produced with considerable differences between individual machines. All variants had substantial four wheel undercarriage with side springing. Warping wings were standard, except for a short period when one aircraft was fitted with a tapered wing.

The second Blackburn monoplane flew at Filey in March 1911.

The triangular-section fuselage was covered around the cockpit with polished veneered wood; all aft of this was covered with fabric. A variety of engines, initially an uncowled Isaacson radial, was fitted, but later there were radial and rotary powered machines with ring cowlings and finally a curved cowling and decking back as far as the cockpit. There were three distinct versions of the *Mercury*, including single seater types for racing and school work, identified as Type B.

The third version incorporated many internal changes, particularly the use of tubular steel wing spars to facilitate wing warping. The first of this version was fitted with a Renault vee-eight engine, although rotary and radial engines were mainly used. This version was known as the *Mercury* Passenger Type, although one at least was built as a single-seater.

Type I. One aircraft built

Power: 50hp Isaacson seven-cylinder air-cooled radial

Data		
	Span	38ft 4in
	Chord	7ft 6in
	Length	33ft
	Weight allup	1,000 lb
	Area	288sq ft
	Height.	6ft 9in
	Max speed	60 mph

Blackburn *Mercury* was built in several versions. This is the first type, a two-seater.

Blackburn *Mercury II* (Type B) single-seater with Gnôme engine.

Type II. Mercury B. Two aircraft built

No.1. A single-seater first flown at Filey in July 1911, but destroyed at Brooklands on 22 July 1911.

No.2. A single-seater first flown at Filey 7 July 1911, but converted to a two-seater in August 1911. Damaged at Eastbourne 23 March 1912 and rebuilt as a single-seater. Used at the Filey School from April 1912 to September, then transferred to Hendon School, where used until June 1913, when the school closed. Identified by 'Type B' on rudders and racing number "33" by the cockpit.

Power: 50hp Gnôme seven-cylinder air-cooled rotary driving a 8ft 6in diameter propeller.

Data		
	Span	32ft (36ft second machine)
	Chord	6ft 6in
	Area	200 sq ft (220 sq ft second machine)
	Length	31ft (32ft second machine)
	Height	8ft 6in
	Weight allup	700 lb
	Max speed	70 mph

Type III. Six aircraft built

No.1 60hp Renault eight-cylinder vee air-cooled. First flight at Filey 6 December 1911 but wrecked there on 6 December 1911. The original parallel chord wing had been replaced with a tapered wing.

No.2 50hp Isaacson seven-cylinder air-cooled radial. First flight May 1912, then used at Hendon School until June 1913. Fitted with a raised top rudder.

No.3 50hp Gnôme seven-cylinder air-cooled rotary. First flight at Brooklands 25 December 1911. Taken over by Lt. Spenser Grey RN, but damaged at Weymouth 10 January 1912. Reflown after repair at Eastchurch 21 February 1912, then to Hendon School in September 1912.

No.4 50hp Gnôme seven-cylinder air-cooled rotary. First flight at Filey in March 1912. Wings with cutaway roots.

No.5 50hp Anzani six-cylinder air-cooled radial.

No.6 50hp Gnôme seven-cylinder air-cooled rotary. First flight at Filey June 1912. Fitted with wings with cutaway roots and redesigned undercarriage.

Blackburn *Mercury III* with Isaacson radial engine.

Data		
	Span	32ft
	Chord	6ft 3in
	Area	195 sq ft
	Length	31ft
	Weight allup	800 lb
	Height	8ft 6in
	Speed with Renault	75 mph

Blackburn Type E. Military monoplane of 1912.

BLACKBURN Type E monoplane

This type was designed to meet the first official British specification for a military reconnaissance two-seater. The basic structure was of steel tube and the fuselage was covered with aluminum panels. The tail surfaces were mainly of light steel tube with fabric covering. The warping wings were constructed on two tubular steel spars, as for the Mercury III.

The first machine was built as a single-seater to a private order. The second of the two machines had a fairing behind the pilot which continued down to the dorsal fin and an extra streamlined fuel tank between the undercarriage struts. A developed version remained as a project only.

No.1 60hp Green four-cylinder inline water-cooled. First flown at Filey in April 1912 and supplied to Lt. Lawrence of the Indian Aviation Co. Ltd. at Brooklands in May, identified as '*No.1 L'Oiseau Gris*'. After further tests this machine was abandoned.

No.2 70hp Renault eight-cylinder vee air-cooled. When tested at Knavesmire, Yorkshire, in June 1912 the machine failed to rise. Except for being shown on the ground at the Cockburn High School, Dewsbury Rd., Leeds, late in 1912, no further use was made of this machine, which it had been planned to enter in the Military Trials, to be flown Lt. Spenser Grey and Capt. R. Gordon.

Data		
	Span	38ft 4in
	Chord	7ft 6in
	Weight allup	950 lb
	Area	290 sq ft
	Length	31ft 2in
	Max speed	80 mph
	Endurance	4hr (1st m/c) 5hr (2nd m/c)

Blackburn 1912 single-seater monoplane. One still flies with the Shuttleworth Collection.

BLACKBURN monoplane 1912 single-seater type

This machine was built to the private order of C.E. Foggin and it flew at the end of 1912 in the hands of Harold Blackburn at Leeds. After initial tests it was taken over by the owner early in 1913. Later the hooked undercarriage skids were replaced with more usual curved type and the wingtips were rounded off. A new owner, M.F. Glew, took over but crashed the machine at Wittering in 1914 and it was not rebuilt before the war began. R.O. Shuttleworth acquired the remains in 1938, but rebuilding was delayed by the second outbreak of war. Nevertheless the machine was rebuilt and re-flown on 17 September 1949 at Henlow.

It remains with the Shuttleworth Collection at Old Warden as the oldest flyable British designed aircraft.

Power: 50hp Gnôme seven-cylinder air-cooled rotary.

Data	Span	32 ft 1 in
	Chord	7ft
	Area	236 sq ft
	Length	26ft 3in
	Height	8ft 9in
	Weight	550 lb
	Weight allup	980 lb
	Max speed	60 mph
	Endurance	2 1/2 3 hr

BLACKBURN monoplane Type I

This was the two-seater development of the 1912 machine with more power and the size increased to accommodate the passenger in the front cockpit. Three of the type were built, the first to the order of a private owner, with minor variations between each machine. There were visible differences to the engine cowlings, where cutaways were made to improve the cooling and air intake supply, and later when no cowling was fitted to the radial engine. One machine had a single kingpost and was a single-seater with freight compartment in place of the front cockpit. The original version had one large cockpit for both crew, but this was modified to make two separate cockpits.

The third machine was known as the 'Improved Type I', and this was later modified into a twin-float seaplane with tail float. This was known as the 'Land/Sea' monoplane since the main floats were clamped direct to the original undercarriage structure and the machine could readily be converted back for land use.

Type I. Two machines built

Power: 80hp Gnôme seven-cylinder air-cooled rotary driving a 7ft diameter propeller.

No.1 Built for Dr. M.G. Christie as a two-seater and first flown on 14 August 1913 at the Yorkshire Aerodrome, Loft house Park, Leeds. It was used extensively until the outbreak of war, when it was commandeered, after which there is no record of its use.

No.2 This was a single-seater with freight compartment and single kingpost. First flown by Harold Blackburn on 14 December 1913. Damaged at York in May 1914. Acquired by W.R. Ding of the Northern Aircraft Co. and converted to a twinfloat seaplane at Cockshot Point, Windermere. It was found to be unsatisfactory and not flown.

Blackburn Type I two-seater monoplane of 1913.

Improved Type I. One machine built

Power: 80hp Gnôme seven-cylinder air-cooled rotary, later replaced by a 100hp Anzani ten-cylinder air-cooled radial when converted to Land/Seaplane. The only machine built was a two-seater shown at the Olympia Aero Show in March 1914. It was commandeered in September 1914 but, being of no military

Blackburn Improved Type I with Sydney Pickles at West Auckland.

Blackburn Type I seaplane with Anzani radial engine. W.R. Ding at Lake Windermere.

value, was returned to the makers. It was then sold to W.R. Ding of the Northern Aircraft Co. in 1915 as the Land/Sea monoplane. It was first flown on floats on 26 October 1915 at Bowness-on-Windermere, where it was used as a trainer until written off on 1 April 1916.

Data		
	Span	38ft
	Chord	6ft 8in
	Area	252 sq ft
	Length	28ft 6in
	Seaplane length	29ft 6in
	Weight (Type I)	950 lb
	Allup (Type I)	1,500 lb
	Weight (Land/Sea)	1,124 lb
	Allup (Land/Sea)	1,733 lb

BLACKBURN Type L seaplane

After preparing a seaplane project in 1913, Blackburn proceeded with the design of a large 'hydrobiplane' for entry in the 1914 Seaplane Circuit of Britain. This was Blackburn's first biplane and the first to be built at the Olympia Works. When the contest was abandoned the Type L was commandeered by the Admiralty and moved in August to Scalby Mills, where it was tested and flown by freelance pilot Sydney Pickles. It was operated for a period on coastal reconnaissance, armed with a machine gun. Early in 1915 the machine crashed into a cliff at Speeton in poor visibility, when being flown by Rowland Ding from Scarborough to Killingholme.

Blackburn Type L seaplane on Scarborough beach.

The Type L was a conventional biplane design of the period, although it was the first Blackburn machine with a square-section fuselage. During the course of its development, the longspan, parallel ailerons on the top wing were replaced by shortspan, tapered units. Other changes included the removal of the engine cowling and the moving of the radiators to the rear center section struts.

A proposed larger development of the Type L was not proceeded with.

Power: 130hp Salmson (Canton-Unné) nine-cylinder water-cooled radial.

Data		
	Span top	49ft 6in
	Span bottom	35ft
	Chord.	6ft
	Area	481 sq ft
	Length	32ft 6in
	Height	12ft 6in
	Weight	1,717lb
	Weight allup	2,475 lb
	Max speed	85 mph
	Ceiling	11,000 ft
	Range	445 miles

1913 Project.

Power: 80hp Gnôme seven-cylinder rotary or 100hp Anzani ten cylinder both air cooled.

Data		
	Span top	44ft.
	Span bottom	36ft
	Chord	5ft
	Area	410 sq ft
	Length	33ft
	Weight.	1,250 lb
	Max speed	estimated 65 mph

1914 Project.

Power: 200hp Gnôme Monosoupape nine-cylinder air-cooled rotary.

Data	Span top	62ft
	Span bottom	47ft 6in
	Weight	1,450 lb
	Weight allup	3,000 lb

BLACKPOOL & FYLDE AERO CLUB glider (Sec. Jack Kemp)

Under construction in February 1910.

Blackpool & Fylde Aero Club glider of 1910.

BLAKE monoplane (Blake, Magheramorne, County Antrim, Ireland)

This machine was advertised for sale for £120 in *The Aero* in August 1910. It was powered by a 12hp JAP engine driving a Spencer propeller, The wing span was 32ft.

BLAND Mayfly biplane (Miss Lilian E. Bland, Carmoney, Belfast, N. Ireland assisted by S. Girvany of Ballymore)

This machine was built as a glider and tested as a kite and in free flight in early 1910. Various changes were made as a result of these tests before the conversion to power was made. Miss Bland sent a series of letters to '*Flight*'' and reported that the aeroplane was flown under power by September 1910. She later described in detail the construction of the machine in '*Flight*' of 17 December 1910 (p.102-157).

Bland *Mayfly* powered version of 1910.

MAYFLY glider

Data	Span	27ft 6in
	Chord	5ft
	Area	200 sq. ft
	Weight	200lb

MAYFLY biplane

The final version was a pusher biplane with open braced booms front and rear, mounted on a tricycle subframe with skids. Unbalanced ailerons were fitted just below the top wing and the front elevators were also capable of being operated differentially. Rear elevators and a small fixed tailplane were employed. For ease of transport the booms were removed and the subframe turned through 90 degrees.

By March 1911 Miss Bland was flying a scale model of a proposed 30ft span Mayfly 2 which was not built, for she married later and disposed of the dismantled aircraft at the end of the year. The engine was bought by the Windsor Model and Gliding Club.

Power: 20hp Avro (Edwards) two-cylinder horizontally opposed air-cooled two-stroke driving a 6ft 6in diameter Avro propeller.

Data	Span	27ft 7in
	Chord	5ft
	Area	250 sq ft
	Length	23ft
	Height	9ft 10in

Bland *Mayfly* glider flown as a kite.

Bland *Mayfly* the final version of 1911.

BLÉRIOT AERONAUTICS (Louis Blériot, Belfast Chambers, 156 Regent St., London, W. Flying School & Works at Hendon and Brooklands)

After Blériot's Channel crossing in July 1909 in his Type XI monoplane, a number of similar machines were made by various firms and individuals, mostly without approval by Blériot. A Blériot School was established at Hendon on 1 October 1910, and in 1914 sheds were taken at Brooklands and enlarged to provide offices and manufacturing facilities, all under the control of M. Norbert Chereau, Blériot's manager for Great Britain.

The RFC used various types of Blériot monoplanes operationally prior to and in the early months of the war, but these were soon relegated to training duties. The majority of Service aircraft were probably French made, but the British company received contracts for the Type XI single-seater, the Type XI-2 two-seater and the single-seat parasol version of 1914. Deliveries of these continued well into 1916, by which date the works were in the process of being transferred to new premises at Addlestone, Surrey.

The quantities of Type XI and XI-2 aircraft made in Britain is uncertain, but it is recorded that at least fourteen parasols Serial Nos.575-586 and 2861-2862 were made at Brooklands.

Blériot Type XI monoplane

Power: 50,70 or 80hp Gnôme seven-cylinder air-cooled rotary.

Data		
	Span	29ft
	Length	25ft
	Area	161 sq ft
	Height	8ft 2in
	Weight	585 lb
	Weight allup	915 lb
	Max speed	65 mph (50hp)
	Endurance	3 hr

Blériot Type XI-2 monoplane

Power: 70 or 80hp Gnôme seven-cylinder air-cooled rotary.

Data		
	Span	34ft
	Length	27ft 6in
	Area	205 sq ft
	Height	8ft 2in
	Weight	738 lb
	Weight allup	1,290 lb
	Max speed	75 mph (80hp)
	Endurance	3 1/2 hr

Blériot XI-2. The two-seater.

Blériot XI single-seater. The earlier version was the inspiration for many British constructors.

Blériot Parasol monoplane used by the RFC was built at Brooklands and Addlestone.

Blériot parasol type monoplane

Power: 80hp Gnôme seven-cylinder air-cooled rotary.

Data	Span	29ft 4in
	Length	25ft 8in
	Area	194 sq ft
	Weight	615 lb
	Weight allup	925 lb
	Max speed	68 mph

BONE monoplane (CA Bone)

A letter in '*Flight* ' of 23 July 1910 (p.586) gave a sketch of a monoplane said to be under construction. '*Flight* ' recommended a 20hp motor. Nothing more was heard of this machine.

Data	Span	17ft 6in
	Chord	3ft 9in
	Length	15ft 9in

BONNARD biplane (L.H. Bonnard, International Aero Co., Huntingdon, Cambridgeshire)

This small single-seat pusher biplane was built on Farman lines except for the position of the ailerons, which were hinged to the front outboard interplane struts. At the time the machine was described and illustrated in '*The Aero* ' on 28 December 1910 (p.506) no engine was fitted.

Data	Span	26ft
	Chord	4ft 3in
	Area	221 sq ft
	Tailplane	28 sq ft
	Elevators	17 sq ft each

Bonnard biplane of 1910.

BOULTBEE monoplane (Harold E. Boultbee, Holy Trinity Vicarage, Burton-on-Trent and James Gardiner, Woolton, Liverpool)

This was a single-seater, midwing monoplane, with a fuselage constructed of steel tubing. The longerons curved down to the nose, where a bearing supported the propeller shaft. The rotary engine was set back in the fuselage, where it was mounted on a cruciform frame carried on the four longerons. The wing was constructed of wood and warping was used for lateral control. The control system was covered by patent No.14990/1909. The under-carriage, fitted with long coil springs, was raked well forward as a precaution, owing to uncertainty about the fore and aft balance. It was to be retractable in flight and was covered by patent No.17291/1909. A tall strut with tail wheel was fitted.

Boultbee monoplane under construction in 1909. It was not completed.

The partners separated and work on the machine was stopped at an advanced stage. Boultbee later held various senior design positions at Bristol and Handley Page and for a time, between the wars, had his own company, the Civilian Aircraft Co.

Power: 12hp Scott two-cylinder horizontally-opposed rotary driving a 6ft Hollands two-bladed steel propeller.

Data	Span	36ft

BOUSTEAD glider (Cedric Boustead, Wimbledon Park, London)

The machine was a biplane with curved lower wing brought up to join the tips of the upper wing. It was a canard controlled by the front elevator. The main structure was of ash with a bamboo elevator. It was flown briefly but successfully at Wimbledon Park.

Data	Span	22ft 6in
	Chord	6ft 6in
	Area	285 sq ft
	Span of elevators	13ft
	Weight	150 lb

Boustead canard glider flown at Wimbledon in 1910.

BRADSHAW monoplane (Granville Bradshaw, Wolverhampton, Staffordshire)

This machine was designed and built by Granville Bradshaw at Wolverhampton when he was a member of the Midland Aero Club, before he joined the Star Engineering Co. A JAP motorcycle engine fitted with an adjustable pitch propeller was used. The machine was present at the Midland Aero Club Show at the park in Sutton Coldfield on 13 November 1909.

BRAGG-SMITH biplane (G.P. Bragg-Smith, Mitcham, Surrey)

Bragg-Smith, a model maker, began construction of a full size canard biplane in 1911, but it was not completed. The machine was shown in model form at Olympia in April 1911 and won first prize for workmanship at the display organized by the RAeC, the AA and the Motor Union.

Bragg-Smith model of automatic stability biplane. The fullsize machine was started at Brooklands in 1911 but was not completed.

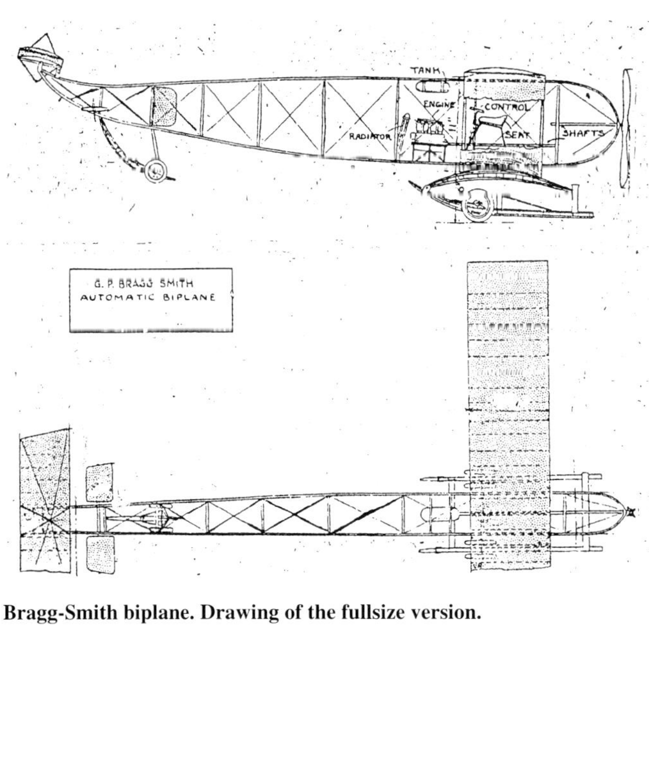

Bragg-Smith biplane. Drawing of the fullsize version.

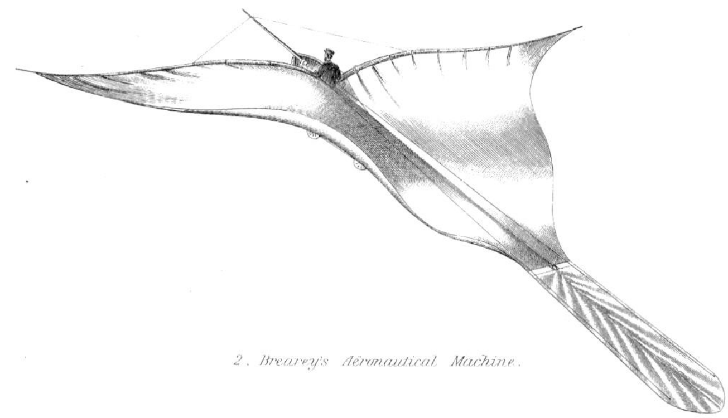

Brearey glider. One of a number of designs by this early pioneer.

BREAREY glider (F.W. Brearey)

The real founder of the Aeronautical Society in 1866 was Fred Brearey, who was to be its Hon. Secretary for its first thirty years. When the first enthusiasm in the Society started to flag it was Brearey who kept it alive until his death in 1896; by then there had been a great revival of interest in aviation matters.

Brearey was an ardent advocate of heavier-than-air flight and designed at least one glider. This was an elegant-looking machine, with birdlike wings and tail. It is not known if it was ever built.

In 1879 Brearey filed a patent (No.2376) for a flying machine described as 'an elongated body pointed at both ends contains the requisite machinery and the passengers Flexible lever arms extend on either side, and a flexible spar extends from the tail end of the body. Silk or other suitable fabric is extended from the arms and along the spar of the tail, thus giving a large supporting surface, and vibrations are imparted to the area which propels the machine by a wave like motion'.

BREGUET AEROPLANES Ltd. (Works, 5 Hythe Rd., Cumberland Rd., Willesden NW. Head Office, 1 Albemarle St., Piccadilly, London.)

This subsidiary of the French company of Douai was established in mid 1912, with a view to obtaining orders for military aircraft from the British government. In the event it would seem that no more than seven Breguet aircraft were used by the British services and the majority of these were imported from France.

The Breguet biplanes had a number of original features. They had an all steel primary structure, the forward fuselage incorporating four channel section longerons, while from the pilot's seat aft a large diameter steel tube continued to the tail and this was encased by an aluminum shell forming the skin of the fuselage. The cruciform tail unit was mounted on a universal joint and moved in two directions.

The unequal span two bay wings were each built on a tubular spar on which the ribs could be rotated against spring pressure, allowing flexibility for warping the top wing for lat-

Breguet Type U.1 biplane. One of a small number built by the British factory.

eral control and serving to alleviate strain induced by gusts. The incidence of the wings could be adjusted on the ground and they could be folded very compactly for transport.

The steel tube undercarriage incorporated main oleopneumatic spring struts with three, later four wheels, the forward of these being steerable by the rudder control wheel.

Various types of engine were fitted, including a horizontal installation of the radial water-cooled Salmson. Of the seven various Breguet aircraft identified as operated by the RFC, the known British-built aircraft, probably serial number 310 with 85hp Salmson is representative. All of the Breguets were withdrawn from service by January 1914.

BREGUET biplane Type U.1

Power: 85hp Salmson (Canton-Unné) seven-cylinder water-cooled radial driving an 8ft 6in diameter propeller.

Data		
	Span top	45ft
	Span bottom	39ft
	Chord	5ft
	Gap	6ft 6in
	Area	400 sq ft
	Height	13ft.
	Length	29ft
	Weight	1,550 lb

Breguet with horizontal Salmson engine. British-built and generally similar to other current Breguet types.

BRISTOL AIRCRAFT (British & Colonial Aeroplane Co. Ltd., Clare St., Filton, Bristol

The company was established at Filton early in 1910 and later that year acquired sheds at Brooklands and Larkhill for test flying and training schools. In order to establish itself as a manufacturer the company obtained a license from the Société Zodiac of Paris for its biplane, an example of which it was arranged to exhibit at the Olympia Aero Show in March, together with a monoplane, which did not arrive. The biplane was not a success, and thereafter the company produced its own designs, the first of which became well known as the Boxkite. It was similar to a Farman and was most successful. The future of the enterprise was assured, and it expanded to produce a succession of types, making the company the largest producer of aircraft in the country in the early years of the industry.

The Flying Schools at Brooklands and Larkhill were equally successful and were responsible for training the highest number of pilots of any such establishment between 1910-1914.

BRISTOL ZODIAC biplane Type 52B

This machine, designed by Gabriel Voisin for the Société Zodiac, was supplied to Filton incomplete for exhibition at Olympia in 1910. After the show it was taken to Filton for completion and then to Brooklands, where it made a brief hop on 28 May, after new wings with more camber were fitted and the side curtains removed. A change to a lighter engine was also reported, but the machine would still not fly and after damage on 15 June was finally abandoned. A batch of five Zodiacs in work at Filton was scrapped.

Power: 50hp Darracq four-cylinder inline water-cooled. Reportedly replaced by a 40hp Gregoire four-cylinder inline water-cooled.

Data		
	Span	33ft 3in
	Area	525 sq ft
	Weight	1,000 lb
	Length	39ft 3in
	Speed	35 mph

Bristol license-built *Zodiac* biplane of 1910. The company's first machine was not a success.

BRISTOL VOISIN monoplane

The only record of this machine, which was designed and partly built by Voisin Frères, was an illustration and brief details in *Flight* of 19 August 1911 (p.727). It had been hoped to acquire a Zodiac monoplane for exhibition in March 1910 and although the report is much later, this may have been the machine, since Voisin had already been engaged as a consultant to work for Bristol. This machine is seldom featured in published records of Bristol aircraft.

Power: 70hp Gnôme seven-cylinder air-cooled rotary.

Data		
	Span	36ft 1in
	Area	215 sq ft
	Length	26ft 3in
	Weight	882 lb

Bristol Voisin monoplane of 1910-1911 was the sole example.

BRISTOL BOXKITE biplane

Bristol took the easy way to establish itself, after the failure of the Zodiac, by copying, with some improvements in both design and workmanship, the Farman pusher biplane. The engineer was G.H. Challenger.

The prototype was unable to fly with the Gregoire water-cooled engine at first fitted. This was replaced by a 50hp Gnôme and when taken to Larkhill flew extremely well on 29 July 1910. Gnôme engines, some of increased power, were mainly fitted to the subsequent aircraft, which reached a total of seventy-six when production was completed in 1915. Of this number, twenty-two were exported to nine different countries.

Bristol Boxkite with extended upper wing.

The Boxkite was basically a two-seater largely employed for pilot training at the Bristol Schools and elsewhere. Apart from the use of various engines, other differences of note on individual machines were the use of three rudders, protection for the crew and double surfaced wings. Upper wing extensions were fitted to the 'Military' type.

Two special single-seat racing machines were also constructed; the first was No.44 flown on 30 May 1911 for M. Tetard to fly in the Circuit de l'Europe (Racing No.3) in which it failed to finish due to engine trouble. This machine had single bay wings with upper wing extensions and a nacelle. The second was a redesign by Gabriel Voisin, with reduced gap and front booms and elevator removed and with single rudder and tailplane. It was at Larkhill in February 1912 in this form, but was rebuilt as a standard school biplane.

Power: 40/50hp Gregoire four-cylinder inline water-cooled initially in prototype (No.7)
60hp ENV type F eight-cylinder water-cooled vee (two aircraft only Nos.8 and 16)
60hp Renault seven-cylinder water-cooled vee. (one aircraft only No.39)
70hp Gnôme seven-cylinder air-cooled rotary. (six aircraft Nos.31, 32, 42, 60, 79 and 139)

Bristol Boxkite with Gnôme rotary engine. The majority built were of similar type.

Bristol Boxkite No.44. One of several variants built for racing.

Data	**Standard Type**	
	Span	34ft 6in
	Area	457 sq ft
	Length	38ft 6in
	Height	11ft 10in
	Weight	800 lb
	Weight allup	1,050 lb
	Speed	40 mph
	Production	15
Data	**Military Type**	
	Span	47ft 8in or 46ft 6in
	Area	517 sq ft
	Length	38ft 6in
	Height	11ft 10in
	Weight	900 lb
	Weight allup	1,150 lb
	Speed	40 mph
	Production	61
Data	**Racing Type No.44**	
	Span	35ft
	Area	350 sq ft
	Length	38ft
	Height	11ft 10in
	Weight	800 lb
	Weight allup	1,000 lb
	Speed	50 mph
	Production	1
Data	**Racing Type No.69**	
	Span	32ft 8in
	Area	420 sq ft
	Length	30ft 9in
	Height	9ft 6in
	Weight	800 lb
	Weight allup	1,000 lb
	Speed	50 mph
	Production	1

BRISTOL glider

This biplane glider was designed by Challenger for Sir George White to present to the Bristol and West of England Aero Club, of which he was President. It was similar to the Boxkite in layout, and designed to carry two people and was later to be fitted with a 30hp engine, which was never installed. It was flown for the first time on 17 December 1910 by Challenger at Keynsham, Somerset, and continued in use until 1912, the inevitable damage being repaired from time to time by the company.

Data	Span	32ft 4in
	Length	33ft 10in
	Height	6ft 8in

The glider built for the Bristol & West of England Aero Club.

BRISTOL Racing biplane. (Challenger-Grandseigne 1911)

This single-seater, unequal span biplane was designed by Frenchmen, M.M. Grandseigne and Versepuy, under the supervision of G.H. Challenger and first appeared at the Olympia Aero Show in March 1911. It was wrecked on its first attempt to fly at Larkhill in April.

Power: 50hp Gnôme seven-cylinder air-cooled rotary.

Data	Span	27ft
	Area	210 sq ft
	Length	25ft
		one aircraft built, works no. 33.

Bristol-Grandseigne No.33 racing biplane of 1911.

BRISTOL monoplane. (Challenger-Low 1911)

This single-seater was the first monoplane designed at Bristol, and two were built, the first arriving at Larkhill in February 1911. The tests were curtailed to return the machine to Filton to prepare it for exhibition at Olympia in March. The second machine was shown at St. Petersburg, Russia in April. The type was abandoned after unsuccessful attempts to fly at Larkhill by Versepuy. The machine had a triangular-section fuselage and warping wings.

Power: 50hp Gnôme seven-cylinder air-cooled rotary.

Data		
	Span	33ft 6in
	Length	31ft 6in
	Area	215 sq ft
	Weight	580 lb
	Weight allup	760 lb
	Speed	estimated 55 mph

Two aircraft built works Nos.35 and 36.

Bristol-Challenger-England-Low No.35 monoplane of 1911.

BRISTOL biplane Type T. (Challenger-Dickson)

This single-seat racing biplane was a new design based on the Boxkite experience and recommendations by Capt. Dickson. Five of the type were built and a sixth partly finished.

The first, Works No.45, was flown by Tabuteau in the Circuit de l'Europe in June-July 1911, in which he finished ninth. Works Nos.51, 54 were built for a team to compete in the *Daily Mail* Circuit of Britain Contest held in July 1911, but none achieved success in this.

The first four machines had 70hp Gnômes, but No.54 was fitted with a 60hp Renault; No.78, the last machine, was not completed, but would have had a 100hp Gnôme; No.51 was sold to a private owner and was fitted with a 50hp Gnôme, but was soon involved in a fatal crash. One of the remaining machines was converted to become the Challenger-England biplane No.59 (*q.v.*).

Bristol Type T No.45 raced by Maurice Tabuteau in 1911.

Bristol-Challenger-Dickson Type T No.52. One of a batch of racing biplanes based on the Boxkite.

The first machine had twin rudders set close together, but on the later machines the rudders were moved into line with the booms. Other changes were made to the nacelle and flying controls.

Power: 70hp Gnôme seven cylinder air cooled rotary. Works Nos.45, 51, 53.
60hp Renault eight-cylinder air-cooled vee. Works No.54.
100hp Gnôme seven-cylinder air-cooled rotary. Works No.78 not completed.
50hp Gnôme seven-cylinder air-cooled rotary. Fitted before sale in No.51.

Data		
	Span	35ft
	Area	350 sq ft
	Length	24ft 6in
	Weight	800 lb
	Weight allup	1,000 lb
	Speed	58 mph

Five aircraft built, one aircraft partly built.

BRISTOL biplane. (Challenger-England)

This was the conversion to tractor configuration of one of the Type T pushers. The work was mainly done by Gordon England, a pilot with no formal design qualifications but with a good practical background.

Bristol-Challenger-England No.59. A conversion of a Type T to tractor type.

The front booms and elevator were discarded and a new single-seater nacelle was built on the lower wing. A single rudder, centrally mounted tailplane and divided elevator replaced the originals.

The machine flew from Larkhill in November 1911 and, although not very successful, was flown at intervals by advanced pupils until 19 May 1912, when it was taxied into spectators, one of whom was killed. The machine was then dismantled.

Power: 60hp ENV type F eight-cylinder water-cooled vee.

Data

Span	35ft
Area	350 sq ft
Length	23ft

Bristol-Prier-Dickson monoplane No.73. This is a two-seater short fuselage version.

BRISTOL monoplanes. (Type Prier P1, Prier-Dickson and Prier-Coanda)

Following the earlier unsuccessful monoplane design, the company engaged Pierre Prier, chief instructor at the Blériot School at Hendon and a capable engineer, to prepare the next monoplane design. The first P1, Works No.46, was not ready for the Gordon Bennett Cup Race on 1 July 1911 but flew soon after. The second and third Nos.56, 57 intended for the Circuit of Britain Contest later that month, also failed to start, No.56 being crashed by Prier before the race and No.57 having a pilot problem.

Thereafter production of both single-seater and two-seaters of five versions proceeded to a total of thirty-four aircraft, when production terminated in December 1912, with two major rebuilds recorded in addition. Of the total, fourteen were exported including those used at the Bristol schools in Germany and Italy. Only two aircraft were bought by the British government owing to the ban on the use of monoplanes by the military.

Bristol-Prier-Dickson monoplane No.98. Advanced trainer with short fuselage.

The original machine had a front mounting to the rotary engine, which was not used on subsequent aircraft. All types had warping wings and sprung undercarriage skids, but a fixed tailplane and hinged elevators were introduced and the fuselage lengthened by 2ft 6in for the majority of machines from No.82 onwards, when the type became known as the Prier-Dickson because of the latter's involvement.

After Prier left the company Henri Coanda designed a side-by-side two-seater, of which three were built.

Power: 50hp Gnôme seven-cylinder air-cooled rotary. Three P1 aircraft Nos.46, 56 and 57. Eleven short fuselage two-seaters Nos.58, 71-76, 83, 84, 90 and 94.
70hp Gnôme seven-cylinder air-cooled rotary. Ten long fuselage two-seaters and three side-by-side Nos.82, 85, 89 and 107-109.
35hp Anzani three-cylinder air-cooled fan type radial. Seven single-seaters Nos.68, 81, 95-98 and 102
40hp Isaacson seven-cylinder air-cooled radial. Replacement for 50hp Gnôme. One single-seater No.56.

Data **P1 single-seater**

Span	30ft 2in
Area	166 sq ft
Length	24ft 6in
Height	9ft 9in
Weight	640 lb
Weight allup	820 lb
Speed	68 mph

Bristol-Prier-Dickson monoplane long fuselage two-seater.

Bristol-Prier school monoplane. A low-powered single-seater.

Bristol GE 1 biplane Works No.64 at Larkhill 1911-1912.

Data	**School single-seater**	
	Span	30ft 2in
	Area	166 sq ft
	Length	24ft 6in
	Height	9ft 9in
	Weight	620 lb
	Weight allup	780 lb
	Speed	58 mph
Data	**Short fuselage two-seater**	
	Span	32ft 9in
	Area	185 sq ft
	Length	24ft 6in
	Height	9ft 9in
	Weight	650 lb
	Weight allup	1,000 lb
	Speed	65 mph
Data	**Long fuselage two-seater**	
	Span	34ft
	Area	200 sq ft
	Length	27ft
	Height	9ft 9in
	Weight	660 lb
	Weight allup	1,080 lb
	Speed	65 mph
Data	**Side-by-side two-seater**	
	Span	35ft 6in
	Area	200 sq ft
	Length	27ft
	Height	9ft 9in
	Weight	660 lb
	Weight allup	1,080 lb
	Speed	65 mph

BRISTOL GE 1 biplane (Gordon England)

After the conversion of the Type T biplane, Gordon England was responsible for the design of a two-seater, side-by-side, two bay biplane for military use; it was to be easily transportable. The machine was tested in May and June 1912 and passed to the German Bristol Co., but was found to be unsuitable for school use. It was returned to Filton in September and scrapped.

The fuselage had a front mounted radiator and a hinged bonnet and was plywood covered at the front. The lower wing passed below the fuselage. Lateral control was by warping with dual control by wheel. The long fin was later deleted and one as on the GE.2 fitted. One aircraft, Works No.64, was built.

Power: 50hp Clerget four-cylinder inline water-cooled driving the propeller at half engine speed.

Data		
	Span	33ft 8in
	Area	320 sq ft
	Length	29ft
	Speed	65 mph

BRISTOL GE.2 biplanes

Two aircraft, developed from the GE.1, were built for entry in the Military Trials of August 1912. Both machines had increased span and rounded wing tips and the fuselage was mounted above the lower wing. One aircraft was fitted with an air-cooled rotary, the other retained a front radiator for a water-cooled engine with geared down propeller. Two aircraft were built, Works

Bristol GE 2 biplane Works No.103 of 1912. No.12 in the Military Trials.

Bristol GE2 biplane Works No.104 with Daimler engine at Larkhill.

No.103 with Gnôme was No.12, and Works No.104 with Daimler-Mercedes was No.13 in the Military Trials.

Power: 100hp Gnôme fourteen-cylinder air-cooled rotary driving a 9ft 6in diameter propeller.
70hp Daimler-Mercedes four-cylinder inline water-cooled driving a four-bladed propeller at half engine speed.

Data **No.103**

Span	40ft
Chord	5ft
Area	400 sq ft
Length	31ft
Weight	1,080 lb (No.104, 1,100 lb)
Weight allup	1,980 lb (No.104, 2,000 lb)
Speed	68 mph (No.104, 62 mph)

BRISTOL GE.3 biplanes

The GE.3 was the final development from the GE.1 and was intended for the Turkish government. The seats were in tandem and the fuselage was faired to circular section. The wings, tail surfaces and detail features were derived from the GE.2. Turkey, at war with Italy, was blockaded by the Italians and delivery was prevented. Problems also existed with weakness of the wing structure and the design was abandoned. Two aircraft were built Works Nos.11, 23.

Power: 80hp Gnôme seven-cylinder air-cooled rotary driving an 8ft diameter propeller

Data

Span	39ft 4in
Chord	5ft
Area	387 sq ft
Length	26ft 3in
Weight	1,096lb
Weight allup	1,996lb
Speed	65 mph

BRISTOL COANDA monoplanes. (Henri Coanda)

After working in France, Henri Coanda, son of the Rumanian War Minister, began work as a designer at Bristol in January 1912. After carrying out some redesign of the Prier-Dickson monoplanes, he followed with a tandem two-seater of similar type. The prototype, Works No.77 went to Larkhill for testing in March 1912 and was followed by a batch of five more for Italy and Rumania. In March a side-by-side version, Works No.80 was flown followed by a further six, of which Italy purchased two and Rumania three.

Two of a variant for the British Military Trials were produced later in the year, Works Nos.105-106 and both were entered. After achieving joint third place, both were bought by the War Office for operation by the RFC as serial Nos.263 and 262 respectively. The former machine crashed with fatal results to the crew on 10 September 1912, and this initiated a ban by the War Office on the use of monoplanes, which was to have far-reaching effects, although the Admiralty continued with their use.

A single machine built with new wings, Works No.111, was tested but was not a success. The final version of the Coanda monoplanes was affected by the accident and although twenty were built, most were converted to biplanes. The overseas sales, mainly to Rumania and license manufacture in Italy, were badly affected, resulting in only one machine being built by Caproni.

All machines were built as two-seaters except for Works No.183, known as the SB.5, construction of which for Italy was discontinued. The fuselage was used later in the construction of the Scout Type A.

Power: 50hp Gnôme seven-cylinder air-cooled rotary. Six school aircraft, Works Nos.77, 132, 186-189. Seven side-by-side, Works Nos.80, 110, 164-166 and 176-177.
80hp Gnôme seven-cylinder air-cooled rotary. Two Military Trials aircraft, Works Nos.105-106
70hp Daimler four-cylinder inline water-cooled. One aircraft, Works No.111.

Bristol Coanda monoplane Works No.105 of 1912. No.14 in the Military Trials.

80hp Gnôme seven-cylinder air-cooled rotary. Twenty aircraft approx., Works Nos.118, 121-123, 131, 142, 154 and 196 plus one built by Caproni

Data **School**

Span	40ft
Area	275 sq ft
Length	27ft
Height	7ft
Weight	770 lb
Weight allup	1,100 lb

Data **Side-by-side**

Span	41ft 3in
Area	275 sq ft
Length	27ft
Height	7ft
Weight	770 lb
Weight allup	1,100 lb

Data **Military Competition**

Span	40ft
Area	242 sq ft
Length	28ft 3in
Height	7ft
Weight	1,000 lb
Weight allup	1,710 lb

Data **Daimler**

Span	39ft 4in
Area	260 sq ft
Length	30ft 9in
Height	7ft
Weight	1,200 lb
Weight allup	1,850 lb

Data **Military**

Span	42ft 9in
Area	280 sq ft
Length	29ft 2in
Height	7ft
Weight	1,050 lb
Weight allup	1,775 lb

BRISTOL COANDA hydro-biplanes

A central float two-seater seaplane, attributed to Coanda but similar in some respects to the GE.3 land-plane, was sent to the Saunders works at Cowes for trials early in 1913. The float, designed by Gnosspelius, was too heavy and the machine could not lift off from the water by pilot Harry Busteed. A Saunders-made Consuta float was fitted, and on 15 1913 the aircraft took off successfully, but engine failure necessitated an emergency alighting, the force of which burst the float and nearly resulted in the drowning of Busteed. The aircraft, Works No.120, was destroyed and a replacement for the Admiralty (Serial No.15) was provided by conversion of one of the Coanda monoplanes (Serial No.121). This was tested initially as a land-plane at Larkhill but needed a new fuselage. After the rebuilding it was tested as a seaplane in September 1913 at Dale in Pembrokeshire. It was further rebuilt as Works No.205 and delivered again early in 1914. Further improvements were needed and the machine was again largely rebuilt with staggered wings with ailerons and a new fin, identified as Type TB.8H.

Bristol Coanda seaplane at Cowes for testing by Harry Busteed.

Power: 80hp Gnôme seven-cylinder air-cooled rotary.

Data **Works No.120 (No type number)***

Span	38ft 8in
Area	436 sq ft
Length	27ft 10in

*As a land-plane before conversion to No.205.

Data **Works No.121 (TB.8)**

Span	37ft 8in
Area	450 sq ft
Length	29ft 2in
Weight	970 lb
Weight allup	1,665 lb
Speed	65-75 mph
Endurance	5hr

Data **Works No.205 (TB.8)**

Span	37ft 8in
Area	450 sq ft
Length	30ft 6in
Weight	970 lb
Weight allup	1,650 lb
Endurance	4hr

BRISTOL COANDA Type BR.7 and Daimler biplanes

This long-range, two-seater biplane was designed as a result of interest by Spain and Germany, and was first seen at Olympia in February 1913. There was a divergence of requirements, for the Spanish type had a Renault engine, whilst a Daimler was specified by Germany. Only one of the latter machines was eventually built at Halberstadt by the German Bristol works. This had extended wings, tapered ailerons and an extra rudder below the tail.

At Bristol seven BR.7's were built, of which five for the Spanish government were not accepted. The prototype, Works No.157, flew in March 1913, but was kept at Larkhill as an

Bristol Coanda Type BR7 of 1913 for the Spanish government. Works No.157.

advanced trainer. Works No.158 was destroyed after a fire in the air, Nos.160, 163 were rarely used. A vee type undercarriage was tested on No.163 and the final machine, No.178, was built with increased span, but remained at Filton.

Power: 70hp Renault vee eight-cylinder air-cooled
90hp Daimler vee eight-cylinder water-cooled

Data	**70hp Renault**	
	Span	38ft
	Area	440 sq ft
	Length	27ft
	Weight	946 lb
	Weight allup	1,826 lb
	Speed	63 mph
	Endurance	5hr

Seven aircraft built. Works Nos.157-158, 160-163 & 178.

Data	**90hp Daimler**	
	Span	57ft
	Area	570 sq ft
	Length	27ft 5in
	Weight	1200 lb
	Weight allup	2,100 lb
	Speed	65 mph
	Endurance	5hr

One aircraft built at Deutsche-Bristol Werke, Halberstadt, Germany

BRISTOL COANDA biplane Type BC.2

This project for a large seaplane was prepared for the Admiralty between April and June 1913 before the TB.8H, Works No.205, was accepted. It was to be fitted with a 200hp Clerget V8 water-cooled engine with two-speed reduction gear. In the spring of 1914 the Admiralty again pursued the purchase of two large seaplanes with the two speed gear. Works Nos.1478 were allocated, but these were eventually built as TB.8 aircraft.

BRISTOL COANDA biplane Type TB.8

The first of these two-seater land-planes for the Admiralty was made by the conversion of one of the redundant monoplanes

Bristol Coanda Type TB 8 prototype at Larkhill in 1913.

(No.144). It flew on 12 August 1913 at Eastchurch for the first time. It was rebuilt after a crash as Works No.225 with a vee type undercarriage and served in the RNAS from April 1914, still with its original serial No.43.

The TB.8 was most successful, and ten of the monoplanes, previously supplied to Italy Germany and Rumania, were bought back for conversion. After this new biplanes were built from Works No.197, fitted with various engines to a total of fifty-three aircraft, mainly for use by the RNAS. The delivery of the last aircraft took place in early 1916. An exception was

Bristol Coanda Type TB 8 H. The prototype converted to a seaplane.

Bristol Coanda Type TB 8 in service with the RNAS.

Bristol Coanda Type GB 75 in service with the RFC as No.601 (should be 610).

Works No.218, the only side-by-side TB.8 converted from the monoplane No.177 and used at Larkhill.

Power: 80hp Gnôme seven-cylinder air-cooled rotary.
60hp and 80hp Le Rhone seven-cylinder air-cooled rotaries.
50hp Gnôme seven cylinder air-cooled rotary.
100hp Gnôme Monosoupape nine-cylinder air-cooled rotary.

Data		
	Span	37 ft 8 in
	Length	29 ft 3 in
	Area	450 sq ft
	Weight	970 lb
	Weight allup	1,665 lb
	Speed	65-75 mph
	Endurance	5hr

Aircraft built. Works Nos.118, 121, 143-4, 147-153, 196-198, 218, 225, 227-228, 331, 342, 870, 893

BRISTOL COANDA biplane Type GB.75.

This two-seater biplane was designed as an improvement on the TB.8, and was largely the work of Frank Barnwell. It was ordered by the Rumanian Prince Cantacuzene and was flown for the first time at Larkhill on 7 1914. Trials continued into June to overcome cooling and balance problems. The order was canceled at the end of June and, after an engine change, the aircraft was taken over by the RFC as serial No.610 or 601. It was exhibited at the Olympia Aero Show in March 1914

Power: 75hp Gnôme Monosoupape seven-cylinder air-cooled rotary.
80hp Gnôme seven-cylinder air-cooled rotary.

Data		
	Span	37ft 8in
	Area	450 sq ft
	Weight	970 lb
	Weight allup	1,650 lb
	Speed	65-75 mph
	Endurance	5hr

One aircraft built. Works No.223.

BRISTOL COANDA biplane Type PB.8

This two-seater pusher biplane was a Boxkite replacement for the Brooklands school. It was delivered to Brooklands in July 1914 after a long period on low priority in the works, but was never flown as the engine was immediately requisitioned on arrival, by the War Office.

Bristol Coanda Type GB 75 Works No.223 at Olympia in 1914.

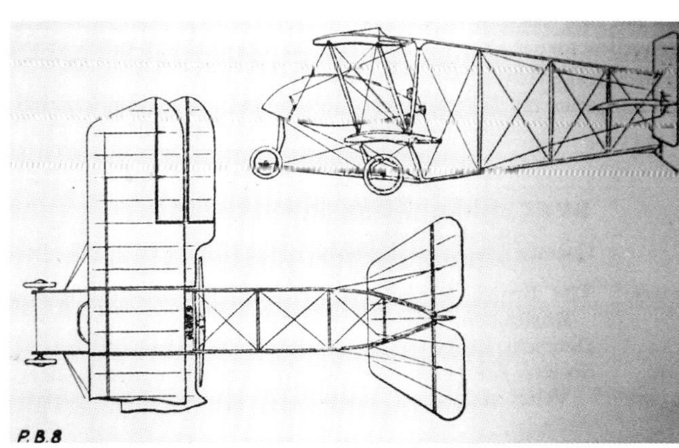

Bristol Coanda Type PB 8 Works No.99. Completed but never flew.

Power: 80hp Gnôme seven-cylinder air-cooled rotary.

Data	Span	27ft 6in
	Length	27ft 6in

One aircraft built. Works No.199.

BRISTOL COANDA biplane Type RB

A project for a two-seater tractor biplane with staggered seats, believed to be for Prince Cantacuzene and to be fitted with the 75hp Gnôme Monosoupape that he had supplied for the GB.75, was prepared in May 1914.

BRISTOL BURNEY hydroplanes

A series of three of these machines with buoyant hulls, buoyancy bags in the first case, and hydrofoils on legs for lifting the machine from the water, were built. The original biplane suggested by Lt. Burney RN was not constructed, as Barnwell had reservations and he proposed a larger monoplane, two versions of which were both built and tested. The design required the use of both air and water propellers.

BRISTOL BURNEY X.1 biplane

This was to be a two-seater biplane based on the GE.1 design, to which it was proposed to add five flotation bags below the wings and fuselage. Three hydroped legs carrying hydrofoils were immersed when the machine was at rest and two of these carried underwater propellers, driven from the single engine. The transfer of power from water to air propeller was effected by engaging and disengaging clutches on a countershaft mounted between the cylinder banks. The X.1 was not built.

Power: 60hp ENV type F eight-cylinder water-cooled vee.

Data	Span	34ft
	Area	325 sq ft
	Length	30ft

BRISTOL BURNEY X.2 monoplane

This second design was built as a flying boat with a planked and sealed hull. It was a side-by-side two-seater monoplane with dual controls and wing warping for lateral control. A similar layout to that of the X.1, was used, with three hydropeds and hydrofoils and water propellers on the two front legs.

Trials were carried out at Dale in May 1912, but damage to the hydrofoils occurred during taxiing before the airscrew could be engaged. During later tests with both air and water propellers engaged the engine proved to be insufficiently powerful. On 21 September 1912, during towing tests with engine removed and equivalent ballast, the aircraft rose and sideslipped into the water and the resulting damage was not repaired.

Power: 80hp Salmson (Canton-Unné) seven-cylinder water-cooled radial.

Data	Span	55ft 9in
	Area	480 sq ft
	Length	30ft 8in

One aircraft built Works No.92.

Bristol Burney X 2 monoplane Works No.92 of 1912.

BRISTOL BURNEY X.3 monoplane

A larger replacement for the X.2 was begun in May 1913. The hull framework was made at Filton, but was made lighter by being covered with 'Consuta' wire-sewn plywood, this work being carried out by SE Saunders Ltd. at Cowes. The lateral control was by ailerons. In addition to the hydrofoil legs a central mounting for contra-rotating water propellers was adopted to cancel out torque reaction and improve stability.

Bristol Burney X 3 monoplane Works No.159 of 1914.

Initial trials in 1913 were carried out with an 80hp Gnôme engine installed and with the wings replaced by a temporary structure with floats at the tips. The main hydropeds were fitted with both water rudders and elevators. An elevator just behind the airscrew was operated in conjunction with the clutch, when the airscrew was engaged.

The final aircraft was ready for testing by Harry Busteed in June 1914, but it grounded on a sandbank and was damaged. Further work was not approved and the machine was stored until 1920, when it was scrapped. One aircraft built Works No.159.

Power: 200hp Salmson (Canton-Unné) 2.M.7 fourteen-cylinder water-cooled radial.

Data		
	Span	57ft 10in
	Area	500 sq ft
	Length	36ft 8in

BRISTOL GB.1 biplane

A design was prepared by Barnwell for a single-seat racer to compete in the 1914 Gordon Bennett Contest, to be held in September, which was canceled by the war. The Board had earlier decided not to compete in the event and the machine was not built. It was similar in layout to the contemporary Scout, but with a fin, and was to be fitted with a 100hp Gnôme Monosoupape engine.

BRISTOL S.S.A. (single-seat armored) biplane

This Scout type was designed by Coanda for a French government requirement with which Breguet was to be involved. It first flew on 8 May 1914. at Larkhill piloted by Sippé. After being damaged at Filton when being flown by Busteed, it was sent to the Breguet works in the damaged condition.

The forward fuselage to aft of the pilot's seat was a riveted steel monocoque enclosing the engine, tanks and pilot. The front of the engine was covered by a large steel spinner with cooling slots. The undercarriage featured castor wheels and rearward extended skids, which obviated the need for a tail-skid. The tail units were similar to those of the Scout biplane. One aircraft built Works No.219.

Power: 80hp Clerget seven-cylinder air-cooled rotary.

Data		
	Span	26ft 3in
	Weight	913 lb
	Area	200 sq ft
	Max. speed	106 mph (estimated)
	Length	18ft 9in
	Endurance	3hr

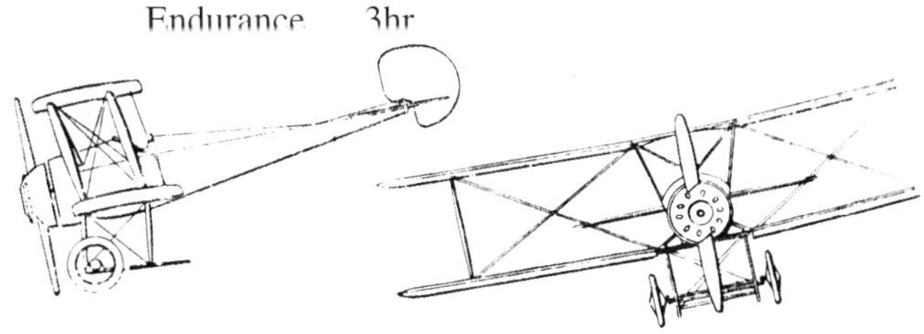

Bristol SSA biplane. This armored Scout was built for a French requirement, but only one example was built.

Bristol Scout Type A of 1914.

BRISTOL SCOUT biplane Types A and B

The redundant fuselage of the monoplane SB.5 (Works No.183) was the basis on which Barnwell designed a small single-seat biplane (Works No.206), which arrived at Larkhill for testing by Busteed in February 1914. After initial trials it was required for display at the Olympia Aero Show opening on 16 March 1914. Soon after the Show the Scout was fitted with increased span wings at Filton and was re-flown in May.

In June Lord Carbery bought the airframe for £400 and installed his own Le Rhone. In the race from London to Manchester and back he damaged the machine during an intermediate landing. After repair at Filton he used it for a race from London to Paris and back on 11 July 1914, but ran out of fuel on the return leg and came down in the Channel. Carbery was saved but the Scout was lost.

Two additional machines were already in hand (Works Nos.229, 230) with some improvements to the cowling and wing bracing and became known as the Scout Type B. The machines, which had not flown when war was declared on 4 August 1914, were requisitioned by the War Office and identified with serial Nos.633 and 648 and then flown to Farnborough on 21 and 23 August 1914. Both saw service in France, and 633 at least was fitted with fixed rifles firing either side of the airscrew.

The performance of the Scout was sufficiently good to result in further orders for improved versions, the Types C and

Bristol Scout Type B of 1914.

D, which were ordered in considerable numbers and delivered from 1915 onwards.

Power: Scout A and B 80hp Gnôme seven-cylinder air-cooled rotary. Scout A also had 80hp Le Rhone seven-cylinder air-cooled rotary.

Data	**Scout A**	
	Span	22ft & 24ft 7in
	Area	161 sq ft and 198 sq ft
	Length	19ft 9in
	Height	8ft 6in
	Weight	617 lb and 750 lb
	Weight allup	957 lb &1,100 lb
	Max speed	95 and 100 mph
	Initial rate of climb	800ft per min
	Endurance	3 and 5hr

One aircraft built Works No.206.

Data	**Scout B**	
	Span	24ft 7in
	Area	198 sq ft
	Length	20ft 8in
	Height	8ft 6in
	Weight	750 lb
	Weight allup	1,100 lb
	Max speed	100 mph
	Initial rate of climb	1,000ft per min
	Endurance	2 1/2 hr

Two aircraft built Works Nos.229, 230

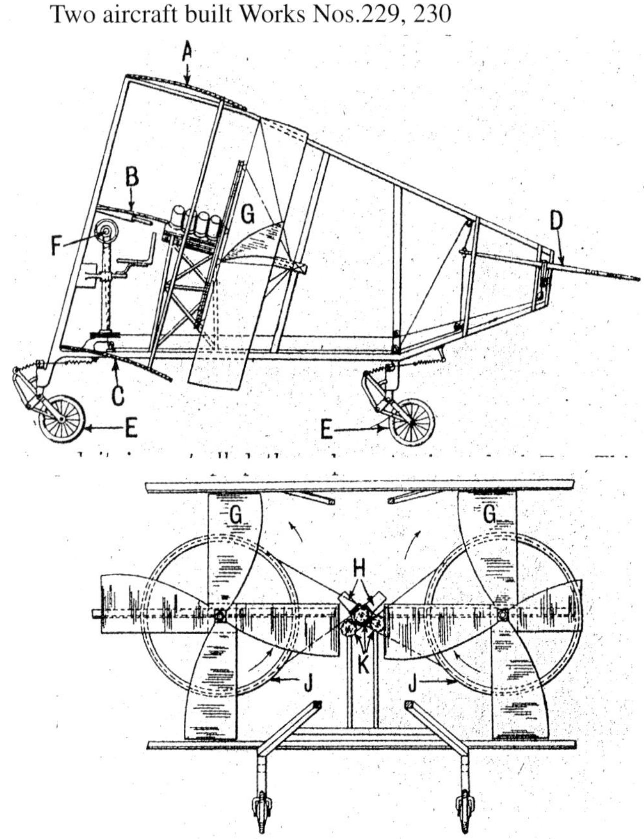

Britain triplane was flown at Wembley by B. Seamer in August 1909. This is the patent drawing.

BRITAIN triplane (Seamer Brothers, Enfield, Middlesex)
This machine was designed by Mr. Henry Britain and was tested successfully by B. Seamer on 30 August 1909 at Middleton House, Forty Hill, Wembley, Middlesex. It rose to a height of about two feet but lack of space prevented a longer flight. A 40hp engine was fitted. Patent No.21952 of 1908 refers.

BRITISH CAUDRON Co. Ltd. (Hendon Aerodrome and Edgware Rd., Cricklewood, from January 1913)
This company, originally W.H. Ewen Aviation Co. Ltd., was established as a flying school and later undertook aircraft manufacture, having acquired the license to manufacture Caudron types in April 1912. Ewen had a number of Caudron biplanes built by Hewlett & Blondeau at Clapham under contract, and other firms are believed to have made machines of similar type. W.E. Ewen Aviation exhibited at Olympia in March 1913 and earlier that year advertised premises at Lanark and Glasgow. Six biplanes were reported to be in production at Lanark, but may not have been completed as these addresses were soon dropped from the firm's advertising.

A Caudron type biplane was built by the James brothers, ex-pupils of the school, at Narberth, Pembrokeshire, in the latter part of 1913, and was reconstructed as a two-seater after a crash (*q.v.*); the machine was taken to Hendon and used as a trainer at the Ruffy-Baumann School.

In April 1914 Ewen left the company and the name was changed to British Caudron, with Messrs. Ramsey and Hunter in charge. The works in the Edgware Rd. were acquired later in the year.

BRITISH CAUDRON biplane Type C
The French Caudron biplane was unusual for a tractor machine in having a short fuselage with outrigger booms to carry the tail unit. The machine was of light construction with considerable flexibility of both wings and tailplane. In both cases about half the chord could be warped, and the tail unit warped in unison with the wings. In these areas one layer of fabric on the underside with the ribs in pockets was used.

Single and twin rudders were fitted, originally almost square in shape, but with curved profiles from 1913. A four-wheel sprung undercarriage was used with two front skids continuing rearwards as the lower longerons to the tail, on which the machine was supported when at rest. Various engines were

British Caudron biplane Type C. Typical of machines constructed by the British company and several other firms.

fitted and the machine was strengthened for aerobatics and developed in several versions, including both single and two-seater versions. In the latter case the nacelle was extended and the passenger housed in the front cockpit behind an upswept decking. The single-seater with low powered engine was mainly used by pupils for 'straights'.

Power: 30hp Anzani three-cylinder air-cooled fan type semi-radial with 6ft 10in diameter propeller.
45-60hp Anzani six-cylinder air-cooled radial.
60 and 70hp Gnôme seven-cylinder air-cooled rotary.

Data **single-seater with 30hp engine***

Span top	30ft.
Span bottom	23ft 3in
Chord	4ft 6in
Gap	4ft 7in
Length	21ft 3in
Area	224 sq ft
Area tailplane	53 sq ft
Rudders	12 1/2 sq ft

*Source *Flight* 30 November 1912, but *Flight* 15 March 1913 gives alternative figures

Data **Type C single-seater with 45hp Anzani**

Span top	34ft
Span bottom	22ft 6in
Area	272 sq ft
Length	22ft 3in
Price with 60hp engine	£660. (18 July 1912)

BRITISH CAUDRON biplane Type G

A larger and stronger version of the Caudron design appeared as an amphibian in the summer of 1913. Twin sprung floats, with apertures housing the wheels, were fitted on a new chassis. The lower longerons were raised at the front to connect to the intersection of the wing rear spars and to the intermediate interplane struts. A pair of cylindrical wingtip floats and two triangular shaped floats, enclosed by skids, supported the tail on land. A pair of small fins and curved rudders were fitted.

Several of the type were used by the RNAS, but these were mainly of French origin, although Hewlett & Blondeau was reported to have one in hand for Ewen in August 1913. A three-seater version was advertised by Ewen in July 1912 at £960.

British Caudron biplane Type G3 used by the RFC. This is A1893.

The RFC acquired a number of the land-plane version, known as the G.3, mainly from France, orders still being placed as late as September 1917, although at least fifty were supplied from the Cricklewood factory. The aircraft was used for reconnaissance in early operations, but was employed mainly for training.

For the land-plane version the longerons reverted to their original low position with front extensions to serve as skids. The original engine was a 80hp Gnôme, but in wartime a variety of engines was used. The data is for a Type G amphibian with 80hp Gnôme engine.

Power: 80hp Gnôme seven-cylinder air-cooled rotary.
100hp Gnôme Monosoupape nine-cylinder air-cooled rotary.
80 & 100hp Anzani ten-cylinder air-cooled radials.
70hp Renault eight-cylinder air-cooled vee.

Data **Amphibian**

Span top	42ft
Span bottom	28ft
Chord	5ft 1in
Area	350 sq ft
Length	26ft 8in
Area tailplane	50 sq ft
Area rudders	16 sq ft

Data **G.8. Land-plane.**

Span top	43ft 6in
Area	304 sq ft
Length	22ft 7in
Height	8ft 6in
Weight	959 lb
Weight allup	1,565 lb
Speed	65 mph
Climb to 6,500ft	27 min
Ceiling	10,000ft

BRITISH DEPERDUSSIN AEROPLANE Co. Ltd. (Originally 'Syndicate') Works at Mildmay Ave, Stoke Newington, and Elthorne Rd., Highgate, London Head Office at 39 Victoria St., Westminster, London SW

The company was formed in April 1912 with Lt. J.C. Porte RN as one of the directors. The chief designer was Dutchman Frederick Koolhoven, and the intention was to design and manufacture monoplanes of the general type produced by the Deperdussin Company in France. The first British-built machines were entered for the Military Trials in August 1912. Both a British and a French built Deperdussin performed well and were taken on by the RFC.

The French machine was to crash with fatal results to the crew within a month, leading to a ban on the use of monoplanes by the RFC, although the RNAS continued with their use.

A British Deperdussin seaplane of original design was exhibited at Olympia in February 1913 and was tested on the River Blackwater in the spring.

The British company was in receivership by August 1913 and, after a financial scandal, the parent company in France was absorbed by Blériot in 1914 and his associated British company was later reconstituted as Blériot and SPAD.

BRITISH DEPERDUSSIN monoplane (Military Trials Type)

This two-seat tractor type differed from the French Deperdussins by the deletion of the undercarriage skids, and inverted vee-shaped cabane struts were used instead of the early single, or later triple struts, of the French machine. Apart from the two machines for the Trials, further machines of the type with various engines, probably a further four, were built at Highgate for the RFC.

Power: 100hp Gnôme fourteen-cylinder air-cooled rotary driving 8ft 6in diameter Rapid propeller.
100hp Anzani ten-cylinder air-cooled radial driving 8ft 6in diameter Rapid propeller.
60hp Anzani six-cylinder air-cooled radial.

Data	Anzani	
	Span	41ft 6in
	Area	270 sq ft
	Length	24ft
	Weight	1,200 lb
	Weight allup	2,000 lb
	Speed	70 mph
	Climb to 1,000ft	4min 45sec
Data	**Gnôme**	
	Span	39ft 6in
	Area	236 sq ft
	Length	24ft 6in
	Weight	1,226 lb
	Weight allup	2,037 lb
	Speed	68 mph
	Climb to 1,000ft	3min 45sec

One of each of above types built for the Trials. No.20 with 100hp Anzani not accepted by the RFC; No.21 with 100hp Gnôme became No.259. Also Nos.280, 100hp Gnôme; 421, 60hp Anzani and Nos.436-437 engine type not known.

British Deperdussin. Military Trials No.20 of 1912.

British Deperdussin *Seagull* of 1913 on test on the River Blackwater.

BRITISH DEPERDUSSIN SEAGULL

This was an original design by the British branch of the company, of which the Admiralty ordered two. The fuselage was of monocoque construction and the wings, which were quite unlike any earlier Deperdussin, had an external tubular steel bracing structure below and no overhead bracing. The spanwise members of this structure were broad fairings of aerofoil section intended to contribute lift, and enclosed the aileron control cables.

One aircraft, exhibited at the Aero Show at Olympia in February 1913 was fitted with a broad central step-less float, to which springing was to be added. Small floats of streamlined shape were fitted at the extremities of the bracing girder and a broad float of aerofoil section, with water rudder, below the tail. The machine tested later on the River Blackwater was fitted experimentally with twin step-less main floats and no wingtip floats.

The crew sat in tandem, with the observer in the front, and provision was made to start the engine from the front cockpit. Full chord, wingtip ailerons were pivoted between the wing spars. The engine was a 100hp Anzani.

Although the Admiralty had ordered two machines these were canceled when the prototype proved to be unsatisfactory and the British company failed. A 100hp Anzani-engined landplane, which became No.885, and may have been one of these machines, was impressed in August 1914.

Data		
	Span	42ft 6in
	Chord	6ft 8in tapering to 7ft 2in
	Length	30ft 6in
	Area	280 sq ft
	Area tailplane	30 sq ft
	Area elevators	16 sq ft
	Area rudder	7 sq ft
	Weight	1,300 lb
	Weight allup	1,980 lb
	Speed	55 mph
	Price	£1,500

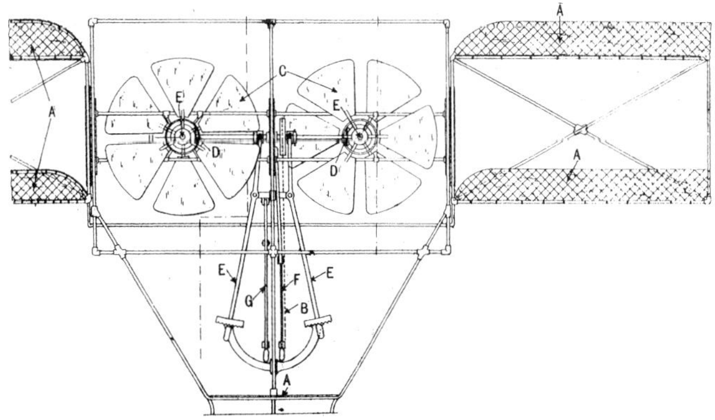

British Matchless convertiplane of 1908.

BRITISH MATCHLESS convertiplane (Arthur Phillips, Market Drayton, Shropshire)

This machine was the invention of Phillips, an engineer and cycle maker, and was covered by Patent No.28 119/1908. It consisted of a rectangular tubular structure, which tapered down to the base where the operator was placed. The biplane wings were mounted on large bearings and could pivot through 90 degrees in unison with the four fanlike propellers. The propeller shafts and wing chord line would be vertical for takeoff and turned to horizontal for forward flight. The propeller shafts were fitted with universal joints to enable the thrust to be deflected for control purposes.

The machine was built in scale form with a Douglas motor cycle engine and was flown on several occasions tethered by a rope. A seven-cylinder two-stroke engine of 12hp was to be fitted and this, together with the structure of the machine, are still to be seen in the Shuttleworth Collection. Phillips was said to know the Wright brothers and to be esteemed by Moore-Brabazon and other enthusiasts.

Data for model.

Span	14ft 6in.
Chord	3ft 6in.
Gap	2ft 6in.
Length	6ft 2in.
Height	6ft 6in.

BROCKLEHURST monoplane. (Major Brocklehurst, Windcrmcrc & Eastchurch.)

This machine appeared at Eastchurch in January 1912 after construction at Windermere. It was described as being designed by Major Brocklchurst, with wings of Etrich style to have automatic stability. The wings were pivoted to allow them to lift under pressure and were interconnected to the undercarriage, through springs, to effect their return. It is believed to have been fitted with an Anzani engine and was based on Patent 24902/1909 for such a wing. FB Fowler, who later ran the Eastbourne Aviation Co., was probably responsible for the design of the landing chassis and other aspects of its construction.

The first mention of this machine refers to a mishap at Eastchurch on 13 January 1913. (*The Aeroplane* 25 January 1913 (p.85), when it was towed back to its shed across the airfield at 25 mph by the Jezzi biplane, from the point where it was stranded.

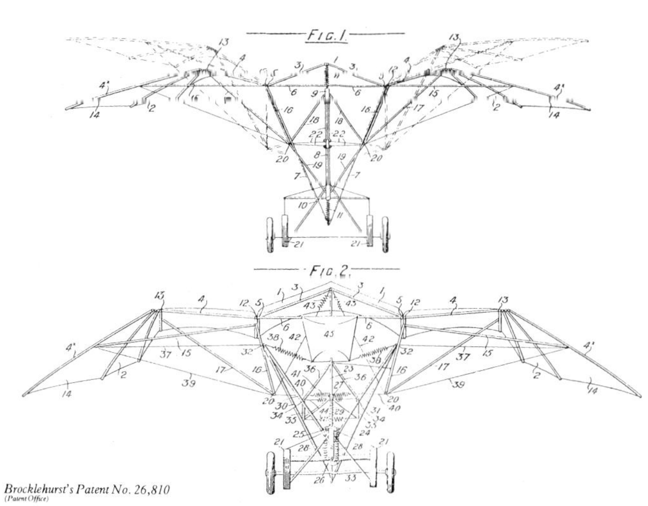

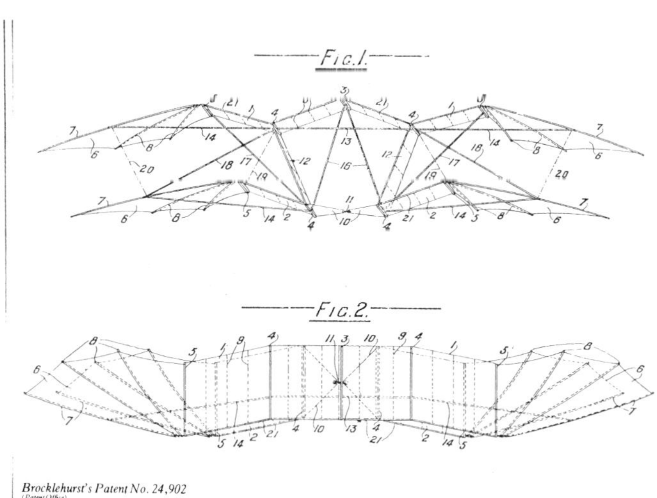

Brocklehurst monoplane. Diagrams from Patent No.26 810/1911.

The machine was transferred to Windermere and fitted with 'roller floats' possibly of the type covered by Patent 6999/1909 granted to Kitchen and Storey. The last reference to the aircraft was in *The Aeroplane* dated 10 April 1913 (p.431), where it was stated that 'Major Brocklehurst's hydromonoplane with Bat's wings' was under construction.

Brocklehurst's earlier Patent No.5662/1909 was for a scheme to provide lateral and pitch control by twisting the tail of an aircraft. His later Patent No.26810/1911, was for 'Batlike' wings in either biplane or monoplane form, which could be folded.

BROCKLEY monoplane (Messrs. Norrie, Duval & Mitchell, Brockley Jack Inn, London, SE.4)
Tests were made of this machine at Abbey Wood, Kent, but damage to the undercarriage and propeller prevented the machine from flying. The construction of the machine, which began in 1908, was of bamboo and steel tubing. Various JAP engines, all of motorcycle type, were tried, starting with a single-cylinder 2.5hp, then a 6hp twin and finally a 16hp racing vee-twin.

Brockley monoplane.

BROCKWAY glider (Brockway, 1 College St., Cleethorpes, Lancashire)
Reported 4 November 1911, but no details given.

BROMLEY monoplane (built at Bromley and tested at Orpington, Kent)
A machine with two two-cylinder engines totaling 24hp, driving one 7ft propeller was reported in November 1909. It was a monoplane with upswept wing tips. The weight with pilot was less than 400 lb.

BROOK aeroplane (Thomas F. Brook, South Parade, Huddersfield, Yorkshire)
Reported in March 1912 with a fuselage of steel tubing and a 45hp JAP engine.

BUCKLE biplane glider (Buckle, Lexton, Colchester, Essex)
Reported in March 1911 and described as being of Voisin type. Also reported was a monoplane of 30ft 10in wingspan and powered by a 45hp Anzani.

Bush glider No.3 of 1910 built by the three Bush brothers.

Bush glider No.4 of 1910 built at Cambridge by Eldon Bush.

BUSH biplane glider (Eldon, Gilbert and Jon Bush, Keynsham, Bristol)
These three brothers designed and built seven biplane gliders between 1909 and 1912.

No.1 was constructed of ash, was of Wright type, but was heavy and did not fly.

No.2 was No.1 reduced in weight and, although flown, was unstable longitudinally.

No.3 was of Curtiss type and built in 1910. It had ailerons hinged to the trailing edges of the bottom wings. Span 30ft, length 24ft, area 260 sq ft. This machine flew for short distances.

No.4 was of Farman type and built at Cambridge by Eldon Bush in 1910. It was fitted with a wheeled undercarriage.

Nos.5, 6 and 7 were modifications of No.4 with the front elevator omitted.

No.7, known as *'The Chocolate Soldier'* had Blériot type elevators but was smashed while under tow by a car.

BUSH MOTORPLANE biplanes Nos.8 and 9
No.8 was built at Bath by the brothers in 1912. It was similar to a Caudron and was a single-seater intended to be powered by an inline engine.

Bush Motorplane No.8 of 1912 built at Bath by the brothers.

No.9 was No.8 with the nacelle widened to accept a 50hp Gnôme rotary engine, the propeller shaft of which broke on test. The machine was not fitted with a replacement as Eldon Bush went to Canada, returning in 1915 to join the RNAS in which service he was killed in 1917.

Bush Motorplane No.9 a two-seater version of No.8 which did not fly.

BURTON monoplane glider (Horace Burton, Grantham, Lincolnshire)

Little is known of this machine with a biplane tail, which was reported in the Press on 23 July 1910.

Data	Span	24ft
	Chord,	5ft,
	Area,	120 sq ft,
	Tailplane,	6ft 6in by 3ft 6in.

BUTLER AND EDWARDS aeroplanes (J.W. Butler and E. Edwards)

This team in 1867 designed a model paper dart shaped monoplane with a small steam generator which propelled the aircraft by a jet of steam. Patent No.2115 of 1867 refers

They also designed a delta wing biplane with a three-bladed propeller unit driven by exhaust steam through jets at the tips of the propeller blades. The wings were to be made of silk covered frames or thin wood or of hollow metal to serve as condensers for steam. Neither of these two interesting designs progressed to full size aircraft.

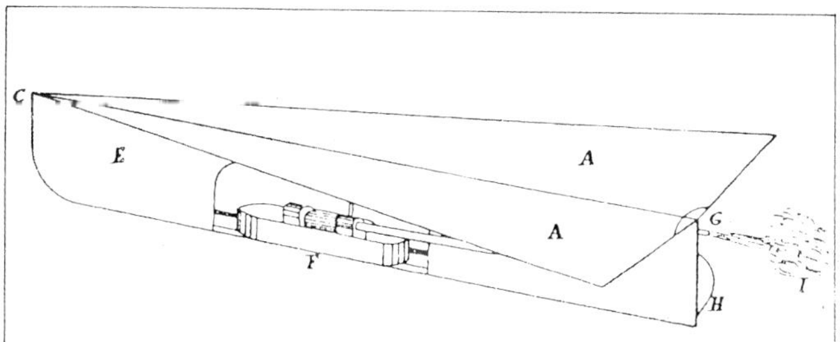

Butler & Edwards monoplane. A design for a monoplane to be propelled by a jet of steam.

BUTTEMER aeroplane (R.W. Buttemer, St. Mary's, Shackleford, Godalming, Surrey)

Robert Buttemer was a scientist and a founder member of the RAC and the Brooklands Automobile Racing Club. In 1897 he had the second motor car in Surrey, a 3.5hp Benz.

An aeroplane of unknown type, fitted with a 25hp Anzani of 'cross Channel type' was built in late 1911 and was offered for sale in *Flight* of 25 February 1912 as being new and unused, for £75. It was fitted with an Handley Page propeller.

It evidently remained unsold and was still at Shackleford in 1919. By then, the design was obsolete and the machine was scrapped.

C

CALEDONIAN MOTOR & CYCLE WORKS (Leith, near Edinburgh, Scotland)

This company built machines identified under the name Gibson (*q.v.*). See also Cooper.

CAMBRIDGE monoplane (Cambridge Automobile & Engineering Co. Ltd. Cambridge)

The following advertisement appeared in *Flight* on 1 October 1910. ' Monoplane, Blériot type, Continental double surfaced, wheels and skids, landing chassis, copper tanks, gauges, lubricators, all fittings, with 25hp two-cylinder water-cooled horizontal motor, tractor screw; everything complete and ready to use; full control; brand new, £75. No other details have been found.

CAMPBELL BRITON monoplane (built for Malcolm Campbell, Bromley, Kent by C.F. McGeorge at Orpington)

Campbell, who became famous as a record breaker on both land and water in later years, built a monoplane in 1909-1910. Inspired by Blériot's Channel crossing, the machine resembled

Campbell *Briton* monoplane of 1910. Malcolm Campbell later became a famous racing and record breaking motorist.

a Blériot in many respects, but was fitted with a 10hp JAP vee-twin motorcycle engine of insufficient power. Damage was caused after it stalled on takeoff, but after repairs a larger 40hp engine was fitted and hops were achieved on several occasions in the summer and autumn of 1910. However, lack of funds precluded continued testing and the machine was finally sold by Friswells.

Candler *Stella* monoplane of 1910 constructed at Farnborough Kent.

Data	Span	39ft

CANDLER *STELLA* or *MONOFOIL* monoplane (Hilda Vale Rd., Farnborough, Kent)

This single-seat tractor monoplane, similar to a Blériot, was being built in 1909-1910 for H.A. Wagstaff Candler of London. The construction was carried out by the Farnborough Aviation Works in Kent, assistance being provided by C.H.M. Alderson (*q.v.*). The *Stella*, also referred to as the Monofoil, was offered in an incomplete state to the Science Museum in 1926, but was not accepted.

Data	Span	30ft
	Area	247 sq ft
	Length	26ft
	Weight	250 lb

CAMERON monoplane (Donald 'Dhonal' Angus Allan Cameron, Castlewood College, Rathmines, Dublin)

Built by an eighteen year old pupil and illustrated in *Flight* 14 May 1910 (p.369), it appears to have been built as a glider, perhaps for a motor to be fitted later. Dhonal Cameron was of Scottish birth but later moved to England, where between 1950 and 1965 he worked as a draughtsman tracer with Vickers at Weybridge.

Cameron monoplane built as a glider by a deaf Scottish youth in Dublin.

Carter biplane of 1911.

CARTER biplane

No details are available of this single-seat biplane, which arrived at Salisbury Plain early in January 1911. The power plant was a 60hp eight-cylinder air-cooled motor of X-form made by the Nonpareil Fitting Co. of Birmingham. The machine made a number of short straight flights only by the end of January in the hands of the owner. There were no further reports.

CAYLEY GLIDERS (Sir George Cayley, 1773-1857. Brompton Hall, near Scarborough, Yorkshire)

Sir George Cayley was a scientist and inventor in a number of fields and included aeronautics among his studies. In articles in various journals he described the principles and configuration of the aeroplane. Much of his work is recorded in his notebooks, which have survived, and he is internationally regarded as the inventor of the aeroplane in its basic form. Apart from his designs for gliders, he published a scheme to a 'borrowed' theory for a double rotor helicopter in 1842. He realized the need for a source of lightweight power other than manpower and experimented with a gunpowder engine, but none of the craft actually constructed was fitted with an engine.

His known designs of full size gliders, realized over a period of forty-two years were flown mostly with ballast, but later a boy and finally a man, his coachman, were carried aloft.

CAYLEY monoplane glider No.1, 1809

This machine, which had a wing area of 300 sq ft and a loaded weight of 140 lb, was fitted with a tail unit with adjustable rudder. The glides were initiated by towing and, in some, ballast was carried, although in certain conditions the lift generated was sufficient to carry the man or boy towing the machine into the air.

CAYLEY triplane glider No.2, 1849.

A sketch of this triplane glider was found by C.H. Gibbs-Smith in 1961. It was operated mainly with ballast, but carried a boy on a few occasions for short glides.

The triplane wing structure was extended aft to carry a cruciform tail unit and a boat shaped nacelle was carried on struts below. The whole machine rested on three lightweight

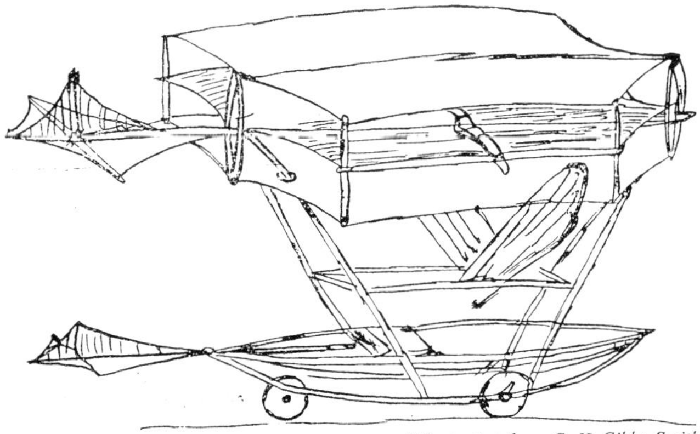

(Illustration from C. H. Gibbs-Smith)

Sir George Cayley's 1849 triplane glider flown briefly with light loads.

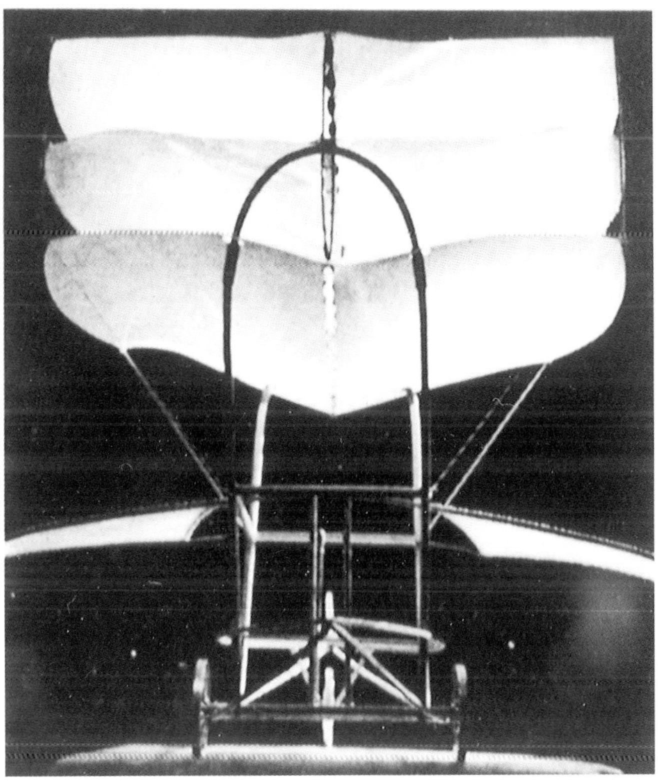

wooden wheels. Control was provided by a second cruciform tail unit on a pole at the rear of the nacelle and by adjustment of the upper elevator by cords from the nacelle. Two wings of 6 ft span were also fitted for control by the pilot, so that Cayley could observe their effect on the glide angle.

Data	Wing area	338 sq ft
	Weight	130 lb

Mechanics' Magazine,

MUSEUM, REGISTER, JOURNAL, AND GAZETTE.

No. 1520.] SATURDAY, SEPTEMBER 25, 1852. [Price 3*d*., Stamped 4*d*.

Edited by J. C. Robertson, 166, Fleet-street.

SIR GEORGE CAYLEY'S GOVERNABLE PARACHUTES.

Fig. 2.

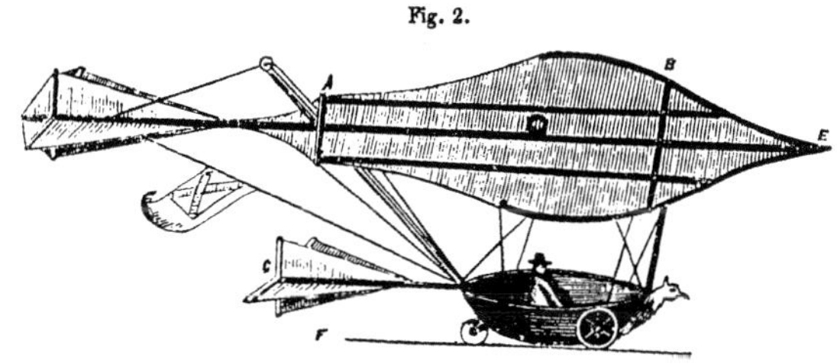

Fig. 1.

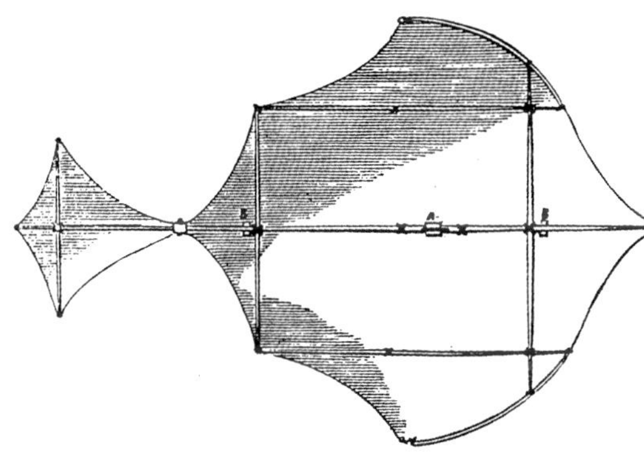

Cayley's 'Governable Parachute' monoplane glider of 1852-1853 which made a manned flight of 500 yards.

Below: Cayley experimental moving wing glider.

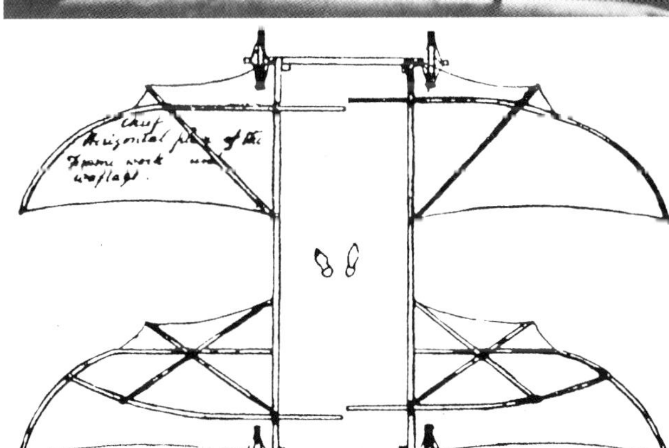

FIGURE 22. Experimental moving wing glider.

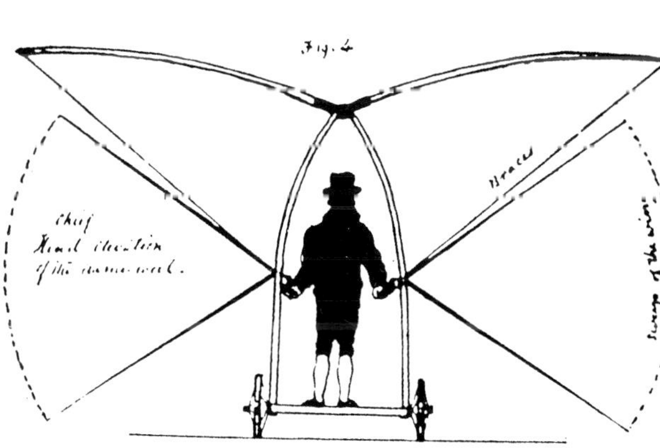

CAYLEY monoplane glider No.3, 1853

Also referred to as possibly a biplane or triplane, this monoplane design was described as 'a Governable Parachute' and information on it was published in *Mechanics Magazine* of 25 September 1852. This is believed to be the version tested in 1853 which made the first known manned glider flight of some 500 yards across Brompton Dale. Apart from the use of a monoplane lifting surface, the general layout was similar in most respects to the 1849 glider. The empty weight was 165 lb.

Subsequent to the finding of Sir George's notebooks in the 1960s, a reproduction of this machine was constructed. This was towed or launched by bungee when flown for filming purposes, prior to being housed in the Museum of Science and Industry, Manchester.

CAYLEY helicopter-aeroplane, 1842-1843

This was a project described in the Mechanics Magazine, which was not constructed, no doubt because of the lack of a suitable motor so its practicability was not proved. The configuration can be seen in the drawings, which are those that accompanied the article and show certain similarities with the No.2 and 3 gliders. The four rotating wings opened out into eight blades for lifting, but became fixed surfaces for forward flight, with propulsion by two four-bladed propellers.

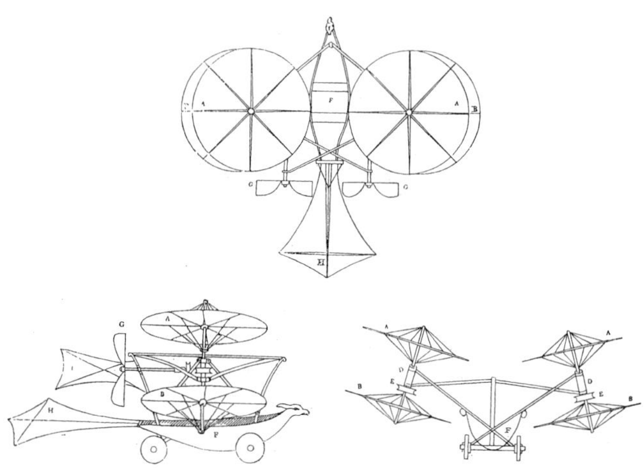

Cayley schemed this helicopter-aeroplane in 1842-1843 but it was not built.

Cayley appears to have been inspired by a letter dated 25 July 1842, written to him by Robert Taylor, proposing a similar scheme some nine months before Cayley's article.

CENTRAL AVIATION Co. monoplane (John Oliver and Joe Bland, Wilmslow, Cheshire)

A monoplane was reported flown at Row of Trees, Wilmslow, and Trafford Park. The machine had a 30hp Alvaston engine.

CENTRAL NOVELTY Co. biplane (99 Snow Hill, and 45 Newhall St. Birmingham)

This model making company advertised in Flight in November 1910 that they were able to offer a full size biplane with *Demoiselle* tail unit and aileron control. The fuselage was finished and wired up and ready for an engine and the wings were covered with fabric. The price was £50 and it was suggested that it would make a first class club glider.

In the 1910 Kelly's Directory they are described as "Aeroplane fittings dealers".

CHANNON biplane (Ernest Huddleston and Ralph Channon, Dorchester)

The glider was based on the Wright-type with front biplane elevator and single rudder on the rear boom. It was later converted to a powered pusher type with twin booms and rudders. The ailerons were transferred from the struts to the top wings. The wheels were removed and it was tested at Maiden Castle using starting tracks in September 1910, when it flew for 100 yards. The machine was under-powered, finance was short, and the project was discontinued. The machine was destroyed by a fire at the works.

Power: 23hp White and Poppe six-cylinder inline water-cooled.

Data Span 50ft.

Channon biplane in its original form as a glider.

Channon pusher biplane tested from a starting rail in September 1910.

Chanter monoplane built at Hendon but soon destroyed in a hangar fire at Shoreham in February 1912.

Chittenden-Robinson biplane No.1 of 1909.

CHANTER monoplane (M. Chanter, Hendon and Shoreham)
This machine was built at Hendon and moved to Shoreham in November 1911, when Chanter transferred his school. It was a single-seater monoplane based on the Nieuport and fitted with a 35hp fan-type Anzani. It flew for the first time at Shoreham on 1 January 1912, but was destroyed in a hangar fire on 29 February 1912.

CHESTERTON biplane (C. Wallis, C. Knightly, P. Booth and E. Muller, High St., Old Chesterton, Cambridgeshire)
These four friends were scientific instrument makers who built a glider in early 1910 which was reported ready for testing in May. The triplane wings were of 35ft span and 5ft chord with control by warping. The machine was intended to lift off from a launching trolley after being towed downhill. A site at Royston was to be used and, if successful, an engine of 25/30hp was to be fitted. No later reports have been traced.

CHITTENDEN-ROBINSON biplane No.1 (J.P. Chittenden and L.H. Robinson, Chiswick, London W)
This was a single-seat, pusher, canard biplane built in 1909, and was fitted with biplane control surfaces at the front and rested on a tricycle undercarriage.

CHITTENDEN-ROBINSON biplane No.2
The second CR biplane was a single-seat tractor with uncovered fuselage structure. The wings, of two and a half bays, were

Chittenden-Robinson biplane No.2 of 1910.

parallel with square tips and with single acting ailerons hinged behind the top wings. In a letter to *Flight*, 19 November 1910 (p.960), Robinson said the machine had not flown owing to the limited space at the ground, although it had lifted some eighteen inches from the ground during tests.

Power: 30hp Alvaston two-cylinder horizontally opposed, water-cooled driving a 7ft diameter Lascelles propeller.

Data Span 30ft
Chord 4ft

Chittenden-Robinson monoplane of 1910.

CHITTENDEN-ROBINSON (SEYMOUR) monoplane

This single-seater tractor monoplane was built at Chiswick for A.W. Seymour, Motor and Aeronautical Engineers of Rugby, in which name it was hoped to continue the business which seemingly did not happen.

Power: 40hp Lascelles four-cylinder air-cooled, fan-type radial driving a 7ft diameter Weiss propeller.

CLAPHAM AERO CLUB monoplane glider (Manor St., Clapham Rd., London SW)

No information on this reported machine.

T.W.K. CLARKE & Co. (14 Union St., later at Crown Works, 22a High St. Kingston-on-Thames and also High St., Hampton Wick, Kingston Bridge)

T.W.K. Clarke was very active from about 1906 in model making and flying, and, having a learned approach to the problems of flight, gave a number of lectures on the subject. His business originally sold models and materials and progressed to full size gliders. Later, powered machines were built, mainly to other people's designs, and propellers were a specialty. He advertised gliders from £40 and aeroplanes from £200, less engine, in 1909 and although the quality of the company's products was highly praised, there is evidence of only a few machines being constructed. There was little publicity given to clients' machines which may account for the few types recorded.

From at least September 1911 to February 1913 he operated from Crown Works, the original boathouse of Alfred Burgoine, but when he moved to Hampton Wick the company appears to have supplied materials, components and models only. It is understood that Clarke joined the staff at Farnborough during the war. The premises at Kingston have long since disappeared.

TWK Clarke glider of 1906-1907 with original elevator, being flown as a kite at Cooper's Hill.

Clarke airborne with the modified glider.

CLARKE biplane glider 1906-1907

This rather frail-looking biplane was originally fitted with a plain rectangular front elevator but later triangular extensions were added. It appears that a launching rail was first tried, but later testing was carried out as a kite, before Clarke himself accomplished glides. These took place on Cooper's Hill, now known as Telegraph Hill, near Manor Rd. South, Hinchley Wood, at the time when this was an open area. Subsequent trials took place near Aldershot.

Data Span upper 39ft
Span lower 31ft
Chord 5ft 3in

CLARKE biplane glider 1909-1910

This equal span biplane with tail mounted elevators and rudder was probably inspired by one of the Chanute types. The operator rested on pads under his arms, on the rails of the main frame, and stood in the cutaway in the center section of the bottom plane and could perhaps transfer his weight to the rear

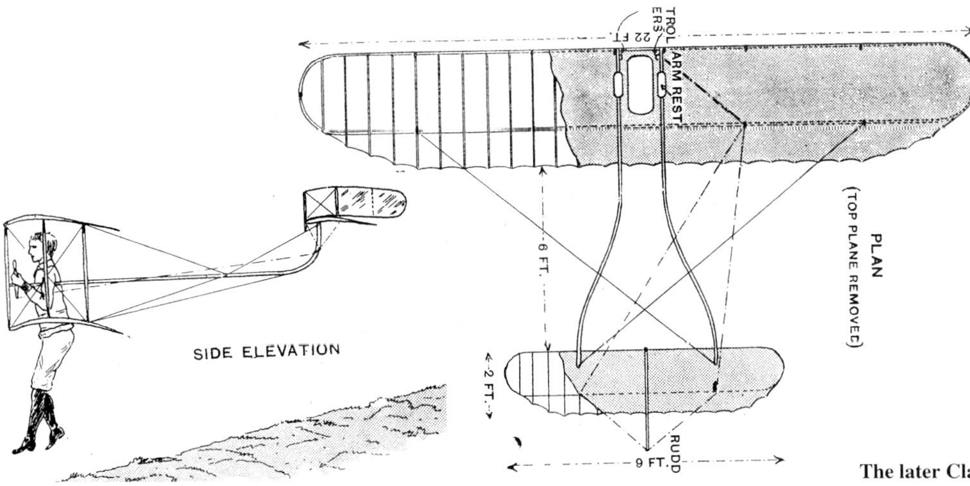

The later Clarke Chanute-type glider of 1909-1910.

spar when in flight. Two levers controlled the rudder and elevators, which could also be warped independently.

The glider was first shown at the Stanley Show at the Agricultural Hall, Islington in November 1909 and has survived in the Science Museum's National Collection. It has recently been exhibited at the RAF Museum, Hendon.

Data	Span	22ft
	Chord	3ft 9in
	Length	10ft

CLARKE glider 1910

A simple tailless glider was designed by Clarke especially for *The Aero*, which published working drawings on 12 December 1910 to enable their readers to build the machine. It was an equal span biplane with a forward mounted elevator carried on double booms. There was no fin or rudder and no ailerons or provision for wing warping.

The aviator picked up the glider and ran down a hill with it until flying speed was reached. He then lifted himself bodily on to the machine with his feet on the skids. His right hand operated the elevator by means of a cable mounted on a wing strut.

Data	Span	24ft
	Chord	4ft 3in
	Length	14ft 1in
	Height	5ft 11in
	Weight	c.120lb

CLARKE-WRIGHT glider

This biplane glider was ordered by Alec Ogilvie and his partner, T.P. Seawright, in order to obtain practice before delivery of a Short-Wright biplane. Delivery took place in August 1909, after only four weeks under construction and gliding took place at Camber Sands thereafter. The design was based on the Wright No.3 glider, but with a biplane front elevator with fixed fin and upright seating. Launching was by weight and pylon as with the Wright machine. A glide of 343 yards was the most achieved.

Data	Span	32ft 8in
	Area	318 sq ft
	Length	18ft

CLARKE monoplane glider 1918

During 1918 a tailless monoplane glider with considerable sweep-back and tip mounted ailerons was photographed in the Farnborough area. It is not known if this had official backing or was a private venture and may even have been a survivor from Clarke's prewar activities.

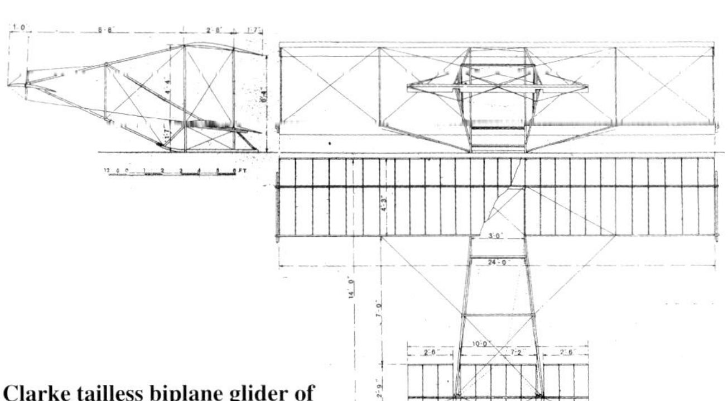

Clarke tailless biplane glider of 1910.

Clarke swept wing monoplane glider photographed in 1918.

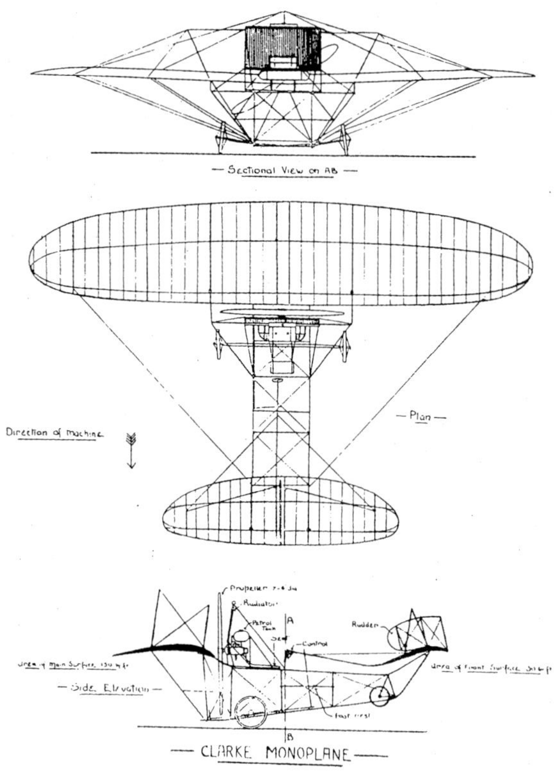

The Clarke prize-winning monoplane of 1910.

CLARKE double triplane

This machine, with front elevator and biplane tail, was under construction in September 1911 when P.K. Turner of *The Aeroplane* inspected the Kingston works. No further reports of its existence can be traced.

CLARKE monoplane

This is the only Clarke powered aircraft of which information exists, as it was described in *The Aeronautical Journal* of October 1910, having won a prize for its design in a competition. It was for Clarke's own use, and was reported to be nearly ready for trial, probably at Brooklands. No reports of trials or successful flights appeared. The machine was a canard with a monoplane foreplane serving as elevator and with a front mounted rudder. Lateral control was by warping the wing.

Power: 25/30hp Thames four-cylinder horizontally opposed water-cooled driving a 7ft 6in diameter Clarke propeller.

Data		
	Span	30ft
	Foreplane Span	14ft
	Mean chord	5ft
	Mean foreplane chord	3ft 6in
	Area	150 sq ft
	Foreplane area	50 sq ft
	Length	20ft
	Weight allup	750 lb

Clayton biplane. An ambitious design which came to nothing.

CLAYTON Amphibious Flyer (C.D. Clayton and Commercial Aero Club)

C.D. Clayton designed a direct lift flyer and floating aircraft to carry three or more persons and which would rise from water vertically.

The machine was a biplane and had four pairs of motors which powered eight aerial wheel-lifting propellers which gave direct lift and also a forward propulsive thrust.

The Flyer was to have been built in cooperation with the Commercial Aero Club and a scale model was completed; nothing further is known of this project.

CLAYTON biplane ('Capt.' Cecil Clayton, Midland Counties Aviation Syndicate, Mansfield, Nottinghamshire)

There were various reports in 1910-1911 of 'Capt.' Clayton, a former balloonist and parachutist, and his aircraft. In April 1910 *The Aero* referred to a biplane which he had just completed, weighing 700 lb and fitted with a 60/70hp rotary engine referred to elsewhere as an Empress. This was replaced by a 50hp Alvaston four-cylinder horizontally opposed, water-cooled type after a crash at Mansfield. The aircraft was flown from the Duke of Portland's estate and referred to as '*The Clipper of the Clouds*'. In 1910-1911 Clayton was using a standard Blériot, which was involved in a fatal accident at Worcester on 9 June 1910, when being taxied by his assistant Berrisford *(q.v.)*.

Data		
	Span	30ft
	Chord	5ft

CLEMENS biplane glider (Charles Clemens & Bros., Malvern, Worcestershire)

This was a Wright type glider built of bamboo, which was wrecked in a gale in March 1911 after two and a half year's work. Span 24ft.

CLEVELAND biplane (J. Cleveland Jones, Newcastle-on-Tyne, and T. Taylor, Pelaw, Co. Durham)

This was a tractor biplane with an uncovered, square section, wooden girder fuselage, tapering in both planes to a vertical post at the rear. The forward bays back to the pilot's seat were cross braced with steel tubes, aft of which cable bracing was employed. The sternpost supported a biplane tail with hinged elevators and twin, square shaped, all moving rudders between the tips of the fixed tailplanes. The undercarriage consisted of

Cleveland biplane built in 1910. It is not known if it flew.

pairs of separate vee shaped units made from streamlined section steel tube and aluminum sockets, braced by cables. Twin skids with upturned tips carried pairs of wheels on axles sprung by flexible bands. The engine was mounted low down in the front of the fuselage and drove a two-bladed propeller by chain at about half engine speed. Long shallow radiators were mounted below the top longerons.

The wings, which had marked camber, were built with the front spar and interplane struts along the leading edge. Three pairs of interplane struts were provided, dividing the wing structure into two outboard bays of equal size, with a smaller center section bay. Ailerons were fitted to both top and bottom wings which were of the single acting type.

The Cleveland biplane, which was designed by Jones and built by Taylor, was completed but there were no reports of tests or of successful flights.

Power: 35hp Thames four-cylinder horizontally opposed water-cooled, driving a two-bladed propeller by chain reduction gear (2:1 approx.)

Data

Span	30ft
Length	26ft
Weight	800 lb

Clout helicopter. This 1903 design was abandoned through lack of official interest.

CLOUT helicopter (George Clout, Durrington, near Worthing, Sussex)

This machine, for which patent application No.13 075/1903 was made, was offered to the War Office, presumably for observation purposes and was abandoned after lack of interest.

The fuselage was a tubular structure with a circular wooden frame on top at the front. On this were mounted the motor, triangular wings and two large fanlike propellers, the whole being free to rotate on the mounting ring. Control was by altering the balance of the car and by the use of 'the steering tail'. The frame of the fuselage could be 'enveloped in descent, converting it into a 13 ft diameter parachute' in an emergency. The photograph shows a scale model built of oak, bamboo and steel tube.

Power: 3.75hp

Data

Rotating wings	30ft diameter
Propellers	10ft diameter
Weight	350 lb

Aircraft Annual 1968 (p45) &1970 (p623)

COCHRANE biplane (William Cochrane & Co., 26 Clarges St., London W1)

Reported in the *Motor Car Journal* of April 1908; to be fitted with a 15hp three-cylinder air-cooled engine. No record of completion. Cochrane continued to produce models and components, including small petrol engines and made a specialty of model and full size propellers with corrugated blades.

Data

Span	30ft
Length	26ft
Weight	800 lb

Flight 21 August 1909 (p.510) 27 November 1909 (p.762)

COCKBURN biplane (G.B. Cockburn, Salisbury Plain)

Cockburn went to France in June 1909 and learned to fly a Farman at Chalons. Later he assisted with the instruction of the Army pilots of the Air Battalion on Salisbury Plain. He was reported to have brought out 'his own constructed miniature Farman' on 11 January 1912 on the Plain, when it was flown by him and others. *The Aeroplane* 18 January 1912 (p.60)

CODY AIRCRAFT

Samuel Franklin Cody, real name Cowdery, was engaged by the Army's Balloon School, and based with the Balloon Factory at Aldershot as Chief Kiting Instructor in February 1905. A set of his kites had been ordered by the War Office, and these had been made by Cody in his workshop at the Crystal Palace. He had already supplied four sets to the Admiralty, which were also made there or at his earlier premises at Alexandra Palace.

During 1905 the Balloon Factory moved to Farnborough and Cody continued his work there although he retained his own premises at the Crystal Palace, where he constructed and tested a large kite like glider. This was moved to Aldershot and later to Farnborough, where it was flown successfully until September, when it crashed heavily, seriously injuring his son Vivian. The method of operation was to fly it as a kite to the required height and the pilot, lying prone on the lower wing, controlled the resulting glide to earth.

There followed a period of effort with the *Nulli Secundus* airship before the next heavier than air machine appeared. This was the Motor Kite, and was generally operated on a wire between poles, although it may also have been flown freely.

Thereafter Cody turned his attention mainly to the aeroplane which became British Army Aeroplane No.1 in which he made the first officially recognized flight of a powered heavier-than-air machine in Britain on 16 October 1908.

The government reviewed the current state of aeronautics with economies in mind and decided to discontinue work on aircraft at Farnborough, as a consequence of which Cody and his contemporary in the field, Lt. J.W. Dunne, were discharged at the end of March 1909. Cody was allowed to keep his aeroplane, but not the engine, although this was loaned to him for a period of time and he was allowed to build a shed on Laffan's Plain on military ground and to operate in the area. British Army Aeroplane No.1 went through many changes and some crashes during its life and in Cody's ownership became known as Cody No.1. It was followed by six more aeroplanes of five basic types, although most were modified in various ways in the course of their lives. Cody designed and redesigned his aircraft, experimenting and improvising and showing great ingenuity and determination, oft en in most difficult circumstances. Successive types up to No.VI all displayed improvements, which enabled Cody to win a number of prizes, including that for the Military Trials of 1912, with No.V, of which two were built and were the only aircraft that Cody sold. His last machine, No.VI, was that which carried him and his passenger to their deaths on 7 August 1913. At the time Cody was planning to build a machine for a transatlantic flight.

The type and subtype numbers used are a convenient means of identification and may not have been in use when the aircraft were current.

Cody biplane glider built at Crystal Palace in 1905.

CODY biplane glider

In 1905 Cody built this large glider, which had kite like construction and was designed to be easily transportable. It was subject to a number of changes in the course of its use until September when it crashed, badly injuring Vivian Cody.

The basic strength of the glider was provided by two 26ft lengths of bamboo, joined at the center, which could be separated to collapse the wings, together with six 10ft removable vertical poles. The outlines of the wings were encircled by cables which braced the fabric when the machine was assembled. The vertical struts fitted into pockets, forming fins above and below the wings and the central spar fitted into triangular wingtip panels, and with bracing wires in each bay, formed a rigid but extremely light structure when assembled. A tailplane and rudder were fitted and, at one stage, a forward elevator. Diamond shaped ailerons were fitted below the wing, later above the top wing, and may have been operated as elevators also. The pilot lay in a cradle in a cutout in the lower center section operating the control surfaces by cords.

In addition to Cody and his sons, a number of Army men flew the glider when trials took place at the Long Valley, Aldershot.

Data	Span	51ft
	Area	807 sq ft
	Weight	116 lb

Cody motorkite biplane of 1905.

CODY Motor-Kite biplane

After the airship *Nulli Secundus* was deflated in August 1907, Cody used the redundant tail surfaces and incorporated them into a sesquiplane with boxkite center section. Bamboo booms extended aft to carry twin vertical fins with a warpable horizontal tail between. The undercarriage consisted of two bicycle wheels and a continuous curved tail-skid between the booms. Single interplane struts passed through the top wing to form kingposts for bracing wires. The engine, mounted in front of the center section drove a pusher propeller by belt and a long shaft. In this form it was tested running along a wire strung between two poles with the wing warping controlled by ropes from the ground. Subsequently front outriggers with biplane elevators and a prone pilot position, with the controls operated by cords, were provided. In this condition it is possible that Cody raised the Motor-Kite a few inches from the ground for very short distances in the vicinity of the Factory.

Power: 12-15hp Buchet three-cylinder fantype air-cooled radial driving a 8ft diameter metal propeller.

Data	Span	40ft approx.
	Width of center section between wing panels	12ft.

CODY biplane glider

In October 1907 the War Office approved an alternative to the Dunne machine to be manufactured at Farnborough. To test features for the new machine, Cody was empowered to make a simple glider of 36ft wingspan, within a cost limit of £50. The glider was similar in layout to the Wright biplane, but avoided the simultaneous wing warping and rudder control, which was the basis of the Wright brother's claim to originality. The machine was flown as a kite with control of warping by wires and bridle from the ground.

Cody Wright-type glider of 1907-1908 being flown as a kite.

British Army Aeroplane No.I was largely Cody's work and flew for 200ft on 29 September 1908.

CODY British Army Aeroplane No.1 and 1A

Construction of this machine began in October 1907 and it was substantially complete in the Airship Shed, minus a suitable engine, by January 1908. The Antoinette engine did not become available until July, and in the meantime some minor work was carried out, including fitting a temporary source of power, perhaps an electric motor, to activate the propellers and their drive. For a time, from May, the machine was transferred to a small shed to allow work to proceed in secure conditions. By September the aircraft was back in the Airship Shed and the engine installation was complete enabling it to appear outside for engine and taxiing tests on the 19 September 1908. From this date several hops and minor damage occurred, but on 16 October 1908 a sustained flight of 1,390ft, reaching a height of some 30ft, was achieved. Although this resulted in a crash, with considerable damage, this is officially recognized as the first powered, sustained and controlled flight in Britain.

The BAA No.1 was basically a twin pusher biplane with front elevator, a boom mounted tail rudder and wingtip ailerons near the leading edge. The pilot was seated behind the engine and the machine rested on a two wheel, narrow track undercarriage with a front buffer wheel and a tail-wheel, to-

Cody/BAA No.I flew for 74 yards on 16 October 1908.

Cody/BAA No.I A flying with streamers attached on 19 January 1909.

gether with wheels at the wingtips. A flexible triangular panel, 'the fantail', extended from the center section of the top wing to the top of the rudder. A further addition was a 'top rudder' above the wing after which the ailerons were discarded. The radiators were moved outboard to the first interplane struts. In this condition the first flight was accomplished.

Changes were made during the rebuild, and the aircraft appeared again as No.1A on 9 January 1909 to fly with streamers attached to check the airflow. Further damage occurred on 20 January 1909 after a flight of 1,200ft but tests resumed again from mid-February until the end of March, when the machine, less engine, was given to Cody upon his discharge from the Factory.

When the machine emerged at the beginning of the year the 'top rudder', introduced to provide lateral control, and the 'fantail' were removed. The front elevator had been increased in span to 25ft, and in addition ailerons were mounted outboard of this, supported on triangular bamboo booms back to the wingtips. The radiators were moved aft and changes were made to strengthen the undercarriage and wingtip wheels. The machine was lengthened by extending the rear boom. In this condition, a forty yard flight, when a height of 10ft resulted in a stall and a heavy landing, proving the center of gravity being too far aft.

Before its next appearance on 20 January 1909 the ailerons were transferred to trail on struts mounted on the rear interplane struts and braced by wires. The 'top rudder' had earlier been mounted within the front booms. However, a short flight resulted in a crash when the front booms folded backwards. Changes were made during the subsequent repair, the ailerons being removed and made into a biplane tail, fitted within the rear booms. Triangular flaps were fitted to the front lower wingtips to act as 'spoilers', probably in conjunction with warping. The Factory drawings of March 1909 show most of these changes and illustrate the general configuration of the aircraft when presented to Cody.

Power: 50hp Antoinette eight-cylinder water-cooled vee driving twin pusher propellers by flat belts.

Data	**No.1**	
	Span	52ft
	Chord	7ft
	Gap	8ft
	Area	790 sq ft
	Length	32ft
	Height	17ft
	Weight	2,260 lb
	Weight allup	2,540 lb
	Speed	40 mph
Data	**No.1A**	
	Area	857 sq ft including ailerons
	Length	44ft
	Height	17ft

CODY biplane No.1B and 1C

No.1B. After March 1909, when the aeroplane became his property, Cody continued with its development, still using the Antoinette engine on loan and with the machine still housed at the Factory until his shed on Laffan's Plain was ready in May. He had earlier removed Capper's biplane tail, and on 14 May 1909 was testing a new rudder mounted above the front eleva-

Cody No.I B after reconstruction on 14 May 1909.

tor, instead of within the booms, coupled to twin rear rudders. Lateral control was still being resolved by repositioning the ailerons; at one stage they were again mounted on the rear outboard interplane struts with wire bracings, and also on booms trailing behind the wings on the second and outboard struts.

Cody flew for over a mile at up to 30ft high in the morning, but crashed in the afternoon while attempting to repeat the performance. In the course of being rebuilt the machine reverted to a single rear rudder and the tailwheel was replaced by a long curved wooden skid referred to as the 'kangaroo tail'. Shortly after the front elevator was increased in span and divided into two separate flaps, capable of differential movement, to provide lateral control.

In June Cody flew distances up to two miles, in July four possibly six miles, and was contemplating a more powerful engine.

Cody No.I C. Although suffering a mishap at Doncaster, the much modified machine was flying strongly to the end of 1909.

No.1C. At this stage a major reconstruction of the aircraft took place with the Antoinette being replaced by a French-made ENV and the pilot's position being moved to in front of the engine. The propellers, now driven by chains, were lowered and the radiators moved to the front of the engine.

Cody flew the reconstructed aircraft on 11 August 1909 for the first time, and considered it a great improvement. A seat was arranged behind the pilot's position and on 14 August 1909 he carried his first passenger. On the 8 September 1909 he flew for 1hr 3min, covering forty miles and reaching a height of 600ft, a record for a flight in Britain.

Cody took the machine to the Doncaster Meeting in October, but after one short flight it overturned on soft ground and was damaged. Cody, now a British subject, was forestalled in his entry for the *Daily Mail* £1,000 prize for a circular flight of one mile, which he could easily have achieved, by Moore-Brabazon on 30 October 1909.

He subsequently moved to Aintree to compete for the *Hartley* prize of £1,000 for a flight from Liverpool to Manchester, in which he was unsuccessful after several abortive attempts. He abandoned further flights with the onset of winter, and the last use of this machine was made in early January.

At the Doncaster Meeting the machine had acquired the nickname *'Cathedral'*.

Power: No.1B 50hp Antoinette eight-cylinder, water-cooled vee driving twin propellers by flat belt.
No.1C 60-80hp ENV type C eight-cylinder, water-cooled vee driving twin propellers by chains, French-made to British parent company's design.

Data		
	Span	52ft
	Length	44ft
	Chord	7ft
	Area	790 sq ft plus
		67 sq ft when ailerons fitted
	Height	13ft
	Gap	9ft
	Weight	2,200 lb
	Weight allup	2,500 lb

CODY biplane No.IIA, B, C, D and E (British Michelin Trophy Type)

No.IIA. For 1910 Cody planned a smaller machine with single pusher propeller, but much in the style of the last version of his first machine. He required more power and planned to fit a ten-cylinder 100hp Phoenix rotary engine, which did not materialize. The design was changed to accept two 60hp Green engines, but only one was fitted initially. On its first appearance the radiators were above the engine and no horizontal tail surface was fitted. In this form it flew two circuits of Laffan's Plain on 23 June 1910, at about 50ft but crashed on landing, injuring Cody.

In the course of the rebuild a fixed tailplane was added. Cody took the machine to the Wolverhampton and Bournemouth meetings in June and July, but his flying was limited and no prize money was won. In August he arrived at Lanark with the machine, but he received a second engine on loan from the Green Company. This could not be synchronized, with the other so both were removed. Cody's visit to Lanark was most unsuccessful.

No.IIB His original French ENV was then fitted and this remained in use until late October. During September he carried nine passengers on various flights, flying 113 miles in the first

Cody No.II. This smaller biplane, built for use in 1910, survived until April 1912 in various forms. No.IIb illustrated with British ENV Type F engine.

Cody No.IIC and D. The 1910 British Empire Trophy No.1 winner with Green engine at Olympia in March 1911.

week and 126 miles in the third. On 30 September 1910 he flew for 1hr 15min and then for 50min. The French-made engine could not be used for the Michelin Contest, for which a £500 prize and trophy were offered. In late October he refitted the Green on a new single mounting and on 4 November 1910 made an attempt lasting 2hr 25min. He then installed a British ENV, which was on loan, in the first week of November, but after a minor accident on 12 November 1910, this was removed and returned to the ENV company.

No.IIC. Cody reverted to a single Green and used this until the end of the year, winning the British Empire Michelin Trophy for a flight of 4hr 47min in a closed circuit, calculated as 185.46 miles, on his third attempt on the last day of the year and of the Contest.

No.IID. Cody exhibited his aircraft at the 1911 Olympia Aero Show, held between 24 March and 1 April. Apart from the various engine changes, to all appearances it had changed very little from its early condition and, by the standards of the day, had proved to be very successful.

The lateral control was by ailerons aided by differential elevators, operated by sideways movement of the control column. A hand-wheel on the column controlled the rudder. The throttle control was by foot pedal. The undercarriage had spring loaded main legs and buffer wheels at the front; the tail-skid was a long wooden 'kangaroo tail' on which the machine normally rested. Cody could see through the honeycomb radiator, and derived warmth from this and small radiators by his feet. A passenger could be carried on a second plough type seat mounted behind and higher than the pilot.

No.IIE. (Omnibus) Cody's earlier Michelin Trophy machine was brought back into use at the end of 1911. It flew fitted with an Austro-Daimler engine giving the power that Cody had long been seeking, on 3 January 1912. Twin booms and tail surfaces, reportedly transferred from No.III, were fitted. The performance of the aircraft was greatly improved and later carried Cody and up to four passengers on agricultural type seats mounted in a line at midgap behind him. The aircraft crashed in April 1912 and was not rebuilt.

Power: No.IIA, B and C 60hp Green four-cylinder inline, water-cooled driving a single pusher propeller by Brampton chain. Two engines were fitted briefly but not flown.

No.IIB. 60-80hp ENV Type C eight-cylinder, water-cooled vee, French-made.

60hp ENV Type F eight-cylinder, water-cooled vee, British-made driving a 10ft 2in diameter propeller by Brampton chain.

No.IIE. 120hp Austro-Daimler six-cylinder, inline water-cooled.

Data		
	Span	46ft (49ft including aileron overhang)
	Chord	6ft 6in
	Area	540 sq ft
	Area front elevator	116 sq ft
	Area ailerons	101 sq ft
	Gap	8ft 6in
	Length	38ft 6in

CODY biplane No.III (Circuit of Britain Type)

Cody's third aircraft was again smaller, but built on similar lines to its predecessor. No ailerons were fitted, lateral control being achieved by the use of wing warping and differential front elevators. An obvious change was the use of twin booms and rudders with fixed circular tailplanes, which could be folded against the mainplanes for storage in the confined space of his shed, now needed to house two aeroplanes. As in his previous aircraft, bamboo was used extensively for the booms and control rods, spruce and hickory being used elsewhere.

The *Daily Mail* Circuit of Britain Contest started at Brooklands on 22 July 1911 and was over a course of 1,010 miles for a £1,000 prize. It was won by a Blériot flown by Lt. Conneau in four days of flying. Cody had problems including poor weather, which extended his time to fourteen days, nevertheless his biplane was the only British-built machine to fin-

Cody No.III Circuit of Britain biplane won the British Empire Michelin Trophies Nos.1 and 2 of 1912 but was unsuccessful in the circuit contest.

Cody No.III Circuit of Britain biplane in flight.

ish and he was placed fourth. This machine proved to be Cody's most successful aircraft to date and went on to win both of the British Empire Michelin Prizes of 1911.

Trophy No.1. 29 October 1911. 5hr 15min, a flight of 261 1/2 miles, £500.

Trophy No.2. 11 September 1911. 3hr 6 1/2 min, a circuit of 125 miles at 40mph, £400.

The machine continued in use for instruction, demonstration and passenger work into 1912, but was crashed badly on 3 June 1912, when being flown by a pupil, Lt. Harvey-Kelly. It was cannibalized to provide parts for a later machine, No.V.

Power: 60hp Green four-cylinder, inline, water-cooled driving a pusher propeller by chain.

Data		
	Span	40ft
	Area	450 sq ft
	Length	30ft
	Height	11ft
	Weight	1,750 lb
	Weight allup	2,500 lb
	Speed range	40-58 mph
	Ceiling	5,000 ft
	Range	350 miles

CODY monoplane No.IV

Cody began testing this new machine, intended for entry in the Military Trials, in the middle of June 1912. Its life was brief, however, for on 8 July he was forced to land after the engine failed and in the resulting collision with a cow, the machine was damaged beyond repair.

Cody No.IV. The monoplane intended for the Military Trials of 1912 but damaged beyond repair before the competition.

The aircraft was a shoulderwing monoplane which rested on a two wheel undercarriage with sprung struts and curved hickory wood struts supporting a central skid with twin front wheels. The skid was extended aft to form a tail support. The front fuselage comprising the engine bay and the side-by-side seating was a spruce structure with cabane for the anchorage of wing bracing wires. The fuselage aft of the cockpit was an unconventional structure of four bamboo longerons, faired with fabric covering tapered to a point. The longerons as they extended aft were splayed out to provide mountings for a fixed cruciform tail, plus pivot points for twin rudders and elevators.

The engine drove, by chain, the propeller mounted above on a shaft at two thirds engine speed. The radiator was mounted in front of the engine, but behind the chain drive. The exhaust was fitted with a silencer, which doubled as a heater and emerged from the top of the starboard fuselage side. Fuel and oil tanks were housed in the decking in front of the cockpit, which had transparent sides. Although the engine was exposed it is probable that cowlings were to be added later.

Lateral control was by wing warping, all control movements being made from a central control column between the two occupants. The elevators were operated by a bamboo rod by fore and aft movement and the rudders by cable by a wheel on the column.

Power: 120hp Austro-Daimler six-cylinder, inline, water-cooled driving by Brampton chain a 10ft 2in diameter Chauvière propeller at two thirds engine speed.

Data		
	Span	43ft 6in *44ft
	Chord	6ft 9in
	Area	260 sq ft
	Elevator area	46 sq ft
	Rudder area	46 sq ft
	Length	37ft *41ft
	Height	12ft 6in *13ft
	Weight	2,400 lb
	Weight allup	3,100 lb
	Speed range	58-83 mph
	*from other sources	

CODY biplane No.VA, B and C (Military Trials Type)

Cody entered both his monoplane and biplane in the Military Trials, but both were damaged about three weeks before the start of the trials. By a great effort he managed to build a new biplane basically using No.III, which had first flown in June 1911, and the engine and parts from the monoplane. He began testing this on 23 July, flying after dusk on some evenings, to have the aircraft ready for the Trials, which began on 4 August, when he flew to Larkhill. He performed so successfully that he was declared the winner of the £5,000 prize.

Cody's success resulted in a War Office order for this machine, which became RFC No.301, and a similar one, No.304. No.301 broke up in the air on 28 May 1913, and Lt. Rogers-

Cody No.VA and B. The winner of the Military Trials of 1912 became No.301 in the RFC. A second machine No.304 is still on display at the Science Museum.

Cody No.V C was fitted with a 100hp Green engine and then won the British Empire Michelin Trophy No.2 of 1912.

Harrison was killed when the front elevators, dating from June 1911, failed. No.304 served briefly with the RFC, but after being damaged, was passed to the Science Museum in November 1913, where it is still displayed.

Some variations existed between the two machines initially, but later they were made identical. The notable differences included the change from anhedral to dihedral setting of the wings; the placing of the rudders closer together and the raising of the elevators. In appearance the Cody V resembled the Cody III except for the shape of the rudders and tail surfaces. The crew area was partly enclosed with fabric.

After winning the Trials, Cody changed the engine for a 100hp Green and proceeded to win the British Michelin Trophy No.2 of 1912 and a prize of £600 for a flight of 186 miles in 3hr 26min. After this the Austro-Daimler was reinstalled before the machine was handed over to the RFC at the end of the year. It was exhibited by the War Office at the Olympia Aero Show in February 1913.

The machine was basically a two-seater but four seats were fitted to VA at one time.

Power: No.VA and B. 120hp Austro-Daimler six-cylinder inline water-cooled driving a 10ft 4in diameter Cody propeller by Brampton chain at 1.75:1 reduction. No.VC. 100hp Green six-cylinder inline water-cooled.

Data		
	Span	43ft
	Chord	5ft 6in
	Gap	6ft. 6in.
	Wing Area	430 sq ft
	Area Elevators	60 sq ft
	Rudders	30sq.ft
	Length	37ft 9in
	Height	11ft 6in
	Tail damper planes	10 sq ft
	VA &B. Weight	1,850 lb allup 2,850 lb
	VC Weight	2,000 lb allup 2,800 lb
	Speed range	48-72 mph

Cody No.VIA for the Seaplane Circuit of Britain was flown only with land undercarriage. This is the machine in which Cody was killed.

CODY biplane No.VI A and B (Water-plane)

The last Cody aeroplane to be constructed was built to compete in the *Daily Mail* Seaplane Circuit of Britain, due to start from Calshot in August 1913. It was completed in July and was taken to Eelmoor Flash on the Basingstoke Canal for flotation trials. Thereafter the floats were removed and a wheel undercarriage fitted for flight trials. These included flying the machine to Brooklands on 3 August 1913 for the meeting on Sunday before the Bank Holiday. Cody planned to fly the machine to Calshot on Thursday 7 August for the floats to be fitted. However, in the morning he arranged to take up passengers and while he was flying with W.H.B. Evans, the Hampshire cricket captain, the machine broke up over Ball Hill near Cove and both Cody and his passenger fell to their deaths.

No.VI was the largest Cody aircraft, and had warping wings and a single front elevator. A large diamond shaped rudder pivoted between the booms and carried a small fixed triangular tailplane. Two seats were fitted in tandem, the pilot's controls and the chain driven propeller were similar to those on previous types. An enclosure to protect the crew was provided after the machine's early appearances.

No.VIA The narrow track land undercarriage incorporated a central skid serving as a front buffer and rear support. Sprung skids pivoted at the wing leading edge below the intermediate

Cody No.VIB. *The Waterplane*, fitted with floats by Harmsworth of Ash Vale, was flotation tested but not flown.

interplane struts prevented contact of the lower wingtips with the ground.

No.VIB The machine never flew as a seaplane, but the flotation gear consisted of a large three stepped central float with fabric decking. The balancing floats were fitted below the first set of interplane struts. These three floats were made by Harmsworth's Boatyard of Ash Vale, constructors of barges and boats, and were tested by being towed along the canal.

Power: 100hp Green six-cylinder inline water-cooled driving by chain a 10ft 8in diameter four-bladed Garuda propeller. A four-bladed Integral was due to be fitted.

Data		
	Span	59ft 6in
	Chord	6ft 10in
	Gap	8ft
	Area	770 sq ft
	Area Elevators	75 sq ft
	Area Rudders	40 sq ft
	Area Damper tailplane	25 sq ft
	Length	40ft 9in
	Height	12ft 9in
	Speed range	45-70 mph

CODY PROJECTS

In 1913 Cody considered establishing his own company for producing aircraft in greater numbers. A catalogue of types was produced in which were listed the No.III biplane and No.IV monoplane, which had been built. In addition there were a smaller and a larger version of No.III, of which there were no sales and these remained unbuilt. The following is the estimated data extracted from the catalogue.

Data	**30-35hp**	
	Weight	750 lb without fuel or pilot
	Lifting surface	400 sq ft
	Width	35ft
	Depth, fore and aft	25ft
	Height overall	9ft 6in
	Test flight	1hr
	Speed range	35-45 mph
	Gliding angle	1 in 7
	Weight carrying	300 lb

Angle of ascent after 100 yard run on level ground

	First 50 yd	1 in 7
	Next 100 yd	1 in 10
	Next mile	1 in 30
	Price	£750

Data	**50-60hp**	
	Weight	1300 lb without fuel or pilot
	Lifting surface	450 sq ft
	Width	40ft
	Depth, fore and aft	30ft
	Height overall	11ft
	Test flight	1hr
	Speed range	40-58 mph
	Gliding angle	1 in 6.5
	Weight carrying	800 lb

Angle of ascent after 100 yard run on level ground

	First 50 yd	1 in 6
	Next 100 yd	1 in 9
	Next mile	1 in 25
	Price	£1,000

Data	**100-160hp**	
	Weight	1850 lb without fuel or pilot
	Lifting surface	500 sq ft
	Width	45ft
	Depth, fore and aft	35ft
	Height overall	12ft 6in
	Test flight	1hr
	Speed range	50-75 mph
	Gliding angle	1 in 5
	Weight carrying	1,000 lb

Angle of ascent after 100 yard run on level ground

	First 50 yd	1 in 4
	Next 100 yd	1 in 6
	Next mile	1 in 20
	Price	£1,200 to £1,800

The above details referred to the ordinary types of 'Touring Machines'. Specifications of 50-60hp and 100-160hp 'Special Racing Types' were available on application.

CODY monoplane No.VII (Transatlantic)

In April 1913 the *Daily Mail* offered a £10,000 prize for a Transatlantic flight, a challenge which Cody intended to accept. He planned to build a monoplane of 120ft span, mounted on twin floats and with a crew of three. For this venture he required a 400hp engine. He prepared a specification for a large twelve-cylinder engine and passed drawings and £600 deposit to a French company to manufacture. This was abandoned after his death.

COLE low power machine (J.W. Cole, 295 Wellington St., Grimsby)

No information can be traced on this machine. A glider was also flown in 1910.

Cole Tandem monoplane with folding wings at Olympia in 1911.

COLE Tandem monoplane (W. Cole & Sons Ltd., 92 High St., Kensington, W14)

A tandem monoplane designed by a Frenchman, M. Magnodex, was exhibited in an unfinished state at the Olympia Show of March 1911. The two wings of equal span were separated by a gap equal to the chord, the pilot being seated between and above the level of the wings. The engine intended to be a 120hp rotary or radial was to drive two propellers mounted on the ends of a single shaft driven by chain. A rudder was pivoted above a fixed tailplane, pitch control being by means of a flap below the pilot's seat. It is believed that the machine was not completed and did not fly.

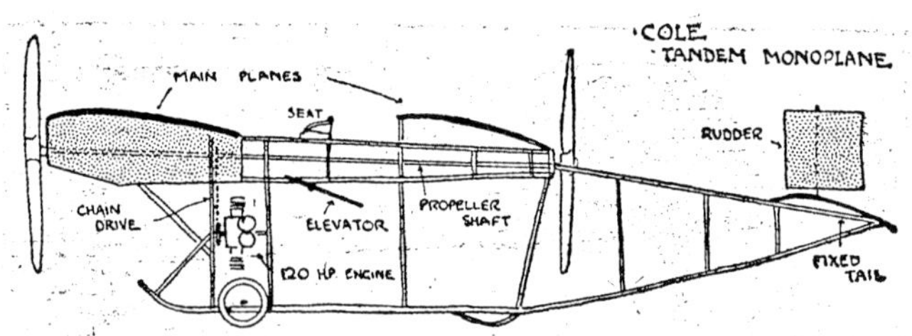

Sketch showing the configuration of the Cole Tandem monoplane.

COLLIER monoplane (H.A. and C.R. Collier, Plumstead, London, SE)

This machine was tested by the Collier brothers on Plumstead Marshes in early 1910, achieving a flight of half a mile on 7 March 1910. It was made by the family concern that later became famous for Matchless motorcycles, and was very similar to a Blériot. The fuselage was of ash and the wings, elevators and rudder were of spruce. It was taken to Brooklands in September 1910, but was damaged during testing by F. Conway Jenkins when the machine was reported fitted with a more powerful engine.

Power: 20hp JAP four-cylinder air-cooled vee driving a 6ft 8in diameter Chauvière propeller
40hp JAP eight-cylinder air-cooled vee fitted later.

Data

Span	30ft
Chord	6ft
Area	180 sq ft
Length	26ft
Weight	500 lb
Weight allup	650 lb

Collier monoplane of 1910 was virtually a Blériot.

COLLINS-HANCOCK monoplane

This machine, illustrated in *Flight* on 24 January 1914 (p.102), was designed as a lightweight military scout. It could be easily transported by road, the fuselage being detachable and the wings could be folded; the dimensions were thus reduced to 20ft by 7ft.

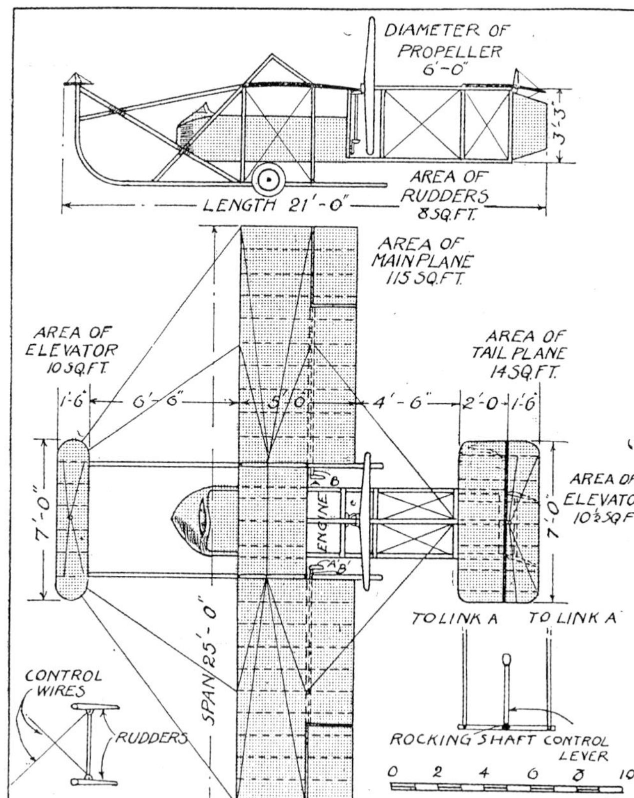

Collins-Hancock monoplane. A design for a lightweight military scout.

The ailerons and elevators were controlled by rods and links. The propeller revolved round the central top longeron and was driven by bevel gears by an engine of unknown type set low at the rear of the nacelle. The pilot, seated at the front of the nacelle would have had a very good all-round view.

This interesting and potentially useful machine appears not to have been built.

Data		
	Span	25ft
	Chord	5ft
	Prop. diameter	6ft
	Area	115 sq ft
	Area: Rudder	2 x 8 sq ft
	Area Tailplane	24.5 sq ft
	Length	21ft
	Front elevator	10 sq ft

Collyer-Lang monoplane. Collyer's first machine of 1910 before he teamed up with BH England.

Collyer-England biplane in 1911 with Alvaston engine.

COLLYER-ENGLAND biplane (Collyer and B.H. England, Shoreham, Sussex)

This machine, built in 1911, was an equal span tractor biplane with three bay wings having squarecut tips. Ailerons were fitted to both top and bottom wings. The undercarriage consisted of two pairs of wheels, each with a central skid, below the inboard interplane struts.

The original engine was a 30hp Alvaston, later replaced by a 35hp Green. Both engines were water-cooled with a radiator mounted below the top wing, the Green requiring a larger unit.

The machine is believed to have been abandoned after damage at Shoreham on 3 May 1912.

Collyer-England biplane in 1912 with Green engine.

COLLYER-LANG aeroplane (Nicknamed 'Otasell' and 'Hellhound')

This single-seater monoplane arrived at Brooklands during November 1910 and testing continued until February 1911 without much success, only brief hops being achieved. In its original form it was a pusher with twin rear booms, with the pilot seated ahead of the wings, between the two front skids. Inset single acting ailerons provided lateral control, and twin rudders within the booms and a tail mounted elevator comprised the other control surfaces. The engine was a 30hp WLA (Adams) motor.

In March 1911 the machine reappeared at Brooklands converted to a tractor biplane with pilot and engine positions interchanged and described as a 'one and a bit plane', with a pair of smaller planes fitted above the originals. No subsequent reports appeared.

Colston glider was built in Glasgow by A. Miller in 1914.

Conisbrough Club. The *Demoiselle* type glider being built in 1911 by Wright and Wilson at Doncaster.

COLSTON glider (A. Miller, Bishopsbriggs, Glasgow)
A canard monoplane glider was built in 1914 by Colston Gliding Club members and crashed in 1915.

Data	Span	24ft,
	Chord	6ft,
	Length	24ft.

COMET Aerocycle (Portman Academy of Motoring and Aviation, Little George St., London W)
As the name implies, the basis of the machine was a bicycle to which a superstructure was added to mount a monoplane wing. This was braced to a cabane above and to the cycle frame and had considerable taper and dihedral. The rider sat below a covered center section the pedals driving a tractor propeller in addition to the rear wheel. The machine was exhibited at the Aero Show at Olympia in March 1910 in the Model Section.

Data	Span	23ft
	Length	13ft
	Optional extra	5 1/2hp engine.
	Price	£160.

Comet Aerocycle on display at Olympia in March 1910.

CONISBROUGH gliders (Conisbrough & District Aeroplane Society, 18 Church St., Conisbrough)
This mainly model building club was presented with a full size glider by the Sheffield Model Aero Club in May 1911. In the course of rebuilding it was reduced in span from 42ft to 24ft. In this form it was only capable of lifting a light person to a height of 20ft when towed manually. The machine consisted of a pair of two bay wings with no stabilizing surfaces. Gliding in this manner was dangerous and was soon discontinued. A second glider of *Demoiselle* type was under construction by club members F.J. Wright and G.N. Wilton at Doncaster in the second half of 1911, but may not have been completed.

COOKE monoplane (W.E. Cooke, Burnley Motor Bus Co., Burnley, Lancashire)
See Howard Wright.

COOPER glider No.1 (G.T. Cooper, Charterhouse School, Godalming, Surrey)
During his last year at Charterhouse School Cooper built a biplane glider between February and July 1911, which was tested under tow.

Data	Span	24ft,
	Chord	5ft,
	Length	18ft.

COOPER glider No.2 (G.T. Cooper, 41 Drumsheugh Gardens, Edinburgh)
After settling in Scotland, Cooper built a second glider in 1911, which he later presented to the East of Scotland Aero Club. It was an unequal span biplane with single-acting ailerons hinged on the trailing edge of the top plane extensions. Bamboo booms carried a horizontal tail and divided elevator.

Below: Cooper glider No.1 built in 1911 while at Charterhouse school.

Cooper was a civil engineer and the Hon. Secretary of the Edinburgh Aeronautical Society and was a driving force behind aviation in Scotland.

Data Span, 22ft,
Length 18ft.

Cooper biplane made several successful flights at Cramond in 1913.

COOPER biplane (G.T. Cooper, Edinburgh)

Cooper designed a biplane in 1913 and it was partly built by John Gibson (q.v.), to whom it was known as his Caledonia XI. It was fitted with a 35hp Humber engine and made several flights in September 1913 between Cramond and Cramond Sands. The wingspan was 30ft.

CORDNER monoplane No.1 (Joseph Cordner, John St., Londonderry, N. Ireland)

The aeronautical Press seem to have missed the work that Corder was carrying out in Northern Ireland until *Flight* published an article on his latest machine in 1915. This was the result of several years of work which included this, and one other full size aircraft, the first of which was tested on the foreshore of Lough Swilly at Lisfannon, near Buncrana, County Donegal in late 1909 and 1910.

The layout was fairly conventional, but with an original type of wing. The fuselage was a triangular section braced wooden girder, originally open, but later covered with fabric. The main struts of the undercarriage were continued up as a

Cordner monoplane No.1 built in Northern Ireland in 1909-1910 with wing with patented tunnels.

pylon for the wing bracing cables, forming a substantial A-frame; a large nosewheel, mounted on tubular struts, completed the chassis; a tail-skid was also fitted.

The tail unit consisted of a one piece tailplane mounted on the top longerons probably moveable for control in pitch. There was no fin or rudder. The wing incorporated the feature of the design covered in Patent No.2521/1911, consisting of a series of triangular section tunnels across the chord, allowing air to enter below the leading edge and to exit above the trailing edge. The system was claimed to 'grip' the air and prevent sideslip. Similar tunnels were fitted to the tailplane.

Power: 40hp JAP eight-cylinder air-cooled vee.

CORDNER monoplane No.2

The second Cordner monoplane still embodied the tunnels in the wing, but was considerably different in other respects. Although it was not reported by *Flight* until 1915, it is believed to have been tested in 1911-1912 at Bond's Field, The Waterside, Londonderry.

The fuselage, apart from the engine bay, was fully covered and was mounted on a two-wheeled undercarriage with twin skids, the front struts being well forward and connected to the engine mounting. The pylon for the bracing cables consisted of four struts on the top longerons just ahead of the pilot. The tailplane on the top longerons, extended well forward and carried a divided elevator. The triangular rudder with rounded

Cordner monoplane No.2. The improved version of 1911-1912.

corners was pivoted on the rearmost fuselage member; there was no fin.

The top surface of the wing was uncovered for about one third of the chord, revealing the exits of the tunnels. Lateral control was by flaps near the tips, acting as spoilers, in an upward direction only.

Power: 45, 50/60 or 80hp Anzani six-cylinder air-cooled radial.

COTT biplane glider (Cott, 21 Malcolm St., Cambridge)
Reported in November 1909. Wing area 180 sq ft.

COVENTRY ORDNANCE WORKS biplanes
This company took over the Howard Wright works at Battersea and hangar No.32 at Brooklands in late 1911, and manufactured two biplanes for entry in the Military Trials, due to start in August 1912. Howard Wright and W.O. Manning were responsible for the design and manufacture at Battersea, although Howard Wright resigned in the autumn when the first aircraft was complete. Although there was some similarity, there were also considerable differences between the two machines.

COW Military Trials biplane No.10
This was a side-by-side two-seater with rotary engine, with fuselage sides parallel in plan view to the tail. The fuselage was mounted above the lower wing, to which it was connected by struts enclosed in a fairing housing the main fuel tank. There was a wide gap between the top and bottom wings, which were connected by four pairs of interplane struts. The wings were tapered, with well-rounded tips and with considerable overhang to the top wing, which was braced to kingposts above the outer pair of interplane struts. The tail unit was unconventional, consisting of small twin fins and rudders and large fixed surfaces on either side of the fuselage. The two rudders and elevators were of similar part circular shape with aerodynamic balance areas. The undercarriage, with central skid, was not sprung and relied on the tires to absorb shocks. The top wing center section was cut away at the trailing edge after the machine's initial appearance.

COW biplane No.10 (first version) with cutout in upper wing center section.

Sopwith tested the machine at Brooklands from 5 May 1912 including the first week of the trials at Larkhill, when he was committed to leave to race in America for the Harmsworth Trophy for power boats. On one flight at Brooklands Sopwith carried three passengers, causing the machine to be nicknamed 'Wombus' (W.O. Manning's omnibus). Raynham flew the aircraft subsequently in the trials at Larkhill, where it was forced to withdraw for various reasons after a poor performance.

After the return of the machine to Brooklands, Manning, still with the company after Wright had left, redesigned the machine which flew in its modified form on 13 January 1913 for the first time. New fourbay wings of increased span and parallel chord were fitted and, as the gap was reduced, the fuselage rested directly on the lower wing. Inversely tapered ailerons replaced wing warping and larger elevators were fitted. A lengthened undercarriage maintained the propeller ground clearance. Later the chain drive was discarded and a smaller two-bladed propeller was fitted. The machine continued in use to the end of 1913.

Power: 100hp Gnôme fourteen-cylinder air-cooled rotary, driving by Renolds chain and shaft at 2:1 reduction a 11ft 6in diameter two-bladed propeller.

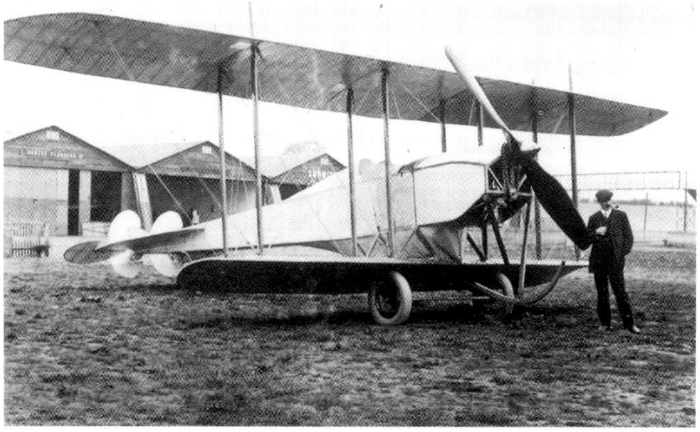

Coventry Ordnance Works biplane No.10 (first version).

COW biplane (second version) with parallel chord wing.

COW biplane No.10 (third version) with direct propeller drive.

Data	1st Version	2nd Version	3rd Version
Span top	40ft	-	56ft
Span bottom	24ft 8in	-	-
Chord top	6ft tapering to 4ft 6in	-	-
Chord bottom	6ft tapering to 5ft 2in	-	-
Gap	8ft	-	-
Wing area	350 sq ft [1]	-	630 sq ft
Tailplane span	-	8ft 2in	-
Elevators span	-	11ft 9in	14ft 3in
Area tailplane	-	35 sq ft [2]	-
Area elevators	-	14 sq ft [3]	27.4 sq ft
Area fin	-	4 sq ft	-
Area rudder	-	14 sq ft [4]	-
Height rudder	-	4ft.	-
Length	33ft 3in	-	-
Height	12ft 8in	-	-
Weight	1,200 lb	-	1,100 lb
Weight allup	1,950 lb	-	1,900 lb
Speed range	20-60 mph	-	-

[1] Also reported as 337 sq ft.
[2] Also reported as 31 sq ft.
[3] Also reported as 17.3 sq ft
[4] Also reported as 15.6 sq ft.

COW biplane No.11. The second machine for the Military Trials with the unsatisfactory Chenu engine.

COW Military Trials biplane No.11

The second machine for the trials was a tandem two-seater with a water-cooled engine, and although of generally similar conception, varied from the first in many features. The fuselage was narrower and shorter but was still parallel in plan and was mounted higher in the wing gap. A similar center skid undercarriage had the addition of two skids behind the wheels. The wings were reduced in span and parallel with rounded tips and with skids under the wingtips. A single fin and rudder were fitted and both these and the tailplane and elevators were large and curvaceous in shape. The Chenu engine had a gear reduction drive and was cooled by radiators on either side of the fuselage.

This machine arrived late at Larkhill for the trials and did not fly owing to engine problems. It seems to have been abandoned soon after.

Power: 110hp Chenu six-cylinder inline, water-cooled driving by gearing at 2:1 reduction a 11ft 6in diameter four-bladed propeller.

Data

Span top	35ft
Span bottom	22ft
Tailplane span	10ft
Elevators span	13ft
Chord	5ft 6in
Gap	8ft
Area	290.5 sq ft
Tailplane area	35.8 sq ft
Elevators area	24 sq ft
Rudder area	10.8 sq ft
Length	31ft 3in
Rudder height	5ft 3in
Height	13ft 2in
Weight	1,250 lb
Weight allup	2,050 lb
Max speed	68-70 mph

COW biplane seaplane

A tractor biplane seaplane was ordered from COW under Contract CP40688/13, to be fitted with a 160hp Gnôme. It was allotted the RN serial number 54, but was never delivered, although it was still on order in September 1915. It was also listed to have an 80hp Gnôme.

Creese-Dederich monoplane at Blackpool in 1909. Creese in the cockpit and Dederich by the propeller.

CREESE-DEDERICH monoplane (A.E. Creese and W. Dederich designers)

Although this machine arrived at the Blackpool Meeting in October 1909 there were no reports of any successful flights taking place there or subsequently. The aircraft was a single-seat tractor monoplane with lateral control by sliding panels to increase the wing area, and consequently the lift, as selected. The undercarriage was sprung pneumatically. These features were covered by Patent Nos.23757-8/1909.

Power: 30hp Anzani three-cylinder air-cooled fan type semi-radial.

Data		
	Span	25ft increasing to 28ft
	Normal area	168 sq ft
	Length	24ft
	Weight	450 lb
	Speed	30mph (estimated)

CROMPTON monoplane (H.D. Crompton, Walton-on-the-Hill, Surrey)

Nothing further was heard after *Flight* published a photograph and details of this single-seat pusher monoplane with front elevator on 4 November 1911 (p.965). The engine was for sale in September 1912.

Power: 30hp Alvaston two-cylinder horizontally opposed water-cooled.

Data		
	Span	30ft
	Area	160 sq ft
	Length	28ft
	Weight allup	600 lb

Crompton monoplane at Walton-on-the-Hill, Surrey in 1911.

D

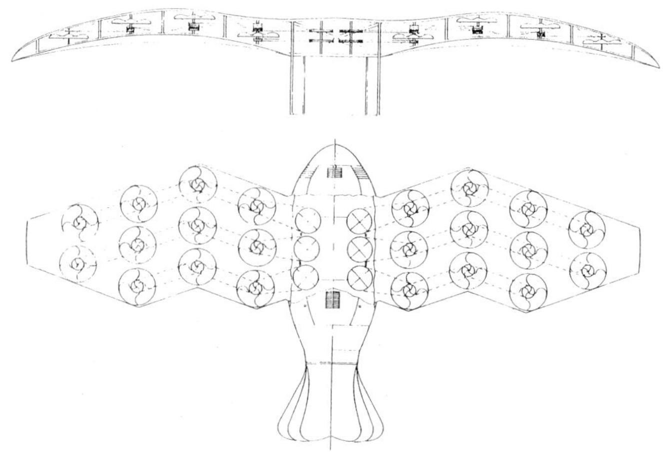

Davidson Air-Car monoplane. The 1897-1898 version with twenty-two lifters.

DAVIDSON Air-Car monoplane 1897-1898 (George Louis Outram Davidson, Davidson Air-Car Construction Syndicate Ltd., Banchory, Grampian, Scotland)

Davidson's interest in flight began in 1883, and he intended his machine to operate in a series of lifts, followed by horizontal flights, like a bird; the airflow being regulated by flaps in the wings, as appropriate. Some of his work is covered by patent Nos.12469/1896 and 13207/1889. Model tests took place at Banchory in 1897 and lift fans were tested by a London consultant, followed by the preparation of a design by a constructional engineer W.L. Hamilton. Finance for construction was sought into 1897, when the project lapsed.

The machine was a high wing monoplane, with a double-decker fuselage for twenty passengers and crew. The lift was obtained from 22 propellers of 6ft diameter in the wings, with their axes set vertically. These were driven by belts from the powerplant in the center section. Automatic stability in all planes, was to be provided by a pendulum arrangement, controlling the tail surfaces, and directional control by nose rudder or 'beak'.

Data		
	Span	100ft
	Length	48ft
	Height	14ft
	Fuselage width	10ft
	Lift generated	10 tons

Davidson Air-Car monoplane. The 1906-1908 version partially built in America.

DAVIDSON Air-Car monoplane 1906-1908 (Head Office, Jermyn St., London. Construction Site, Nichol's Castle, Denver, Colorado, U.S.A.)

A second version of the project was revived by Davidson in 1906, working from offices in London, where he engaged A.V. Roe as a draughtsman. In April the project was moved to the U.S.A., where Davidson had mining interests. British patent No.1960/1907 was taken out in January to protect the design.

The machine was still a monoplane with deep arched wings, but the lift and propulsion system now consisted of a single large lifter in each wing, the outer half of which protruded from the wing. The vertical axis could be tilted to control the direction of flight, and was a means of overcoming the undulating flight path of the first scheme.

The incomplete center section, with its rotary lifters, was constructed at Montclair, near Denver, but suffered a mishap when tested on 6 May 1908, with just one Stanley steam engine of only 10hp. No further progress was made with this machine and activities returned to Britain.

Power: Two 50hp Stanley steam engines driving two 110 blade, 27ft 8in diameter rotary 'lifters'.

Data		
	Span	67ft
	Lift generated	7 tons
	Length	60ft
	Height	13ft
	Fuselage width	8ft
	Distance between axes of lifters	40ft

DAVIDSON Gyropter biplane 1908-1911 (Davidson's Gyropter Flying Machine Ltd. Amerden Bank, Taplow, Berkshire)

Davidson returned to Britain to build his next machine named 'Gyropter', which he entered for the *Daily Mail* prize of £10,000, for a flight from London to Manchester. The first plans were for a machine weighing four tons, carrying twelve passengers, which would do the flight in three hours. The machine was shown in model form at Olympia in 1911, and was a biplane with three pairs of wings in tandem with lifting fans, as on the previous machine, fitted at the center pair. The fuselage was again a double-decker, with a 'beak' nose rudder, and tail surfaces. The two steam engines in the fuselage drove the lifters through shafts and bevel gearing. By June 1910 the wooden structure was well advanced, but funds were short, so Davidson issued a new company prospectus. By February 1911 both lifters and one engine had been installed. Further funds were required and a new brochure now gave a weight of seven tons, a lifting power of ten tons, twenty passengers and speed in excess of 100 mph. The finance was not forthcoming and the project faded away.

Power: Two 60hp Stanley steam engines driving two 26ft 10in diameter 'lifters'.

Data		
	Span	76ft
	Length	66ft

DAVIES biplane gliders (Walter Davies, Dudley, Worcestershire.)

The first of Davies' two gliders was built in 1911, and his report on his first experience under tow was conveyed in a letter to *Flight* of 12 August 1911 (p.708). Later, when describing his second glider, he explained that the lack of a suitable ground and local disinterest caused him to dismantle it and build a more readily transportable machine. A drawing of the No.2 glider and notes on his trials appeared in *Flight* on 6 December 1913 (p.133-178). Both towed and free flights were made, the longest being 130ft.

Data		**No.1**	**No.2**
	Span top	30ft	30ft 1in
	Span bottom	20ft	20ft 1in
	Chord	5ft	5ft
	Area	283.75 sq ft	285 sq ft
	Length	-	21ft
	Weight allup	-	285lb

Davidson Gyropter tandem biplane of 1908-1911. This model was at the Olympia Aero Show in 1911 and construction was well advanced at Taplow.

Davies glider No.1 built in 1911 at Dudley.

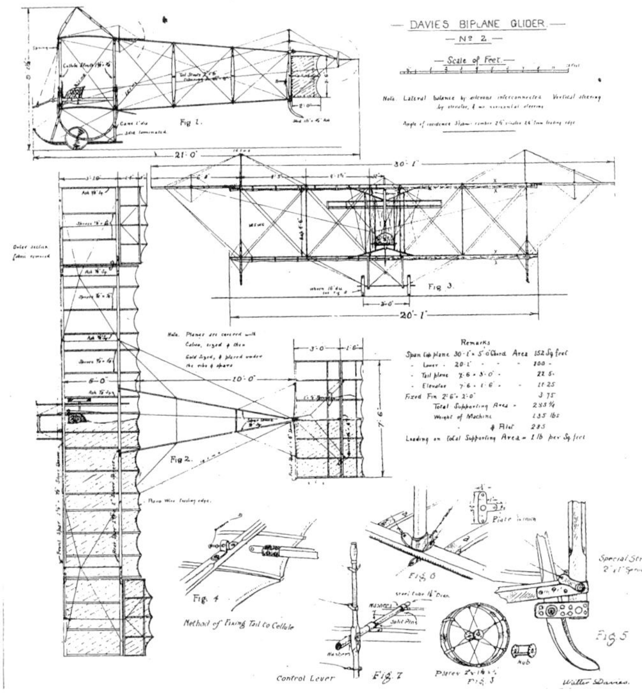

Davies glider No.2 of 1913.

DAWSON glider (Mr. and Mrs. Charles E. Dawson (née Miss Gertrude Robins) Naphill, near High Wycombe, Buckinghamshire)

This unequal span glider was designed by Dawson and made by Mulliners Coachworks Ltd., Long Acre, London and Northampton in 1911. It carried the stage name of his actress wife, Gertrude Robins, under the front elevator.

Dawson glider built by Mulliners Coachworks Ltd. for actress Gertrude Robins (Mrs. Dawson).

DAWSON & de PAVILLET monoplane (Canterbury Motor Co., The Pavilion, Rhodaus Town, Canterbury, Kent)

H.A. Dawson was the driving force behind the construction of this aircraft, his partner, H.F. de Pavillet, being more concerned with the business aspects of the motor manufacturing and sales side.

The machine, which resembled a *Demoiselle*, was constructed of bamboo and steel tube, and was fitted with a 30hp Alvaston twin-cylinder, horizontally opposed, water-cooled engine. This had sufficient power for only a few hops when the machine was tested by Dawson in June 1910 at Old Park, St. Martin's Hill, Canterbury. The engine was changed for an Antoinette, probably a vee four, water-cooled type of 24hp,

Dawson and de Pavillet monoplane constructed by the Canterbury Motor Co. in June 1910.

which proved to be too heavy. An Anzani, reportedly a horizontally opposed engine, was then fitted, but the aircraft was still not successful. It was sold to Mr. Dan Sherrin of Whitstable, less the engine, which was used to drive a compressor at the works for many years.

DAY monoplane (F.W. Day, 334 Commercial Rd., Portsmouth, Hampshire)

Also referred to as the Hampshire Aero Club monoplane. The machine was at Fort Grange in April 1910, no engine was fitted, and it was offered for sale in May 1910.

Data		
	Span	35ft
	Chord	6ft 1in
	Area	220 sq ft
	Length	22ft
	Weight	320 lb
	Area tailplane	30 sq ft
	Area rudder	6 sq ft

Day monoplane built by a member of the Hampshire Aero Club in 1909-1910.

De BOLOTOFF triplane (Prince Serge de Bolotoff, Brooklands)

This young Russian aristocrat was for a time the owner of a Voisin in France in 1909, but later moved with his mother to West Norwood, where his large triplane was built. It had Voisin-type side curtains and box tail and, with the narrow track undercarriage, it was necessary to provide skids at the tips of the wings and the tail in addition to the tail-wheel. The two-bladed tractor propeller was driven by a 120hp Panhard eight-cylinder vee water-cooled engine. It was taxied at Brooklands on 22 November 1913 by Harold Barnwell, the Bristol pilot, but the

De Bolotoff triplane which finally appeared in 1913.

undercarriage was damaged on rough ground. Little more was heard of the machine, which was already an obsolete type and was under-powered.

Data		
	Span	35ft 9in
	Length	39ft
	Gross Wt	about 2640 lb

De HAVILLAND biplane No.1 (Geoffrey de Havilland, Bothwell Rd., Fulham & Newbury)

The design and manufacture of this machine began in 1908 and continued throughout 1909 at Fulham until November, when it was moved to a shed at Seven Barrows near Newbury on Lord Carnarvon's estate for final assembly. This was completed and initial tests were carried out by December, when, during a fast run, de Havilland attempted to take off. This resulted in complete failure of the wing spars, made from weak pinewood. de Havilland was assisted with the construction by F.T. Hearle, later a director of the DH company, and his wife.

The design was a three bay biplane with twin pusher propellers. The fuselage was an open girder tapering to the front, which mounted a divided elevator, and to the rear where the rudder was pivoted between the girder and a boom mounted on the top wing. Single acting ailerons on the top wing provided lateral control. The undercarriage had front and rear wheels in addition to the main pair, supplemented by wingtip skids.

The engine was also designed by Geoffrey de Havilland and was made for him for £250 by the Iris Motor Co. of Willesden, where his elder brother Ivor had been chief designer before his early death.

Power: 45hp Iris four-cylinder horizontally opposed water-cooled driving two aluminum adjustable pitch propellers through shafts and gearing.

Data		
	Span	36ft
	Chord	6ft
	Gap	6ft
	Area	408 sq ft
	Area elevators	36 sq ft
	Tailplane	25 sq ft
	Rudder	7 sq ft
	Length	29ft
	Height	10ft
	Weight	850 lb

De HAVILLAND biplane No.2

The second de Havilland machine used the original engine, but was an entirely new design resembling a Farman in layout. The Iris engine was cooled by a radiator mounted horizontally above the fuel tank between the wings. The wooden structure of twin booms, parallel in plan, carried at the front a single elevator, and the tapering rear booms mounted a fixed tailplane and fin, elevator and rudder. The small fixed tailplane was added

De Havilland No.2 was flown in 1910 and was purchased by the War Office and used for development work at Farnborough identified as FE 1.

De Havilland No.1 was a twin-pusher biplane which suffered structural failure during taxiing trials in December 1909.

to overcome tail heaviness. Lateral control was by ailerons, hinged at the trailing edge of the top wings.

The machine flew for the first time on 10 September 1910, and by November flights up to forty minutes had been accomplished, including some with a passenger. In this month de Havilland applied to Mervyn O'Gorman, Superintendent at Farnborough, for a technical post which he took up in January 1911. In addition, the aircraft was purchased for £400 by the War Office and used at Farnborough for development work, being renamed FE.1 after its acceptance in January 1911. The foreplane and front booms were removed during the course of this work.

Power: 45hp Iris four-cylinder horizontally opposed water-cooled driving a two-bladed wooden propeller.

Data
Span 33ft 6in
Chord 5ft 6in
Gap 5ft 6in
Area 340 sq ft
Weight allup 1,100 lb
Length 40ft
Speed 37 mph

DENDY glider (H. Dendy, Eastbourne, Sussex)
This glider, reported on 4 June 1910, was a biplane built on Henry Farman lines.

Data
Span 24ft
Chord 5ft 3in
Length 22ft

DEWHIRST monoplane (J. Dewhirst, Campbell St., Keighley, Yorkshire)
A monoplane with 20hp JAP engine of which no details are known, was offered for sale in November 1911.

DING-SAYERS monoplane (Rowland Ding and W.H. Sayers, Brooklands)
This single-seat canard monoplane was built during 1911 and was tested at Brooklands. It was fitted with a 50hp Gnôme engine and was based on Sayers' design of a model biplane which won first prize at the 1911 Olympia Aero Show.

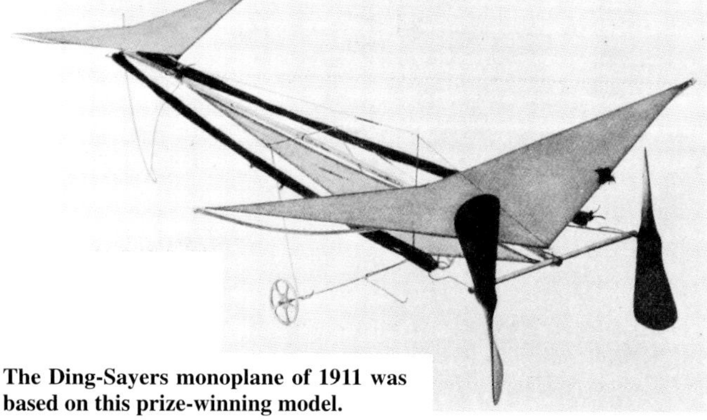

The Ding-Sayers monoplane of 1911 was based on this prize-winning model.

Dixon *Nipper* No.1 monoplane was wrecked on test at Acton in 1911.

DIXON NIPPER No.1 monoplane (H.S. Dixon, 73 Twyford Avenue, Park Royal, W3)
This single-seat, tail first monoplane was built in 1910-1911 and was tested at the London Aviation Ground, Acton, in February 1911. After a number of short, straight flights it was badly damaged, and although Dixon requested assistance for its reconstruction, it was not heard of again. Warping of the wings was controlled by a waist belt and wires by sideways movement. A biplane elevator at the nose and rudders at the wingtips served for pitch and yaw control.

Power: 25hp Advance four-cylinder air-cooled vee driving a 6ft 4in diameter Twining two-bladed pusher propeller.

Data
Span 26ft
Area 210 sq ft
Length 20ft
Weight allup 530 lb

FIG.I.

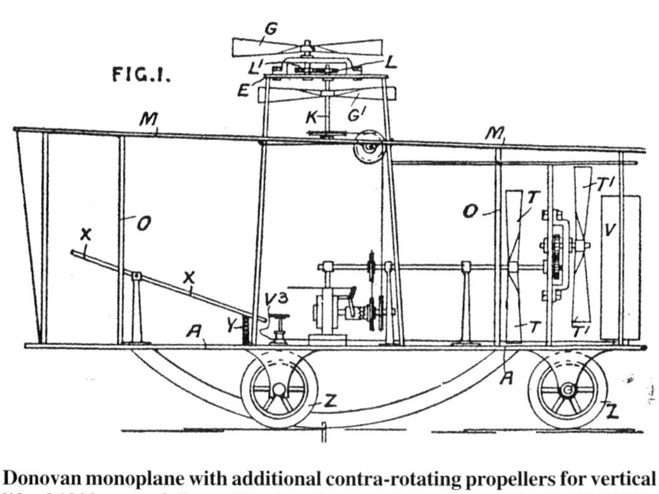

Donovan monoplane with additional contra-rotating propellers for vertical lift of 1909 was a failure. The drawings are from Patent No.21618/1908.

DONOVAN monoplane (Donovan Aeroplane Co., West Hartlepool. Built by the Howcroft Carriage Co., Oxford Rd., West Hartlepool)

This machine was built by the Howcroft Company as a result of the initiative of Capt. Joseph Donovan, who interested a group of friends in the project, and who formed a company in May 1909. It was to be a two-seater based on Patent No.21618/1909 taken out by Donovan.

The engine was designed and tested in two-cylinder form by Mr. Harry Fothergill, a trained engineer of Richardson, Westgarth Ltd. The six-cylinder version was completed by Gale's, a local garage.

The machine was an annular wing pusher monoplane with the additional feature of twin contra-rotating lifting screws, a type later described as a 'convertiplane'. The main framework of struts and spars, made of sycamore wood, contained a central platform, above which was mounted the circular canvas wing. The platform provided the mounting for the engine and the operator's station. Power was taken by belts and shafts to the contra-rotating pusher propellers and, at right angles, by a similar arrangement to the lifting blades. The pairs of screws were not coaxial, but were offset from one another, the final drive being by spur gearing to provide contra-rotation.

A tall rudder behind the pusher propeller was pivoted between the top and bottom members of the frame. An elevator, identified as a 'scaling rudder', ahead of the operator, was to be set at an angle of forty-five degrees for takeoff and, with the lifting screws engaged, the machine was intended to take to the air, by a combination of vertical thrust and wing induced lift. When airborne, a clutch could then release the drive to the lifting screws leaving full power available for propulsion. The machine rested on three wheels, the front pair being mounted on two flexible skids, and were designed to be jettisoned after takeoff to save weight. It was envisaged that the wing would act as a parachute to cushion the landing.

Unfortunately Capt. Donovan and his associates had no real appreciation of the design requirements and the aircraft proved to be excessively heavy. It was housed in a shed at Rift House Farm and when brought out for trials on 25 October 1909, failed even to taxi. It was eventually sold for £35 at auction a year later.

Donovan monoplane model of the unsuccessful fullsize machine.

Power: 30hp Fothergill six-cylinder inline water-cooled.

Data	Span	28ft
	Weight	1,000 lb

Druiff-Neate Cycloplane. A pedal-powered machine made for the designers by C.G. Spencer & Co. in 1909.

DRUIFF-NEATE Cycloplane

Constructed in 1909 for the designers by Messrs. C.G. Spencer & Co. of Highbury, N London, the machine was pedal powered driving a 4ft diameter propeller, although the fitting of a motor was contemplated. The movable wingtips and tailplane were operated by wires from the handlebar.

Data	Span	20ft
	Chord	5ft
	Area	100 sq ft
	Weight	50 lb

The Car 22 December 1909 (p.264)

DUGUE self-propelled monoplanes (Andrew L. Dugue)

Two cycloplanes were designed in autumn 1909. The No.1 machine had a monoplane wing mounted on a bicycle. A rear extension carried a rudder linked to the handlebars and a propeller driven by a system of chains and sprockets linked to the

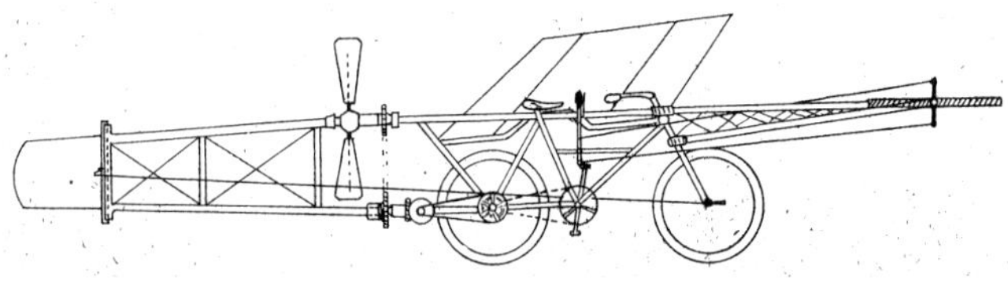

Dugue cycloplane.

bicycle's pedals. A front extension carried an elevator controlled by a joystick. The No.2 machine was similar in layout but had the airscrew mounted in front of the handlebars.

DUNDEE glider (Dundee Aero Club, 3 Baltic St., Dundee)
The club acquired the glider in the spring of 1911 and reported that glides had taken place from a hill near St. Fort station. The club appears to have continued with models only, after the glider was destroyed by a cow in 1912.

DUNGEY monoplane (Hugh Percival Dungey, Middleton's, Cranbrook, Kent)
This machine was built for Dungey, but was fitted with a 20hp Dungey two-cylinder engine driving a Beney propeller. It was described as of Blériot-*Demoiselle* type and was tested in 1911. Dungey became a member of the Aero Club of Great Britain and Ireland in 1909, whose secretary was ABE Cheeseman.

DUNNE aircraft (J.W. Dunne, Lt. and Capt., Balloon/Aircraft Factory, Farnborough 1906-1909. Blair-Atholl Syndicate Ltd., 1 Queen Victoria St., London EC, 1909-1914)
Lt. Dunne was transferred onto halfpay owing to being physically unfit for active service in the Army. He was not a trained engineer but, from 1904, had ideas for aircraft possessing automatic stability. In 1906 he was engaged by Col. Capper to work at the Balloon Factory to develop these ideas, beginning with a glider and later powered aircraft, all relying on swept wings as their main feature. In order to preserve secrecy, when the early machines were ready for trials in 1907-1908, they were taken to Blair Atholl in Scotland with a mainly military team of support personnel.

As a result of economy measures in 1909, Dunne together with Cody, was discharged from Farnborough, being allowed to keep his aircraft, less the engine, when he severed his connection in March. Thereafter the Blair Atholl Syndicate was formed by several titled men of wealth with offices only in the City, the aircraft construction being carried out mainly by Short Bros. at Leysdown or Eastchurch. License arrangements were later concluded to make the aircraft in the U.S.A. by the Burgess Co. of Marblehead. In France a similar arrangement was begun with the Astra Co. but was concluded with Nieuport.

The total number of machines built was small, but included two for the War Office in 1913. It soon became evident that there was no great future for machines of the Dunne type and the Syndicate was liquidated and Dunne withdrew from aeronautical work in May 1914.

Dunne D1A glider at Blair Atholl being manhandled onto the launching trolley.

DUNNE D.1A biplane glider
This was built at the Balloon Factory in 1906-1907 and taken in a dismantled condition to Blair Atholl for tests, which lasted between July and October 1907. Col. Capper achieved a number of short flights from downhill launches, but the machine was eventually damaged. The glider was an unstaggered biplane with constant chord, swept back wings with negative incidence at the tips, and a drooped trailing edge at the intersection of the wings. There were skids below the center section for landing, but for takeoff the machine rested on a trolley. The pilot was seated on the lower wing and appeared to rely on the inherent stability of the design, as there was no evidence of means of controlling pitch and roll, nor was there a rudder for directional control.

DUNNE D.1B biplane
The original glider had been built with a view to later conversion to become a powered aircraft, and towards the end of the work in Scotland, the conversion of the damaged glider took place. The powerplant was two Buchet 12hp engines mounted coaxially on a tubular frame, driving twin pusher propellers through flat belts, similar to the scheme used later on the D.4. The takeoff was arranged using a track of wooden planks, but on the first attempt in October, the machine veered off and was badly damaged. The pilot may have been Dunne himself or possibly Lt. Westland. This and the onset of winter, caused the party to return to Farnborough.

DUNNE D.2 glider
This proposal for a smaller form of the Dunne-Huntington triplane was not constructed.

DUNNE-HUNTINGTON triplane
This unusual machine, variously referred to as a triplane, a biplane or a monoplane, was basically designed by Dunne during the winter of 1907-1908. The layout was passed to Professor Huntington, an active member of the Aero Club, to complete the detailed drawings. The machine was built by Short

Dunne-Huntington triplane. The first version at Eastchurch in 1910 with Wolseley engine.

Bros. at Leysdown, and after a long gestation period, was flying at Eastchurch in April 1910. It went through a number of changes, and was still flying well in April 1913, its original design by this time being quite outdated.

When operating at Eastchurch in early 1910, the open fuselage structure was originally mounted on a three wheeled chassis and skid plus a tailwheel; later the nosewheel was removed and replaced by two semicircular skids at the front. The front wing was mounted at the nose and had considerable sweep back. On the same level, at the rear end, was a wing of parallel chord, with anhedral on the outboard section. This was also split diagonally to form triangular flaps for control in roll and pitch. The top wing was mounted above the central fuselage and braced to it by struts, and to the rear wing by vertical struts at the leading edge; also by raked struts to the front where the anhedral began. The area between the staggered struts was, at one time, filled in to form side curtains. The pilot was seated well forward of the biplane wings in the inner apex of the vee shaped front wing.

Dunne-Huntington triplane still with Wolseley engine but with modified chassis and side curtains.

The engine was a Wolseley water-cooled unit mounted in the center, with the radiator placed well behind. It drove the twin tractor propellers through chains and shafts. A Gnôme rotary replaced this heavy installation in 1912, in which form the machine was reported to reach 43 mph.

Power: 50hp Wolseley eight-cylinder, water-cooled vee driving twin tractor propellers.
70hp Gnôme seven-cylinder air-cooled rotary.

Data		
	Span	59ft
	Chord	10ft
	Length	50ft

DUNNE D.3 glider

This was a biplane of hang glider type with swept wings, which was to test features of the D4 powered aircraft, of which it was a smaller version. The component parts were taken to Blair Atholl and erected with some difficulty, between 2 and 19 September 1908. The glider was flown for the first time on 20 September 1908 by Lt. L.D. Gibbs. This officer did most of the gliding in the time that the machine was in use until 16 October 1908, during which period a longest flight of 157ft was achieved.

The top wings had ailerons serving also as elevators, with an auxiliary elevator at the trailing edge of the wing intersection. Two small triangular flaps were fitted later, hinged to the leading edge on either side of the point of the lower wing. These both had straight leading edges and were coupled to the other controls, but were soon discarded after criticism by Lt. Gibbs of their effect on handling.

Data		
	Area	200 sq ft
	Weight	80 lb

DUNNE D.4 biplane

The wings of D.1 were used to build this machine, in conjunction with a new tubular steel structure incorporating a sprung four wheel undercarriage, also providing the housing for the powerplant. This was a REP fantype semi-radial, driving a pair

The Dunne D4 biplane was built by mounting the D1 glider on a four-wheeled chassis which also incorporated the pilot's position.

of steel pusher propellers through flat belts, one of which was crossed. The transmission incorporated a clutch, and the shafts carrying the belt pulleys were supported at both ends. The propellers, whose blades were adjustable for pitch, were 6ft diameter originally, but were later reduced to 5ft 7in. The wingtip interplane struts were covered in fabric to form side curtains.

The machine seems to have been assembled for the first time at Blair Atholl and trials took place there between 15 November and 4 December 1908. Straight runs only were made reaching a maximum of 40 yards airborne distance.

This was the last machine made by the Balloon Factory and was given to Dunne, less the engine, upon his discharge.

Power: 25hp REP seven-cylinder air-cooled, fantype semi-radial driving twin pusher propellers

Data		
	Wing area	504 sq ft
	Weight allup	1,035 lb

DUNNE D.5 biplane

Dunne went on to prepare the design of the first machine after he had left Farnborough. He made arrangements, with C.R. Fairey as his assistant, for Short Bros. to make the machine at Leysdown, from whence it was taken to Eastchurch to fly in the spring of 1910. Dunne flew the machine himself for a distance of 2 1/4 miles on 27 May 1910. It continued in use until the following year, when it was wrecked by an inexperienced pilot.

The wing design of D.5 followed the line of development pursued by Dunne to achieve automatic stability, incorporating negative incidence and reducing gap towards the tips. The wings were divided into four bays, with radiators on the inboard, and side curtains on the outboard interplane struts. The hinged flaps at the tips of the top wing combined all the control functions.

The single-seater fuselage was extended well aft and was covered along its length, and was constructed around the tubular structure comprising the engine and outboard propeller shaft mountings. The machine rested on a substantial undercarriage with twin main wheels and a tailwheel with skid, all sprung independently. Twin nosewheels controlled the initial attitude on takeoff, but were later removed.

Dunne D5 biplane with nosewheel undercarriage was built by Short Bros. at Leysdown in 1909-1910.

Dunne D5 at Eastchurch in 1910 after modifications to the undercarriage and removal of the nosewheel.

Power: 60hp Green four-cylinder, water-cooled inline driving two 7ft diameter propellers through chains, sprockets and shafts.
60hp ENV type F eight-cylinder water-cooled vee was also reported fitted

Data		
	Span	46ft
	Chord	6ft
	Gap	6ft at center
	Angle of sweep	32 degrees at leading edge
	Area	552 sq ft including elevons
	Length of fuselage	18ft
	Overall length	20ft 4 1/2in

Dunne D5 showing the shape of the swept wing.

DUNNE-CAPPER monoplane glider

As early as 1905 Dunne had made a model of a monoplane, the design of which was submitted to the War Office and rejected. A full size glider version of this was made at Farnborough in 1907 under Col. Capper's authority. It is believed to have been tested at Blair Atholl, unsuccessfully by Capper, at the end of August during the first official expedition.

The glider was a parasol monoplane with swept wing and down turned wingtips, which was mounted on a structure of two A-frames, joined by longitudinal members, on which the pilot was supported by his arms, control being effected by weight shift.

The Dunne-Capper monoplane was evolved from Col. Capper's glider at Larkhill and was first tested in January 1911 but could not be made to lift off.

DUNNE-CAPPER monoplane

When designing his early monoplane, Dunne required a 10-12hp engine, but nothing of suitable weight was available, consequently work proceeded on the machine as the glider previously referred to.

It was not until 5 January 1911 that Capper's monoplane with Green engine was reported undergoing tests on Salisbury Plain (Larkhill) handled by Lt. Cammell. In the week ending 14 January further tests were carried out, in which Col. Capper participated. The machine could not be made to lift and Capper decided that further changes were required.

In its new form the fuselage consisted of three A-frames, at the lower ends of which, a four wheeled undercarriage and twin skids were fitted. A third skid was fitted forward of these. The pilot and engine were mounted on a floor formed on the crossbars of the A-frames, with a radiator mounted outboard on the port side. The parasol wing, which was attached at the top of the A-frames, was well braced to these, and by additional cables, to three kingposts.

Parts from the glider, probably the complete wing, were reported to have been used in its construction, which was largely the work of C.R. Fairey and was carried out at Larkhill.

The Dunne-Capper monoplane with twin nosewheels at some stage of its development.

Dunne D6 monoplane built by Short Bros. at Leysdown in 1911.

DUNNE D.6 monoplane

A new version of the monoplane was described and illustrated in *Flight* 24 June 1911 (p.542-545) and was reported to be undergoing trials at Eastchurch. The wing was of similar type to that used earlier on Capper's monoplane, but the fuselage structure was considerably changed.

The main tubular structure was built of two vertical frames 9ft 6in apart, with cross members between, on which an open wooden structure housed the pilot, tanks and engine.

The wing was attached to the four vertical members of the frame, which extended above to form kingposts for the bracing wires. The radiator was mounted vertically above the centerline. The machine was a single-seater and was built by Short Bros.

Power: 60hp Green four-cylinder inline, water-cooled driving a 7ft 3in diameter pusher propeller

Data		
	Span	36ft
	Length	21ft
	Chord	6ft 3in tapering to 5ft
	Height	11ft
	Area	248 sq ft. including ailerons/elevators

DUNNE D.7 Auto-Safety monoplane

This lighter and smaller machine was exhibited incomplete on the stand of the Blair Atholl Syndicate in March 1911 at the Aero Show at Olympia, placarded as being built to the order of Col. J.E. Capper. The general configuration was similar to the D.6, whose appearance it preceded. It was a reconstruction of the Dunne-Capper monoplane by Short Bros., with Fairey acting for the Blair Atholl Syndicate. Capper had decided in February that his machine should be modified and it was still being worked on when the show opened on 24 March 1911.

The D.7. was tested in June 1911 at Eastchurch and was still airworthy in the early part of 1912. On 12 January 1912 Dunne flew the machine to demonstrate its stability to Alec Ogilvie and T. O'Brien Hubbard of the Aeronautical Society, during which he wrote a note while the aircraft was flying 'hands-off' at 60mph.

Dunne D7 Auto-Safety monoplane was a smaller version of D6 built for Col. Capper and here seen exhibited at the 1911 Aero Show at Olympia.

Power 50hp Gnôme seven-cylinder air-cooled rotary driving a pusher propeller.

Data	Span	35ft
	Weight	1,050 lb
	Area	200 sq ft
	Weight allup	1,409 lb
	Speed	60mph

J.W. Dunne in the D7bis monoplane at Eastchurch in 1912 where the reconstruction from D6 was carried out.

DUNNE D.7 bis monoplane

This was the conversion of the original D.6 to a two-seater with air-cooled engine. It followed the general configuration of the earlier machine but had an additional flap, centrally mounted to improve pitch control.

Power: 70hp Gnôme seven-cylinder air-cooled rotary driving a pusher propeller.

Data	Span	35ft
	Area	200 sq ft
	Weight	1,200 lb
	Weight allup	1,728 lb
	Speed	60 mph

DUNNE D.8 and D.8bis biplanes

The first of these was a rebuild of the original D.5 but with a propeller driven direct by the engine, replacing the twin propeller arrangement. Some of the weight saved was offset by a most complex steel tube undercarriage structure with narrow track wheels and front and rear skids, supplemented by sprung wing tip skids.

Dunne was flying the D.8, still with Green engine and the D.7bis monoplane, at Eastchurch in June 1912 after a long illness. Capt. Carden and N.S. Percival also flew the machine in 1912 and 1913. The former obtained his pilot's certificate on it and purchased the machine, although it was subsequently bought back by the Syndicate.

Dunne D7bis monoplane was a reconstruction of D6 converted to a two-seater with air-cooled engine.

Dunne D8 with 50hp Gnôme at Farnborough.

Dunne D8bis with 60hp Green.

The Dunne biplane was present at Larkhill in August 1912 during the Military Trials although not as an entry; it still had a Green engine with a single four-bladed propeller. The indications are that it was a partial conversion from the D.5 before the Gnôme, which became the standard power plant, was fitted.

A second D.8 was assembled at Hendon and flew on 18 October 1913 in the hands of the French pilot, Commandant Felix. It was he who had flown the first D.8 to Villacoublay from Eastchurch on 11-12 August 1912 for demonstrations in France in connection with the Nieuport license, which followed the earlier interest of the Astra company. Nieuport produced a machine with a simplified undercarriage, in time to exhibit it at the Paris Show in December 1912.

The War Office ordered two D.8 aircraft, referred to by Dunne as D.8bis, in March 1913, one was canceled due to late delivery, the remaining aircraft arrived at Farnborough in March 1914 and was identified with the serial No.366. The exact number of the type built is uncertain. There is a reference to a Green-engined machine in 1913, but this may have been confused with the D.8 at Larkhill in 1912.

Apart from the undercarriage changes, on the second machine onwards, the shape of the side curtains was altered, eliminating the rectangular cutout at the top and tapering the outline at both top and bottom. The control surfaces in the top wings of the D.8 were increased to four, although each pair apparently worked in unison, but were still capable of being moved differentially for control of roll and yaw. The later machines also had the nacelle extended further forward.

Power: 50 and 80hp Gnôme seven-cylinder air-cooled rotary driving a 8ft 2in diameter Integrale propeller.
60hp Green four-cylinder inline water-cooled driving a four-bladed propeller.

Data		
	Span	46ft
	Angle of sweep	30 degrees at leading edge
	Chord	6ft
	Area	545 sq ft
	Length	25ft 9in
	Weight	1,400 lb (1,900 lb with Green)
	Weight allup	1,900 lb (2,114 lb with Green)
	Max speed	56 mph

DUNNE D.9 monoplane (Also referred to as the James monoplane)

This was a tailless pusher design of almost delta wing form. It was a single-seater with a protective cage above the cockpit, serving also as an anchorage for the bracing wires. The wheels on the twin skid undercarriage were covered by fairings. The power plant was a 35hp Levis five-cylinder inline two-stroke engine designed by H. Newey, mounted inverted in the nose, and driving the propeller by extension shaft.

Dunne was assisted by C.R. Fairey with the stressing and the machine was constructed in 1913 for a Mr. James by Levis Ltd. of Stechford, Warwickshire, motorcycle manufacturers. The name 'Leonie' and the initials 'AB', probably of Arthur Butterfield, one of the directors of Levis, were painted on the wheel covers.

Dunne D9 monoplane built by Levis Ltd. for R.E. James in 1913 was wrecked on its first takeoff.

The machine was wrecked on its first attempt at flight by hitting an obstruction on a playing field at Castle Bromwich, forestalling the proposal to fit floats for later trials on Edgbaston Reservoir.

DUNNE D.10 biplane

The D.10 was a variant of the D.8 and was a two-seater with 80hp Gnôme built in 1912, but later converted to D.8 standard.

Power: 80hp Gnôme seven-cylinder air-cooled rotary.

Data		
	Span	45ft
	Area	448 sq ft
	Weight allup	2,202 lb
	Max speed	50 mph

DYOTT monoplane (George M. Dyott)

This small single seater was completed in early 1913 by Hewlett and Blondeau Ltd., of Clapham Junction, London SW, to the design of G.M. Dyott. Soon after its initial flight tests he took it to the U.S.A. for a tour lasting six months. Dyott returned with the machine to Hendon in October, and in November it overturned at Beachy Head during the London to Brighton Race and was damaged. It was repaired and in 1914 was taken over by the Admiralty as serial No.1598.

Dyott entered the Royal Navy, but before the war had also designed a large twin-engined machine, two of which were completed in 1916 for the Admiralty by Hewlett and Blondeau.

The monoplane was a handsome machine designed for easy assembly and servicing and it performed well on its small Gnôme engine. The lateral control was by wing warping and the wing bracing wires were carried to a pylon of vee-shape above the fuselage and to the undercarriage below; the latter being a point of criticism. The cockpit was well furnished with instruments, including a control movement recorder of Dyott's own design.

Power: 50hp Gnôme seven-cylinder air-cooled rotary.

Data		
	Span	29ft
	Chord	5ft 10in
	Area	148 sq ft
	Area tailplane	14 sq ft
	Area elevators	12 sq ft
	Area rudder	4 sq ft
	Length	23ft
	Speed	45-75 mph
	Endurance	3hr

Dyott monoplane built by Hewlett & Blondeau in 1913.

E

EASTBOURNE AVIATION Co. Ltd. (F.B. Fowler, Managing Director; E.L. Gassler, Chief Designer, St. Anthony's Hill, Eastbourne, Sussex)

This was one of the small manufacturers, which evolved from a flying training school established by F.B. Fowler in 1911, with himself as instructor. Mon E.L. Gassler was a Swiss pupil, who continued with the company as chief designer and pilot.

Monoplane

This was a single-seat monoplane of conventional design, built during 1913. No fin was fitted and lateral control was by ailerons. It was intended for advanced pupils and private owner use, but only one was built because of lack of demand due to the impending war.

Power: 35hp Anzani three-cylinder air-cooled fan-type radial.

Data		
	Span	29ft.2in.
	Chord	4ft 11in
	Area	135 sq ft
	Area tailplane	14 sq ft
	Area elevator	10 sq ft
	Area rudder	7 sq ft
	Length	21ft
	Height	8ft in flying attitude

Eastbourne Aviation Company monoplane constructed in 1913.

EAC biplane. The second design by E.L. Gassler appeared in 1914.

Biplane

This single-seater was completed in February 1914, to the order of Lt. R.E.B. Hunt and was for exhibition flying. It was in use at Eastbourne until just before the outbreak of war.

The design, by Gassler, was of a simple unstaggered two bay biplane with warping top wing. The overhang of the wing was braced from kingposts, in line with the outboard interplane struts. The front struts were connected to the front spar, which lay along the leading edge. The lower wing was of parallel chord, but the trailing edge of the top wing tapered from a cutaway center section, but changed to become parallel over the warpable length of the span.

The four wheel undercarriage included twin skids, joined by struts to the bottom longerons. There was no fin and the rudder was unbalanced. The engine was a 50hp Gnôme seven-cylinder air-cooled rotary, enclosed in a circular cowling, cutaway at the lower quarter.

Data Cruising speed 65 mph

Military biplane

The Military biplane, designed by M. Gassler, was exhibited at the Aero Show at Olympia in March 1914. Quick assembly and dismantling for transport were a feature, and included hinges enabling the tailplane to fold down against the fuselage. The use of bolts and quick release cable fittings enabled the wings to be removed and packed flat for transport.

The machine itself was a conventional unstaggered biplane, with large gap and top wing extensions. The top wing was carried on four steel struts of streamlined section. These were splayed to clear the cockpit area, which was a single opening, with the pilot in front and passenger behind. A starting handle was provided in the pilot's cockpit.

The engine was totally enclosed, except for the lower quarter, with a rather flat fronted circular cowl. The rear undercarriage struts, also of streamlined steel tubes, were extended forward to form short skids.

Power: 80hp Gnôme seven-cylinder air-cooled rotary driving a 8ft 2in diameter propeller.

Data		
	Span	34 ft 6 in
	Area	245 sq ft
	Weight	950 lb
	Length	25 ft
	Speed range	50-75 mph

Circuit biplane

The *Daily Mail* Seaplane Circuit of Britain, which was due to be held in August 1914, was prevented by the outbreak of war. The EAC aircraft, built to compete as No.5, was to be flown by F.B. Fowler and was also sponsored by Frank Hucks.

The aircraft was a twin tractor biplane with the engine buried in the fuselage. This drove the two propellers, which were mounted on the first pair of interplane struts, through shafts and bevel gearing. The deep fuselage provided accommodation for two crew, seated side by side, in the open cockpit ahead of the wings, and was large enough for one other crew member just behind, with space for a further man beside the engine.

The three bay wings were parallel in chord and of equal span with three degrees of dihedral on the lower. Control was by normal ailerons, elevators and rudder. There was no fixed fin, although provision to increase the side area, both above and below the fuselage, was made if found necessary as a result of trials.

The twin floats, with a single step and air ventilated, were spaced wide apart and of such proportions that wingtip and tail floats were unnecessary.

EAC military biplane of 1914 was exhibited at Olympia.

EAC twin-tractor biplane built for the Seaplane Circuit of Britain abandoned in 1914.

There were reports of problems with distortion of the struts mounting the two propellers and although the machine was still at Eastbourne at the end of the year, there was no indication that it had ever flown, and it was dismantled later.

Power: 100hp Green six-cylinder inline water-cooled driving twin-tractor propellers 8ft 2in diameter by shafts in torque tubes, and bevel gearing.

Data		
	Span	54ft
	Chord	6ft 6in
	Area	700 sq ft
	Length	31ft
	Height	11ft 6in
	Floats length	19ft
	Floats beam	2ft
	Spacing at centerline	12ft
	Endurance	7hr
	Weight	1,850 lb
	Weight allup	2,500 lb
	Speed range	45-65 mph

EAST GRINSTEAD BOY SCOUTS glider (Patrol Leaders Smith and Beard)

This glider, of Chanute type, was constructed by the Scouts in 1912. It had a span of 20ft and achieved flights up to 25ft high and 200ft long.

East Grinstead Boy Scouts Chanute type glider of 1912.

EAST LONDON AERO CLUB glider

This club proposed to build a glider based on the Valkyrie design. A sketch appeared in *The Aero* of 22 March 1911 (p.249), but there were no further reports.

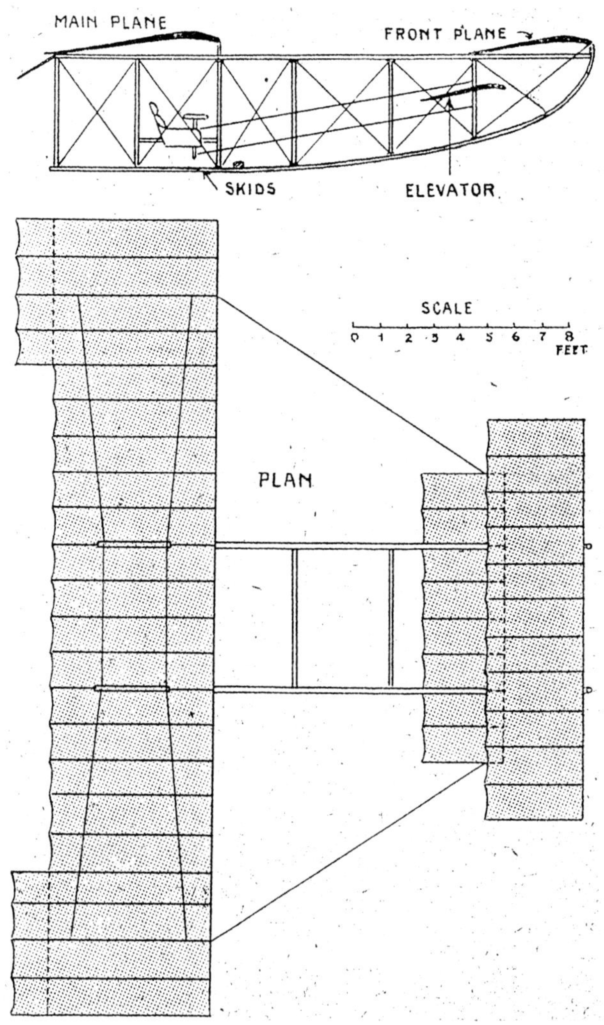

East London glider. A sketch which appeared in 1911.

EDGAR glider (Norman W. Edgar, Bristol)

Flown by members of the Bristol and West of England Aero Club, this biplane glider was built during 1913 with the intention of fitting a light engine later. However the wing area was found to be inadequate and the intention was not realized.

Edgar glider built in 1913 by Norman Edgar of the Bristol and West of England Aero Club.

It was a sesquiplane with twin booms, parallel in plan, carrying the tailplane and rudder, and was mounted on an undercarriage with two motorcycle wheels and skids.

Data	Span	26ft
	Length	21ft
	Area	170 sq ft
	Chord	4ft 7in

EDINBURGH cycle biplane

A letter in *The Aero* of 21 September 1909 (p.298), referred to a self-propelled machine constructed near Edinburgh and also gave a sketch. It was based on a bicycle and was, no doubt, the machine reported earlier to be on test at Colinton Mains.

Data	Span	18ft
	Length	20ft

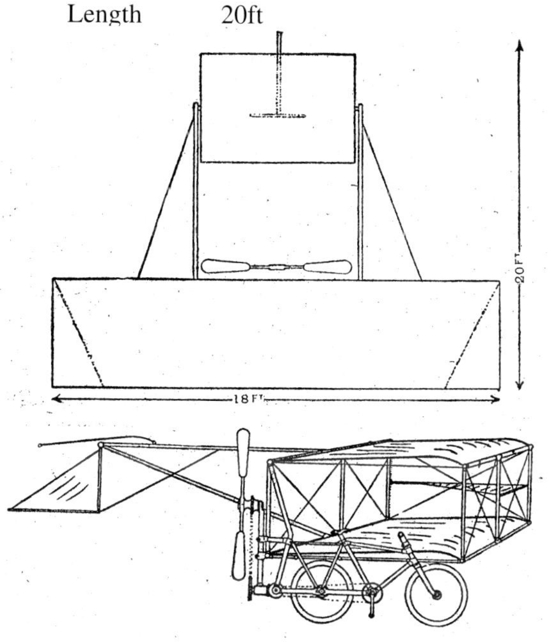

Edinburgh cycle biplane. A sketch which was published in 1909.

EDWARDS Rhomboidal biplane. (Walton & Edwards Aeroplane Co., Shed 37, Brooklands)

First described in *Flight* on 5 February 1910 (p.87-90), and reported to be backed by a large automobile firm, believed to be Humber, the Rhomboidal made its first appearance at Brooklands on 11 March 1911. It was not until June that attempts to taxi or to fly were made. On 27 June 1911 the pilot, named Martin Rucker, lost control and ran across a ditch onto the racing track, causing severe damage to the machine. There were no further reports and it was apparently not repaired.

This most unusual form of annular biplane was built around two triangular section longitudinal girders, built up from wooden sections, and was wire braced. The lower girder provided the mounting for the engine and seat for the pilot, and rested on splayed struts, which were the main undercarriage members. These were wire braced and carried twin skids and two pairs of wheels. A nose wheel was mounted at the extreme front and a skid at the rear end.

Edwards *Rhomboidal* at Brooklands in 1912. An unlikely looking flying machine.

The wings, in plan, were in the form of a rhomboid or open diamond shape. The structure for these consisted of built-up spars of streamlined section, at top and bottom, and at right angles to the main girders, with interplane struts between. The shape of the wings was maintained by tension cables along the leading edges and curved battens across the chord, but the trailing edges were allowed to remain flexible. Bracing wires from the top and bottom wingtips were taken down to the undercarriage structure. The twin tractor propellers were mounted on shafts on outriggers and were extensively braced by cables to the main structure.

Elevators and a rudder were mounted at the tail, but reliance was placed on wing dihedral for lateral balance.

Power: 50hp Humber four-cylinder inline, water-cooled driving by chains twin tractor 8ft diameter propellers. 60hp Green. four-cylinder inline, water-cooled from June 1911.

Data	Span	38ft
	Chord	3ft front areas 9ft rear areas
	Area	1,200 sq ft
	Length	48ft
	Height	14 ft 11in (16ft 10in to top of rudder)
	Weight	1,600 lb
	Dihedral	8 degrees

Eggleton No.1 glider built at Eastchurch.

EGGLETON gliders (R.H. Eggleton, 24 Cranbury Rd., Eastleigh, Hampshire. Flying at North End, Eastleigh)
Eggleton was an apprentice at the LSWR Carriage Works at Eastleigh, and is reported to have made eleven gliders in the years 1911 to 1914. Of these, three can be identified and illustrated, the first and second being illustrated in *Flight* 17 February 1912 (p.159).

The first, a parasol monoplane with ailerons and normal tail surfaces, had an enclosed section for the operator, but was of rather flimsy construction, perhaps accounting for its early replacement.

No.2 was a canard monoplane of Valkyrie type with forward elevator, where the pilot sat in a gap in the wing center section. The span was 28ft and the length 24ft.

No.3 was a biplane of unequal span, rather like a Caudron, with a small nacelle on the lower wing. No rudder was fitted and control was achieved with wings and elevators which could be warped. A wheeled undercarriage was fitted. This machine was the subject of a note and illustrations in *Flight* 18 October 1913 (p.1144).

Eggleton No.2 glider followed in 1912.

Eggleton No.3 biplane glider of 1913.

Few details of any of the other gliders remain, except that the last had a span of 31ft 6in and was 24ft long. Activities were terminated by the outbreak of war.

ELLIS biplane 'Sawfinelle No.2'
It is not clear whether this was a glider or to be a powered aircraft. It was of Farman type with upper span of 26ft and lower of 21ft, chord 4ft 6in. No other information has been traced.

ELLIS-WILLIAMS monoplane (W. Ellis Williams, University College of North Wales Bangor)
See Williams monoplane.

ELSE glider (J.H. Else, Birmingham)
Else was a working carpenter and a member of the Birmingham Aero Club who built a small monoplane glider in 1909 (q.v.) Else was practicing glides at the Midland Aero Club's meeting at Sutton Park on 13 November 1909, but failed to take off.

The glider had a boat shaped fuselage, with a high wing mounted halfway along its length. A small tailplane was mounted at the rear and a large moveable foreplane gave elevator control

The undercarriage comprised of two large bicycle wheels and a tail-skid.

The Aero 30 November 1909 (p.476) and 28 December 1909.

ELSWICK aeroplanes (Elswick Autocars Ltd., Faggs Corner, Bedfont, Middlesex)
Six aeroplanes were in the course of construction when the company, in liquidation, was put up for sale in 1911. The type of machine is not recorded and presumably work on these was not continued.

ELSWORTH monoplane (Oswald Elsworth, Tyneside Foundry Co., Water St., Newcastle)
Elsworth was a member of the Northumberland Aero Club, with flying ground at Bolden Racecourse. His machine was nearing completion in February 1910.

The layout was similar to a Blériot with open, square section girder fuselage, but with a triangulated undercarriage with twin skids, designed to receive the landing loads and provide drag for braking. The wings had squarecut tips and were warped by a wheel control, fore and aft movement of which operated the elevators. A tailplane and rudder were fitted but no fin. There is no evidence that flight was achieved.

Power: 25hp Advance four-cylinder air-cooled vee.

Data	Span	28ft
	Weight	150 lb without engine.

Elsworth monoplane at Bolden racecourse in 1910.

EMPRESS monoplanes Nos.1 and 2 (Empress Engineering Co., 180 Stockport Rd., Manchester. Proprietor Charles A. Fletcher)

The first of Fletcher's monoplanes was built in 1909 and tested on 20 October 1909 at Heaton Park, after attempts to start the engine failed, a week earlier. Several runs were made during which the machine lifted off the ground for 20-30 yards, before damage to the wheels and one of the wings occurred. The machine then appears to have been abandoned, although it was shown at the Manchester Aero Club Show at White City in early March 1910.

***Empress* monoplane outside the Empress factory. Charles Fletcher is on the far right.**

The second monoplane was reported to be larger and was to be tested on 17 January 1910.

However, before Fletcher could take his seat, the machine was blown twenty feet into the air and finished up severely damaged in a clump of trees. Fletcher claimed to have flown 75-100 yards on previous trials, but this was the last reported appearance of this monoplane

The first aircraft was a parasol monoplane and was a mixed structure of steel tube and bamboo with the pilot seated in a wicker chair. The propeller was made of steel tube laced across with wire and covered with canvas. There are no details of the second monoplane except that the same engine was probably used.

The power was supplied by a five-cylinder Empress engine, which Fletcher also designed. It was rated at 50hp and was an air-cooled rotary, which apparently infringed Gnôme patents and later became the subject of a court case.

Data		
	Span	34ft
	Length	30ft
	Propeller	6ft
	Area	345 sq ft (*320 sq ft)
	Weight	365 lb (*330 sq ft)
		*alternative figures

EMPRESS biplane

Fletcher produced his third machine in 1910 and it first appeared at the Manchester Aero Club Show in March, followed by testing on 8 August 1910 at Manchester Race Course, Castle Irwell. One of the 'pilots' who lifted the machine off for short distances was Jack Alcock of later Atlantic fame, who, as an apprentice at the Empress Works, assisted in the construction of all three of Fletcher's aircraft. Under pressure from the racecourse management after various crashes, Fletcher removed the aircraft to Brooklands, but after losing the case over the Gnôme patents, the company went out of business.

The biplane was a pusher on the lines of a Farman and was powered by an Empress seven-cylinder rotary engine. An Edwards four-cylinder two-stroke engine of 30hp was also fitted for a time, and reports indicate that either engine could enable the aircraft to become airborne.

***Empress* biplane at Manchester Race Course in August 1910. Charles Fletcher at the controls.**

ENGLAND biplane (Gordon England, Shoreham, Sussex)
Gordon England was reported to be building a 'small high speed, landgoing, pusher biplane' in July 1913, which was approaching completion but did not materialize.

EVANS monoplane (Evans, Doe Lea, Derbyshire)
This variant of a Blériot type monoplane was built, over a period of sixteen months in 1910-1911, by Mr. Evans, a miner. The power plant was a 12hp Madison motorcycle engine driving a 6ft diameter propeller. The power was insufficient for the machine to fly.

Data	Span	26ft
	Length	24ft

Evans monoplane with the constructor at Doe Lea, Derbyshire in 1911.

EVERETT-EDGECUMBE monoplane (Everett-Edgecumbe & Co., Colindale, Hendon)
E.I. Everett began the design of his monoplane in 1908, and constructed it in the wooden shed used by Paulhan in his flight from London to Manchester in April 1910, the area around which was later absorbed into Hendon aerodrome. The machine was assembled over a long period and C.R. Fairey was one who assisted in his spare time. It was finally ready for testing on 6-7 December 1910, when Bernard Clutterbuck was able to make short hops, also Everett himself in January 1911.

Although described as Blériot-like, there were major differences, in particular the marked dihedral and camber and the end plates of the squarecut wings. The axle for the wheels was moved forward during the trials from the sprung trailing portion of the twin skids to a central position between the mounting struts.

The fuselage was originally covered throughout with fabric, but later was uncovered aft of the trailing edge of the wings. The fixed fin was triangular and mounted on top of the rear fuselage and carried a square shaped rudder the lower edge of which was cut at an angle to clear the one piece elevator. Separate fixed tail surfaces were attached to the lower longerons.

Power: 35hp JAP four-cylinder water-cooled vee driving a 6ft diameter propeller.

Data	Span	25ft
	Area	140 sq ft
	Length	16ft
	Height	9ft 6in
	Max speed	30 mph.

Everett Edgcumbe monoplane built at Hendon in 1910.

EVERSHED biplane (Sydney H. Evershed, Burton-on-Trent, Staffordshire. Flying Ground at Netherseal)
This machine was constructed during the winter of 1909-1910 and attempts to fly were made at Grangewood Farm, Netherseal on several weekends in May-June 1910 without success. The final attempt was on 11 June 1910 when, under pressure from the farmer who had charged for admission, use was made of a slope to gain speed. The machine rose 6ft in the air, but landed awkwardly, damaging a wheel and causing more general damage. Sydney Evershed had overspent on the project and abandoned all further attempts to fly, but went on to develop electrical equipment, later founding the firm of Evershed and Vignoles.

The machine was a small biplane loosely based on the early Wright glider but driven by a single tractor propeller. The wings were parallel and with slight dihedral and camber. The fuselage consisted of three bamboo longerons tapering to a universal joint at the rear to carry the box kite tail, which was the sole means of control. The pilot was seated on the lower center section behind the engine. The undercarriage consisted of two motorcycle wheels and a single cycle type tailwheel.

Power: 10hp JAP two-cylinder air-cooled vee type.

Data	Span	25ft
	Length	23ft

Evershed biplane of 1910 built at Burton-on-Trent.

F

FARBROTHER monoplane (A.M. Jesse Farbrother, 44 Westcott Rd., Wokingham, Berkshire)

The maker of the Wokingham Flyer (*q.v.*) also built a monoplane, which rose about one foot above the ground on its first and only test. The wings were described as having ribs like those of an umbrella. See Wokingham monoplane.

Fardell glider. Built on the Isle of Wight in 1910.

FARDELL glider (Fardell Bros., Morton Manor, Brading, Isle of Wight)

The Fardell brothers built their glider and tested it on Brading Downs in 1910. It probably went the way of similar projects and was finally dismantled after a season of gliding.

The glider was a neat looking, two bay biplane with the pilot seated on the lower center section. The rudder and elevator were carried on tail booms, which were parallel in both plan and elevation; the rudder was tall and was pivoted inside the booms, in front of the elevator. The machine was raised well clear of the ground on twin curved main skids and a double tail-skid. The wings were double surfaced and without dihedral, and were originally fitted with ailerons mounted on the front outboard interplane struts, later removed and fitted to the lower wings.

FAULKNER aircraft (S. Faulkner, 28 Boundaries Mansions, Boundaries Rd., Balham)

A report in the Press of 'a new ingenious flying machine' that did not materialize.

FAVRE monoplane (C. Favre, London)

A monoplane was under construction in a mews garage of the Head Chauffeurs' Club of Great Britain and Ireland in 1910-1911. There is no evidence of its completion.

Favre monoplane of 1910-1911 seems not to have been completed.

FAY-WILKINSON biplane (P.P. Fay & Wilkinson, Withernsea, Near, Hull, Yorkshire)

This machine, of which a crude drawing and description appeared in *The Aero* 27 July 1909 (p.153), was said to be under construction by Messrs. Fay and Wilkinson after years of experiment.

The biplane main wings were each divided into a center section and outer panels, the latter being controllable by pivoting on their main spars. A similar tail unit was also controllable, but presumably was of smaller span, and may have been of monoplane type.

The engine drove two tractor and two pusher four-bladed propellers through chains and shafts and was mounted on the lower main center section behind the pilot. The 70hp six-cylinder engine was made in Leeds and weighed 300 lb. An undercarriage of four pairs of wheels supported the machine. A stabilizer consisting of a weight suspended below the machine was connected into the control system.

Fay-Wilkinson biplane. Sketch of the machine which was partly built in 1909.

FERGUSON monoplane No.1 (J.B. Ferguson & Co., Little Donegall St., Belfast. Later Harry Ferguson Co. Ltd., The May St. Motor Co. Ltd., Belfast)

Harry G. Ferguson was the designer and pilot, whose first aircraft were constructed by his brother's firm of motor engineers. The machine was completed late in 1909 and although a 35hp Green had been acquired it was abandoned after a temporary

Ferguson monoplane No.1 was built in Belfast by Harry Ferguson in four versions. This is the first version as tested in December 1909 with Beedle propeller.

Ferguson monoplane No.1 second version with modified undercarriage.

Ferguson monoplane No.1 third version.

flywheel disintegrated on test, and the aircraft was fitted instead with a 35hp air-cooled JAP.

The machine had an open rectangular section girder fuselage, incorporating an A-frame which formed the upper pylon and continued down as the main members of the complex chassis with central skid. A large wheel supported the tail. The wings were parallel in chord and were fitted with skids in the form of hoops of cane at the tips, and also further inboard. Warping was employed for lateral control. The tail surfaces consisted of a tailplane on the fuselage centerline carrying a triangular elevator, which operated between top and bottom rudders.

The trials at Hillsborough Old Park using a Beedle propeller, resulted in a few short hops only and, greatly disappointed, Ferguson went off to both Rheims and Blackpool meetings, returning with new ideas for an improved version of his aircraft.

The second version was ready for testing at Hillsborough in December 1909. This was fitted with a new wing with tapered trailing edge. The undercarriage was simplified and no skid was fitted; a second pylon for bracings was added. On 31 December 1909, after a few short hops, the machine covered 130 yards reaching about 15 feet in height, in a 25-30mph wind, the first flight of an Irish-built aircraft. This was achieved using a Cochrane propeller, which Ferguson regarded as unsatisfactory, together with other propellers that he had tried, and he said so, in rather blunt letters to *Flight*. He praised a propeller from Clarke of Kingston, which he used later, and with which he was more successful, although changes to the machine no doubt contributed. The flying ground was also unsatisfactory and future tests were to be carried out elsewhere.

The third version was taken to Massareene Park, County Antrim in April 1910. It had an increased span wing, without the large hoop skids, an increased height undercarriage and a long tapered fin. The rear fuselage was covered with fabric and a large AA badge (Aviation Association of Ireland) was mounted at the nose. A Clarke propeller was fitted. Although some flights were made, the location proved unsatisfactory, also further modifications to the machine were indicated.

The fourth version was ready for trials in May 1910 and was taken to a the beach at Magilligan Strand, Lough Foyle,

Ferguson monoplane No.1. The fourth version at Magilligan Strand which was wrecked when flying from there in October 1910. The fuselage nose was modified and a skid was fitted.

County Derry. The aircraft had been considerably modified by the addition of an extra 2ft bay at the nose. The wing had been reduced in span with altered bracings to the raised pylon, now positioned further forward, and without the earlier kingposts. The undercarriage was now fitted with a central skid and the tailwheel was later replaced by a skid. The rudders were triangular and no fin was fitted. Ferguson made a number of flights at Magilligan, increasing steadily to 2 1/2 miles and up to 40ft height by 8 June 1910. On 15 July 1910 the dismantled machine was moved to Newcastle, County Down for a display at Donard Park, but the ground and the weather were unsuitable and he moved to the sands at Dundrum Bay. On 8 August 1910 he was able to cover three miles at a height of 100ft at 35-40mph, for which he won a £100 prize, donated by the Newcastle Recreation Committee.

Further improvements were made to the machine, the design of which was now basically established. The new nose bay was made triangular in section, and then reverted to rectangular. The fuel tank, previously carried above the fuselage, was placed inside and then transferred to a position between the undercarriage legs. The fuselage was completely covered.

Flying continued at Magilligan Point with passengers being carried, including on 10 October 1910, T.W.K. Clarke, the supplier of Ferguson's favored propeller. Later that day the machine crashed badly and was wrecked beyond repair.

Power: 35hp JAP eight-cylinder air-cooled vee.

Data	**Version**	**(1)**	**(2)**	**(3)**	**(4)**
	Span	34ft	34ft	40ft	36ft
	Chord	6ft	7ft tapering to 5ft	7ft 6in tapering to 5ft	
	Length	30ft	-	-	32ft
	Area	192 sq ft	192 sq ft	292 sq ft	240 sq ft
	Weight	-	620 lb	660 lb	-
	Weight allup	-	760 lb	800 lb	900 lb
	Speed	-	32 mph	-	35-40 mph

FERGUSON monoplane No.2

The second Ferguson monoplane embodied the changes made progressively to the first machine, but except for the engine and possibly minor parts salvaged from the wreck of the first, was completely new. It was taken to Newtownards and took off unexpectedly from the beach in windy conditions on 12 June 1910. The propeller and skid were damaged on landing. On 14 June 1910 Ferguson flew with a heavy passenger, who was injured in the landing when further damage occurred. After repairs it was re-flown at Strangford Lough in October and carried several passengers. Ferguson flew the aircraft at Magilligan Strand in 1912, but was concentrating on his motor business and by January 1913 the monoplane was being flown successfully by O.G. Lywood. John Williams, an associate in Ferguson's company, also flew the machine, but abandoned it after various problems. The engine was salvaged, but the aircraft was not rebuilt.

Ferguson monoplane No.2 was a success and was flying from June 1910 to early 1913. A nosewheel was fitted.

The fuselage of No.2 was of triangular section and the aircraft was generally reminiscent of the Antoinette, having triangular tail surfaces, tapered wings, and the pilot seated high above the top longerons. The undercarriage originally had two wheels and a nose skid but was subsequently modified to include a front wheel between the two convergent landing skids; the hollow axle of the unsprung main wheels allowed the warp cables to pass through, and enabled the deletion of the separate underside mounting for the pulleys. Warping was controlled by lever by the left hand, the elevator being operated by a separate stick. A single A-frame was used as the anchorage for the top cables. The fuel tank was carried across the fuselage, just behind the engine, with room for a passenger seated below the pylon.

A replica of this aircraft, fitted with the original engine and seat, was built in 1972-1973 for the Irish Aviation Museum by Capt. J.C. Kelly-Rogers, a well-known ex-BOAC and Aer Lingus pilot.

Power: 35hp JAP eight-cylinder air-cooled vee driving a 7ft diameter Clarke propeller.

Data		
	Span	32ft
	Chord	7ft tapering to 5ft
	Area	180 sq ft
	Area tailplane	26 sq ft
	Area elevator	11 sq ft
	Area rudders	9 sq ft
	Length	27ft
	Height	9ft 9in
	Speed	55 mph

FILEY ornithopter (built at Low Cliffs, Filey, Yorkshire)
Towards the end of 1910 Robert Blackburn had rented a hangar and furnished bungalow from John WF Tranmer, a local engineer and director of Northern Automobile Company. The hangar overlooked a three mile stretch of sand, very suitable for aeroplanes and was initially used by Tranmer to test Blériot monoplanes. The Blackburn Flying School used the Filey hangar until September 1912, when they moved to Hendon.

The hangar was then used by 'a foreign gentleman' (possibly German) who used space in it to construct a small ornithopter, which could be fitted with three different sizes of wings. These were described as resembling a Crow, a Sparrow and a Seagull presumably this meant that they were of low, medium and high aspect ratio. The constructor departed from the area at the outbreak of war in 1914 and the hangar was used as a drill hall by the army. The remains of the ornithopter survived for several years.

FILL monoplane (Vincent Fill, c/o Eastbourne Aviation Co.)
This was one of the 'unofficial' Blériot monoplanes built entirely by a pupil pilot, trained at the EAC and flown from August 1913.
Power: 30hp Anzani three-cylinder Y type air-cooled radial.

Fill monoplane was an 'unofficial' Blériot type constructed at the Eastbourne Aviation Co.

FITCH biplane (A. Fitch & Son, Motor Engineers, 610 Rosemary Rd., Clacton-on-Sea, Essex)
The Laking No.1 biplane, completed in 1909 largely by T.W.K. Clarke, was abandoned by its owner, Guy Laking, when the initial tests were unsuccessful and he went abroad. It was subsequently handed over to Frank Fitch, who reconstructed it with a single 40hp engine.

The aeroplane, still identified as Laking No.1, was taken to a field at Little Clacton on 10 June 1911 for final assembly and testing. It made its first flight on 4 July 1911 in the hands of Mr. Anthony Westlake, who himself had made a monoplane, in 1913 (q.v.). The circular flight covered about 300 yards at a height of 11ft at 20mph and the aircraft returned safely to near the original takeoff point. In the initial euphoria after this successful flight, the Press reported that a flight would take place to Colchester and a passenger could be carried, also a training school was to be established.

The machine was very lightly built and was reported to weigh only 5 1/2 cwt with pilot ready for flight. The major change from the Laking machine was the installation of the single engine said to be 'specially designed' and possibly built by Anthony Westlake. It may also have been the 40hp Lascelles which was auctioned in October 1913 (see Westlake).

The layout of the machine was basically as the original Laking No.1 and was of a typical pusher biplane of the period, with front elevator and rear rudder on booms and presumably wing warping. The pilot, seated on the lower center section, rested his feet on the center of the axle, and controlled the machine by two levers.
Power: 40hp with fuel injection pumps, driving a 7ft 8in diameter pusher propeller.

Data		
	Span	33ft
	Chord	5ft 6in
	Gap	5ft
	Area	330 sq ft (Press reported 165 sq yds!)
	Weight allup	616 lb
	Fuel capacity	6 gal

FLANDERS aircraft (L. Howard Flanders Ltd., Brooklands Shed 33 and Townsend Rd., Richmond, Surrey)
Howard Flanders began working as an assistant to A.V. Roe at Lea Marshes in 1909. In 1910, with financial support from Dukinfield-Jones and others, he established himself in business, initially at Brooklands. He also acquired premises at Richmond in March 1912. Flanders was a Cambridge graduate and an intelligent designer, who expressed serious views on various aviation matters of importance.

The firm began with a series of monoplanes which, unfortunately, were delivered for military service at the period of the ban on monoplanes and so were never used. The only biplane from the company, for the Military Trials of August 1912, was let down by the lateness of the unsatisfactory ABC engine. Flanders, who had a physical disability, was involved in a bad motorcycle accident in the autumn of 1913 and his company passed into bankruptcy. He went to Australia in January 1914 for six months to recuperate, and returned in July 1914 to be engaged as chief designer for Vickers at Crayford.

Flanders F1 twin-pusher monoplane was partly built between October 1910 and May 1911 but was abandoned through lack of the selected engine.

FLANDERS F.1 monoplane

This was to be a twin pusher monoplane with a 120hp ABC engine and variable incidence wing to aid stability. The engine did not materialize and construction was abandoned in May 1911.

Flanders F2 a single-seater of 1911 was the constructors' first machine to be completed.

FLANDERS F.2 monoplane

Flanders' first aeroplane to be completed was a single-seater tractor monoplane of conventional layout. The fuselage was based on a central structure incorporating two heavy ash members in the forward fuselage, which supported the engine and pilot and included front and rear spar sections, to which the wings were attached. Also included were the main kingpost and a beam, to which the upright members of the undercarriage were attached. The external shape of the fuselage was in the form of a conventional box girder, with plywood covering at the front and fabric at the rear end. The center structure was tapered and extended aft to integrate with the box girder.

The wings were tapered and braced by steel tapes from the front spar to the undercarriage structure, and to a post above the fuselage. The rear spar bracing cables also operated the warping control. The undercarriage had a central skid, the radius members being sprung and attached to the horizontal beam in the primary structure. Horizontal springs were fitted to realign the wheels, which could deflect under side loads. The tail unit had no fin.

The Green engine was fitted with a flywheel behind the propeller. Cooling was catered for by two radiators, positioned on either side of the front fuselage partly below the wings. The machine was flown initially by E.V.B. Fisher on 31 July 1911.

Power: 60hp Green four-cylinder inline, water-cooled driving a 8ft diameter Regy propeller.

Data		
	Span	34ft 6in
	Chord	7ft tapering to 5ft 6in
	Area	200 sq ft
	Length	29ft 6in
	Dihedral	4 degrees

FLANDERS F.3 monoplane

This was the F.2 converted to a two-seater with increased length and wing area. The second cockpit was positioned in front of the pilot. The warping control cables were transferred to a post below the fuselage, separate from the undercarriage structure, and the pulley was enclosed in a streamlined casing. The side mounted radiators were taller and shorter, and positioned further back in the gaps between the wing and the fuselage. An additional small radiator, with header tank, was fitted behind the engine, which also carried an exhaust silencer. A small fin was also added.

Flanders F3 was the two-seater version of F2 which crashed with fatal results at Brooklands in May 1912.

The F.3 was first reported flying with a passenger on 22 November 1911 and continued in use until 15 May 1912, when it crashed killing the pilot, E.V.B. Fisher, and his passenger.

During its relatively short career the F.3 was used by Marconi at Brooklands for his ground to air wireless experiments.

Power: 60hp Green four-cylinder inline, water-cooled driving a 8ft 6in diameter propeller.

Data		
	Span	41ft (also 42ft reported)
	Chord	7ft 8in tapering to 5ft 8in
	Area	240 sq ft
	Area tailplane	22 sq ft
	Length	31ft 9in
	Area elevators	13 sq ft
	Area rudder	6.5 sq ft
	Height	8ft 9in (tail down)

Flanders F3 with early Marconi wireless at Brooklands.

FLANDERS F.4 monoplane

The War Office ordered four F.4 monoplanes in late 1911, and these were built in the new works at Townsend Rd., Richmond, before being taken to Brooklands for final assembly and flight trials. The first machine was flying there by 7 July 1912 in the hands of F.P. Raynham, the others being transferred there progressively until 1 January 1913, when the last machine made its first flight.

The earlier type of fuselage structure was discontinued and a conventional type of box girder now constituted the primary structure, and a decking was provided around the two cockpits. The undercarriage of the first machine failed to satisfy the rough field requirement, when tested at Farnborough in September, and the machine was returned for modification.

A strengthened undercarriage structure, with separate legs and half axles, was attached direct to the bottom longerons, and was sprung by four rubber cords on each leg. Larger wheels, which did not castor, replaced the original design on all four aircraft. The engine was now air-cooled, thus dispensing with radiators, and a four-bladed propeller was used.

The general configuration was otherwise similar to the earlier F.3 aircraft except for a pair of parallel chord wings fitted to the fourth machine. Steel tapes, chains and sprockets were used to operate the warp control.

Flanders F4. Four of these monoplanes were built at the new Richmond works in 1912 for the War Office but were discarded after the official ban on monoplanes for the RFC.

Power: 70hp Renault eight-cylinder, air-cooled vee with integral fan and ducting. 2:1 reduction gear driving four blades consisting of two separate 8ft 6in Regy propellers.

Data		
	Span.	40ft 6in
	Chord	8ft 6in tapering to 5ft 9in
	Area	240 sq ft
	Area tailplane	25 sq ft
	Area elevators	18 sq ft
	Area rudder	10 sq ft
	Length	31ft 6in
	Height	9ft tail up
	Weight	1,350 lb
	Weight allup	1,850 lb
	Speed range	41-67mph

Four aircraft built, serial Nos.265, 281, 422 and 439

FLANDERS B.2 biplane

The ABC engine intended for the biplane built for the Military Trials of August 1912, was not ready in time to be fitted, so the machine was towed, without its power unit to Larkhill. The undercarriage of the machine, No.14 in the trials, was damaged on the way. After repairs, the aircraft took part in the assembly test and subsequently the engine was fitted. After a brief test, mechanical trouble developed, described as a 'tied-up camshaft ' caused by a stripped timing gear, and the aircraft was withdrawn.

The bottom of the fuselage was basically of triangular section with a full-length keel, changing to rectangular in the forward area, to a position behind the rear cockpit, where the bottom longerons joined to the central keel. A rounded decking ran the full length of the top.

The wings were of two bays with top wing overhang, braced by cables to kingposts above the outer interplane struts. The front inboard struts were of strong ash to carry the loads from the undercarriage, which was mounted immediately below. The axle was sprung and moved in slots in the vertical members, being restrained by a tubular member with a pivot on a vee strut below the engine, which also served to mount a long curved central skid.

Flanders B2. The firm's only biplane was constructed in 1912 for the Military Trials but was under-powered with its 40hp ABC, fitted when the 100hp ABC failed on test.

Flanders B2 modified and fitted with an Isaacson radial engine in 1913.

Lateral control by warping of both top and bottom wings was operated by cables and pulleys. A conventional tail unit included an aerodynamically balanced rudder of comma shape.

Subsequently a 40hp ABC was installed and the machine flew with this for the first time on 22 December 1912, piloted by Raynham, and later also with two passengers in its capacious fuselage. On 29 December 1912 the aircraft was damaged and the wings were then replaced with ones of slightly less span and altered tip shape. A split rudder to serve as an air brake was tried in the spring of 1913, but was soon discarded. In October 1913 a 60hp Isaacson radial replaced the ABC and an improved performance resulted.

A further engine change was made in June 1914, after it was reported that Lt. R.E.B. Hunt RN had bought the machine, with the fitting of a 70hp Gnôme in a circular cowling, with cutaway lower portion. A fin and new rudder were also fitted. The Flanders biplane was taken over by the Admiralty at the outbreak of war, becoming serial No.918 in the RNAS until deleted in January 1915.

Power: (1) 100hp ABC eight-cylinder, water-cooled vee driving a 8ft 6in dia Regy propeller.
(2) 40hp ABC four-cylinder inline, water-cooled.
(3) 60hp Isaacson seven-cylinder, air-cooled radial driving a 8ft 6in dia Lang propeller
(4) 70hp Gnôme seven-cylinder, air-cooled rotary.

Flanders B2 in its final form with cowled Gnôme rotary in 1914.

Data Version	**(1)**	**(2)**	**(3)**	**(4)**
Span top	43ft	43ft	41ft	40ft
Span bottom	27ft	27ft	27ft	27ft
Chord top	6ft 7in	6ft 7in	6ft 7in	6ft 7in
Chord bottom	5ft 1in	5ft 1in	5ft 1in	5ft 1in
Area wings	400 sq ft	400 sq ft	400 sq ft	400 sq ft
Area tailplane	28 sq ft	28 sq ft	28 sq ft	
Area elevators	14 sq ft	14 sq ft	14 sq ft	
Length	31ft 6in*	31ft 6in	31ft 10in	31ft
Weight	1,250 lb	670 lb	1,000 lb	1,050 lb
Weight allup	2,000 lb	1,100 lb	1,571 lb	1,650 lb
Speed (mph)	65 (est.)	38-56	40-65	40-68

* *The Aero* reported 30ft 6in

FLETCHER aircraft, see EMPRESS

FOLEY biplane (J.S. Foley and F.W. Baker, High House, Oldswinford and New Rd., Stourbridge, Worcestershire)

The machine was described in *The Aero* 29 March 1910 (p.247) as 'in course of construction at the motor works of F.W. Baker & Co. of Stourbridge'. Foley, a Midland Aero Club member, was the instigator and financed the venture.

It was said to be a combination of a biplane with a monoplane, but seems to have been a tractor biplane with a monoplane tailplane, elevators and rudder. At the front, carried on booms, was a divided elevator also serving for lateral control. A mechanical device for securing automatic stability was to be fitted, but no details were given. Ash and oval steel tube were the materials employed. The undercarriage had sprung wheels and incorporated skids. In the local Press, the report stated that it was hoped to complete the machine in a fortnight, i.e. mid-July. The difficulty of finding a suitable 'all British engine' prevented its appearance at the Wolverhampton meeting. There were no further reports on this machine.

Power: 35-40hp air-cooled engine of English make driving a 6ft diameter propeller.

Data		
	Span	35ft
	Length	40ft
	Area	360 sq ft
	Area tailplane	40 sq ft
	Area rear elevators	46 sq ft
	Area front elevons	60 sq ft
	Area rudder	18 sq ft
	Weight	500-600 lb

FORBES & ARNOLD monoplane (Victor F.E. Forbes and Arthur J. Arnold, Leighon Sea, Essex)

This machine was the subject of Patent No.20846/1909, although it was considerably later when details and illustrations appeared in *Flight.*

The upper part was a boxkite unit of diamond shape, with additional surfaces in the form of long flexible sails on either side. The lower part was a four wheeled chassis, with the operator seated on a wicker chair. Pitch and yaw control was to

Forbes and Arnold monoplane of patented design was built in Essex in 1910-1911 and may have lifted off briefly.

Francis *Comet* cycloplane with multiple wings of 1910.

be obtained from the surfaces with marked anhedral, mounted at front and rear, stability being achieved by the pendulum effect. A low powered motor, mounted on the chassis, drove a tractor propeller by belt. The first engine was water-cooled, probably a car engine, requiring a radiator, and gave insufficient power. This was replaced and a corrugated blade propeller fitted, probably a Cochrane, with which a brief flight is believed to have been made at Rochford.

The structure, made of tube or bamboo, appears to be flimsy and the whole concept seems impractical. Nothing further was heard of this machine.

Data	Span	16ft
	Area	350 sq ft
	Length	24ft
	Height	10ft
	Weight allup	450 lb

FOULIS biplane (Wilfred Foulis, Sunbury Mews, Belford Rd., Edinburgh)

Foulis was a motor engineer who went to Brooklands to practice flying at the Lane School, and became an agent for Lane monoplanes in Scotland. He achieved the first flight in May 1910 on a Lane monoplane near Edinburgh. Foulis had built a Farman-type pusher biplane in 1909 with ailerons mounted between the main planes. Few details are available, and it was presumably abandoned.

Power: 80hp Charron eight-cylinder air-cooled driving two 9ft 6in diameter propellers.

Data	Span	31ft
	Area	480 sq ft
	Weight	960 lb

FOX monoplane (E.W. Fox, Hemel Hempstead, Hertfordshire)

Fox is reported to have built a full size Blériot-type monoplane, which was waiting for the loan of a 25hp engine.

FRANCIS 'Comet' cycle aeroplane (B.S. and P. Francis, 97 Huddleston Rd., Tufnell Park, London)

This was a bicycle with two sets of planes, which was exhibited at Olympia in March 1910. Each set comprised six planes placed in vee-form, one set in front of, and one behind the rider. The total surface area was 120 sq ft and the incidence of the planes could be controlled. There was a pusher propeller powered by the rider.

FRANKLIN biplane (H.H. Franklin, A.E. Lowy, H.W. Holt and C.M. Spielman)

This crude machine was built in the 1909 Easter vacation, by four Cambridge undergraduates, at the home of Franklin, in the Chilterns.

The machine was a tandem biplane of large proportions, with front biplane elevator and tail mounted rudder. The pilot sat behind the front wings and behind him was a 12hp Buchet twin-cylinder, air-cooled motor driving through chain, the twin pusher propellers behind the rear wings. For lateral control there were four hinged flaps on each wing, pivoted diagonally to face forward, an unconventional scheme of doubtful merit. Air in compression was used in the undercarriage legs.

Limited time allowed only brief trials which resulted in some damage on the ground and the machine was abandoned.

Franklin biplane of 1909, abandoned after brief trials.

FRISWELL monoplane (Friswell (1906) Ltd., Motor Agents, Albany St., London NW)

This machine began life as a Roe Triplane early in 1909, and was in course of construction at Lea Marshes, when the partnership between AV Roe and George Friswell, broke up. The machine was taken over by Friswell at a stage where the fuselage structure, midwing, center section and tailplane were assembled, and was then redesigned by Capt. T.T. Lovelace as a monoplane. It was not completed and was finally auctioned by Friswell for £5.50.

The engine was to be a 35hp JAP eight-cylinder, air-cooled vee unit and, unlike Roe's other triplanes, the Friswell machine had a rectangular section fuselage.

FRITZ monoplane (Fritz Goetze, designer, H. and D.J. Oyler, 35 New Cavendish St., London W1, constructors)

This machine resembled a Blériot XII or *Demoiselle* in certain features, particularly the positioning of the pilot and engine and the type of undercarriage.

The fuselage consisted of three bamboo longerons, the upper being 2 1/2in diameter and curved, and triangular steel tube frames all suitably braced. The wings were also curved to a gull wing shape and were based on two 2in diameter, and one 1in diameter, bamboo spars with bamboo ribs. Lateral control was by pedal for warping the wings. An aluminum member formed the leading edge, and a steel wire the trailing edge. Fullchord elevators, outboard of the tailplane tips, were controlled by a hand-wheel. The rudder was also controlled by a hand-wheel and was the only vertical surface. A streamlined fuel tank was mounted in the cabane above the wing with coolant radiators just below.

The Fritz monoplane was illustrated and described briefly in *Flight* in August 1911 but was not referred to again, although it was in a completed state at an unidentified airfield.

Power: 40hp ENV type D eight-cylinder, water-cooled vee.

Data		
	Span	32ft
	Chord	7ft
	Area	210 sq ft
	Area tailplane	15 sq ft
	Area elevators	15 sq ft
	Area rudder	9 sq ft
	Length	29ft

Fritz monoplane of 1911 was built by Oylers Ltd.

Frost ornithopter of 1877 was steam-powered and quite impractical.

FROST ornithopters (Edward Purkis Frost, West Wratting Hall, Cambridgeshire)

Frost experimented with flapping wing devices, using actual goose feathers for the wing surfaces. Early trials in 1877, in association with Dr. F.W.H. Hutchison, of a large machine powered by a 5hp steam engine, with a total weight of 650 lb, were a failure when attempts were made to lift a man. This device had four sets of wings, with a maximum span of 30ft. Later tests of a small model, with 3 sq.ft of wing area, were carried out on a whirling arm, powered by a 2hp electric motor to produce the flapping motion, and this induced rotation of the arm.

A full scale machine, smaller than the 1877 machine, followed in 1905, powered by a petrol engine from a 3hp BAT motor cycle. With a span of 20ft and weighing 232 lb, it was constructed on a tubular frame on four wheels, with a platform on which the operator stood. It was said to have risen up to two feet in the air and tended to move forward. This machine had only one pair of wings. This work was reported to the Cambridge University Engineering Society and recorded in *The Automotor Journal*.

Frost ornithopter of 1905 was able to 'jump' under the power of its motorcycle engine.

E.P. Frost was president of the Aeronautical Society from 1908 to 1911, until the more progressive members forced a reorganization of the Society and he retired. The 1905 machine exists at the Shuttleworth Collection and the model is preserved in the Science Museum.

FULFORD monoplane No.1 (W.H. Fulford, Mills-Fulford Ltd., Coventry)

This very small *Demoiselle* type monoplane was exhibited at the Stanley Cycle Show in the Agricultural Hall, Islington in November 1909. It was in the nature of a demonstration rig, with very short wingspan, for space considerations at the show. It was also referred to as the Mills-Fulford.

The machine was powered by a four-cylinder FN air-cooled motor cycle engine driving a 6ft diameter tractor propeller by chain, and had a front elevator with differential action, as well as a combined tail elevator and rudder. The structure was of steel bicycle tubing and was mounted on three wheels with curved strip springs.

Fulford monoplane on exhibition at Islington in 1909.

G

Gaskell-Blackburn biplane was built from parts of three other machines in 1913-1914.

GASKELL-BLACKBURN biplane (Vivian Gaskell-Blackburn, Brooklands)

Gaskell-Blackburn learnt to fly at the Bristol School at Brooklands, receiving certificate No.617 dated 10 September 1913. He then constructed a biplane, that first appeared for engine run on 27 February 1914. First flight was on 2 April 1914 at Brooklands, where it was used until the outbreak of war.

The biplane was a conventional single-seat, two bay, tractor somewhat resembling the Wong Tong Mei. *The Aeroplane* reported at the time that a few parts from that machine, albeit adequately strengthened, were actually incorporated in the Gaskell-Blackburn. A much later record stated that parts of three aeroplanes, namely a Champel, the Parsons biplane and a Sommer of the Pashley brothers were used in its construction.
Power: 40hp ABC four-cylinder inline, water-cooled.

GAUNT Cycloplane (John Gaunt, Gargrave, near Leeds, Yorkshire)

John Gaunt first appeared on the aeronautical scene at the Olympia Aero Show in March 1909, with an attachment to a bicycle to provide lift to aid the cyclist. It could also be fitted to a boat.

GAUNT biplane No.1 (Apperley Bridge, near Bradford, Yorkshire and Southport Aerodrome Hesketh Rd., Southport, Lancashire)

Gaunt soon discarded the Cycloplane and proceeded to design and construct a twin pusher biplane. The machine was tested in April 1910 at Apperley Bridge, near Bradford, but on 27 April 1910 a petrol explosion occurred, which damaged the machine and badly burned John Gaunt.

The machine was rebuilt and was out again in August, but Gaunt had now moved across to Southport, and rented a hangar adjacent to the sands. After engine problems, the machine

Gaunt Cycloplane, a lift-aid for cyclists.

appeared to be ready to lift off, until damage occurred through the undercarriage sinking into the soft sand on 30 August 1910. It was probably at this stage that Gaunt decided to save the weight of the heavy chain drives and other fittings by converting the machine to a single propeller tractor.

Power: 30hp Alvaston two-cylinder horizontally opposed water-cooled.

Data Wing area 200 sq ft

GAUNT 'See-Saw' aeroplane

The following letter, with sketch, appeared in *Flight* 11 March 1911 (p.213) and described an unlikely looking biplane, which presumably remained unbuilt.

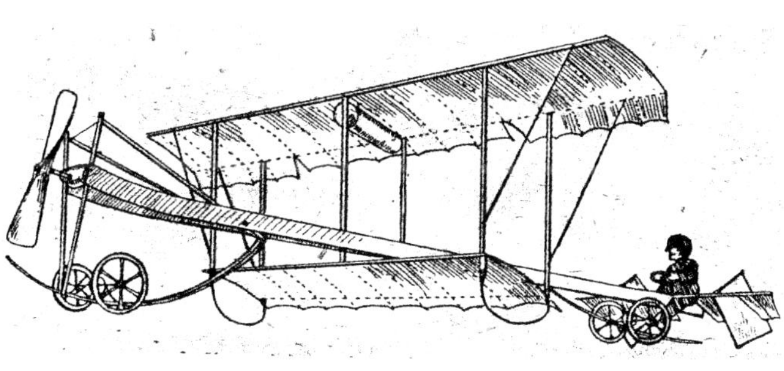

Gaunt See-Saw aeroplane had no future.

'I enclose a sketch of the see-saw type of aeroplane that I have invented and for which I claim that it is easier to learn to fly than any other type. The pilot has the whole of the machine in front of him with direct control and "feel" of both elevating and steering handle, which is of gunmetal and similar to the cycle handlebar in use for steering, but with up and down movement for balancing.

Primarily, the invention is to secure a better sense of direction in beam winds. The present loading of aeroplanes is a central disposition of carried weights, which in a beam wind acts as a pivot for the aeroplane to veer round.

Features not to be lost sight of are: the compass is a long way from the engine, so is the pilot, and the provision of a hoeshoe or grapnel will be useful to hold back aeroplane whilst starting engine and as an emergency drag in finishing a flight.

The idea is the outcome of another original machine I have designed, built and flown with an Alvaston Engine. John Gaunt.'

GAUNT monoplane

Flight reported on 1 April 1911 (p.296), trials at Southport of a Gaunt monoplane, by the Hon. W.S. Leveson-Gower RN, of which he was part owner. On a third run on 24 March 1911, the machine rose to 10-15ft, but crashed with damage to the wing, undercarriage and propeller. No other information can be traced on this machine.

GAUNT biplane No.2 '*Baby*'

First reports of the '*Baby*' biplane appeared in *Flight* of 8 July 1911 (p.592), when Gaunt made a straight flight of two miles. Straight flights up to 100ft high were reported a week later, and these increased progressively up to 300ft high, and to four miles and return. Gaunt apparently flew regularly at Southport in the summer of 1911, and started to instruct Leveson-Gower on 7 September 1911. The machine was still flying in August 1912, but the end seems to have come on 22 August 1912, when it was damaged in an accident.

The machine was an unequal span biplane, with the same Alvaston engine and radiators fitted on either side. Control was by wing warping and conventional tail surfaces. The undercarriage was fitted with two wheels and a central skid; a tail-skid and wingtip skids were provided. The machine was constructed

Gaunt achieved some success at last with his second biplane.

in the conventional style of the time, except for the wings, which were covered with thin French plywood; the separate sheets were fitted with eyelets along the edges, and were laced together with cord.

Power: 30hp Alvaston two-cylinder, horizontally opposed, water-cooled driving a 7ft diameter Gaunt propeller

Data		
	Wing area	208 sq ft
	Weight	600 lb
	Speed	35-50 mph

The George and Jobling biplane was flown at both Eastchurch and Gosforth.

GEORGE & JOBLING biplane (A.E. George & Jobling, Smith St., Newcastle-on-Tyne, Motor Engineers)

A.E. George, a racing motorist, acquired the Voisin 'Bird of Passage' but abandoned it after a crash at Shellbeach. He subsequently decided to build a machine of his own design, which he was ready to exhibit at the Olympia Aero Show in March 1910. On 13 May 1910 he flew straights up to 130 yards at Eastchurch and, by the end of May, had made twenty flights. At the end of August George had passed the tests for his pilot's certificate on it at Eastchurch. By October the machine was moved north, but on 6 October 1910 at Gosforth Park, Newcastle damage occurred. The machine had flown 300 miles, with a longest flight of eighteen minutes and, although reported to have been repaired and ready to fly again, there was no further report that this took place.

This well designed pusher biplane used bamboo for the booms, which were to be replaced by steel tube. It had front and rear elevators, a single rudder and side panels in the front booms. Midgap ailerons were fitted to the front outboard interplane struts, together with warping extensions to the top wings for lateral control. Small side curtains were also fitted. All controls were operated by the hand-wheel and movement of the control column.

The undercarriage was made of steel tube, and the twin skids could be lowered to aid the passage of the machine over rough ground when alighting. A steerable tailwheel and wing tip skids were fitted.

George and Jobling biplane modified with side-curtains on front booms.

Power: 60hp Green four-cylinder inline, water-cooled driving through chain a 9ft dia. G&J propeller at 2:1 reduction on a shaft above the engine, the whole being mounted on a subframe. The radiators were of surface type forming part of the lower wing.

Data		
	Span	30ft
	Chord	5ft 6in plus 1ft 3in warpable
	Gap	5ft
	Area	325 sq ft
	Area tailplane	50 sq ft
	Area elevators	36 sq ft
	Area rudder	22 1/2 sq ft
	Area ailerons	27 sq ft
	Length	31ft
	Weight	662 lb
	Weight allup	862 lb
	Max speed	48mph

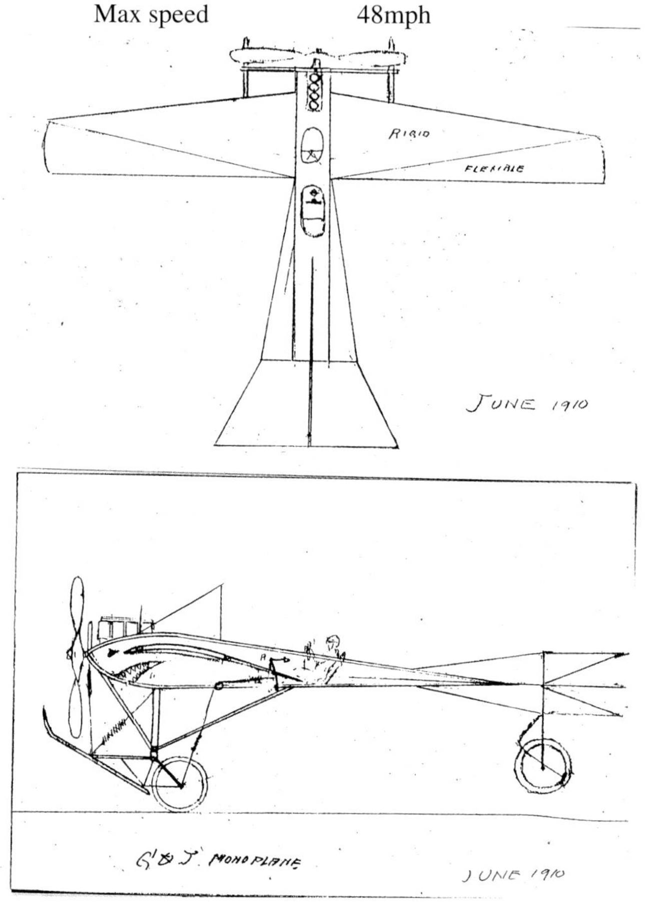

Sketch of the proposed G & J monoplane.

GEORGE & JOBLING monoplane

In June 1910, George was preparing the design of a new monoplane. This had the means of retracting both the main and tail wheels for landing on the skids, to reduce the length of run. The machine was a tandem two-seater on the lines of an Antoinette, with a tapered wing capable of being warped over the area, from the intersection of the fuselage at the trailing edge, to the leading edge at the wingtip, namely 32% of the area. The fuselage was parallel in plan, with elongated fins and horizontal surfaces extending forward to the wing junction. The wing area was 200 sq ft and a 90hp motor was estimated to give a speed of 95mph. There is no evidence that construction ever commenced.

Gerrard monoplane wrecked at the London Aviation Ground, Acton in December 1910.

GERRARD monoplane (F.A. Gerrard, The London Aviation Ground, Acton)

Little is known about this aircraft except that which can be deduced from the illustration in *The Aero* 28 December 1910 (p.509), which shows it inverted after crashing on 18 December 1910 at Acton.

Although similar to certain Howard Wright designs, the machine had original features The fuselage girder was deep towards the tail and the strong cabane had a lateral crossbar. This, together with a vertical post for bracing wires above the tailskid, no doubt enabled Gerrard to leave the aircraft unhurt, also preventing damage to the propeller.

The rudder and elevator were on a universal fitting, with a separate tailplane fixed to the lower longerons. The wing tips were rounded, the wings, without ailerons, were presumably warpable.

A gust of wind is reported to have caused the accident and was sufficient to wreck Gerrard's aspirations, for there were no further reports of him or his aircraft.

Gibson biplane No.1 built at Leith in 1909 was unsuccessful.

GIBSON biplane No.1 (Caledonian Motor & Cycle Works, Leith, near Edinburgh, Scotland. J. and G.T. Gibson, proprietors)

This pusher biplane was designed and built during 1909 by the Gibsons, father and son, at their Caledonian Motor and Cycle Works. It had the typical layout of a machine of the period, with front and rear monoplane elevators and a fixed tailplane and single divided rudder. These units were apparently supported by lower booms only but were presumably braced also by wires. The chassis consisted of short curved skids carrying two pairs of sprung wheels on separate axles.

The wings were of equal span, and single acting ailerons were fitted to both top and bottom. The pilot sat on the lower center section, with the engine, fuel tank and radiators on either side. The machine was not capable of flight.

Power: 30hp Alvaston two-cylinder horizontally opposed, water-cooled.

Data Span 30ft

Gibson biplane No.2 was a reconstruction of No.1. A further batch of nine was built.

GIBSON biplane No.2

The second machine was a reconstruction of the first and was completed in the following year, and flown in August 1910 at Beautland Farm near Balerno.

The modifications included changing the front monoplane elevator to a biplane unit with twin rudders. The attitude of the machine on the ground was altered by lowering the tail, and rigging the machine with slight stagger. The front and rear booms were full depth strut braced units with additional support at the front, provided by the extended, curved Sommer-type skids, which met at the front elevator attachments.

The Gibsons built a further nine biplanes, one of which was shown at the Scottish International Exhibition in 1911. During 1913-1914, they were building a machine of 40ft span and had started to design and build an engine. This work was discontinued when war broke out. Of the eleven machines, four were destroyed during testing, one was burned at an exhibition in Brussels, the rest were sold. No detailed information on the design of these has been found.

Power: 30hp Alvaston two-cylinder horizontally opposed, water-cooled.

Data Span 29ft
Length 30ft

Gibson biplane No.3. A Farman type with twin tractor propellers.

GIBSON biplane No.3

A letter from John Gibson published in *Flight* on 19 February 1910 illustrates the third biplane which they had made and sold. This was a Farman type machine with twin pushed propellers driven by a single engine of unknown make. There were twin fins and rudders and ailerons set at mid gap. There was a conventional undercarriage with two main wheels and a tail wheel. No further details have been found.

Gibson mentions that they were making a fourth machine of 29ft span and that a fifth was being made for display at the Olympia Show.

GIBSON ornithopter (William Gibson, Outram St., West Hartlepool)

This machine was shown at the exhibition at Crystal Palace, organized by the Aeronautical Society in 1868. It consisted of a framework, to which four wings were attached that could be operated by a treadle, so that when two were up, the other pair were down. The machine did not lift off, and Gibson proposed to build a larger one with more area, of which there is no information.

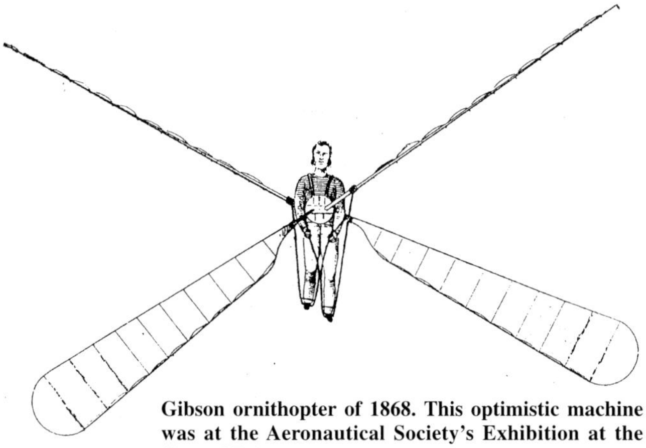

Gibson ornithopter of 1868. This optimistic machine was at the Aeronautical Society's Exhibition at the Crystal Palace.

Data		
	Length of wing	12ft
	Chord	1ft 6in tapering to 1ft
	Area of two wings	37 sq ft
	Weight	41 lb
	Weight allup	188 lb (5 lb per sq ft approx.)

GLASGOW glider (Glasgow Model Aeroplane Club, 101 Vincent Rd., Glasgow)

It was reported in Flight of 9 April 1910 (p.274) that the club had a monoplane glider designed and built by Mr. W. Crebar, president of the club, which had flown successfully.

Gloucester monoplane an unsuccessful venture of 1910.

GLOUCESTER military monoplane (Gloucester Aeroplane Co., Barton St., Gloucester. Campbell A. Ping and J.A. Barnet)

This newly formed company made a tractor monoplane, to the design of Campbell Ping, in the summer of 1910. It was of conventional layout, but with a special arrangement, whereby the rear wing spars could be individually moved, by separate pedals, to control the angle of incidence. The wings were single surfaced with fabric, except for a section of the leading edge, tapering from 1ft 6in to 1ft along the upper surface, which was also covered. The control units were single surfaced as well. The machine was low powered and lightly constructed.

There is no indication that the machine was ever flown and it later appeared for sale with the effects of W.W. Peet, in bankruptcy, on 20 May 1913, together with a part made tandem monoplane (almost certainly the Webb-Peet monoplane). Previously a 'new type biplane' was said to be under way.

Power: 20-24hp Phoenix five-cylinder air-cooled rotary driving a 6ft dia. propeller.

Data		
	Span	32ft
	Chord	6ft 9in tapering to 5ft
	Weight	305 lb

GLOVER direct lift aircraft (J.P. Glover)

The only information is that this machine was built in 1912 by the Blackburn Aeroplane Co. at 6 Balm Rd., Leeds.

GLYDOPLANE glider (Threlfall & Leigh, 30 Withy Grove, Manchester)

No details are available for this glider reported built in 1909-1910.

Gnosspelius No.1 hydro-monoplane was tested, but not flown at Windermere in 1910.

GNOSSPELIUS hydro-monoplane No.1 (Oscar Theodore Gnosspelius, Silver Holme, Windermere and Borwicks Boatyard, Bowness, Windermere)

Gnosspelius was of Swedish origin, but of English upbringing, and had been involved in civil engineering works abroad. His aviation interest stemmed from the time of the Blackpool Meeting in 1909.

His first machine was much like a Blériot with a rectangular section open fuselage with pyramid pylon for bracing and warping cables. It was fitted originally with twin floats, with which it would not lift from the water, It was then refitted with a single float 14ft long and 4ft wide sprung by rubber rings.

The machine was under-powered and neither version of the hydroplane was successful when tested between August and November 1910.

Power: 20hp Alvaston two-cylinder horizontally opposed, water-cooled.

GNOSSPELIUS hydro-monoplane No.2

The second Gnosspelius machine was designed with a triangular section fuselage, to save weight, and was fitted with a more powerful engine. It was again built at Borwick's at Bowness and was fitted with a central float, together with small wingtip floats of streamlined shape. Trials in this form on Windermere in the spring of 1911 indicated the need for ventilation of the float, to enable it to unstick. In June the float from No.1 was fitted with vanes at the bow. An attempt at flight with this float was made on 25 November 1911 which ended with the machine upside down in the water.

Gnosspelius No.2 hydro-monoplane was tested in 1911 but could not be flown until 1912 after changes to the floats.

The machine was subsequently rebuilt with box type wingtip floats and No.2 flew for the first time on 13 February 1912, after taxiing trials the previous day. Gnosspelius was the pilot for the first and a number of subsequent flights, becoming the first man in the area to design and fly his own aircraft. The machine continued to fly successfully in the hands of R.C. Kemp and Lt. J.F.A. Trotter until the spring of 1914.

Power: 40-50hp Clerget four-cylinder inline, water-cooled.

Data		
	Area	190 sq ft
	Weight	650 lb
	Speed	55 mph
	Central float	(first) 12ft long by 5ft wide
		(second) 14ft long by 4ft wide

Gnosspelius hydro-biplane was flown for a short time only in 1913.

GNOSSPELIUS hydro-biplane

After receiving Aviators Certificate No.360 in March 1912, the first to be issued to a pilot trained on a hydro-aeroplane, Lt. Trotter ordered a two-seater biplane to be designed by Gnosspelius. The machine was launched by the 8 September 1913, but did not takeoff until 20 September 1913, when Trotter alighted heavily immediately after lifting off and damaged the chassis and propeller. After repairs and a change of main float, only limited flying took place due to various problems, the machine apparently being flown for the last time on 11 November 1913.

The four-bladed propeller was geared down by chain and sprockets from the 100hp Green, six-cylinder inline, water-cooled engine, which was not powerful enough for the heavy aircraft that Gnosspelius had designed. A large radiator was mounted alongside the passenger seat on the port side. The three bay wings had ailerons on the top plane only. No fin was fitted and the rudder post extended down to carry a water rudder.

GOLDMAN Crucifer biplane (L. Goldman)

This ambitious streamlined biplane was designed by Goldman in 1909 and was built as a model; a full size machine may have been commenced, but there is no evidence that it was ever completed

Goldman *Crucifer* biplane. An ambitious design which was not built.

The swept back wings were mounted on the torpedo like body on a ball bearing collar so that if a sudden gust of wind struck underneath one wing it moved about the central body without imparting motion to this since it was mounted on the sliding collar.

The engine was mounted within the body and drove twin propellers, mounted in midgap behind the wing, by chains or shafts and gears.

The wings could be folded back over the body and the engine was connected by suitable gearing to the wheels so that the aeroplane with its wings folded could be driven along a road. Lack of finance prevented further development of this clever design.

GOODDEN Dragonfly monoplane (Frank Widenham Goodden, Elmthorpe Rd., Wolvercote, Oxfordshire)
A gale in November 1911 destroyed the Imperial Aero Club, at Port Meadow, Oxford owned by J. Bett & Co., where Gooden was employed. He purchased the remains and built a new shed, in which he built a tractor monoplane. He flew this single-seater from the middle to the end of 1912, when he sold out and moved to Hendon. The *Dragonfly* was destroyed by fire, with the buildings, on 6 August 1913.

The fuselage was of circular section, plywood covered at the front and faired to match at the rear, with stringers and fabric covering. The machine was of conventional layout and rested on its main wheels and front skid.

The wings were cut back to the rear spar alongside the fuselage and were fitted with trailing ailerons, and braced by many cables to a pylon and the undercarriage structure. The flowing lines of the tail surfaces perhaps inspired the name.

Goodden *Dragonfly* built and flown in 1912 by Frank Goodden, the famous Farnborough testpilot killed in the SE.5 prototype.

Power: 35hp JAP four-cylinder air-cooled vee.

Data		
	Span	30ft
	Area	168 sq ft
	Length	25ft
	Weight allup	650 lb
	Max speed	60 mph

GOODMAN monoplane (T.L. Goodman, 29 Oxford Street, Manchester)
This Blériot type monoplane was advertised for sale in April 1911 without engine or propeller. No other details are known.

Mr. GORDON'S Flyer
Built in c.1909 with a two-cylinder JAP engine.

GOULD biplane & monoplane (Gould & Co., London Inn Square, Exeter, Devon)
A pusher biplane was offered for sale in February 1910 and was illustrated in a partly built state in *The Aero*. A Blériot type tractor monoplane designed by George Weeks and fitted with an Anzani engine and Cochrane propeller was also built in the same year.

Gould biplane under construction at Exeter in 1910.

Gould monoplane was an 'unofficial' Blériot type.

GRAHAME-WHITE AVIATION Co. (Claude Grahame-White, Hendon Aerodrome)

Grahame-White operated a successful motor business and, after a brief period of ballooning, turned to heavier than air machines in 1909, after the developments in France and Blériot's Channel crossing. He learnt to fly at the Blériot School at the end of the year, receiving Brevet No.30 of the Aero Club de France in January 1910. Thereafter he set up his own flying school at Pau, later transferring this temporarily to Brooklands. In May 1910 *The Aero* reported that he rented ground at Park Royal, near the Plumes Hotel, but this report was premature. The report also referred to 'half a dozen monoplanes of Blériot type', built at his workshop at Walham Green, under the supervision of R.W.A. Brewer.

The school was moved to Hendon in 1911, where Grahame-White also ran many flying displays, which made Hendon famous. He also established his manufacturing company there, early in 1911, and this operated into the early post-war period, but was forced into receivership in 1922. From 1913 to 1915, J.D. North, later of Boulton and Paul, was in charge of design.

Farman and Blériot machines from the French factories had been used by Grahame-White since 1909, but when his school commenced operations the number of machines was increased. The aircraft in use in 1911-1912 were often reported to be 'British-built Farmans' but the number actually made is uncertain. To indicate the scale of operations, reports at times referred to biplane No.10 and monoplane No.6. A photograph in *Flight* 11 February 1911 (p.114) showed five Blériots, one Farman and one G.W. Baby already available. The manufacture of aircraft developed from Farman types, to those of original design, and the production capacity was enhanced by a large addition of plant and machinery in October 1911.

GRAHAME-WHITE Sommer type biplane

A single propeller, pusher type biplane with seats for two was exhibited at Gamages store from 6 to 11 June 1910, with which Grahame-White was hoping to fly to Paris from London. It was described as 'British-built', and was constructed before the works at Hendon were established. It was made by the Windham Detachable Motor Body Co. Nothing more was heard of this machine at the time.

This Sommer-type biplane was built for Grahame-White in 1910 by Windham's motor body company.

In some respects like a Sommer biplane, the machine had front booms and long curved skids, to which the front elevator was hinged. Rear booms carried a fixed tailplane but no rudder. The front spars of the wings were at the leading edge. Large single acting ailerons hung from the rear spars of the top wings, with additional fixed trailing edge surfaces inboard of these.

A later reference states that a Sommer biplane was dismantled at Brooklands by Compton Paterson, at that time a GW pilot, and in November it was being reassembled with a Gnôme engine for the use of the pupils. Although not confirmed, this is presumed to refer to this machine.

Power: 40hp Humber four-cylinder inline water-cooled.

Data		
	Span	33ft
	Gap	5ft 9in
	Length of tail, rear of wings	23ft
	Area	450 sq ft
	Weight allup	800 lb

GW *Baby* at Olympia in March 1911 was built for GW in America by the Burgess company.

GRAHAME-WHITE *Baby* & *New Baby* biplane

While competing in America, in the autumn of 1910, Grahame-White arranged for an aircraft to be constructed to his requirements by the Burgess Co. of Marblehead, Mass. It is recorded here although not strictly British. The machine arrived at Hendon in January 1911 and made its first flight on 11 February 1911 in the hands of the owner.

The *Baby* was a scaled down Farman-type, with front and rear elevators carried on booms and with ailerons on all wings, and was capable of carrying a passenger. The platform, for the crew and the engine, was tilted to increase propeller ground

GW *New Baby*. This is probably the second machine also built by Burgess, which arrived at Hendon in March 1911.

clearance, due to the low overall height. The crew were completely exposed in the aircraft as received, but a primitive nacelle was added before the machine appeared at the Olympia Aero Show in March 1911.

A much modified version for racing had rear booms tapering down to a single tail plane, with semicircular split elevators. The front boom and elevators were removed and a ring cowling fitted round the engine. An undercarriage with two wheels and nosewheel was used.

A note in *Flight* 11 November 1911 (p.985), refers to a 50hp Indian-engined version dispatched to California for use during Grahame-White's second U.S. tour. The Indian engine was a modified version of the Gnôme made by the Hendee Manufacturing Co. However most use was made of a Nieuport monoplane during this visit.

In 1911 the Grahame-White Aviation Co. were advertising as manufacturers of the *Baby* aircraft, but it is not clear how many, if any, were British built. A crate from Burgess, presumably delivering another *Baby* arrived at Hendon in March 1911.

Power: 50hp Gnôme (£950) or 70hp (£1050) seven-cylinder air-cooled rotary driving a 7ft 6in diameter propeller mounted ahead of the engine.
50hp Indian. Gnôme made by Hendee.

Data		
	Span	27ft
	Chord	4ft 9in
	Gap	4ft 10in
	Area	235 sq ft
	Area front elevator	19 1/4 sq ft
	Area rear elevator	10 1/2 sq ft
	Area tailplanes	37 1/2 sq ft
	Area rudders	12 1/2 sq ft
	Length	32ft 3in
	Height	8ft 6in
	Weight	420 lb
	Weight allup	655 lb
	Max. speed	55mph improved to 60-65mph in July 1911
	Endurance	1 1/2hr

GW *Baby Racer* a much modified version at Hendon.

GW Blériot with ENV engine installation.

GRAHAME-WHITE Blériot monoplanes

The machines reported to be under construction at Walham Green were probably standard type Blériots destined for the school. However one machine used at Hendon had an original engine installation, namely a 40hp ENV type D, with a frontal radiator. This flew for the first time on 24 January 1911 piloted by Clement Gresswell, the chief instructor of the school at Hendon.

GW School biplane with Gnôme engine and single rudder.

GW School biplane with Gnôme engine became RFC No.309.

GW Type VIIc known as the *Popular*.

GRAHAME-WHITE Boxkite type biplanes

These Farman type biplanes, used from 1911 onwards for training purposes at Hendon, contained variations between individual machines. Most had a single rudder and were powered by a 50hp Gnôme rotary engine. One early machine had been built at the end of 1910 to compete in the Baron de Forest Contest to be flown by Gresswell, but this was prevented by damage on the ground at Dover, in a gale. To qualify, it was fitted with a 60hp ENV type F eight-cylinder, water-cooled vee engine made at Willesden. After repair it flew at Hendon as School Biplane No.1, from 21 January 1911 until February, when a 60hp Wolseley was substituted, the ENV being reinstated in September 1911.

One of the school biplanes with 50hp Gnôme was bought by the War Office in 1913, identified as serial No.309, and based at Farnborough.

GRAHAME-WHITE Military biplane Type VI

This two-seater pusher biplane was shown at Olympia in February 1913 and was one of the earliest attempts to produce an aircraft with offensive capability.

The engine was front mounted driving the propeller, through-shaft and chain, at the rear of the nacelle. A Colt machine-gun above the engine, with a wide field of fire, could be operated by a gunner in the front cockpit.

An unusual feature was the triangular section rear boom consisting of tubular longitudinal members, the top one protruding forward through a bearing in the propeller boss, enabling the control wires to pass through to the tail.

The undercarriage incorporated wide skids, which carried the wheels on sprung mountings. The wheels each had two rims fitted with separate tires to prevent canting of the wheels on bumpy surfaces. Roll control was by wing warping.

The basic design was attributed to Horatio Barber, with J.D. North responsible for the final design work.

Power: 90hp Austro-Daimler or 120hp Austro-Daimler six-cylinder inline water cooled driving a 10ft diameter Chauvière propeller through shaft and Brampton chain.

Note: The engine fitted at the Olympia Show was of 90hp. It is doubtful if the 120hp Austro-Daimler was ever fitted for little was heard of this machine after the Show.

Data		
	Span top	42ft 6in
	Span bottom	23ft
	Chord	6ft 1in
	Gap	7ft
	Length	33ft 9in
	Wing area	435 sq ft
	Area tailplane	50 sq ft
	Area elevators	25 sq ft
	Area rudder	20 sq ft
	Weight empty	2100 lb
		2200 lb with 120hp engine
	Weight allup	2750 lb
		2850 lb with 120hp engine
	Speed	50-70 mph
	Endurance	6 hours

GW Type VI Military biplane was never used by the RFC.

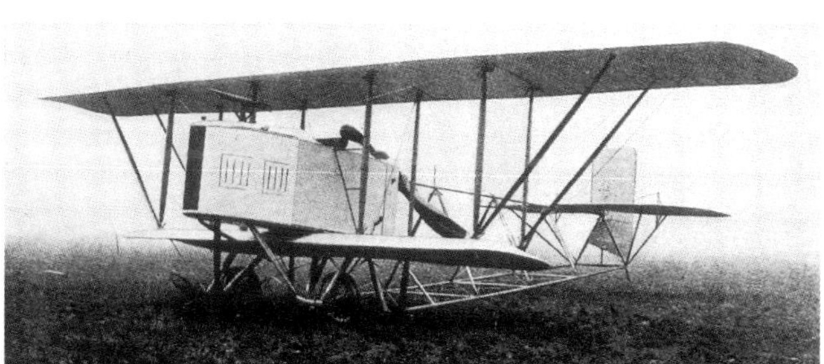

GW Type VII was acquired for the RFC as No.283 in 1913.

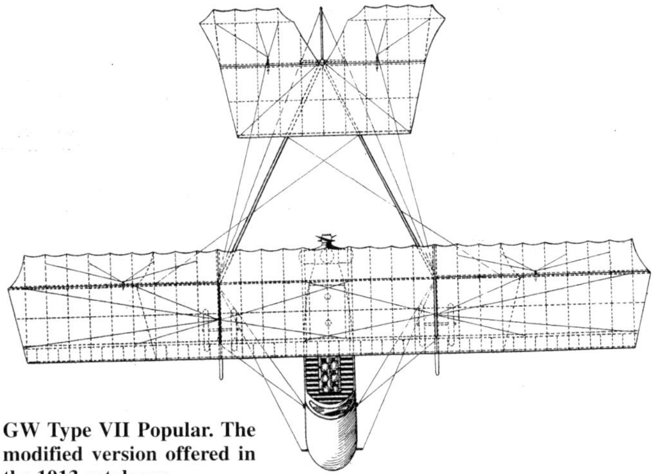

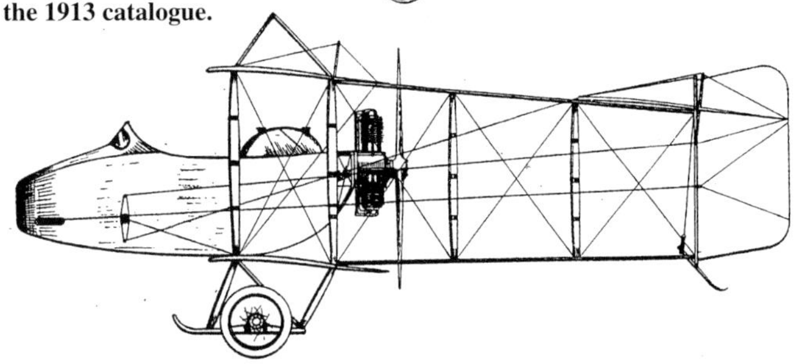

GW Type VII Popular. The modified version offered in the 1913 catalogue.

GRAHAME-WHITE *Popular* biplane Type VII

The single-seat *Popular*, designed by J.D. North, appeared at Hendon in the last week of January 1913. It was also offered as a two-seater with 50hp Gnôme, which was apparently not built. The single aircraft built was acquired by the War Office as No.283 in March 1913.

The wide spaced booms tapered to the tail in side view and, together with the very short span lower wing, gave the machine a gawky appearance, not enhanced by the wide track undercarriage and squarish nacelle. The top wing was fitted with ailerons and there was a full width elevator and divided rudder. The machine rested on its four wheels and rear portions of the main skids; there was no separate tail-skid.

Power: 35hp Anzani three-cylinder air-cooled Y-type radial driving a 7ft diameter GW propeller

Data		
	Span top	28ft.
	Span bottom	14ft
	Chord	5ft 3in
	Area	205 sq ft
	Area tailplane	20 sq ft
	Area elevator	14 sq ft
	Area rudder	10 sq ft
	Length	23ft 6in
	Speed range	4060 mph
	Price	£400

Power: 70hp Gnôme seven-cylinder air-cooled rotary driving a 8ft 6in diameter propeller.

Data		
	Span	38ft
	Chord	6ft 3in
	Gap	5ft 10in
	Area	475 sq ft
	Area tailplane	40 sq ft
	Area elevator	9 1/2 sq ft
	Area rudder	14 sq ft
	Length	26ft 10in
	Height	9ft 10in
	Speed	45 mph
	Endurance	4hr

GRAHAME-WHITE *Popular* passenger biplane Type VIIC

Derived from the Boxkite, rather than the Popular Type VII, this appears to be a variation evolved in the works under Mr. Bill Law, the superintendent, prior to the appointment of J.D. North as chief designer.

The obvious differences were the removal of the front booms and elevator, and the fitting of a projecting nacelle. The tail booms were apparently raked downwards. This machine was acquired by the War Office in March 1913 and identified as serial No.354 at Farnborough.

Right: GW Type VIII hydro-biplane as exhibited in seaplane form in March 1913.

GRAHAME-WHITE *Popular* biplane Type VII (1913 version)

A considerably different version of the Type VII appeared in the Grahame-White catalogue of 1913. This is the version that would have been built if orders had been received. The main difference from the version flown was in the tail structure. The tail booms tapered in plan to a point at the rear and were virtually parallel in side elevation. The nacelle was of an improved rounded shape and was lowered onto the bottom wing. The tailplane, with divided elevator, was fitted on the top longerons, together with a tall, one piece rudder and sprung tail-skid, which enabled the main skids to be reduced in length.

Power: Type A 35hp Anzani three-cylinder Y type radial driving a 7ft diameter GW propeller
Type B 50hp Gnôme seven-cylinder air-cooled rotary driving a 8ft diameter propeller.

Data		
	Span top	29ft 2in
	Span bottom	17ft
	Chord	5ft
	Gap	5ft
	Area	230 sq ft
	Area tailplane	28 sq ft
	Area elevator	20 sq ft
	Area rudder	12 sq ft
	Speed Type A	45-50mph
	Speed Type B	50-60mph
	Endurance	4hr
	Price Type A	£400
	Price Type B	£660

GRAHAME-WHITE hydro-biplane Type VIII

Intended for the private owner, the Type VIII twin float seaplane, another design by J.D. North, was shown at Olympia in February 1913.

The machine was a conventional tractor biplane of unequal span. The passenger was seated well forward, with a fuel tank between him and the pilot. There were three pairs of interplane struts, the inboard set, adjacent to the fuselage, in lieu of center section struts. The 10ft overhang of the top plane, which carried the ailerons, was braced to triangular kingposts

GW Type VIII with land chassis as acquired for the RFC in April 1913.

GW Type IX monoplane with monocoque fuselage and 35hp Anzani was under-powered. A 50hp Gnôme was fitted later.

and could be folded down for storage. The floats had a single step and a concave rear underside and were air ventilated. The Type VIII was also flown at Hendon as a land-plane with a cumbersome undercarriage, with four main, and two nose wheels, plus a tail-skid.

The machine was acquired by the War Office in April 1913 in land-plane form. It was possibly to be serial No.287, but was struck off charge on 26 November 1913.

Power: Type A 60hp Anzani six-cylinder air-cooled radial driving a 8ft 6in diameter propeller
Type B 70-80hp Gnôme seven-cylinder air-cooled rotary

Data		
	Span top	42ft 6in
	Span bottom	24ft *
	Chord	5ft
	Gap	5ft
	Floats	15ft. long, 1ft. 9in. beam, 1ft.3in. deep
	Area	380 sq ft *
	Length	25ft
	Height	12ft
	Weight	850 lb
	Weight allup	1,300 lb
	Speed Type A	45-55 mph
	Speed Type B	50-65 mph
	Endurance	3 1/24 hr
	Price Type A seaplane	£900
	Price Type B	£1,250

*Also quoted as 22ft 6in and 335 sq ft respectively
Above data from GW catalogue

GRAHAME-WHITE monoplane Type IX

W. Rowland Ding was the designer of this small single-seater with monocoque fuselage, which flew straights at Hendon for the first time on 7 September 1912 in the hands of Marcel Desoutter. The machine was a conventional braced monoplane with warping wings and twin skid undercarriage; a silencer was fitted to the exhaust system. The aircraft was under powered, and the 35hp engine was reported replaced by a 50hp

GW Charabanc fitted with Austro-Daimler engine.

Gnôme, but even this did not solve the problems of the Type IX, and there was no further news of this machine.

Power: Type A 35hp Anzani three-cylinder Y type air-cooled radial with silencer driving a 7ft diameter propeller
Type B 50hp Gnôme seven-cylinder air-cooled rotary driving a 7ft 10in propeller

Data

	Type A	Type B
Span	32ft	
Chord	6ft 6in	
Area	208 sq ft	
Area tailplane	18 sq ft	
Area elevators	15 sq ft	
Area rudder	11 sq ft	
Length	21ft	
Height	9ft	
Speed	50-55 mph	65 mph
Endurance	3 1/2hr	4 hr
Price	£550	£850

GRAHAME-WHITE *Charabanc* biplane Type X

This was a further J.D. North design for passenger carrying, seating four passengers in wicker seats behind the pilot; it was later reported flying with eleven persons aboard. In appearance it resembled a scaled-up Henry Farman biplane with two and a half bays, top wing extensions and with triple rudders.

The main fuel tank was under the seats supplying a gravity tank on one of the inboard interplane struts. An oil tank of similar size was fitted on the opposite strut. A horn and exhaust silencer were fitted.

The machine flew initially with a 90hp Austro-Daimler on 30 August 1913, piloted by Louis Noel with two passengers, the load was progressively increased when the more powerful engine, acquired from Cody's estate, was fitted in September. The machine was reported to show a great increase in speed with this. Early in November a 100hp Green was fitted to qualify the machine as 'all British', in order to enter the contest for the British Empire Michelin Trophy No.1. This duly happened, when on 5 November 1913 RH Carr flew the required 300 miles, flying continuously between Hendon and Brooklands, to win the prize.

A further use was made of the *Charabanc* to carry out the first parachute descent from an aeroplane in Britain on 9 May 1914. The aircraft still retained the Green engine, when exhibited, less wings, at Olympia in March 1914. Only one aircraft was built.

Power: 120hp (Initially 90hp) Austro-Daimler six-cylinder inline water-cooled
100hp Green six-cylinder inline water-cooled with 9ft 6in diameter Lang propeller

Data

Span top	62ft *
Span bottom	43ft 8in
Chord	7ft 6in
Gap	6ft 2in
Area	790 sq ft *
Area tailplane	128 sq ft
Area elevator	39 sq ft
Area rudders	36 sq ft
Length	37ft 6in *
Weight	2,000 lb
Weight five up	3,100 lb
Speed	30-55 mph

(Averaged 51mph in Michelin Contest)

**The Aeroplane* quoted 62ft 6in, 849 sq ft and 38ft 6in respectively

GW Charabanc of 1913-1914 was designed to carry four passengers but flew with eleven aboard. Fitted with a Green engine.

GW biplane '*Lizzie*'. First version of 1913, a Morane fuselage with wings of a Popular.

GW biplane '*Lizzie*'. Second version with two bay wings in 1914.

GRAHAME-WHITE biplane, nicknamed '*Lizzie*' or '*Tea Tray*'

Following demonstrations of looping by Pegoud at Hendon in September 1913, a strange single-seater biplane was constructed for racing and acrobatic work. This machine was made from the fuselage of a Morane-Saulnier monoplane and a set of biplane wings of a Popular. It flew for the first time on 22 November 1913. In this form it flew successfully until some time in 1914, when the lower wings were replaced with wings of near equal span to the top ones, and additional struts converted the machine into a two bay biplane.

Its demise is not recorded, but a photograph of the Blackburn Type L seaplane in the Olympia Works at Leeds shows '*Lizzie*' in the background.

Power: 50hp Gnôme-cylinder air-cooled seven rotary driving a 7ft 6in diameter propeller

Data		
	Span top	28ft 6in
	Span bottom	14ft
	Chord	5ft 6in
	Gap	6ft 3in
	Length	21ft 10in
	Area	220 sq ft
	Area elevators	15 sq ft
	Area rudder	10 sq ft
	Weight allup	850 lb
	Max speed	65 mph

GW Type XI Military biplane. The only one of the type of 1914.

GRAHAME-WHITE Military biplane Type XI

J.D. North designed a new aircraft, which was first seen at Olympia in March 1914, unflown. It was tested at Hendon on 9 May 1914 by Louis Noel, but was reported to be nose heavy, requiring a longer tail and larger tail surfaces. The location of the cockpits for carrying a gun also needed to be reversed. It is believed that the machine was soon abandoned.

The Type XI was a typical pusher biplane of the period, with two bay wings of near equal span. The tail booms tapered in plan and were readily detachable for transport. The rudder was balanced, and a small fin was mounted above the tailplane, which carried divided elevators. Ailerons were fitted to top and bottom wings and were connected by balance cables.

The engine bay was covered by metal cowling panels, cooling being effected by leaving the underside exposed. The engine itself was mounted low down and drove the propeller on a separate shaft mounted above.

Power: 100hp Gnôme Monosoupape nine-cylinder air-cooled rotary driving a four-bladed Integrale propeller through chain and sprockets at a reduction of 5:3 approx.

Data		
	Span top	37ft
	Span bottom	35ft
	Chord	5ft
	Gap	6ft
	Length	26ft 6in
	Area	358 sq ft *
	Area tailplane	26 sq ft
	Area rudder	15 sq ft
	Area elevators	23 sq ft
	Weight	1,000 lb
	Weight allup	1,550 lb
	Speed range	42-80 mph

**The Aeroplane* reported 358 sq ft

GRAHAME-WHITE 'Circuit of Britain' biplane Type XIII

The Type XIII was entry No.4 for the 1914 contest, canceled by the outbreak of war. It was being tested from Southampton Water late in July, when it capsized, giving J.D. North, its designer flying in the passenger seat, a ducking. The machine was reported to be ready, less engine, after repair in late Au-

GW Type XIII Circuit of Britain biplane in seaplane form.

gust, but was then fitted with a wheel undercarriage for testing. It failed to interest either the naval or military authorities as a reconnaissance aircraft.

The fuselage was said to resemble that of a Morane-Saulnier monoplane, owing to the horizontal knife-edge termination at the tail. The rather bulbous engine cowling was cutaway at the lower segment, and faired to a point at the sides of the flat sided fuselage.

The single bay wings were heavily staggered with N type interplane struts, an original feature at the time, and carried ailerons on the top wing only. No tailplane was fitted, only balanced divided elevators, with conventional fin and balanced rudder, the lower portion of which acted as a water rudder, and was cropped when a tail skid was fitted.

The twin wooden floats were short, with a single step and were mounted on a steel tubular chassis. The tail float was streamlined and was made of metal.

Power: 100hp Gnôme Monosoupape nine-cylinder air-cooled rotary

Data		
	Span	27ft 10in
	Chord	5ft 6in
	Gap	5ft 6in
	Floats length	14ft 10in
	Floats beam	1ft 9 1/2in.
	Floats depth	1ft 2in
	Area	290 sq ft *
	Weight	1,000 lb
	Weight allup	1,800 lb
	Max speed	85 mph
	Endurance	5 1/2hr

**The Aeroplane* reported 300 sq ft

GRAHAME-WHITE monoplane Type XIV (Morane-Saulnier Types G&H)

An example of the French Morane-Saulnier Type G arrived at Hendon in May 1913, and was sufficiently impressive for the manufacturing rights to be acquired by Grahame-White. One British-built machine was supplied to the RNAS, serial No.1242, in December 1914, and a batch of twelve to the RFC, serial Nos.587-598, in 1914.

GW Type XIV was the Morane-Saulnier Type H produced under license.

GW Type XIII was later tested with a wheel undercarriage but no production ensued.

The fuselage was parallel in plan, finishing at a horizontal knife-edge; the rather bulbous engine cowling was faired to the flat sides. The pilot and passenger were seated close together in a single open cockpit. The wheels were fitted on a sprung divided axle, the halves pivoting at a vee-shaped center pylon, to which the wing bracing wires were attached. A similar pylon provided a mounting for the warp actuating gear. The upper pylon for the warping pulley was hinged to the top longerons and was itself braced against the pull of the warp cable, by a cable passing through a pulley at the front, enabling it to be hinged down when the aircraft was dismantled.

Three of the single-seat version Type H were built before the war and were impressed into the RFC as Nos.623, 627 and 629.

Power: 80hp Gnôme seven-cylinder air-cooled rotary driving a 7ft 10in diameter Integrale propeller

Data		
	Span	30ft 6in
	Chord	6ft
	Length	21ft 6in
	Area	160 sq ft
	Area elevators	20 sq ft
	Area rudder	6 sq ft

GRAHAME-WHITE pusher type biplanes from 1912 leading to Type XV

In 1912-1913, a Boxkite with twin rudders, the *bi-rudder 'bus*, still with front elevator and exposed crew was used extensively at Hendon. On 27 November 1913, flown by Marcus D. Manton, it was used to demonstrate the use of a Lewis machine gun for firing at ground targets at Bisley. A similar machine to this, RNAS No.1600, was supplied to the Admiralty early in 1915. This became the prototype for further batches, supplied in wartime to the RNAS and later to the RFC, to a total of 135 aircraft.

The early deliveries were fitted with a front elevator, which was discarded at some stage, and a nacelle for pilot and passenger, with full dual controls, was fitted. The top wings could be fitted with extensions, and the ailerons were fitted with balance cables. A variety of engines was fitted, including the 60hp Le Rhone, the 70 and 80hp Gnôme air-cooled rotaries and the 60hp Green water-cooled type.

Despite the considerable differences between the early and later types, still being delivered in 1917, these aircraft were generally known as the Type XV.

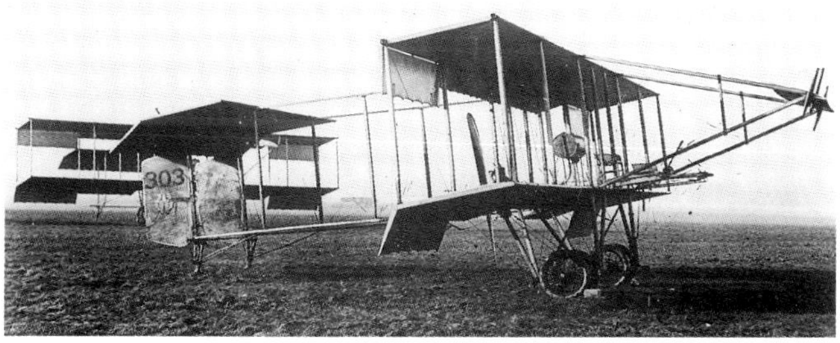

GW Type XV pusher biplane with front elevator.

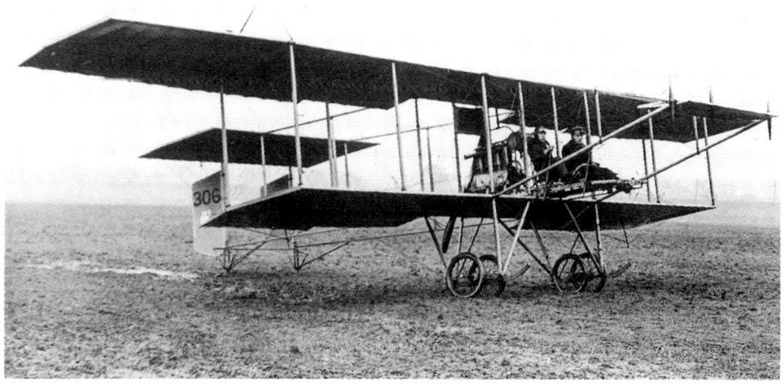

GW Type XV in the second form with dual control and wing extensions. This is a GW School machine with Green engine.

GW Type XV. The first of many supplied to the RNAS and RFC fitted with a nacelle.

GRATZE monoplane (Eugene V. Gratze Ltd., 446 Whitfield St., London W1)

This machine, originally known as the '*Dirigoplane*', was taken to the Blackpool Meeting in October 1909. It did not fly, as the wings had been lost in transit, and temporary wings made there were unsuitable.

The system of control was unusual and complicated, involving the use of three hand wheels and a pedal. One wheel operated a tail elevator and the incidence of the wings; the second operated the rudder portion of the cruciform tail together with horizontal movement of the propeller axis, which was mounted on a universally jointed shaft; the third wheel enabled the sweep angle of the wings to be changed, together with the upward alignment of the propeller. The foot pedal was to move the flexible wing tips upwards, in unison, with the object of increasing the dihedral angle to improve stability.

During the winter months, changes were made to the machine, which was at Canewdon, near Southend, in the middle of May 1910, where it was christened '*Daisy*' in the traditional

Gratze monoplane was at the Blackpool Meeting of October 1909 in this form.

Gratze *'Daisy'* monoplane in much modified form in 1910.

manner. The alterations included the addition of a shortspan plane, above the center section, the center portion of which served as an elevator. This operated in conjunction with flaps on the leading edges, near the wing tips, which could be raised vertically. A larger cruciform tail unit was fitted, and the rounded ends of the wings squared off. It is not clear how much of the original control system was retained, but presumably the flexible wing tips were no longer used. Although perhaps 'flown' in November 1909, no further reports of flight were published. Finally, in November 1911, the machine, which had cost the owner £800, was advertised for sale at £80.

Power: 40hp JAP eight-cylinder air-cooled vee.

Data		
	Span	40ft
	Chord	5ft 6in
	Length	28ft 6in

Gray monoplane of 1910-1911.

GRAY monoplane (William Edrington Gray and E.L. Gray, Lake District area)

The brothers were reported to have built a monoplane in 1910-1911 and a biplane with Humber engine in 1914-1915. There is some confusion, for W.E. Gray recorded in 1966, details of a monoplane built and flown in March 1917; presumably this should be 1911.

The layout was of a conventional tractor monoplane, which was mainly constructed of yellow pine, with corner blocks glued and wrapped with glued fabric. The elevator and rudder were made of steel tube.

The fuselage was fully covered with fabric, and terminated with a tail unit with small triangular fixed surfaces on either side, to which was hinged a large one piece elevator. The shield-shaped rudder pivoted on a post above the fuselage and was provided with a large balance area. Roll control was by ailerons. The undercarriage consisted of separate half-axles pivoted on inverted vees and a central skid.

There is no information on the biplane, which was probably not completed after the outbreak of war.

Power: 35hp Humber three-cylinder air-cooled fantype radial (referred to as from the Windermere monoplane) driving a Chauvière propeller

Data		
	Span	32ft
	Length	21ft 6in
	Area	189 sq ft
	Area tailplane	10 sq ft
	Area elevator	23 sq ft
	Weight	650 lb
	Weight allup	800 lb
	Speed	35-40 mph
	Materials cost	£ 50

Green biplane. A powered kite of Cody pattern built in 1908-1909.

GREEN biplane (R.W.L. Green)

Built a small, power driven Cody-like kite in 1908-1909.

GROHMANN monoplane (Alfred Grohmann, Laurel Bank Terrace, Feniscowles, near Blackburn, Lancashire)

Grohmann was a carpenter of German origin working in Lancashire, who built a man-powered machine mounted on a lady's bicycle in 1910. This was made to test the wings and

Grohmann monoplane built near Blackburn in 1910.

controls, before being rebuilt as a powered machine, of which no more was heard. A 4ft 6in diameter tractor propeller was driven through chain and bevel gearing.

Data	Span	24ft
	Area	36 sq ft
	Area of elevator	9 sq ft
	Weight	100 lb

GROOMBRIDGE & SOUTH multiplane (C. Groombridge and W.A. South)

This large aeroplane had two tiers of five wings fore and aft and six propellers carried three on either side of the connecting framework. A seventh screw was mounted on top of the whole structure.

Two 20hp engines were fitted and these drove the propellers via a central main shaft. The propellers were vanes which worked in the manner of oars with a driving stroke in the vertical position and a return stroke with the blades feathered.

The machine was fitted with four road wheels linked to the engine so that a speed of 20 mph could be attained before the propelling vanes were brought into use.

The photograph shows a full size mockup made of wood in 1903 and was rather heavy; the final version would have been built of lighter materials. Patent Nos.13241 1900 and 25723 of 1902 refer.

Data	Span	60ft
	Length	80ft

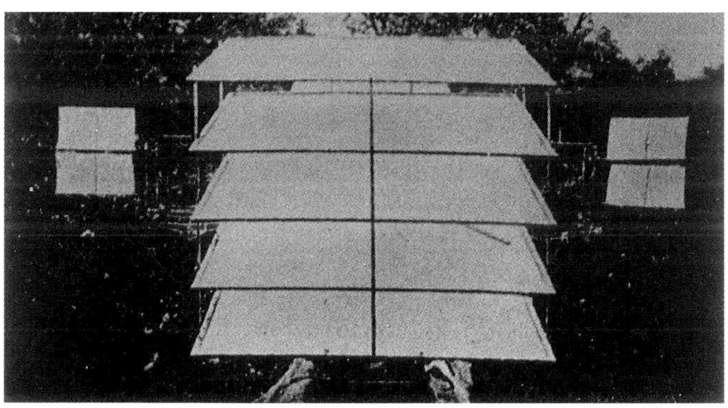

Groombridge & South multiplane. A fullsize mockup of a project of 1903

Grose and Feary monoplane, built by apprentices at Windham's works at Clapham Junction, was tested at Oakington, Cambridgeshire. in 1910.

GROSE and FEARY monoplane (A.M. Grose and N.A. Feary, Oakington, Cambridgeshire)

Construction of the tractor monoplane, designed by the partners, was carried out at the works of the Windham Detachable Motor Body Co., at Clapham Junction, London, SW11. It was ready for testing at Oakington in April 1910, but appears not to have flown.

The machine had a triangular section fuselage with sprung undercarriage, and was of conventional layout. The wings could be adjusted for incidence, which may have been related to the 'patented stability device'. The long chord, shortspan ailerons hung from the rear spars and were unbalanced.

Power: 25hp Advance four-cylinder air-cooled vee

Data	Span	26ft
	Length	25ft
	Area	160 sq ft

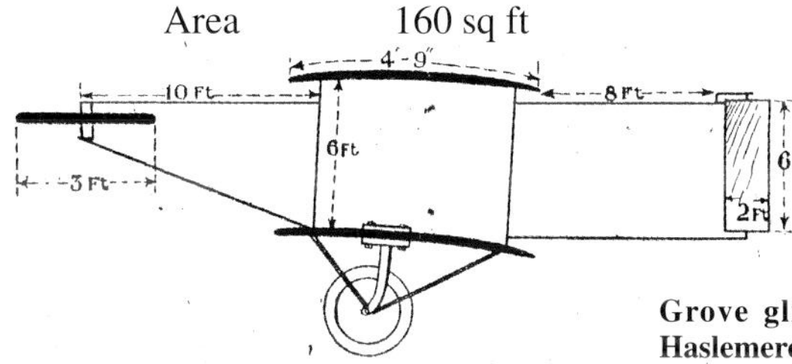

Grove glider built at Haslemere during 1909.

GROVE glider (Arthur T.M. Grove, Haslemere, Surrey)

This was a biplane glider with front elevator and rear rudder, and was mounted on a two wheel undercarriage. It was built in 1909, but was smashed before flight at Henley-on-Thames, by a gust of wind. A reference of July 1909 gave slightly different details to those below and stated that a 25-32hp Antoinette motor was to be fitted later with the engine and pilot side by side. The biplane was taken to Shellbeach for testing in September 1909. Grove stated in January 1910 that he had begun construction of a monoplane glider, of which nothing further was heard.

Data	Span	25ft
	Chord	4ft 9in*
	Gap	6ft *
	Length	22 ft
	Area	119 sq ft
	Area front elevator	18 sq ft
	Area rudder	12 sq ft *
	Weight	500 lb

*Also reported as 4ft 6in, 6ft 1in and 9 sq ft respectively

Guillon and Clouzy biplane, built in 1907 by two Frenchmen, was tested on Epsom Downs.

GUILLOU & CLOUZY biplane (Henri Guillou, or Guillon, de Puirajou, Epsom, Surrey)
Attempts to fly this machine were made by Frenchman, M. Guillou, on Epsom Downs on 11 April 1907, without success, although the claim was made that it was 'just lifting' at 20 mph, and more power was needed.

The machine had a tricycle frame, on which the pilot lay prone, and to which was attached the two bay wing structure, made of bamboo with calico covering. A triangular front elevator protruded from the top wing and there was a fixed tailplane.

A 20hp Buchet four-cylinder engine was mounted low in the frame, and drove a 5ft diameter, broad bladed propeller by chain. The 370 lb weight of the machine was deemed to be excessive.

H

HADDON gliders Nos.1 and 2
See Birmingham Aero Club.

HALL biplanes (Hall Aviation Co., London Aerodrome, Hendon)
J.L. Hall operated a flying school at Hendon from 1913, using a variety of machines, Avro, Blériot, Deperdussin and Caudron. In 1914 the company began the manufacture of Caudron machines, of both single and two-seater types, the latter with dual control. These were powered by 35hp and 45hp Anzani engines.

One machine used Caudron wings, but with a fuselage, instead of a nacelle and was powered by a 50hp Gnôme. It was commenced in 1914 and eventually flew in June 1915 and continued in use as No.6 of a large fleet of training machines of Caudron type, operated until 1918. It was assembled in Hall's shed, but was built by another company, possibly Ruffy-Baumann.

The Hall biplane of 1914-1915 used Caudron wings.

Hall-built Caudron G2 biplane at Hendon.

HAMBLE hydrobiplane HL.1 (Hamble River, Luke & Co., Hamble, Hampshire)
Constructed by an established boat building company, to the design of Frank Murphy, late of Bristol, the machine was shown incomplete at Olympia in March 1914. Changes were made later and the machine was tested by Gordon England, but costs were excessive, and the machine and the premises were put up for sale by July 1914.

The aircraft was a large two-seater pusher seaplane with twin wooden main floats and twin metal tail floats, moving with the rudders. There were four pairs of interplane struts each side, with radiators mounted on the innermost pairs. Ailerons

Hamble HL1 hydro-biplane at Olympia in March 1914 was abandoned when the company was liquidated later in the year.

were fitted to both top and bottom wings, the latter having pronounced dihedral.

The spruce tail booms, to be replaced by steel tubes later, extended aft from the second interplane struts. The tailplane and elevator were carried on the top booms, the pair of rudders pivoting on posts below. The pointed nose nacelle, was covered with two layers of cedar, laid up diagonally over the structure, and was mounted on short struts between the wings. The main floats were of similar construction, but rendered watertight, and were mounted on a steel tube chassis, intended to be sprung later.

Power: 150hp NAG (British-made) six-cylinder inline, water-cooled driving a Normale pusher propeller direct.

Data		
	Span top	60ft
	Span bottom	53ft
	Chord	6ft
	Gap	6ft
	Length	30ft
	Area	678 sq ft
	Area tailplane	39 sq ft
	Area elevator	33 sq ft
	Area rudders	32 sq ft
	Weight	1,300 lb
	Weight allup	2,550 lb
	Speed range	32-65mph
	Endurance	5hr

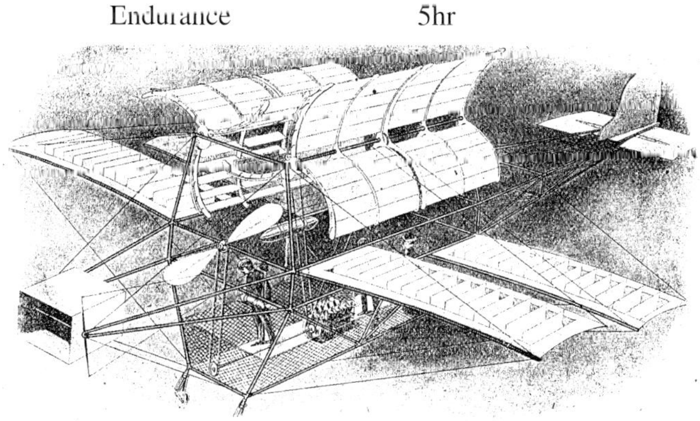

Hammant. This ambitious scheme for a direct-lift machine was tested in model form only after being patented in 1908.

HAMMANT direct lift machine (William Hammant, London)

A design with two long revolving paddles was patented by Hammant, and was illustrated in *The Aero* of 29 March 1910. Slots in the paddles were opened and closed by cams, as they revolved. Experiments and model tests carried out by Braun and Co., Pentonville Road, North London were reported to be very satisfactory.

HAMMOND biplane (E.V. Hammond, 88 Louisville Rd., Balham, London SW)

Hammond carried out experiments with helicopter rigs in 1906-1908, but came to the conclusion that the power required for vertical lift was not available at that time. He then turned to the construction of a pusher biplane at Brooklands, which made its appearance at some time in 1909. He later claimed to be the sole experimenter at Brooklands throughout the spring, summer and autumn of 1909.

The rather crude looking single-seater, pusher biplane was not reported upon in the Press at the time, but a surviving photograph shows the general configuration. Unusual features were the large interplane ailerons and the double boxkite-like rudder, which seemed to be carried on swinging tail booms.

The machine was presumably abandoned after the trials, which were of limited success, although Hammond was among the first dozen British pilots to get off the ground in a British-built machine.

Hammond was at Brooklands in the Spring of 1909 with this crude biplane.

HAMMOND triplane

The successor to the biplane appeared at Brooklands for engine testing at the beginning of January 1911. Its general construction was reported to be flimsy and the commentators of the time were doubtful of its potential. On the 14 January 1911 it was taxied at speeds up to 10 mph by Howard Pixton but, although further brief appearances were made, it was presumably abandoned soon after. Hammond claimed that it made several short flights with a 14hp engine but that it was underpowered.

The layout was similar to other types, usually biplanes, of the period but, unusually, it was a twin tractor with belt drive to the propellers, and the pilot behind the engine. Lateral con-

Hammond triplane. This was Hammond's second attempt at flight at Brooklands in 1911.

trol was achieved by means of unbalanced ailerons on the top plane. The machine was mounted on a four wheeled chassis, with skids under the wing tips to aid balance on the ground. No further details of the machine are recorded.

HAMMOND monoplane

A single-seater tractor monoplane was built by Hammond at Brooklands in Shed No.2 during the summer of 1913. It was a simple design with two parallel steel tubes forming the fuselage. The engine was a 30hp Advance four-cylinder water-cooled vee type, already an old type, probably used in his earlier machines. The wing was based on a single spar made of 2 1/2in diameter by 22swg steel tube.

Hammond was constructing this monoplane at Brooklands in 1913.

Hampshire Aero Club glider resulted from work carried out at the United Services College at Windsor sponsored by P.Y. Alexander.

HAMPSHIRE AERO CLUB glider

Patrick Y. Alexander, who had devoted his time and fortune to the advancement of aeronautics since about 1890, was the president of this club. Among his previous work was support for the United Services College (later the Imperial Service College) at Windsor where, in 1909-1910, he encouraged the making of models by the students, leading to a full size glider. It was this machine that he transferred to Gosport for the use of the Hampshire club.

The glider consisted of a braced biplane structure with four pairs of interplane struts each side and no other stabilizing or control surfaces. The operator stood on the front spar of the lower wing and was towed into a suitable breeze by the helpers on the ground. Some degree of control may have been possible by weight shift by the operator.

The glider was first flown at Fort Grange, Gosport in April 1910 and also later that year.

HANDLEY PAGE glider (Frederick Handley Page, Creekmouth, Barking, Essex)

Handley Page, who later became famous as the major manufacturer of large aircraft, experimented with flapping wing models, but became actively interested in fixed wings in 1906-1907. He became acquainted with José Weiss in 1908, through the Aeronautical Society, thereafter making use of the Weiss

Handley Page glider. The maker famous for large aircraft had small beginnings in 1909.

patented wing form, with curved leading edge and flexible tips, in a number of his own designs.

The first of these was a glider, with both main wing and foreplane of Weiss shape, mounted on a tricycle chassis, which carried the seated pilot. The attempts to fly from slopes at Barking Creek in 1909 were unsuccessful. The glider was reported sold to a Mr. Rodez.

HP Type A (HP1) The '*Bluebird*' of 1910 with Weiss-type wing.

HANDLEY PAGE monoplane Type A. HP.1. (Handley Page Ltd., Barking)

A small tractor monoplane, named *'Bluebird'*, with Weiss type wing, was shown incomplete at Olympia in March 1910, priced at £375. The machine left the ground after the undercarriage had been moved forward, and a small tailplane was added. The flight took place on 26 May 1910 but the aircraft was damaged by side-slipping due to lack of lateral control.

Power: 20-25hp Advance four-cylinder, air-cooled vee driving a 6ft 6in HP propeller

Data		
	Span	32ft 6in
	Chord	6ft
	Length	20ft 6in
	Area	150 sq ft
	Weight	300 lb
	Weight allup	450 lb
	Speed	35 mph
	Price	£375

HP Type B (HP2) built for Planes Ltd. of Freshfield.

HANDLEY PAGE biplane Type B. HP.2

This machine was built at Barking in 1910 for WP Thompson of Planes Ltd., Freshfield, but sustained damage on trial and was further wrecked, when the shed, in which it was housed, collapsed in a gale. It was subsequently rebuilt as the Planes biplane (*q.v.*).

HP Type C (HP3) was a reconstruction of the Type A with Alvaston engine.

HANDLEY PAGE monoplane Type C. HP.3

This was a reconstruction of the Type A, carried out in 1910. Wing warping was introduced together with a large rudder which was still of cruciform type on a universal mounting. A twin strut pylon replaced the single kingpost of the original. A 25hp Alvaston water-cooled engine, with a radiator mounted edge on and to one side, was fitted, but this was of insufficient power and was replaced by a 50hp Isaacson radial.

The Type C was abandoned in favor of the Type D and was transferred to the Northampton Polytechnic Institute as an instructional airframe.

HP Type C (HP3) with Isaacson radial engine.

Power:	25hp Alvaston two-cylinder horizontally opposed, water-cooled	
	50hp Isaacson seven-cylinder air-cooled radial	
Data	Span	30ft
	Length	21ft
	Area	150 sq ft
	Weight	250 lb
	Weight allup	450 lb
	Speed	35mph

HANDLEY PAGE monoplane Type D. HP.4

The Type D single-seater was shown at Olympia in March 1911, priced at £450, but remained unsold. It was fitted with a 35hp Green engine, with radiators each side of the fuselage. The engine was on loan and was to be replaced by the Isaacson radial from the Type C, after the lack of success at the show.

The machine at Olympia had a monocoque fuselage, long skids, warping wings, a rudder with fixed fin below the large elevator and a small fixed tailplane. When it came to fit the radial engine, it was found impractical to mount it on the monocoque fuselage, which was replaced on a second machine with a new girder type, fabric covered fuselage.

The aircraft was due to fly in the Circuit of Britain Contest to be held on 22 July 1911, with R.C. Fenwick, who also contributed to the design, as pilot. Unfortunately he crashed the machine on its first flight on 15 July 1911 and was immediately dismissed by Handley Page As a result of this, the second machine was not ready in time to compete in the Contest, being flown later by Edward Petre at Fairlop in Essex. The color of its new doping scheme caused it to receive the nicknames *'Yellow Peril'* and *'Antiseptic'*.

Power:	35hp Green four-cylinder inline water-cooled	
	50hp Isaacson five-cylinder air-cooled radial	
Data	Span	32ft
	Chord	6ft
	Length	22ft
	Area	156 sq ft
	Weight	420 lb

HP Type E (HP5) in original 1912 form as a two-seater.

HANDLEY PAGE monoplane Type E. HP.5. Also identified as E/50. (Handley Page Ltd. Barking & Cricklewood)

The Type E was flown on 26 April 1912 at Fairlop, by Edward Petre, for the first time and was the first successful Handley Page aircraft, although it was damaged when landing on the. rough ground when it returned to the factory at Barking in June. On the 27 July 1912 Petre flew the 55 miles from Fairlop to Brooklands along the Thames, the first such flight across London. The aircraft remained at Brooklands during August, while the factory was moved from Barking to Cricklewood, and then moved to Hendon. Modifications were made, and a further flight took place on 1 February 1913 by Sydney Pickles, the machine then going on exhibition at Olympia from 14 February 1913.

HP Type D (HP4) was at Olympia in March 1911 with Green engine.

HP Type E (HP5) in 1913 with modified fin and ailerons instead of warping.

The Type E retained the Weiss type crescent wing, braced to fuselage-mounted pylons and kingposts towards the tips, which were flexible for warping, although ailerons were fitted later. The fuselage was a shallow, braced girder, tapering to a vertical post at the rear, with a deep fairing below and decking above, built up with formers and stringers and was fabric covered. The tail unit consisted of a long tapered tailplane with semicircular elevators, a fin of low, parallel shape and a tall curved rudder. The central skid, split axle undercarriage embodied spring loaded struts with rubber cord shock absorbers; the tail was supported by a tall springy skid below the rudder post.

The pilot flew the aircraft from the front cockpit and was protected by a cowl covering the upper half of the engine. The aircraft was well constructed and smartly finished, with a blue fuselage and yellow wings and tail, and again became known as the *'Yellow Peril'*.

During repairs in early 1913, the fin was changed to a more elegant one of triangular shape. In 1914, before the aircraft was sold to the Beatty School of Flying, it was converted to a single-seater, with twin skid, cross axle undercarriage. However the machine was requisitioned by the War Office, who retained the engine, returning the airframe to Handley Page, where it remained until disposed of in 1940.

Power: 50hp Gnôme seven-cylinder air-cooled rotary driving a 8ft 3in diameter HP propeller

Data		
	Span	42ft 6in
	Chord	7ft 8in max
	Length	27ft 6in (27ft single-seater)
	Area	220 sq ft
	Area tailplane	32 sq ft
	Area elevators	18 sq ft
	Area rudder	12 sq ft
	Weight	800 lb
	Weight allup	1,300 lb
	Speed	60mph
	Endurance	3hr
	Price	£850

HP Type E (HP5) in 1914 as a single-seater school machine at Hendon.

HP Type F (HP6) built for the Military Trials of 1912 but withdrawn with engine trouble.

HANDLEY PAGE monoplane Type F. HP.6. Also identified as F/70

Alongside the Type E, a second monoplane was designed for the Military Trials, for which the previous machine was unsuitable, and was entered as No.28. Handley Page outlined the design to meet the specification, and Henry Petre prepared the drawings and was nominated as pilot. A crescent wing was to be used, which Henry Petre distrusted from experience of flying the Type E, which was unstable laterally in level flight, and he produced drawings for an alternative pair of straight wings, which were made without authority. These were fitted for the first flight at Barking, where the machine was built, but were removed when the machine was dismantled for dispatch to Larkhill, without being seen by Handley Page. Henry Petre refused to fly the machine with crescent wings and it was taken over by his brother Edward who flew it for the first time on 21 August 1912 in this form. Its performance at Larkhill was dogged by engine trouble, which eventually caused a forced landing and damage. The machine was withdrawn and taken to Hendon for repair and flew again on 9 November 1912, this time with Wilfred Parke as pilot. He was flying from Hendon to Oxford on 15 December 1912 with HP's manager, A. Arkell Hardwick as passenger, when the machine crashed at Wembley and both men were killed. It was not rebuilt.

The deep crescent wing, similar to the that of the previous Type E with warping control, was retained after the first flight, but was braced to a single point on a pylon above the fuselage. The pylon consisted of two hinged vee-struts which could be folded down for transport.

The fuselage was deep and flat sided, with curved top and bottom fairings, and seated two side by side. The observer, on the port side, was provided with a transparent panel also serving as an entry hatch. A center skid undercarriage, with divided axle and spring struts, was intended to support the tail also but was supplemented with a tail support of rattan cane There was no fin, but a large semicircular tailplane, divided elevator and balanced rudder with scalloped edges constituted the tail unit.

The engine was enclosed in a circular front cowl with center air intake entry and slotted exits for the cooling air below the fuselage. The machine was of solid construction, well finished and bore the company name, under the wings, and initials on the fuselage.

Power: 70hp Gnôme seven-cylinder air-cooled rotary

Data		
	Span	43ft 6in
	Length	30ft 2in
	Height	10ft 6in
	Area	250 sq ft
	Area tailplane	40 sq ft (including elevators)
	Area rudder.	14 sq ft
	Weight	850 lb
	Weight allup	1,450 lb
	Speed	55-58 mph
	Price	£1,050

HP Type G (HP7) as No.892 in RNAS service 1914-1915.

HANDLEY PAGE biplane Type G. HP.7. Also identified as G/100

The second biplane made by Handley Page was also of his own design, and was inspired by the monoplane ban, and a desire to improve on the BE.2, a small contract for which was in hand at Cricklewood, where the Type G was also constructed. Piloted by E.R. Whitehouse, it flew for the first time on 6 November 1913 at Hendon and, after company trials, was purchased by the new owner of the Lakes Flying Co., Lindsay Bainbridge, to be flown by Rowland Ding, also a director of that company.

It was intended to convert the machine later to a twinfloat seaplane, but it was flown on tour as a land-plane after being taken over by Ding in May, until a mishap on 27 July 1914, when it was damaged and returned to the factory by rail. The repair was completed after the declaration of war. The machine was then taken over by the RNAS and continued to fly at Hendon and Chingford as No.892, until it was crashed in August 1914, the engine being bequeathed to Ding by Bainbridge, who was now deceased.

The Type G had a fuselage and tail unit similar to those of the Type E monoplane, with the pilot in the rear seat and room for one or two passengers in front. The engine was covered by a cowl at the top and had a continuous exhaust ring in front, exhausting through outlets below the bottom wing, to which a silencer was fitted later.

The top wing derived from the Type E, and the lower wing was a smaller version of that of the Type F. The wings had three bays, the outer bay having a single strut only at the rear spar, with short struts connected to both top and bottom leading edges.

The fuselage was mounted above the lower wing, to which it was attached by pairs of struts on either side. With the first type of undercarriage, which was of twin skid type, long struts were connected to the lower longerons just behind the engine. After the crash on 2 May 1914 a modified undercarriage with vee struts connected to the lower longerons was fitted, together with a smaller fin and the silencer. During its time with the RNAS the fuselage under-fairing was extended aft.

HP Type G (HP7) biplane of 1913 in original form with twin-skid undercarriage and single long cockpit.

Power: 100hp Anzani ten-cylinder air-cooled radial driving a 8ft 6in diameter Integral propeller. The 100hp Green originally planned was not fitted.

Data		
	Span	40ft
	Chord	6ft 6in max
	Length	27ft later 25ft 1in
	Gap	6ft 6in
	Area	384 sq ft
	Area tailplane	32 sq ft
	Area elevators	19 sq ft
	Area rudder	12 1/2 sq ft
	Weight	1,150 lb
	Weight allup	1,775 lb
	Speed range	35-73 mph
	Climb to 3,000ft	10 1/2min
	Endurance	4hr
	Price	£1,500

HP Type G (HP7) with modified undercarriage and two cockpits.

HANDLEY PAGE monoplane Type H/70 & H/110

The first of these projects by H.A. Petre was to be based on the Type F and fitted with a 70hp Gnôme. The second, an improved type with 110hp Anzani and tandem seats in an elongated cockpit, was drawn by G.R. Volkert in early 1913.

HANDLEY PAGE biplane Type K/35.

A proposal for a single-seat, scaled down version of the Type G for training purposes was drawn by Volkert in December 1913; intended for use by the Beatty School at Hendon, it was not built.

Power: 35hp Anzani Y-type three-cylinder radial

Data	Span	30ft
	Area	225 sq ft

An alternative Type K was described in the Handley Page Catalogue of 1913. This specified a two-seater with 80hp Gnôme. This type apparently remained on paper only.

Data	Span	42ft
	Area	250 sq ft
	Weight	900 lb
	Speed	65mph
	Endurance	4 1/2hr
	Price	£1,100

HANDLEY PAGE biplane Type L/200. HP.8

This large single-engined, tractor biplane was designed by Volkert in December 1913 to compete for the *Daily Mail* Prize of £10,000 for an Atlantic crossing. It was ordered by Princess Ludwig of Lowenstein-Wertheim, who planned to accompany Rowland Ding on the flight, and it was nearing completion in the summer of 1914. The seats were side-by-side with dual controls. The machine was apparently a scaled-up version of the Type K/35 and was abandoned, although offered to the RNAS with straight wings.

Power: 200hp Salmson (Canton-Unné) fourteen-cylinder water-cooled radial.

Data	Span	60ft
	Area	900 sq ft
	Length	41ft
	Weight	2,800 lb
	Weight allup	6,000 lb
	Speed range	43-80mph
	Endurance	23hr

HANDLEY PAGE biplanes Types M/200 & MS/200 HP.9

In August 1914 when the 200hp engine for the L/200 was taken over by the Admiralty, a version was offered with two 100hp engines in the wings and then the M/200 with engines, mounted nose to nose in the fuselage, driving tractor propellers on the wings by chain. Both versions were rejected, as was the seaplane version MS/200, with twin main and tail floats. A crew of four was to be carried.

Power: Two 95hp Salmson (Canton-Unné) seven-cylinder water-cooled radials.

Data	M/200 land-plane.	
	Span	70ft
	Area	950 sq ft
	Length	37ft 6in
	Weight	3,000 lb
	Weight allup	5,000 lb
	Speed	75mph

HANDLEY PAGE biplane Type N/80. HP.10

A proposed scout version of the Type K/35, with a parallel center section inserted, and two seats side-by-side, the design of which was discontinued in January 1915.

Power: 80hp Gnôme seven-cylinder air-cooled rotary.

Data	Span.	32ft 6in
	Length	24ft

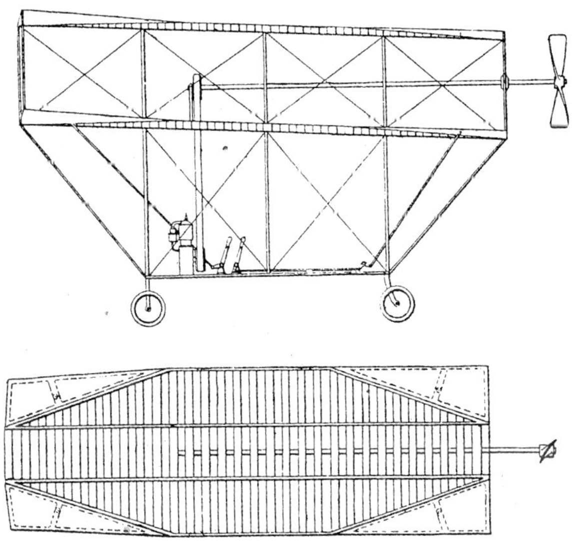

Hare narrow biplane. Sketch of the patented machine of 1908.

HARE narrow biplane (EH Hare)

The subject of patent No.19805/1908, the machine was intended to fly in the direction of its larger axis. The wings incorporated equally spaced slats and triangular flaps at each corner to effect control. A pusher propeller on a long shaft was belt driven from a motor mounted well below the wing structure. It is assumed that the machine was not constructed.

HARPER monoplane (A.M. Harper, 2 Edbrooke Rd., St. Peter's Park, London W9)

This two-seater side-by-side tractor monoplane, designed by Harper, was being built in 1912 by Weston-Hurlin & Co. of Paddington, and was entered as No.25 in the Military Trials, held at Larkhill in August. It did not take part and was probably not completed.

Power: 60hp Green four-cylinder inline water-cooled.

Data	Span	35ft
	Length	27ft
	Speed	60mph

HARRIS biplane and HARRIS & FAGAN glider (L.H. Harris, Clapham, London)

There is no detailed information on these projects, reported in the Clapham Observer of 12 and 19 January 1912 and 29 March 1912

HARRISON glider (J. Harrison, Peverell, Plymouth, Devon)

It was reported that this glider was flown at the Old Martello Tower, Marsh Mills, Plymouth in 1910.

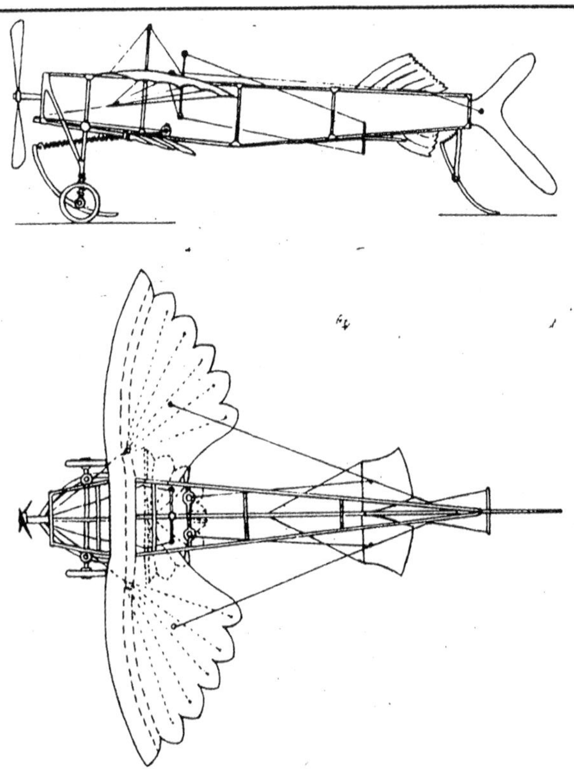

Side and plan views of the Hart Aeroplane

Hart patented wing was tested on a Blériot monoplane at Hendon in 1912.

HART monoplane (William Henry Hart, Thornton Heath, Surrey)

Patent No.1372/1911 was taken out by Hart in conjunction with Carl von Buchs for 'Improvements in Aeroplanes'. A company named Hart Aeroplanes & Waterplanes Ltd. was formed on 12 November 1912 and this was joined briefly by Howard T. Wright, before he moved to J. Samuel White & Co. Trials of a set of wings fitted to a 35hp Blériot at Hendon, under the supervision of Grahame-White Aviation Co., were satisfactory, but not of interest to the industry. The Hart Co. was dissolved on 21 June 1916 having achieved nothing.

The invention consisted of a form of aeroplane wing construction, equivalent to that of a flying fish, with an arrangement of closing, balancing fans to which the balancing ends of the wings were connected, and a special forked rudder with fins above and below the fuselage to maintain the balance of the machine. The patent drawings convey the general arrangement.

HARTILL monoplane (A.E. Hartill, Cleveland St., Wolverhampton)

This machine was of *Demoiselle* type, built by Hartill for Dr. Hands. It was at Dunstall Park in 1910, but whether it flew is uncertain.

Power: 20hp Alvaston two-cylinder horizontally opposed, water-cooled.

Data	Span	26ft
	Length	24ft
	Weight	450 lb

Hartill monoplane was of *Demoiselle* type and was at the Dunstall Park Meeting Wolverhampton in 1910.

HAWKINS-OGILVIE triplane (Capt. A.C. Hawkins, Bertram Ogilvie & C.I. Nairn of Napier, New Zealand based at Winchester, Hampshire and later at Brooklands)

The partners built three machines between 1907 and 1910 in New Zealand. A pusher biplane powered by a 10hp Rex vee-twin motorcycle engine was mounted on a ramp for takeoff, but failed to sustain itself in flight. The 1910 machine, which had large ailerons below the bottom wing, was seen by Lord Kitchener on 11 March 1910, during a visit to New Zealand and he encouraged the experimenters to transfer to England for the greater opportunities, and to compete for the Baron de Forest Prize.

A model of a triplane was made during the voyage and the full size machine was constructed by Handley Page, but after the end of 1910 and thus not in time to compete for the Prize. It was ready for testing near Winchester in early March 1911, when it was reported 'rolling'. Ogilvie was the pilot with Hawkins providing the financial support.

There was a modest degree of success during the trials at Winchester and later at Brooklands before the finance was exhausted and the group returned to New Zealand. The machine was by this time of dated design.

The aircraft was an inelegant tractor triplane mounted on a high chassis with long skids and a tall tail-skid. The two bay wings were staggered and joined by long pairs of interplane

Hawkins-Ogilvie triplane built by Handley Page for partners from New Zealand was tested near Winchester and at Brooklands in 1911.

struts, the forward struts being positioned at the leading edge, and the outers at the extreme tips of the wings. The tips of the wings were squarecut, and were fitted with duplicated, curved cane skids and differential ailerons fitted to all three pairs. The ailerons were separate rectangular flaps, not inset into the main wings, but were hinged at the trailing edges.

The fuselage girder was fabric covered from the pilot's cockpit aft to the tall sternpost, which supported the tail-skid below and carried a rhomboidal-shaped rudder above. There was no fin, only a strut to support the rudder post. The tailplane was rectangular, mounted on the top longerons and carried a one piece rhomboidal-shaped elevator, hinged along its narrow edge. Longitudinal stability was said to be automatically regulated by the action of the wind on the planes, the mechanism for this being covered by patent No.8212/1910.

Power: 50hp Alvaston four-cylinder horizontally opposed, water-cooled.

Data Span 24ft

HAYNES monoplane (Haynes, Thatcham, Berkshire)

No information on this machine reported in 1911.

HELICOPLANE direct lift machine (Helicoplane Flying Machines Ltd., Wembley. C. Lorenzen, Chief Engineer)

Lorenzen patented his scheme for a helicopter, No.13809/1908, and later formed a company to develop his ideas for variable pitch lifting screws. Ground tests of these, driven by an electric motor were carried out, and tested by the independent authority at Faraday House.

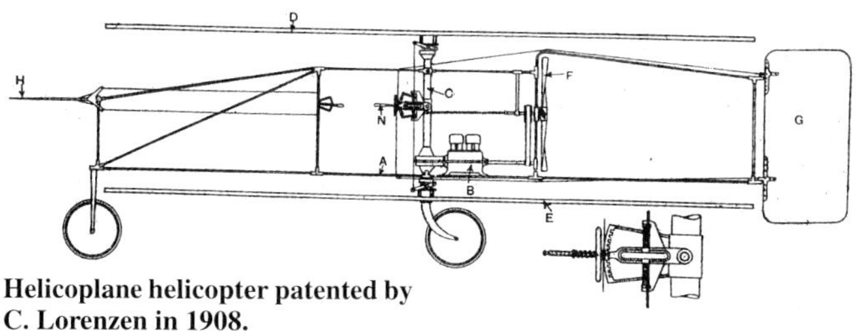

Helicoplane helicopter patented by C. Lorenzen in 1908.

The Helicoplane was not built, but the company went on to produce propellers, including large two, three and four blade propellers up to 18ft diameter, some being supplied to the German government in 1911 for airship use.

HENSHAW cycloplane (Jack Henshaw)

This was reported to have been tried downhill by the young constructor, but required larger wings to enable it to take off. Nothing further reported.

Henshaw cycloplane of 1910 had insufficient wing area to achieve gliding flight.

HENSON monoplane (William Samuel Henson, Chard, Somerset)

W.S. Henson was an inventor who began experimenting in 1840 with model gliders, and in 1842, his patent No.9478 described a monoplane of advanced conception, to be powered by a steam engine, from which 25-30hp was expected. Earlier, he had become acquainted with John Stringfellow, a fellow inventor and designer of machinery, including light steam engines for driving lace making machines. It is believed that Stringfellow contributed his knowledge to further the design of the aeroplane.

From 1843 Henson and Stringfellow experimented with various steam powered models, with wing span up to 20ft, near Chard. In association with others, a company known as 'The Aerial Steam Transit Co.' was formed in 1844. There was considerable publicity, but the project was premature and collapsed. Henson finally withdrew from aeronautical work and emigrated to America in 1848, but Stringfellow worked on after 1848 achieving short flights with small models. His interest revived in 1866, and he won a prize of £100 at the Crystal Palace for a successful flight. Stringfellow died in 1883 and certain of his models are exhibited at the Science Museum. Henson died in 1888 at Newark, New Jersey, U.S.A.

The proposed Henson aeroplane was to be a large monoplane, with passengers housed in a cabin below the wing, and was mounted on a tricycle undercarriage. The machine was to be launched down an incline and then propelled by two six

Henson monoplane. The fullsize version was part of an ambitious scheme in the middle nineteenth century and was to be steam powered.

bladed propellers driven by belts from the engine, which with its large boilers was positioned above the cabin.

Power: 25-30hp Stringfellow two-cylinder steam engine.

Data	Span	150ft
	Chord	30ft
	Wing area	4,500 sq ft
	Tail area	1,500 sq ft

HEWITT ornithopter (S.R. Hewitt, 39 Lonsdale St., Belfast)

This machine was constructed in 1908, and had two planes of 32ft span and 7ft chord, of which 2ft was flexible trailing edge. The machine was mounted on a tricycle undercarriage and had a front elevator and biplane tail.

Hewitt ornithopter of 1908 built at Belfast.

Hewitt glider built in 1909 in North Wales was replaced by a Bleriót in 1910.

HEWITT glider (V.V.D. Hewitt, Trefnant, N. Wales)

Hewitt's glider was a biplane with single surfaced wings, the fabric covering being on the underside. The published illustration showed a biplane front elevator; there was a tailplane, not illustrated, and midgap ailerons. The fitting of an engine did not proceed. He took delivery of a Blériot monoplane in early 1910 and flew to Ireland in it, in April 1912, but gave up aviation in 1914.

Data	Span	30ft
	Chord	6ft

HEWLETT & BLONDEAU aircraft (Mrs. Hilda B. Hewlett and Gustave Blondeau)

After running a flying school at Brooklands, the partners began to manufacture aircraft and components, moving for the purpose in 1912 into the former Mulliners Coachworks factory at 216 Vardens Rd., Clapham Junction, SW, which became known as the Omnia Works.

A variety of aircraft types, to other people's designs, was built, including, in 1912, at least three Hanriot single-seat monoplanes, fitted with various engines, for the French company. A number of Caudron biplanes, of Types C and G for W.H. Ewen Ltd., were also built in 1913. In early 1913 the firm completed a private order for the Dyott monoplane (*q.v.*) and later received a British government order for two BE 2a's Nos.49-50, which served with the Navy. This was followed by an order for twelve

Hewlett & Blondeau Farman-type biplane for the Pashley Brothers.

Caudron G2 biplane built by Hewlett & Blondeau.

Hanriot monoplane built by Hewlett and Blondeau.

BE 2c's Nos.976-987 and a further six Nos.1189-1194, also for the Navy, while the firm was at the Clapham Junction works.

The company had built the Farman biplanes used by its own school at Brooklands, and also for other users. In 1912 a batch of three Farman biplanes was made for Vickers, for use at their flying school at Brooklands, the first of which was in use by August 1912. These were later referred to as Vickers Boxkites and numbered 19-21 in the Vickers fleet, the first two had 50hp Gnôme engines and No.21 a 70hp Gnôme. Two of these machines were taken over by the RFC after the outbreak of war becoming serial Nos.639 and 642. The dimensions of the latter machines was as follows:

Span	51ft 3in (without extensions 35ft)
Length	39ft.
Wing area	433 sq ft (without extensions 394 sq ft)
Gross Wt	835 lb

A Farman type biplane with a nacelle with two seats was built for the Pashley Brothers and used extensively at their Shoreham Flying School.

Hewlett & Blondeau Farman-type biplane for Vickers Flying School.

Hewlett biplane glider possibly with Francis Hewlett at the controls.

HEWLETT glider biplane

A photograph taken in about 1912 possibly shows Francis Hewlett, son of Hilda Hewlett, at the controls of a small biplane glider. Although quite well designed, it may have been built as a 'toy' at the Hewlett & Blondeau works. The photo may have been taken in the grounds of a church in Northwick Terrace, Maida Vale, London NW, close to the Hewlett home.

HIGGINS glider (Henry Higgins, Cheltenham, Gloucestershire)

This glider was designed and built in 1910, and was tested successfully at Cleeve Hill, near Cheltenham. It was a canard monoplane with the pilot seated below the wing, above the axle of the wheels. Control was provided by a front elevator, and by warping the wing, with a foot bar. A movable 'blinker' was fitted in the front bay of the single girder structure.

Data		
	Span	26ft
	Chord	6ft
	Length	20ft
	Area	160 sq ft
	Area elevator	16 sq ft
	Weight	94 lb

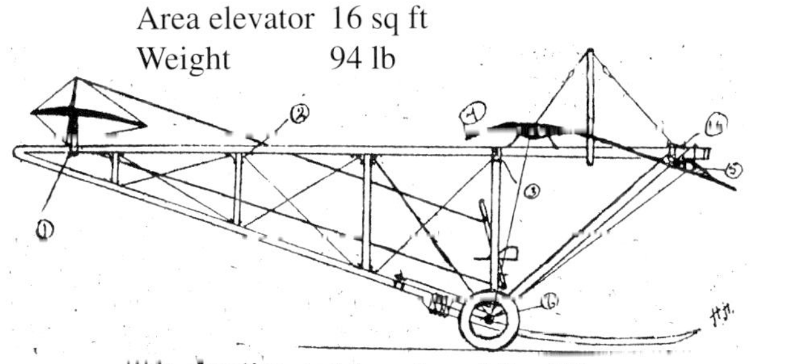

Side elevation and key view of the glider.

Higgins glider of 1910 on Cleve Hill, Cheltenham was fitted with an engine later without success.

HIGGINS monoplane

The glider had been built with the intention of fitting a low powered engine later, and one was installed as a pusher in 1911. A two-cylinder JAP of 10-12hp was fitted below the wing, driving a 7ft 2in diameter propeller by chain. A tail was fitted 12ft behind the cg; a fixed plane below the elevator increased the lifting surface to 196 sq ft and the weight to 280 lb. In this form the machine was unsuccessful.

In 1912 further changes increased the span to 30ft, the lifting area to 240 sq ft and weight to 300 lb. There were no further reports and it is presumed that the machine was not a success.

Hill monoplane built by a firm in Bury, Lancashire. was based on the Blériot XI.

HILL monoplane (Hill & Co., Bury, Lancashire)

This company built a single-seater monoplane, similar in layout to a Blériot XI, but with a modified undercarriage and three large bicycle type wheels. There is no evidence that the machine actually flew.

HILL glider (Geoffrey T.R. Hill and Roderic M. Hill)

This biplane glider, with boat-like hull, was built by the Hill brothers in 1913-1914, and was flown off the slope of Firle Beacon in the South Downs. It was intended to fit an engine and wheels, but this was prevented by the outbreak of war. Geoffrey was to become famous as the designer of Pterodactyl aircraft, built by the Westland company; Roderic became an Air Marshal.

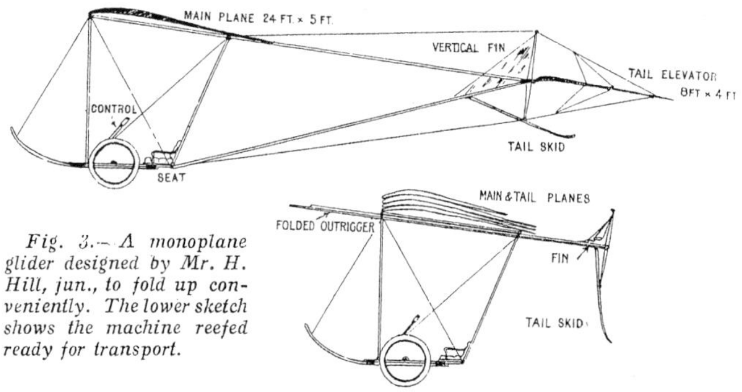

Hill folding glider designed by a member of the Paddington Aero Club in 1912.

HILL glider (H. Hill, Paddington Aero Club)

This was a monoplane glider, of the *Demoiselle* type, designed to fold easily for transport. It was proposed by a club member in 1912 and may have been evolved from the Hill glider mentioned below.

Data		
	Span	24ft
	Chord	5ft
	Area	120 sq ft
	Tail elevator	32 sq ft

HILL monoplane glider (E. King Hill, Paddington, London)

This glider of *Demoiselle* type was partially complete in December 1910.

HOLLAND & HOLLAND monoplane. See Nicholson monoplane.

HOLMES glider (D.C. Holmes, Thames Ditton, Surrey)

Holmes and his friends made four gliders between October 1909 and the end of 1910.

No.1. A Chanute-Voisin type which was too small and achieved hops of only 10-12ft.

Data		
	Span	15ft
	Chord	3ft 6in.

Hill glider was built by the brothers at their Sussex home in 1913-1914. Later they became well known in aviation circles.

No.2. This was modified from No.1, with increased chord and gap and a larger tail. Glides of 60ft were made.

Data	Span	15ft
	Chord	5ft.

No.3. The bamboo frame of the earlier glider was converted to a Wright type, fitted with a controllable elevator and a fixed fin. In this form glides of 180ft up to 10ft high were made.

No.4. This was a new triplane of Wright type, which was smashed by wind and horses before it could be flown.

Data	Span	27ft
	Chord	6ft 6in
	Length	18ft
	Wing area	300sq ft
	Weight	80lb

HOLT aeroplane (H.W. Holt, King's College & University Engineering Laboratory, Free School Lane, Cambridge)

Described as a 'Dihedral, triplane, biplane tandem' this small machine had a triangular section fuselage, with apex uppermost. It was of tractor type and was being assembled in the premises of Mr. Harry Williams at Victoria Park, Chesterton, near Peterborough, in September 1911. Nothing more was reported on this venture.

Power: 10-12hp JAP two-cylinder, air-cooled, driving a 6ft 2in diameter Cochrane propeller by Renolds chain at a reduction of 2.75 to 1

Data	Span	18ft
	Area	150 sq ft
	Length	12ft
	Height	6ft
	Weight allup	400 lb
	Speed	30mph
	Endurance	2hr

Hopkirk and McCormick glider built in Ireland in 1912.

HOPKIRK & McCORMICK glider (F.C. Hopkirk and W.E. McCormick, Ireland)

This biplane glider was built in 1912, and was towed down a slope, achieving several short hops, culminating on 14 December 1912, with a good flight in gusty weather, after which the machine was wrecked on landing. The glider had a front elevator, ailerons and 18ft long skids fitted with wheels.

Data	Span	20ft
	Chord	5ft
	Gap	5ft 8in
	Area	200 sq ft

Hornstein biplane with Green engine built by the Thames Bank Wharf Co. in 1909-1910. The first machine with JAP engine was tested at Upper Halliford near Shepperton.

HORNSTEIN biplanes (N.A. Hornstein, a Russian domiciled in London)

Two biplanes designed by Hornstein, with contributions by Livet de Loriere of the Thames Bank Wharf Co., were constructed in 1909-1910. The first of these, with a JAP engine, was flown very successfully by Hornstein at Upper Halliford, near Shepperton, Middlesex on 25 and again on 27 March 1910, when he landed in a ploughed field, without injury. Further flights were also made by Ernest de Loriere between 17 and 24 April 1910 when distances of 300-400 yards were flown, rising at times to fifty feet.

Both machines were of similar design, being pusher biplanes with front elevator and the tail carried on open booms. The undercarriage had three pairs of bicycle wheels, the front pair on mountings, with projecting skids. Ailerons were fitted to top and bottom wings and were operated by the pilot, seated on the lower wing, by sideways movement of the body, the back of his seat being arranged to pivot for this purpose.

The second machine had a Green engine and was of approximately similar size. Hornstein also referred to an 80hp Italian engine to be given a trial on a later machine. The Hornstein machines soon faded from the aviation scene.

See the Thames Bank Wharf Co.

Power: 35hp JAP eight-cylinder air-cooled vee
35hp Green four-cylinder inline, water-cooled
Both engines drove 6ft 11in diameter Chauvière propellers

Data		**JAP**	**Green**
	Span	32ft 10in	36ft
	Chord	5ft 9in	
	Area	378 sq ft	414 sq ft
	Length	34ft	
	Weight	696 lb	700 lb

HOWARD-WINDHAM monoplane

Referred to in *Flight* of 28 August 1909 (p.530) See Windham.

HOWARD WRIGHT aircraft (Howard T. Wright Bros. Ltd., Belgravia Chambers, 72 Victoria St., SW1. Works at Prince of Wales Rd., Battersea)

Howard Wright became involved in aeronautical work when working for Hiram Maxim, between 1899 and 1904. In July 1905, together with his brothers Warwick and Walter, the above company was formed, which included the goodwill of Maxim's company. An enquiry from an Italian inventor, Federico Capone, intended for the defunct Maxim company, was received in early 1907 for a helicopter to Capone's own design. The order was concluded and was the start of manufacture of flying machines by Howard Wright.

The manufacture of the helicopter began in November 1907 at the premises of Warwick Wright Ltd. at 110 High St., Marylebone, but was transferred to Battersea under railway arch No.80. Growth of aviation work necessitated further premises and arches Nos.79 and 82 were rented later. The ground used for testing conventional aircraft was Brooklands, where Shed No.9 was in use throughout 1911, until the business was closed. Howard Wright then became the manager of the Coventry Ordnance Works Aviation Dept at the beginning of the following year.

In addition to aircraft of his own design, Howard Wright produced others, bearing their owner's or designer's name, notably the Barber, ASL and Poynter monoplanes. Apart from responsibility for the actual construction, Howard Wright was no doubt involved with some design aspects of these, and probably bore most of the design responsibility in the case of the Cooke, Lascelles and Scottish Aeroplane Syndicate monoplanes. In the five years of its existence the company built approximately thirty-five aircraft, Howard Wright being joined in December 1908 by W.O. Manning, who from then on undertook much of the design responsibility.

HOWARD WRIGHT gliders

A triplane glider was reported in October 1908 and a biplane glider in course of construction in June 1909 at Battersea. No information on these is available.

HOWARD WRIGHT-CAPONE helicopters

Capone had made, in Italy, an unmanned helicopter to prove his design of rotors. He then made the enquiry received by Howard Wright for the manufacture of a second machine, which was to be piloted. Patent application 7129, with drawings by Howard Wright, was made in 1908. The machine was completed in March 1908 and taken to Norbury golf course, before being dispatched to Italy. The machine proved to be unsuccessful and a third machine of much lighter weight was commissioned, which was completed and sent to Italy in January 1909. Unmanned tests were made at Naples in October/November 1909, but were inconclusive.

A fourth Capone helicopter was started in June 1909 and was delivered to Italy that year. There is some evidence of a fifth Capone helicopter and an ornithopter being made by Howard Wright.

Howard Wright built several Capone helicopters. This is the second version in 1909.

The helicopters designed by Capone consisted basically of a structure of light gauge steel tube, welded to form a framework with outriggers. At the tips of the outriggers were mounted two-bladed lift rotors, driven through shafts and bevel gearing from the engine mounted across the framework, the pilot being seated immediately behind.

The top surface of the framework was covered with fabric to form a lifting surface in forward flight. Hinged triangular flaps at the tips of the hinged tailplane were to give yaw control, and triangular flaps below the rotors, lateral control, as well as stabilizing the machine on the ground. Large rectangular surfaces, within the main framework, were given a flapping motion by a drive from the engine, to produce a movement of air below the body and assist stability.

Forward motion was imparted to the machine by two four-bladed tractor propellers, driven by bevel gearing from the main rotor shafts. The machine was mounted on four wheels, one of which could be coupled to the rotor drive shaft by chain, to cater for taxiing on the ground. There were six small auxiliary wheels to aid stability on the ground.

The second and third helicopters were similar except that considerable weight was saved in the construction of the latter machine, through the use of a lighter air-cooled power unit and other changes of design. The fan to cool the engine was used for forward motion also, and the undercarriage was reduced to three main wheels.

CAPONE No.2 helicopter

Power: 50hp Antoinette eight-cylinder water-cooled vee driving two 4ft 8in diameter propellers

Data	Span over rotors	48ft 2in
	Length over rotors	27ft
	Wing span/rotor axis	28ft 10in
	Max wing chord	11ft 8in
	Tailplane span	13ft 2in
	Tailplane chord	8ft 2in
	Control area	81 sq ft inc. rudders
	Wing area	160 sq ft
	Rotor diameter	19ft 4in
	Weight allup	1,250 lb
	Rotor blades	6ft 4in by 2ft 4in
	Gliding speed	20mph

Howard Wright Capone helicopter was tested at Naples.

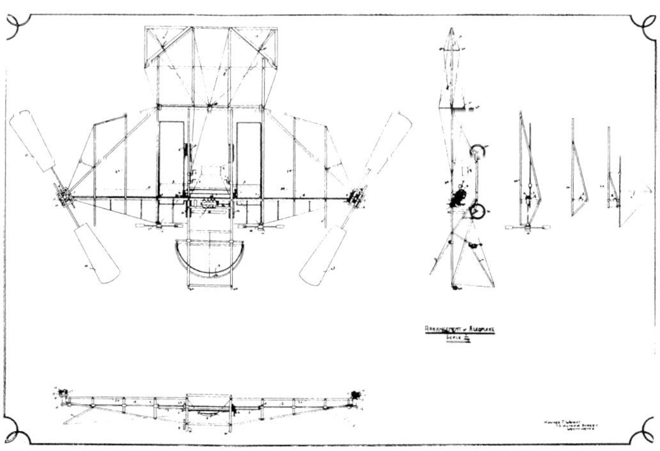

CAPONE No.3 helicopter

Power: 30hp REP seven-cylinder air-cooled, fantype semi-radial

Data	Span over rotors	56ft
	Wing span/rotor axis	30ft
	Weight allup	600 lb
	Rotor diameter	26ft
	Gliding speed	15mph
	Rotor blades	6ft 6in by 3ft

The first Howard Wright biplane was shown at Olympia in 1909.

HOWARD WRIGHT biplane 1909 type

A machine constructed early in 1909, to the order of M.H. Seton-Karr, was shown at Olympia in March. It was damaged in transit to Fambridge, Essex, where it was taken for trials after the show. Further damage occurred during the trials, resulting from the propeller bursting and also the shed collapsing. Short hops of some 30ft in length were achieved by the end of May on the unsatisfactory ground. Later trials were transferred to Camber Sands, where flight was achieved, the machine continuing in use until the end of the year.

Another machine of the same type, already constructed, was reported to have been ordered by W.E. Cooke of Burnley in September 1909. This was not delivered to Cooke, who received the first Howard Wright monoplane instead.

This rather ungainly pusher biplane, although well constructed and incorporating advanced features at the time, was not a great success.

The structure of the nacelle and tail booms was of welded light gauge steel tube, some of streamlined section, the nacelle was fabric covered and accommodated two persons. The machine had a single track undercarriage, consisting of one main wheel and a smaller tailwheel, supplemented by large wheels at the wingtips, on one of which the machine normally rested.

The wings were of wooden construction, fabric covered. The gap was maintained by three pairs of interplane struts, the outboard pair covered by fabric side curtains. Small ailerons, extending slightly beyond the trailing edge, were fitted to top and bottom wings.

The front elevator was a biplane structure with side curtains and the whole unit pivoted on the nose of the nacelle. The tail unit consisted of a fixed biplane structure with side curtains and central rudder.

During the course of the trials the lower corners of the side curtains of the tail were cut away to provide ground clearance. The members of the undercarriage, for the single main wheel, were strengthened and an additional pair of small wheels were fitted on a new mounting, below the rear of the nacelle. These wheels were intended to aid the leveling process on take off, as this did not take place at a sufficiently low speed.

The power was transmitted through a reduction gearbox to contra-rotating propellers, the front blades of which were of greater chord, and intended to absorb two thirds of the power. The blades were of wood on tubular steel shanks, but after the propeller disintegrated at Fambridge a single wooden propeller was fitted.

Power: 60hp Metallurgique four-cylinder inline, water-cooled driving through a 3 to 1 reduction gearbox a patented, contra-rotating 5ft diameter pusher propeller

Data		
	Span	40ft
	Chord	6ft 6in
	Gap	6ft 6in
	Area	520 sq ft
	Length	43ft
	Weight allup	1,600 lb
	Weight	1,100 lb
	Speed	35mph
	Price	£1,200

HOWARD WRIGHT monoplane 1909 type

This was the only one of its type, and was delivered to W.E. Cooke of the East Lancashire Aeroplane and Motor Co. on behalf of the Burnley and District Aero Club instead of the biplane originally ordered. It was dispatched by rail to Burnley where it arrived on 9 October 1909 and was immediately assembled for exhibition at the Athletic Ground, Brunshaw and was taxied, after some difficulty starting the engine. It was also displayed at Accrington and Blackburn, where the propeller flew off and was damaged. Thereafter the machine soon passed into other hands and finally to H.K. Hales of Burslem, the man who donated the Hales Trophy for the Blue Riband crossing of the Atlantic by ocean liner. Hales attempted to make the first flight of the machine at Keele Racecourse on 10 July 1910, where it crashed badly after twenty-five seconds. It was rebuilt and advertised for sale in *The Aero,* 12 October 1910, but this was the last heard of the machine.

Howard Wright's first monoplane of 1909, built for W.E. Cooke of Burnley, Lancashire, was not a success.

There was some similarity between this and the Barber monoplane, which had been delivered shortly before. It had a welded light gauge tubular steel fuselage structure, parallel in depth throughout its length, apart from a slightly increased depth of the engine bay. Three large spoked wheels, in forks, were each sprung by a telescopic strut, the radius arm pivoting on tubular mountings below the fuselage.

The wings were tapered and braced to the undercarriage uprights, and to a cabane above the fuselage, in which the streamlined fuel tank was hung. Lateral control was by warping the wing using foot pedals. A structure for mounting the warping gear hung below the fuselage. End type elevators were fitted to a fixed tailplane below the fuselage girder; a simple triangular rudder completed the tail unit.

The engine was cooled by long narrow condensing radiators on each side of the fuselage and could be started from the pilot's position, which was placed well back along the fuselage. The machine was intended to hold three people.

Power: 50hp Antoinette eight-cylinder water-cooled vee.

Data		
	Span	32ft
	Mean chord	7ft 6in
	Area	240 sq ft *
	Length.	27ft
	Weight	750 lb
	Endurance	3hr
	Price	£1,000

**Flight* reported 200 sq ft

HOWARD WRIGHT monoplanes (1909-1910 'Standard' type and Lascelles Ornis)

In early December 1909 *The Aero* described and illustrated the latest machine at Battersea, still minus its Lascelles engine, and indicated that two similar aircraft had been made for customers and several others were in the course of construction. A machine of the type appeared at Olympia in March 1910 on the stand of R. Lascelles & Co. of 13 Greek Street London, identified as the Lascelles *Ornis*, but this had a Blériot-like rudder, instead of the triangular type shown earlier in *The Aero* 7 December 1909 (p.493). The *Ornis* was sold to A.G. Power, who housed the machine at Brooklands. The monoplanes produced later from Battersea were of somewhat different types.

The description as a 'standard' type may well have applied only to certain details of the construction, for there were no large quantities of the type made. Reports of the time gave variations of dimensions for the aircraft being described.

The fuselage was based on a wooden, braced girder made of ash, with steel angle fittings at the junctions and spoke and nipple type bracings. The undercarriage was of steel tube, as in the 1909 Cooke type monoplane, and three equal sized wheels were fitted.

The wings were parallel in chord and incorporated control by warping. End elevators were fitted to the tailplane and the rudder shape was triangular, when first seen, but rectangular on the Ornis, making the machine appear more like a Blériot.

1909-1910 monoplane

Power: 35hp Lascelles four-cylinder air-cooled, fantype, semi-radial.

Data		
	Span	28ft
	Area	160 sq ft
	Length	27ft
	Weight	500 lb

Howard Wright 1909-1910 monoplane was similar to a Beriót XI but with triangular rudder.

Ornis monoplane

Power: 35hp Lascelles four-cylinder air-cooled, fantype, semi-radial driving a 8ft diameter Weiss propeller

Data		
	Span	28ft
	Chord	6ft
	Length	28ft
	Area	160 sq ft
	Weight	400 lb
	Weight allup	600 lb
	Speed	30mph
	Price	£450

Howard Wright built the '*Ornis*' monoplane shown on the Lascelles stand at Olympia in 1910.

Howard Wright *Avis* monoplane of which a small number was built in 1909-1910. This early machine was fitted with a fantype Anzani engine.

HOWARD WRIGHT monoplanes *Avis* Type 1910

The first of these aircraft was delivered at Brooklands to the Scottish Aviation Syndicate in which the Hon A.R. Boyle was the leading figure and pilot. The machine was named *'The Golden Plover'* and he tested it early in 1910, but it was underpowered with the original 25-30hp Anzani and so was returned to the works for a more powerful engine to be fitted. Other changes were also carried out, including reducing the length of the rudder. It was returned later to Brooklands where it was flown successfully in March and was then renamed *Avis*.

Four *Avis* machines were built for the SAS and were fitted with various engines. After the early Anzanis, JAP and ENV engines were also fitted; a report of a 35hp Green being used in June 1910 may have been incorrect.

There were several changes of design from the earlier type monoplanes. The undercarriage was completely redesigned and now consisted of a tubular structure with twin wooden skids and two pairs of wheels on separate axles, sprung by rubber cord. The tailwheel was smaller and was sprung by coil spring.

The fuselage girder was reduced in depth towards the tail to carry a universally jointed tail unit of cruciform type. This was similar to the *Demoiselle* type and had triangular shaped balance areas forward of the hinge on both horizontal and vertical surfaces. The actual surfaces were of light construction with wire edges giving a scalloped effect. On Avis III, at least, more area was added, which eliminated the pointed trailing edge tip.

Power: 25-30hp and 30-35hp Anzani three cylinder fantype, air cooled semi-radial driving a 6ft diameter propeller
35-40hp JAP eight-cylinder air-cooled vee (*Avis* II & III)
35hp ENV type D eight-cylinder water-cooled vee (*Avis* I)

Data		
	Span	28ft
	Chord	6ft 4in
	Length.	27ft.
	Area	160 sq ft
	Area rudder	14 sq ft
	Area elevator	22 sq ft
	Weight	430 lb
	Weight allup	630 lb
	Speed	35-40mph
	Price	£370-£490

Howard Wright monoplane 1910 type had a Blériot-type tail and ENV engine.

HOWARD WRIGHT monoplane 1910 Type

An alternative type of monoplane was also constructed in 1910, one of which, for Warwick Wright, appeared on the company's stand at Olympia in March. The first of the type to fly was the second machine for Capt. Hinds-Howell, which became airborne at Brooklands on 26 March 1910. The third was the machine bought by Tom Sopwith, on which he carried out his first flights in October and was used to test the first 40hp ABC engine, in April 1912, by R.L. Charteris.

The structure of the aircraft was similar to that of the Avis, but a more conventional tail unit with end elevators and a rectangular rudder, as on the Ornis, were obvious changes. Water-cooled engines were used on all three machines, the radiators on the first two being across the full width of the wing, just below the surface. Sopwith's machine had one radiator positioned below the fuselage, one vee-shaped pylon above the fuselage and a tail-skid instead of a tail wheel, also full span elevators.

Power: 35hp ENV type D eight-cylinder, water-cooled vee driving a 6ft diameter propeller

Data		
	Span	27ft
	Chord	6ft 6in
	Length	29ft
	Area	190 sq ft *
	Area tailplane	18 sq ft
	Area elevators	18 sq ft
	Area rudder	7 sq ft
	Weight	455 lb*
	Weight allup	655 lb*
	Speed	35-40mph
	Price	£650

*Various figures quoted

Sopwith began his flying career on this ENV powered Howard Wright monoplane in 1910.

HOWARD WRIGHT Monoplanes, various types

In January or February 1910, when the representative of *The Aero* visited the works at Battersea, several aircraft were under construction, whose completion and utilization cannot be established. They were as follows:

'Santos Dumont' type

A slightly larger version of the *Demoiselle* with a triangular section fuselage of bamboo. It is believed that this machine was completed in 1910.

Power: 30hp Darracq two-cylinder horizontally opposed, water-cooled with 6ft diameter propeller

Data		
	Span	19ft
	Area	120 sq ft
	Weight allup	255 lb
	Speed	48mph
	Length	21ft

'Antoinette' type

A two-seater enlarged version of the *Avis* was under construction. The wing was completed in a crate and the engine already installed in the fuselage. No record of completion of this machine has been traced.

Power: 60hp Green four-cylinder inline, water-cooled
Data Span 42ft
Area 320 sq ft
Length 40ft
Weight allup 800 lb

'Blériot XII' type
This machine was described as such from the position of the engine low down in the frame, with the propeller above, driven by a chain. The pilot and passenger sat below the wing behind the engine. The monoplane tail had end elevators and the multi-sided rudder was similar to that on the Avis.
Power: 60hp ENV type F eight-cylinder water-cooled vee
Data Span 35ft
Area 360 sq ft
Length 46ft
Weight allup 1,100 lb

Racing Type biplane
Power: 50hp Gnôme seven-cylinder air-cooled rotary.
Data Span 36ft
with outer lower panels removed 22ft
Chord 6ft
Gap 6ft
Area 432 sq ft could reduce to 348 sq ft
Length 37ft

HOWARD WRIGHT biplane 'Curtiss' type
It was reported that a machine was built in 1910, conforming generally to the design of the American pilot constructor, Glenn Curtiss. Little is recorded, except that the engine was of 35hp driving a pusher propeller, the span was 33ft and the area 270 sq ft.

HOWARD WRIGHT biplane
More of the 1910 type biplanes were made than of any other Howard Wright types. In basic design the type resembled a Farman, although the curved ends of the tail booms, an original feature, were changed later. The structure of the aircraft was mostly of mahogany and ash with steel fittings. It had both front and rear interconnected elevators, with a single tall divided rudder. The wings were fabric covered on the lower surfaces only and single acting ailerons were fitted to the top and bottom wings and to the top wing extensions, which were an alternative fitment to some aircraft. These extra ailerons were coupled to the existing controls and could be folded down with the extensions for storage purposes. The undercarriage consisted of twin wheels on separate axles, sprung by rubber cords to the skids, similar springing being used for the tailskid.

Howard Wright biplane of 1910 without wing extensions.

Tom Sopwith used his Howard Wright biplane very successfully in Britain and America.

The standard machines mostly had the ENV type F water-cooled engine with radiators on each side, the oil tank just above the engine and the fuel tank below the top center section. For long range flights an extra tank replaced the passenger seat, behind the pilot, and both Sopwith and Reynolds, who used the biplane, had fairings fitted in front to protect them from the elements.

The Howard Wright biplane was used by Sopwith for the Baron de Forest Contest at the end of 1910, in which he was successful, with a flight of 169 miles into Belgium.

At the Olympia Aero Show in March 1911, Tom Sopwith's machine was shown on the Aero Club's stand, while the Warwick Wright stand showed a special racing version for Robert Loraine. This had a Gnôme engine and reduced wing chord, with ailerons on the top wings only. The outer panels of the lower wings could be removed back as far as the first pair of interplane struts and the overhanging top wing was then braced by wires.

The first of the machines with ENV engines was owned by E.M. Maitland, and was first flown on 1 August 1910 at Larkhill. This resulted in a crash and injury to Maitland. It was repaired and then flown by Watkins, but was later sold to the War Office and identified as serial No.F3. The second machine went to John Done, known as Jack Dare in the theatrical world. The third machine was bought by Tom Sopwith and was used by him very successfully for competition and instructional work.

The fourth aircraft was exported to New Zealand to the Walsh brothers of Auckland where it survived in modified form until December 1913, and the fifth, with the smaller 40hp ENV type D, was exported to Rangoon, Burma for W.C. England in 1912.

The Howard Wright racing biplane shown at Olympia in 1911 was fitted with a Gnôme rotary instead of the usual ENV.

The eighth machine was flown by Lt. Reynolds in the Circuit of Britain Contest in July 1911, but he was forced to abandon the flight at Harrogate. This machine may have been fitted with a Green engine at some stage.

The ninth machine went to J.L. Longstaffe, a little known pilot, who made an unsuccessful attempt for the British Empire Michelin Trophy No.2 in October 1911. Grahame-White also purchased a Gnôme engined version for use at Hendon.

Power: 60hp ENV type F eight-cylinder water-cooled vee driving a 8ft 4in diameter
40hp ENV type D eight-cylinder water-cooled vee (one aircraft) propeller.
60hp Green four-cylinder inline water-cooled (one or two aircraft)

Data		
	Span	36ft (48ft #)
	Gap	6ft 6in
	Length	36ft 6in
	Area (inc. ailerons)	415 sq ft (504 sq ft #*) (532 sq ft #)
	Area tailplane	48 sq ft (34 sq ft *)
	Area front elevator	27 sq ft (23 1/3sq ft *)
	Area rear elevator	18 sq ft (12 3/4 sq ft *)
	Area rudder	15 sq ft
	Weight	800 lb
	Weight allup	1,200 lb
	Speed	36-45mph
	Endurance	5hr
	Price	£1,000

*Alternatives from various sources
#With extensions

HUBBARD cycloplane and monoplanes (Chas. Hubbard, 224 Bruntisfield Place Edinburgh)

Hubbard began his aeronautical work in September 1909 with a cycle-based machine with a wing area of 240 sq ft. It was pedal driven with a 5ft 3in diameter propeller and weighed 145 lb. This was no more successful than others of a similar type.

In October his second effort was a monoplane of 190 sq ft wing area, powered by a 60hp Rex motor. The machine weighed 260 lb and was presumably too heavy to fly.

His third machine was another monoplane tested at Portobello on 17 June 1911 and was described as of Blériot type with a hammock seat. Two twin-cylinder Rex engines were coupled together to drive a 6ft diameter propeller. A 30hp three-cylinder engine of unknown make was also fitted.

HUMBER monoplane (Humber Ltd., Coventry. Flying ground at Brooklands)

Humbers announced their intention to set up an aviation department in September 1909 to produce fifty aeroplanes at £400 each; this number was not built and by 1912 aircraft manufacture had been discontinued. From early 1910, hangars Nos.4 and 5 were rented at Brooklands for flying operations.

This well established motor company made, as its first product, a monoplane resembling a Blériot in its general layout, although any connection was denied by the company in the Press. There were differences in detail and the machines were fitted with Humber's own design of three-cylinder engine.

The early tests of Humber monoplanes took place at Cannes in the south of France from 5 January 1910, whence the first two machines had been taken to obtain good flying weather in winter. They were flown by Capt. Dawes and J.V. Neale. Dawes took the machines on to Egypt, where one was flown at the Heliopolis Meeting by Frenchman, Le Blon, who gained third place in prize money earnings. In early April Le Blon died in an accident when flying a Blériot at St. Sebastian in Spain.

Two machines were later at Brooklands, but only began trials in May in the hands of novice pilot G.A. Barnes, with the first public flight taking place on 28 May 1910. The machine was wrecked on 18 June 1910, but Barnes continued on a second machine and took his certificate soon after, later flying at various meetings, including Wolverhampton and Bournemouth. A fuselage with a 50hp. engine was also shown at Olympia in March 1910.

Power: 30hp Humber three-cylinder air-cooled, fantype semi-radial.
50hp Humber four-cylinder inline, water-cooled as in the Lovelace type.

Humber monoplane of Blériot-type was flown in France and Egypt in 1910.

Humber monoplane designed by Capt. Lovelace was shown at Olympia in March 1910.

HUMBER monoplane (Lovelace type)

Capt. T.T. Lovelace was engaged by Humber as designer and pilot, and he produced as his first design a machine which was shown at Olympia in March 1910, before it had flown. Later it was taken to Brooklands, where it was tested by Lovelace who carried out straight flights up to 10ft high on 25 May 1910. One of these resulted in a heavy landing and some damage, the machine being reported as tail heavy.

The aircraft was seemingly evolved from the basic Blériot XI design, with some changes introduced. It featured a long central skid behind the undercarriage, carrying the warp control on one of its supporting vee struts. There were large swept tail surfaces with trailing curved elevators and a long rudder with a pointed end. The engine was water-cooled with two narrow, vertical radiators at the front.

Power: 50hp Humber four-cylinder inline, water-cooled driving a 7ft diameter propeller

Data		
	Span	33ft (33ft 3in)*
	Chord	6ft 10in (6ft.9in.)*
	Length	26ft 6in (26ft 8in)*
	Area	210 sq ft (232 sq ft)*
	Area tailplane	36 sq ft
	Area elevators	32 1/2 sq ft
	Weight	500 lb
	Speed	50mph
	Price	£750 (£775)*

*Alternative figures from *The Aero*

Also shown at Olympia in 1910 was this monoplane designed by Frenchman Le Blon.

HUMBER monoplane (Le Blon Type)

Hubert Le Blon was a racing motorist who later took to aviation with Leon Delagrange, but was killed in a crash in Spain on 2 April 1910, soon after the machine he designed for Humber was exhibited at the Olympia Show. It embodied a number of original design features but interest in it died with its creator.

The fuselage consisted of a long tapered tube of light wooden monocoque construction covered with fabric. It was small in diameter and the pilot sat outside with his legs straddling the monocoque to operate the rudder bar. The fuselage hung below the wings, which were joined together by tubular center section spars that were integrated to the cabane, the engine mounting and the undercarriage. The main members of the latter were two large diameter shaped steel tubes joined below by a straight cross member, from the ends of which, short laminated springs carried the cross axle and bicycle type wheels. The tail was supported by a long curved ash tail-skid.

The center section of the wings was completely open. The wings themselves were braced with steel tapes and the entire trailing edge portions were hinged to serve the dual purpose of ailerons and camber changing flaps.

The tail unit was comprised of a long tapering fin, with rudder of equispaced areas above and below the fuselage centerline, with a shape that curved forwards at the trailing edge. The tailplane was of the type with end elevators, mounted below the fuselage and could be adjusted for incidence on the ground. The tailplane was itself supplemented by a long tapering extension carried forward below the fuselage.

Power: 30hp Humber three-cylinder air-cooled fantype semi-radial driving a 6ft 6in diameter (6ft 11in)* Chauvière propeller

Data		
	Span	29ft (29ft 2in)*
	Chord	6ft 6in (6ft 10in)*
	Length	26ft 8in (24ft)*
	Area	192 sq ft (186 sq ft)*
	Area tailplane	16 1/2 sq ft plus extension
	Area elevators	11 sq ft
	Area rudder	6 sq ft (10 sq ft)*
	Weight	490 lb (480 lb)*
	Price	£480

*Alternatives from *The Aero*

HUMBER biplane 1910 type

This tractor biplane shared the stand at Olympia with the monoplanes in March 1910. It was a side by side three seater, with the instructor in the center with overriding control of the pupils controls, which were triplicated.

The uncovered fuselage, which was square in section, tapered down to the tail by curvature of the upper longerons. The structure was of wood and the tail surfaces were similar to those of the Lovelace monoplane.

The wings were spaced by four pairs of interplane struts at the front and rear spars, all being of tubular steel. A widetrack undercarriage, below the longerons, was braced to a pair of

Humber's first biplane was a side-by-side three-seater for instructional work. At Olympia in March 1910.

A modified version of the 1910-1911 Humber biplane in India without the top-wing extensions and the sloping side-curtains.

outboard struts. The spars were jointed for ease of transport and the trailing edge sections of the wing outer portions were flexible for warping. This machine was probably not flown.

Power: 50hp Humber four-cylinder inline water-cooled driving a 6ft 11in diameter Humber propeller.

Data		
	Span	40ft (41ft 6in)*
	Chord	6ft 8in
	Gap	5ft
	Length	33ft (36ft)*
	Area	482 sq ft
	Area elevators	16 sq ft
	Area rudder	12 sq ft
	Speed	50mph
	Price	£1,000 (£1,100)*

*Alternatives from *The Aero*

HUMBER biplane 1910-1911 type

The biplane, shown at Olympia in 1910, was soon replaced by a more normal design of the time based on the French Sommer type, the design being attributed to A.H. Bailey.

The first of these was reported to have flown at Brooklands on 19 October 1910, having been fitted with a Gnôme engine as, no doubt, the Humber engine still required development. The machine was packed, together with a second biplane and two monoplanes, and was dispatched to India by early November for demonstrations during an exhibition at Allahabad, in January and February. The team was managed by Capts. W.G. Windham and G. Dawes, with pilots Henri Pequet and Keith Davies, who also carried out pioneering air mail flights. The team was back by April and Pequet flew the Olympia machine at Brooklands for the first time on 6 May 1911 and the following day made a flight of 1hr 10min. Further flight trials of the engine, in a monoplane, took place throughout the summer, but apart from one in a monoplane sold to the Pashley brothers in August, the Humber engine was not generally adopted.

Humber biplane of late 1910-1911, with original side-curtains at wingtips, was shown at Olympia in 1911.

The machine was a typical pusher biplane with front elevator. It had the long curved skids attached to the front elevator supports, associated with the Sommer type. The tailplane was adjustable for incidence, by hand wheel adjacent to the pilot, an unusual feature at the time. The most obvious departure from standard practice, were the side curtains set at an angle between the two planes. These were warpable and were used in conjunction with small semicircular ailerons, positioned inboard of the outer pair of interplane struts, on the top wing. The top wing extensions could be folded down to reduce the span for storage, the side curtains then being folded upwards. Rudders were fitted above and below the tailplane.

Power: 50hp Humber four-cylinder inline, water-cooled with 7ft diameter propeller
50hp Gnôme seven-cylinder air-cooled rotary

Data		
	Span top.	45ft 8in
	Span bottom	33ft 9in
	Length	40ft
	Chord top	6ft 9in
	Chord bottom	6ft 9in tapering to 5ft 9in
	Area	506 sq ft
	Weight	820 lb

Humber gave up the manufacture of aeroplanes and aero engines in August and sold all their remaining stock by auction on 4 September 1912. The sale included one Blériot type, four biplanes, three Blériot fuselages and two Humber Le Blon fuselages. Also a number of aero engines and finished and unfinished parts for aero engines and woodwork and metal parts.

HUMPHREYS gliders (Jack Edmond Humphreys, Wivenhoe, Essex)

Prior to attempting powered flight, Jack Humphreys built three gliders from 1902 onwards, culminating in man carrying flights of up to half a mile from the cliffs at Coombe Farm, Fowey in Cornwall. He made a study of bird flight and based his highlift wing design on that of birds, confirming by practical experiment the airflow and lifting characteristics.

Humphreys withdrew from aviation after the series of mishaps at Brooklands with his No.3 monoplane, and strong criticism voiced in The Aeroplane by C.G. Grey concerning his piloting ability. He was a qualified dental surgeon and returned to this profession establishing a practice in Harley St., London W1.

HUMPHREYS biplane

The construction of an amphibious biplane was begun in the autumn of 1908 at Forrest's Boatyard, Wivenhoe. It was launched on the River Colne at Rowhedge Ferry Hard in April 1909, but soon sank at its moorings. It had previously been brought to London for exhibition at the Aero Show at Olympia in March but, because of its size, the machine could not pass through the doors. The subsequent trials were unsuccessful and the machine, which became known as the '*Wivenhoe Flyer*' in the Press, never did more than just taxiing on the river at speeds up to ten knots.

The biplane wings were widely spaced and curved in shape, with large triangular tip ailerons. The top center section embodied a tapered 'keel', a theory of Humphreys to assist longitudinal stability. A triangular front elevator and flexible tailplane, with rudder above, were fitted at the rear of the machine. The tailplane was fixed but could be operated differentially in conjunction with the ailerons.

The engine was mounted across the center on bearers, at midgap, and drove twin pusher propellers through shafts and bevel gearing. A four wheeled chassis, which could be folded up, was incorporated, but probably not fitted for the trials on water. The machine floated on an open central hull and tip floats, mounted on the lower wings.

Steel tube was used for the main frame with wood in the structure of the lower wing, tail and supports for the front elevator. The mainplanes and fin were covered with thin aluminum sheet, the rest of the surfaces were fabric covered.

Power: 35hp JAP eight-cylinder air-cooled vee driving two counter rotating 8ft diameter steel bladed propellers at 400 rpm

Humphreys biplane was built at Wivenhoe in 1908-1909 but could not become airborne.

Data		
	Span	45ft
	Hull	12ft 6in by 8ft beam
	Tip floats	6ft by 1ft 6in beam
	Chord	13ft
	Length	38ft
	Area	650 sq ft *
	Weight allup	1,750 lb

*Area uncertain but claimed loading of 1 lb per sq ft does not seem feasible

HUMPHREYS monoplane No.1 (British Aeroplane Syndicate Ltd., 80 Brook St., London W. Works at Wivenhoe.)

In 1909 Humphreys decided to raise funds to form a company, and took full page advertisements in the aviation press for the purpose in May and June. The syndicate formed following this, being advertised for the first time in *The Aero* on 22 June 1909, but the advertising faded away by the end of August. However, Humphreys had somehow financed the building of a monoplane, with the declared intention of competing for the *Daily Mail* prize of £1,000 for a flight of one mile by an allBritish aircraft. This he attempted on unsuitable ground at Rowhedge near Wivenhoe on 16 October 1909 and the machine was wrecked when it hit a drainage ditch.

No data exists for the monoplane, other than that which can be deduced from a surviving photograph. It was a tractor type and had a triangular section fuselage, apparently made of tube, with a pylon above to carry a streamlined fuel tank, bracing wires and warping controls. The tubular undercarriage structure extended rearwards, terminating in a large sprung tailwheel, the structure also providing anchorages for bracing cables.

The wings were deeply cambered and curved downwards, and the leading edge curved back to the trailing edge tips. The lateral control surfaces were wide chord flaps hinged behind the trailing edges. The tailplane was of considerable size and probably contributed lift. The pilot was seated well aft, an indication of the effect of the sweep-back and distribution of lift. The rudder, below the tailplane, was rectangular in shape and

Humphreys monoplane No.1, also built by Forrester's boatyard at Wivenhoe, was probably not completed.

emblazoned with the Union Jack over the whole surface. The original engine, a 35hp JAP, was fitted with a clutch and extension shaft for the propeller, but was replaced by the new 50hp Green.

The machine was transported to the Bournemouth Meeting in July 1910, where it arrived dismantled on horse and cart transport, on the fourth day of the meeting, 21 October 1909, but it did not perform.

Humphreys monoplane No.2 at Wivenhoe.

HUMPHREYS monoplane No.2 (Wivenhoe and Brooklands Hangar No.10)

This machine was considerably different from No.1, although the fuselage, which was covered, was still triangular in section and again fitted with the 50hp Green, driving the tractor propeller by clutch and extension shaft. Later these were removed and the propeller was fitted direct to the crankshaft. The original undercarriage was similar to that of an Antoinette, with a central sprung strut, using tension springs. A single front skid was fitted. This was later replaced by one of A type with two pairs of wheels, and a larger elevator was also fitted.

Long tapered fins were fitted above and below the fuselage; a skid supported the tail. The wings had curved leading edges, but less sweep-back to the trailing edges; they were braced by kingposts and did not droop as on No.1. Lateral control was by warping the wings, which were cut away at the center section to allow access to the cockpit.

Humphreys monoplane No.2. Was taken to the 1910 Blackpool Meeting but not flown.

This is the machine described in *The Aero* as the 'earlier version' and presumably is that shown in the only remaining photograph taken at Forrest's Boatyard. The machine, with its wing span of 48ft, was the largest at Brooklands at the time, and became known as the *'Elephant'*. It appeared for a test flight on 9 October 1910 in the hands of Humphreys, but it fell over to one side on take off and was damaged. It was tried again at Abbey Fields Colchester on 5 November 1919 but did no more than taxi.

Humphreys monoplane No.3, nicknamed the *Elephant*. Shown here in its shed at Wivenhoe.

HUMPHREYS monoplane. No.3

A report in February described the naming ceremony of a 'new machine' by the Hon. Mrs. Assheton-Harbord, who christened it *'Mary'* in honor of the new Queen on 4 February 1911 at Brooklands. Although described as new, the machine may well have been a reconstruction of its predecessor.

This machine was totally different from No.2 but still retained the triangular section fuselage and Green engine mounted on the top longerons. The undercarriage was of Antoinette type with a central sprung strut. Long tapered fins and small triangular rudders were fitted above and below the fuselage and a skid supported the tail. The wings had curved leading edges but little sweep-back to the trailing edges; they were braced by kingposts and had little anhedral.

In May and June, Humphreys was reported to be taking lessons at the Hanriot School. The monoplane, described in *Flight* as of 'Blériot type', was not reported out until 30 August 1911, when Humphreys took off, but overcorrected at 10ft from the ground and the aircraft was severely damaged in the ensuing crash.

A major repair took place and subsequently Gordon Bell tested the machine on 12 December 1911 when it was reported to lift rapidly at half throttle, indicating tail heaviness. To correct the balance, flights were made with one and two passengers. By now the undercarriage was replaced by a Farman type structure with 'A' frames and two pairs of wheels and skids. Large diagonal radiators were attached to each side on the un-

Humphreys monoplane No.3 with modified undercarriage and radiators. Was tested at Brooklands and took off with three people aboard.(Photo from Steve St Martin).

dercarriage struts and a larger elevator was fitted. The rear fuselage was now uncovered.

On 16 January 1912 Humphreys' mechanic was seriously hurt in a propeller swinging accident. Dual ignition was fitted and one switch must have remained on. Straight flights were made on the 20-21 January 1912, by a pilot named Hunter, who again flew the machine to test a new Regy propeller fitted to the 60hp Green. On 12 May 1912, after taxiing towards the sheds, Humphreys carelessly crashed into the Hewlett & Blondeau school machine, wrecking it completely.

The No.3 monoplane seems to have been built at Wivenhoe and the photograph of the original version was taken inside a shed at Forrest's Boatyard.

HUMPHRIS direct lift machine (James Humphris, Willow Walk, Wood Green, London N)

No details are available, other than the use of 16 and 30hp Oppermann engines and that patent No.13853/1909 was taken out.

HUNT tandem biplane (A. Castle Hunt, Marlborough Aeroplane Works, Merton, Surrey

A tandem biplane was under construction in January 1910 and was of similar layout to the J.T. Davies machine which was illustrated in *The Aero* of 11 January 1910.

It had two pairs of wings mounted 3ft above the fuselage, the two bottom members of which were curved up in front, forming an outrigger to carry the elevator on a level with the main planes. The front and rear wings were 7ft apart and there was no vertical rudder. The undercarriage consisted of three wheels and skids.

Data	Length	30ft
	Span	28ft
	Chord	5ft

HURD-HALES ornithopter (Milner Hurd and H.K. Hales, Burslem, Staffordshire)

This flapping wing machine was built in 1910 and was not a success. It was fitted with an 18hp engine and had a span of 28ft.

Hutton & Wilson monoplane at Scarborough in 1910.

HUTTON & WILSON monoplane. (Filey, Yorkshire)

A tractor monoplane with 35hp Alvaston engine, built by two Irishmen, was reported ready for testing in August 1910. No further information was recorded.

HYERMAN & WARREN biplane (London Aerodrome, Hendon)

A Sommer type biplane was reported built by Hyerman and Warren at the Grahame-White School at Hendon. It was fitted with a 60hp Aster engine and was brought out for engine running on 8 January 1912. No further reports can be traced.

I

INGHAM biplane (Ralph Ingham, Ballenden Old Mill, Rawtenstall, Lancashire)
This machine was reported to be of Farman type, when built in April 1910.

Data	Span	33ft 6in
	Length	41ft 6in
	Height	12ft

INTERNATIONAL biplane (International Aero Co., Ferrars St, Huntingdon)
This concern built the Bonnard biplane in 1910 (*q.v.*).

J

JAMES Bros., No.1 and 2 biplanes (Henry Howard and John Herbert James, Narberth, Pembrokeshire)
The James brothers built a Caudron type C single-seater biplane with a 45hp Anzani engine, in the second half of 1913. It was the younger brother, Herbert, who was flying the machine on its initial flight on 25 September 1913, when it crashed on test at Clynderwen. Although considerable damage was done, it was repairable and Herbert suffered only minor injuries.

The aircraft, which was built with a single rudder, had carried 'James Bros. No.1' on both the rudder and the lower wings. When the aircraft was rebuilt as No.2 it was a two-seater with the later Caudron-type twin triangular rudders, and small fixed fins and was unmarked. Herbert flew it again in this form on 22 November 1913, but damaged the propeller on landing. It was not until 20 April 1914 that flying recommenced, with increasing success until the outbreak of war.

The brothers had originally become involved in aviation by an apprenticeship with the Gloucester firm of Webb, Peat & Co. which was terminated with the liquidation of that firm. Thereafter they enrolled with the International Correspondence School, to study aeronautical theory, and moved to Hendon for flying tuition, both receiving their Aviation Certificates in October 1912. They continued at the Ewen School as instructors, until returning to Wales, in the summer of 1913, to build the aircraft at Narberth, where, in July 1914, they planned to operate a small aircraft factory. A monoplane was in course of design but this, and the factory scheme, were discontinued with the outbreak of war. The biplane was moved to the Ruffy Baumann School in February 1915, and the brothers spent until the end of 1915 instructing at Hendon.

Howard spent the first half of 1916 with the RNAS, and was then transferred to Farnborough, where Herbert was already engaged as a civilian test pilot, both becoming commissioned in the Special Reserve of the RFC in March 1917. The

The James brothers of Narberth built this Caudron Type C in 1913 and later rebuilt it as a two-seater.

older brother crashed badly in December 1917 and was invalided out of the RAF in October 1918. The younger brother continued to fly after the war, becoming a famous test and competition pilot for Nieuport and General Aviation Co., and later for Gloster and the Air Navigation and Engineering Co. (ANEC). He served in the RAF in a non-flying capacity in World War II, but died in February 1944 while still serving.

JAMES monoplane (Rupert E. James, Birmingham)
See Dunne D9 monoplane for this machine, built by Levis Ltd. in 1913.

JAMES transatlantic monoplane
R.E. James announced, in the press in September 1913, his intention to build a machine to compete for the *Daily Mail* £10,000 prize for an Atlantic crossing. The aircraft was to carry two pilots, with dual controls, and would refuel four or five times at points on the way.

It was to be fitted with a 160hp engine, would have a span of 50ft and reach a speed of 80mph. The scheme was abandoned.

JAMES monoplane (J.M. James, Horn Hill, Lowestoft, Suffolk)
This machine was reported in July 1910 to be of Blériot type with a small seven-cylinder two-cycle rotary engine. It is assumed that it made at least one unsuccessful flight as it was advertised in Flight on 17 September 1910 as complete but with damaged wings.

JAMES & THOMPSON hydroplane (Rupert E. James and F.A. Thompson, Edgbaston, Birmingham)
The partners were members of the Birmingham Aero Club and Thompson ran the Belmont Aviation Co. at Ladywood where, at the latter's premises, the hydroplane, including its engine, was built for James. It was alleged to have flown off a local reservoir with both partners on board in August 1913. The machine had been constructed in some secrecy by a deaf mute carpenter named Pickering.

Power: 10hp Belmont single-cylinder, air-cooled. Weight 38 lb.

Data	Span	33ft
	Length	16ft

JAMOGOCHIAN cycloplane (M.A. Jamogochian)

This cycle plane was designed in 1910 and consisted of a parasol front wing and biplane rear wings mounted on a bicycle. A front mounted propeller was driven by the pedals via a chain and bevel gear. A rudder was attached to an outrigger at the rear and was connected to the front wheel so that movement of the handlebars gave lateral control. The front wing was fitted with an elevator which was controlled by a lever attached to the crossbar.

Data	Front wing	10ft x 4ft
	Top rear wing	10ft x 4ft
	Bottom rear wing	10ft x 3ft
	Total wing area	110 sq ft
	Empty weight	c100 lb

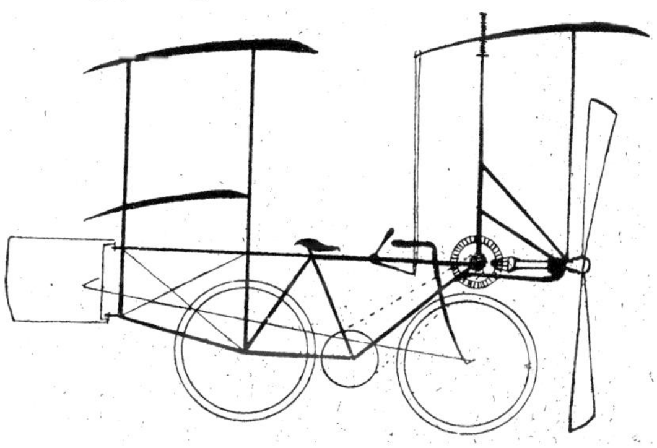

Jamogochian cycleplane of 1910.

The JAP Harding monoplane was similar to a Blériot.

The JAP Harding monoplane was later fitted with warping wings without ailerons.

JAP-HARDING monoplane (J.A. Prestwich & Co., Tottenham, London, N)

This well known manufacturer of motorcycle engines also produced light aero engines and, in 1910, built a monoplane fitted with one for demonstration purposes, to be flown by H.J. Harding, their Paris representative. The machine made its first flight on 10 April 1910 at Tottenham Marshes, but thereafter was taken to Amberieu in France, where Harding learnt to fly it, before returning to England without his pilot's certificate, due to a misunderstanding. He was thus unable to compete at Blackpool or Lanark in August.

The machine was generally similar to a Blériot except for wings fitted with ailerons, hinged on the trailing edge, which were replaced in France with Blériot type warping wings. Harding subsequently obtained his certificate and continued to fly the machine from Tottenham during the rest of the season. It was later handed over to the Science Museum, where it is still exhibited.

Power: 40hp JAP eight-cylinder, air-cooled vee driving a 6ft 10in diameter JAP propeller

Data	Span	30ft
	Chord	6ft 9in
	Length	27ft
	Area	190 sq ft (230 sq ft later)
	Area tailplane	16 1/2 sq ft
	Area elevators	7 1/2 sq ft
	Weight	510 lb
	Speed	50 mph

JEZZI No.1 biplane (P.G. Leo Jezzi, Shawfield Park, Bromley, Kent)

Construction of this machine was begun in 1909 in conjunction with Arthur Cooper, and it was moved to Eastchurch for assembly in January 1910, where successful flights were made by the following August. Aviation was a spare time occupation for Jezzi, who was employed in the City of London.

The machine was similar in layout to the Wright biplane, having twin pusher propellers driven by chains from the engine, which was mounted on the lower center section with the pilot seated in front. A small propeller for cooling was driven from the front of the engine. The three bay wings had warping

Jezzi biplane No.1 at Eastchurch in 1910 in original pusher configuration.

Jezzi biplane No.2 as first flown uncovered and as a sesquiplane.

sections, which extended beyond the chord, and were separated from the main wing by a gap at the inboard end. Fixed fins were built onto the rear outboard interplane struts, which were wide and of streamlined section.

The tail booms, of light tubular construction, carried a fixed, high mounted tailplane with twin rudders below. The front structure of heavier tubular members incorporated twin skids and carried a biplane elevator, with twin rudders between them, coupled to work with those at the back. The undercarriage consisted of two main wheels below the center section.

The biplane, although performing well, was capable of improvement and was converted to a twin tractor type; it flew in this form for the first time on 12 November 1910. The engine had been moved forward and reversed with the pilot repositioned behind. The front elevator and rudder, together with their supporting structure, were removed and a rear mounted elevator was fitted. The balance of the machine having changed, it now rested on a skid behind the tail booms.

Power: 35hp JAP eight-cylinder, air-cooled vee driving twin propellers by chain

Jezzi biplane No.1 in 1910-1911 after conversion to tractor layout.

JEZZI No.2 biplane

Jezzi discarded his first biplane, replacing it with No.2, which appeared in September 1911 at Eastchurch, where he ran it at high speed on the ground to test features of the machine, which was still incomplete and uncovered. On 13 January 1912, it was reported that he used the machine to tow the Brocklehurst monoplane back to the hangar, after it was stranded on the airfield. The machine had been covered by the end of November.

The first flight of No.2 is not recorded, but by 10 March 1912 he had already flown and carried several passengers. The machine had been modified by August 1912, by the addition of extensions to the lower wings, making the wings of equal span. In this form Jezzi used the machine continuously at weekends throughout 1912. As an amateur designer and constructor with limited time and resources, Jezzi made a machine which was fast, very reliable and served him well.

The design was of a simple tractor two-seater, with tapering wooden fuselage girder, mounted on struts between the wings. The top wing had considerable overhang when the machine was originally built as a single bay biplane, with strut and kingpost bracing of the overhanging portion, which was flexible for warping of the inversely tapered trailing edges. The struts and kingposts were removed when the machine was converted to a two bay biplane.

A streamlined headrest was provided behind the pilot, and streamlined fairings were added to the supporting struts for the fuselage and undercarriage. In addition to twin skids on the main two wheel undercarriage, the axle of which was sprung by rubber bands, twin tail skids and skids, below the first interplane struts, were provided, all of which were laminated springs. There was a balanced rudder, but no fin, and a tailplane with divided elevators, all of which were parallel and square ended.

Power: 35hp JAP eight-cylinder, air-cooled vee, driving a 7ft 8in diameter tractor propeller

Data		
	Span top	28ft After conversion 27ft 7in
	Span bottom	14ft After conversion 27ft 7in
	Chord.	4ft
	Gap	5ft
	Length	24ft 6in
	Area	166 sq ft After conversion 220 sq ft
	Area tailplane	10 sq ft
	Area elevators	15 sq ft
	Area rudder	9 sq ft
	Speed	65 mph

Jezzi biplane No.2 in its final form in 1912 at Eastchurch.

JOBLING monoplane (Charles Jobling. Trials carried out at Eastchurch)

A monoplane of somewhat peculiar appearance, designed by Jobling, appeared at Eastchurch on 23 September 1910 for the first time, and carried out straight runs on the ground only. On 19 October 1910 the machine flew 130 yards to a height of thirty feet, before a wire strainer broke and the machine crashed. There were no further reports of this aircraft.

The machine was a large straight-winged monoplane, with curved and rearward swept tapered wing extensions, serving as all controls. These surfaces were extremely flexible, and were operated by wires, attached at the extreme points. The machine was a pusher without tail surfaces and appeared most unstable. The construction was reported to be by Handley Page.

Power: 45hp Iris four-cylinder, horizontally-opposed, water-cooled. Designed by Geoffrey de Havilland and made by the Iris Motor Co. of Willesden.

Data		
	Span	60ft
	Leading edge	40ft plus 10ft extensions each side.
	Weight	1,300 lb

JONES bicycle monoplane (S. Jones, Alfred Place, Dingley, Leicestershire)

This pedal powered machine was reported to have a span of 30ft, was 25ft long and weighed 280 lb, complete with rider.

JONES monoplane (Jones, St. Mabyn, Chudleigh, Devon)

This monoplane was reported in January 1912, and was fitted with a 6ft diameter Cochrane propeller. No further information has been traced.

K

KAUFMAN ornithopter (Joseph Kaufman, Glasgow)

Kaufman exhibited a large flapping wing ornithopter at the Aeronautical Society Show at the Crystal Palace in 1868. A boiler generated steam through a flexible pipe to the test rig and provided some forty pounds of lift. Like many ornithopters it flapped itself to destruction.

KEITH-WEISS Aviette (Alexander Keith and Jose Weiss)

Weiss (*q.v.*) was an artist who had experimented with gliders and powered aircraft since 1902, before he collaborated with Keith, an elderly Scotsman with knowledge of the functioning of birds in flight. The partners designed an ornithopter, the wings of which were actuated by foot pedals. The wings were retained in the normal position by springs, but were given a propelling beat of three feet, when activated.

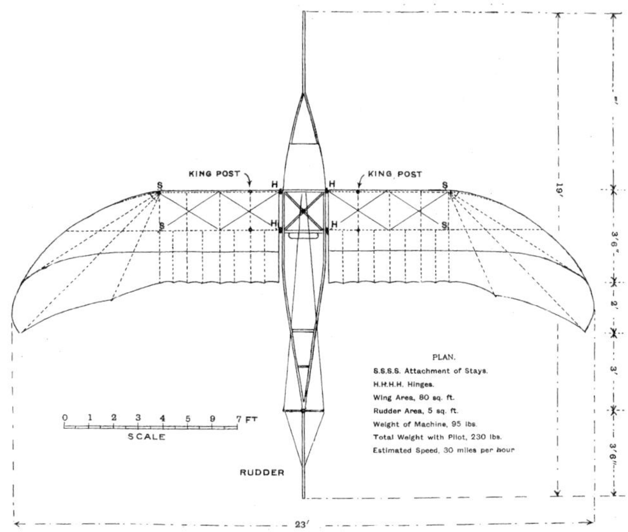

Keith-Weiss *Aviette* was tested as an unmanned glider at Amberley in 1912.

The machine was tested as a glider, with ballast, at Amberley Mount, Sussex in the summer of 1912.

Data		
	Span	23ft
	Length	19ft
	Area	80 sq ft
	Weight	95 lb
	Weight loaded	230 lb
	Est. speed	30 mph

KENSINGTON glider (Kensington Aero Club, L le Senechal, 4 Angola Mews, North Kensington)

At a preliminary meeting on 2 November 1910, it was reported that a member was building a full size glider. Nothing further was reported.

King monoplane built in 1908-1909 at Kessingland flew locally until wrecked during a gale.

Kitchen annular biplane the forerunner of later machines by Lee and Richards.

KING monoplane (Alec or Alick King, Kessingland, Suffolk)

This machine was built in 1908-1909 at Kessingland, and was tested on Benacre Denes, where it lifted off and was then damaged by running into a bank of earth. After repair it was removed, in October 1909, the few miles inland to Beccles Common, where Capt. Sanders proposed to fly and to build a hangar. The King monoplane was housed in a tent but was wrecked when this was blown down during a gale and was not rebuilt.

The aircraft had some similarity to a Howard Wright monoplane, but with a Blériot type undercarriage with cycle wheels. The machine had a fuselage structure made of bamboo, which was parallel in elevation, but tapered in plan to the vertical sternpost. To this was attached the rudder, a light five sided unit with wire boundaries; there was no fixed fin. The tailplane with Blériot-like end elevators, was fixed ahead of the rudder to the lower longerons.

The wings were of parallel section with square tips, and with spars in a diamond pattern. They were braced by cables and kingposts to the undercarriage and fuselage pylon, which also served to provide a pivot point for the warping control cables, which apparently warped the leading edge.

The sides of the center fuselage were fabric covered; the rear fuselage and engine bay were uncovered. The engine itself is believed to have been made by a local man named Talbot and was a horizontally opposed, air-cooled type of four-cylinders, although other reports referred to a three-cylinder engine.

KITCHEN annular biplane (John George Aulsebrook Kitchen, Scotforth, Lancashire)

Kitchen patented ideas for circular winged aircraft, with special aerodynamic features in 1910 and 1912. A biplane based on his theories was built in 1910 and housed at Famine Point near Heysham, Lancashire. The machine may have been tested, but did not fly before it was purchased at the end of 1910 by Cedric Lee, a wealthy engineer and textile manufacturer from Manchester. He was joined by G. Tilghman Richards, an engineer and member of Manchester Aero Club.

A period of testing and modification, to incorporate some of Richards' ideas ended with the machine overturning in a ditch and finally being destroyed, while under repair, on 15 November 1911, when a gale blew down the hangar. Thereafter the experiments were conducted with models and a full size glider, to test the conflicting views of Kitchen and Richards.

The Kitchen biplane was basically of nacelle and boom type, the uncovered nacelle housing the pilot with the motor mounted on the front, was suspended between the front center section of the wings. The two tail booms were unusual, being constructed with curved vertical members, and carried a pair of rudders and a single tailplane, behind the top wing. The undercarriage had a wide track and had twin skids and four wheels.

The circular wings with central aperture, were built with ribs fore and aft and were attached to inner and outer boundary members. The wings were spaced by a pair of interplane struts each side, together with pairs of diagonal struts connected to the lower longerons. Single-acting ailerons hung on the front interplane struts. The engine was a 50hp Gnôme seven-cylinder air-cooled rotary.

KITCHEN-LEE-RICHARDS annular glider

A full size glider biplane, with a circular lower wing and an upper wing of tapered form, was built in January 1912 for testing Kitchen's theories. The trials carried out at Sellet Banks, near Kirby Lonsdale, using a 52ft long launching track and pylon, with weight and cable system, similar to the system employed by the Wright brothers.

Kitchen-Lee-Richards glider version No1 tested near Kirby Lonsdale in 1912.

The third and final version of the Kitchen-Lee-Richards glider.

Lakes *Waterbird* was a Curtiss-type and was built by Avro in 1911.

The pilot was seated in the center of the aperture on a simple fuselage, to which the circular edge members of the wing were rigidly attached. The upper wing was mounted on a pylon above the fuselage and on single pairs of interplane struts. The elevators, inset into the lower wing, trailed behind the wing profile.

Changes were made to the glider, which was tested in three basic forms until the end of 1912, when Kitchen's ideas were discounted and Lee and Richards moved to London to continue their work (*q.v.*).

Data	Span	22ft
	Area	400 sq ft
	Weight	215 lb
	Weight allup	385 lb
	Speed	22mph

L

LAKES FLYING Co. (Capt. E.W. Wakefield of Kendal and H. Stanley-Adams. Hangars at Hill of Oaks and Cockshot, Windermere, Cumberland)

This company, financed by Wakefield, was formed in December 1911 to run a seaplane flying school, for passenger carrying and for aircraft construction. Prior to forming the company Wakefield ordered an aircraft from A.V. Roe & Co. (*q.v.*). This machine became known as the Lakes *Water-Bird*. The company was taken over by the Northern Aircraft Co. on 11 November 1914.

LAKES seaplane *Water-Bird*

The aircraft built by Avro was of typical Curtiss design, with front elevator and single rudder, and tailplane carried on tapering bamboo booms. The wings were of unequal span, the upper carrying four single ailerons, the inboard being of wider chord and with curved trailing edges.

The machine was flown for the first time, as a land-plane at Brooklands, by R.C. Kemp on 19 June 1911, followed by

other pilots, before it was transferred to Windermere. Once there, a three step central float with canvas top, made by Borwick and Sons of Bowness, was fitted, together with cylindrical canvas wing tip floats. In this form it flew in the hands of Stanley-Adams on 25 November 1911.

The long delay before the machine was flown was mainly caused by Wakefield's criticism of the engine, which was secondhand and had to be sent back to France to be reconditioned, but also because of some constructional problems in the airframe. The aircraft was damaged beyond repair at Cockshot Point, when the hangar was destroyed by a gale on 30 March 1912.

Power: 50hp Gnôme seven-cylinder air-cooled rotary driving a 8ft 6in diameter propeller.

Data	Span top	41ft
	Span bottom	32ft
	Chord	5ft
	Length	36ft 5in
	Main float	12ft long 2ft beam 1ft deep
	Wing tip floats	3ft long 8in diameter with springboards below
	Area	365 sq ft
	Weight	780 lb
	Weight allup	1,130lb
	Speed	33-45mph
	Ceiling	800ft

One aircraft built at Manchester by A.V. Roe & Co.

Lakes *Waterhen* was a modified copy of the *Waterbird* built by the Lakes company at Windermere in 1912. Shown here with single float.

Lakes *Seabird* was a reconstruction of the Avro Duigan biplane in 1912.

LAKES seaplane *Water-Hen*

The Lakes company built a copy of the *Water-Bird*, incorporating various modifications, and named it *Water-Hen*. The changes included a single pair of parallel chord ailerons and a tail elevator. To cater for passenger carrying a larger central float with a single step, and with aluminum sheet covering of the bottom and sides, was again made by Borwicks. An enclosed nacelle with pointed front was fitted, to protect the crew, in the winter of 1912-1913.

Adams flew the *Water-Hen* for the first time on 30 April 1912 and it continued in use for passenger carrying and training work until well into 1916.

Later a longer nacelle with curved front was fitted and subsequently, twin widely spaced floats; at the same time the wingtip floats were removed. A water rudder was fitted below the rudder briefly, but was not effective. One aircraft only was built.

Data as for *Water-Bird* except for the following information from *Flight*.

Data	Span top	42ft
	Area	375 sq ft (Quoted incorrectly in *Flight* as 270 sq ft)
	Area tailplane	11 sq ft
	Area rudder	9 1/2 sq ft
	Area rear elevator	6 1/2 sq ft
	Area front elevator	17 1/2 sq ft
	Main float	12ft long 6ft beam 1ft deep

Lakes *Waterhen* was later fitted with twin floats and a nacelle.

LAKES seaplane *Sea-Bird*

The *Sea-Bird* was a reconstruction of the Avro Duigan tractor biplane, carried out under the supervision of Stanley-Adams at the Hill of Oaks base. It flew for the first time on 28 August 1912. It suffered a number of mishaps, with disintegrating propellers and handling by inexperienced people, during its long period in use as a passenger carrying and training aircraft, until finally written off on 3 June 1915. It was not a handsome aircraft and, although fitted with the original secondhand Gnôme from the *Water-Bird*, it served a most useful life on Windermere.

The fuselage of the Avro was used with new three bay warping wings, and was initially mounted on a large two-step central float, with rubber band springing to the chassis. Inflated airbags, on springboards, were used at the wingtips and tail and a water rudder was hung behind the main float. Various changes of float were made, including an unsatisfactory amphibious type at the end of 1912. Twin floats being fitted in the winter of 1914-1915, when the wingtip floats were also discarded.

The engine was mounted high in the fuselage with a semicircular cowl, although this was removed later, probably due to overheating.

Power: 50hp Gnôme seven-cylinder air-cooled rotary driving a 8ft 6in diameter propeller

Lakes *Seabird* was later fitted with twin floats.

Data

Span	39ft 4in
Main float	12ft long 4ft beam 1ft deep
Chord	4ft 7in
Length	29ft 4in
Area	350 sq ft
Area tailplane	22 sq ft
Area elevators	14 sq ft
Area rudder	15 sq ft
Speed	62mph

Lakes (later Northern Aircraft Co.) monoplane was built by Borwicks boat-builders of Bowness in 1913-1914 originally with a central float.

LAKES monoplane (Later NAC monoplane)

This pusher type monoplane seaplane was designed by O.T. Gnosspelius, and was made for the Lakes company by Borwick and Sons of Bowness in 1913-1914. The first flight was made in September 1914 by W.R. Ding, who had taken over the company by this time, and renamed it the Northern Aircraft Co. The aircraft was damaged on several occasions, including having the whole tail section broken off in May 1915 and, although repaired, was finally written off at the end of that month.

The pilot and passenger sat in tandem in a nacelle of rectangular section with a curved prow. The engine was attached at the rear, with the propeller in front of the engine, and the wings joined at the top longerons. The tail surfaces were attached to a braced, outrigger structure of tapering depth, which was carried forward to provide a chassis for mounting the single main float. This was attached to the nacelle by three pairs of struts and to the wing center section by two more pairs. Triangulated pylons, above the center section, provided anchorages for the wing bracing wires. Lateral control was by ailerons. Inflatable floats were fitted under the wings and tail.

The Lakes (NAC) monoplane with twin floats and other changes.

As with previous Lakes seaplanes, the original single step central float was replaced with single step twin floats, covered with aluminum alloy sheet, in which form it flew first on 1 May 1915. The wing and tail floats were deleted and the tail surfaces increased in area at the same time.

Power: 80hp Gnôme seven-cylinder air-cooled rotary fitted with an Integral propeller

Laking biplane No.1 was taken over by Frank Fitch and flown near Clacton in July 1911.

LAKING No.1 biplane (Guy Francis Laking, owner. Components built by T.W.K. Clarke of Kingston-on-Thames. Assembled by A. Fitch & Son, 610 Rosemary Rd., Clacton-on-Sea, Essex)

Guy Laking, the son of Sir Francis Laking, physician to King Edward VII, specified and financed the building of a biplane, largely made by T.W.K. Clarke and assembled by Fitch & Son, under the supervision of Frank Fitch. A shed was erected at Bockings Elm, Little Clacton for the trial flights. The machine was still in the course of manufacture in October 1909 when reported in the Press, but later, when it appeared for the first time, the owner was reported to have gone abroad and the trials were postponed indefinitely. Laking appears to have lost interest and passed responsibility for the machine to Frank Fitch. (See Fitch biplane)

Power: Two 12hp JAP engines driving 5ft 8in diameter pusher propellers

Data

Span	33ft
Chord	5ft 6in
Gap	5ft
Area	330 sq ft (Press reported 156 sq yd!)
Weight allup	504 lb

LAMPLOUGH Orthopter & Lifterplane (Lamplough & Sons, Albany Works, Willesden Junction, London, NW)

This machine was designed to operate on the principles of bird flight and was exhibited incomplete at the Aero Show of March 1909 at Olympia.

It consisted of an elongated framework of ash members covered with Continental fabric and containing the mechanism for operating a pair of biplane wings, with their leading edges facing one another. The propulsion system consisted of two pusher propellers on long shafts driven from the forward part of the machine. A swaying motion was imparted to the lifters, by cranks, and their angle of incidence was controlled by links. Conventional biplane wings, with five pairs of interplane struts, were mounted to the sides of the framework, and biplane elevators with single rudders at front and rear.

Needless to say the Orthopter was a failure and in 1910 was followed by a biplane ornithopter with wings actuated by a 25hp motor, which was probably a modification of the earlier machine.

Data	Span	20ft
	Area	945 sq ft
	Weight	900 lb

Lamplough ornithopter appeared incomplete at the Olympia Aero Show in March 1909.

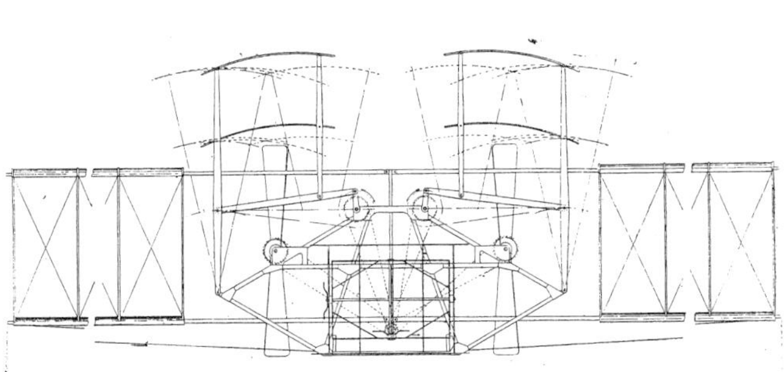

Lane glider operated on the Members Hill at Brooklands in 1910.

LANE glider (Charles Lane, 31 Foley St., Lane's British Aeroplanes Ltd., with works at King's Place, King St., Camden Town, London, N. Flying at Brooklands, Hangar No.1)

A Wright type glider was operated from a starting rail on the Members Hill at Brooklands in the early part of 1910. The short glides, down the slope, received publicity because some were by Mrs. Gavin, an actress, which were claimed to be the first flights by an aviatrix in Britain. The machine was bought by Mr. Gavin and later carried his wife's stage name under the wings. It was taken to their home but little more was heard of it.

The glider was a two bay biplane with single acting ailerons on the top wings, the lower wings being curved up towards the tips. A front elevator was carried on booms and two rudders were mounted in the rear booms, that also carried an elevator.

LANE monoplane

This single-seat monoplane was built on the lines of a Blériot XI, but with a tailplane below the bottom longeron, and divided elevators mounted separately above and slightly behind. The castering undercarriage was sprung by rubber cords, connected by wires horizontally to a point midway down the fuselage, and served the purpose of realigning the wheels, as well as providing sprung suspension. Construction was of wood with a tubular steel undercarriage.

The first report was of trials conducted by Charles Lane on 13 January 1910. Wilfred Foulis, who was to become Lane's agent in Scotland, received a machine in February, which he was reported to be flying near Edinburgh in the spring. A mono-

Lane monoplane No.1 with modified undercarriage.

Lane monoplane No.1 at Brooklands.

Lane monoplane No.2 at Olympia in March 1910. Several of the smaller machines at the rear were made.

plane with the fuselage uncovered was shown at Olympia in March 1910. This machine, and the Foulis aircraft, were certainly fitted with NEC engines. In May, *The Aero* reported two machines in hangar No.1 at Brooklands, by then fitted with non-castering undercarriage with twin skids and four wheels, with which Lane and Astley continued flying throughout the summer.

It appears that the first Lane monoplane was originally fitted with a three-cylinder Anzani engine, which was replaced with the four-cylinder two-stroke NEC. This was an unsatisfactory engine, which did not produce the claimed power, and was finally replaced by a 35hp ENV, with which it was flown on 26 May 1910, receiving a complimentary Press report.

An early Lane monoplane fitted with an Anzani engine was wrecked at Brooklands on 25 July 1911 when flown by Pashley.

Power: 25hp Anzani three-cylinder air-cooled fantype semi-radial
25-30hp NEC four-cylinder inline water-cooled two-stroke driving a 7ft diameter propeller
35hp ENV type D eight-cylinder water-cooled vee.

Data		
	Span	30ft
	Chord	6ft 6in
	Length	24ft
	Area	160 sq ft (201 sq ft)*
	Area tailplane	21 sq ft (26 sq ft)*
	Area elevators	18 sq ft (24 sq ft)*
	Area rudder	5 1/2 sq ft (8 sq ft)*
	Weight	425 lb (450 lb)*
	Weight allup	650 lb
	Speed	30mph
	Price with ENV	£500

*Alternative data from *The Aero*

LANE monoplane

In addition to the single-seater, Lane also exhibited at Olympia an incomplete side-by-side two-seater, similar in layout to the Blériot XII, but with Lane's own type of tail unit and undercarriage. In this case the elevator was in one piece and the rudder was divided, the lower portion serving as a skid.

The engine was mounted low down in the airframe, with the propeller level with the leading edge, chain driven at about a reduction of two to one. The pilot and passenger sat on a bench seat behind the engine, just below the wing trailing edge.

The Blériot XII was not a great success and the similar Lane machine followed it into obscurity. There are no reports of testing and the machine may not have been completed after the Aero Show.

Two Lane monoplanes, a single-seater and a two-seater, were offered for sale by L.N. Palmer of Tooting, London SW at bankrupt prices.

Power: 60hp NEC six-cylinder inline water-cooled two-stroke driving by chain an 8ft 6in (8ft 8in)* diameter propeller at 2 to 1 reduction

Data		
	Span	36ft 6in
	Chord	7ft 10in (8ft)*
	Length	24ft
	Area	250 sq ft (320 sq ft)*
	Area tailplane	30 sq ft
	Area elevators	25 sq ft (26 sq ft)*
	Area rudder	7 sq ft (8 sq ft)*
	Weight	770 lb (675 lb)*
	Weight allup	970 lb
	Speed	30mph
	Price	£800

*Alternative contemporary data

LANE biplane

Lane built a Farman type biplane in 1910, fitted with a 60hp ENV type F engine. It was illustrated in *Flight* on the 17 December 1910 (p.1037) with C.W. Bowles, an ex-pupil of the Lane School in the pilot's seat; thereafter it went to India for the Maharajah of Patiala, for whom it was purchased.

It was a single-seat pusher of typical Farman layout with biplane tail and single front elevator. Single acting ailerons were fitted on top and bottom wings which were equal in span.

Lane biplane at Brooklands in 1910 before dispatch to India.

Lee-Richards annular monoplane No.1 of 1913 before covering.

LANGLEY & ROBERTSON glider (Richard Langley and William Robertson, Middlesbrough)
The partners built a glider in 1913-1914 which was tested at Ormesby Bank in July 1914.

LANGLEY SMITHERS monoplane
This aircraft was built in 1908-1909 and was assembled and tested at Knockholt Cricket ground in Kent. It took off, but crashed on the first attempt and appears not to have been rebuilt.

The fuselage of the monoplane was an open parallel girder, with curved top and bottom members, meeting at both front and rear ends. A tailplane and front elevator were fitted. There was considerable dihedral to the wings, which were braced to a tall pylon of four struts, and could be warped. The unidentified type of motor drove twin tractor propellers, apparently by shafts and bevel gearing.

Langley-Smithers monoplane of 1908-1909 at Knockholt, Kent.

LASCELLES Ornis monoplane
See Howard Wright.

LEE biplane (Lee, "Haven Bank", Formby, Lancashire.)
A Farman-type biplane was advertised in *Flight* 3 August 1912. It was fitted with a 35hp engine of unknown make. The mainplanes needed recovering. It was housed in a shed next to Freshfield aerodrome. This may have been Cedric Lee (q.v.)

LEE RICHARDS annular monoplane No.1 (Cedric Lee and George Tilghman Richards)
The association with Kitchen (*q.v.*), having ended in 1912, wind tunnel tests were carried out at the East London College and at the NPL to establish aerodynamic data for a new design of monoplane. After a false start with Blackburns, the detail design and construction was undertaken by James Radley and E.C. Gordon England, his works manager and test pilot at Shoreham. After problems created by Radley, the machine was completed and ready to be flown by Gordon England on 23 November 1913. After a minimum of taxiing, he flew successfully for the first time, although noting tail heaviness, until the engine cut out through lack of fuel, on the approach to land. In the resulting, crash Gordon England was injured and out of action for some months. The engine and other mechanical parts were salvaged for use in a new machine.

The monoplane wing was a complex structure, which required considerable ingenuity by Gordon England in its construction, involving as it did, a variety of sections at different positions in the wing. This was built in two halves around two steel tube, semicircular spars, spaced by steel tube compression struts at intervals. On to this basic structure were placed numerous wooden ribs, positioned radially, and connected by inner and outer edge members and stringers. At the rear edge, each wing was cut straight across to form flaps, serving as both elevators and ailerons, their trailing edges continuing the outline of the wing.

The fuselage was a tapering wooden girder of four longerons and struts braced by wires, with the addition of a curved top decking. The propeller was mounted on a long extension shaft on the Gnôme engine, which was mounted on a bulkhead, to which the rear spars were also attached. Behind this were fuel and oil tanks, with the passenger seat and pilot behind. An inverted Y-pylon, in front of the passenger's cockpit, provided an anchorage for the numerous bracing wires. A long dorsal fin and semicircular rudder were fitted. The undercarriage consisted of two wheels on swing axles pivoted on the vee-shaped pylon below the fuselage, which also served as the anchorage for the lift wires. A pair of nose wheels was mounted on the central member protruding forward of the pylon. The tail was also protected by a skid.

Power: 80hp Gnôme Monosoupape seven-cylinder air-cooled rotary

Data		
	Span	22ft
	Dihedral	5 deg
	Length	23ft 6in
	Area	280 sq ft
	Weight allup	1,680 lb (2 crew) 1,500 lb (as flown)
	Speed	83-85 mph
	Climb	300ft per min

Lee-Richards annular monoplane No.2 of 1914 was similar to No.1 but with additional elevators above the tail and behind the wings.

LEE RICHARDS annular monoplane No.2

The second aircraft was completed and flown by England in February or March 1914, but after a number of flights totaling some 25 hours, a tendency to spin was not resolved and he resigned. Following this N.S. Percival flew once, followed by Gordon Bell from early April until 25 April 1914, when the machine was wrecked, the pilot having lost elevator control. This monoplane was almost identical in appearance to No.1 except for an additional elevator above the rudders, and the parallel chord elevators behind the wings, which were moved outwards during the course of development. The wing dihedral was reduced from 5 to 3 degrees. A speed range of 30 to 70 mph was reported.

At the end of February, Lee made two entries of machines with Austro-Daimler engines for the Gordon Bennett Race of September 1914, which was abandoned when war broke out. These machines would have been based on No.2, or its successor, would have a wingspan of fifteen feet and were reported to be partly built.

LEE RICHARDS annular monoplane No.3

The third machine incorporated changes to improve control and reduce lateral stability, by a reduction of dihedral to 1 1/2 degrees, and changes of wing section. The improvised biplane tail of No.2 was replaced with enlarged elevons, positioned well out to the side, and faired into the wing outline. A large ventral fin was added with a parallel chord rudder of greater area.

The date of completion is not recorded, but the machine was being flown by Gordon Bell up to the outbreak of war. Cedric Lee, who was an inexperienced pilot, then attempted to fly No.3 in August 1914 and crashed into the River Adur, wrecking the machine, but only suffering minor injuries himself.

Lee-Richards No.3 annular monoplane of 1914, the final version.

LEEMING glider (J.F. Leeming, Southport, Lancashire)

A rudimentary monoplane glider, made and flown in 1910 down sloping ground, reached a height of 1214 feet and covered about thirty-five feet distance. It was controlled by weight shift. A rough sketch and description was given in a letter to *The Aero* of 25 January 1911 (p.65). This referred also to the construction of a biplane of twenty-four feet span.

Data		
	Span	20ft
	Chord	6ft
	Length	14ft
	Tailplane span	10ft
	Tailplane chord	3ft
	Area	120 sq ft
	Tailplane area	30 sq ft

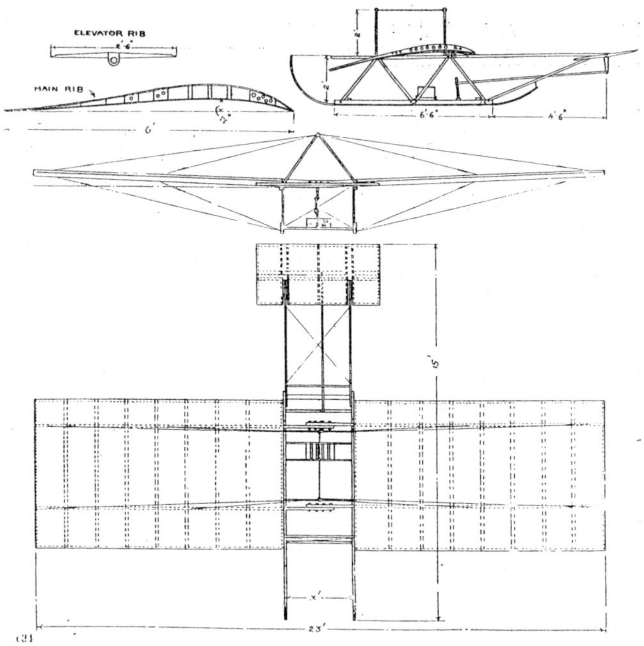

LGD glider made by members of staff of *The Motor* in 1909.

LGD monoplane glider. (LGD of *'The Motor'* magazine, London)

This tail first monoplane glider was made in 1909 by a member of the staff of *The Motor,* who described it in that journal as one of the few gliders with a double surfaced wing, then in being.

The two spars, of the parallel chord wings, were joined at the center, and were attached to the top members of an open structure. The lower members of this structure were skids, supporting a platform, on which the operator was seated. He controlled the flat section front elevator with a short control column. The wings were braced to an overhead cabane and below to the skid members. Spruce and hickory wood were used in its construction.

Data	Span	23ft
	Chord	6ft
	Length	15ft
	Height	4ft
	Area	140 sq ft
	Weight	53 lb

LING monoplane

Jane's all the World's Airships 1909 reported that this machine was being built at Hull, but nothing more was heard of it. This was almost certainly the Newington monoplane (q.v.)

LINOGREN Aerial Yacht (Oscar Linogren)

It was reported that a 'Patent Aerial Yacht' was built on the Thames in 1873.

LIPMAN helicopter (S. Lipman, Manchester Aero Club)

This machine, of which no drawings or photographs have been found, was reported in 1909 to have six helices or lifters driven by a petrol engine in the center of the body and two propellers which worked independently.

Liwentaal glider. Artist's impression of the machine of 1894. (Courtesy Anthony Bryant).

LIWENTAAL glider, the Dittisham Aerostat (Albert or Alexander Liwentaal, Snail Cottage, Dittisham, Dartmouth, Devon)

Liwentaal was a Swiss who worked in the shipyard of Simpson and Strickland at Noss, Dartmouth. Between 1891 and 1911, a number of patents were taken out in the name of Alexander Liwentaal relating to flying machines.

His first attempt at flight seems to have been in February 1894, from a hill above Dittisham Mill Creek. Although the machine lifted six or seven feet for several yards, it was damaged when hit by a gust of wind. A further attempt was made in April from slopes at Bozamzeal, Dittisham which resulted in severe damage and injury to Liwentaal needing treatment in Dartmouth hospital.

The pilot sat on a bicycle saddle astride a single steel tube fuselage, mounted on an undercarriage consisting of a single bicycle wheel and a tail skid. The wing, shaped like that of a gull, comprised a straight main spar and curved leading edge member, and ribs of pine and a wire trailing edge. The tailplane was 'all moving' and there was a rudder for directional control.

LIWENTAAL ornithopter

After the accident with his glider, Lt. Liwentaal proposed, in a letter to the Press, to build a two-seater powered machine with flapping wings, which was said to be in course of construction in December 1894. Nothing further was heard of this and Liwentaal moved to London. His use of military rank probably stemmed from service in the Swiss Army, a largely part-time organization.

Power: 30hp and 5hp. The small engine presumably to operate the wing flapping at a maximum of forty strokes per minute.

Data	Span	54ft
	Length	27ft
	Area	475 sq ft
	Area tailplane	125 sq ft
	Weight	300 lb
	Weight allup	600 lb
	Speed	70mph est.

LIWENTAAL monoplane (Alexander Liwentaal, 92 High St., Islington, London N)

Some years later, when he was living in London, Liwentaal took out patent No.11784/1911. This was for a tractor monoplane with a control system consisting solely of air under pressure directed to valves at the extremities of the machine, where it emerged as jets. The operation of the valves was to be controlled by pendulums. This machine remained as a project only, but the control by air jets was remarkably similar to the reaction control system used in the 1960s in the Harrier jet fighter.

In 1909, back in Switzerland, Liwentaal constructed the unsuccessful Libellan monoplane at Vernier, Geneva and was involved in the promotion of an airship, to be named *'Ville de Geneve'*, which failed to materialize due to lack of funds. In wartime he is reported to have had an aircraft factory at Courbevoie, near Paris, which was bombed by a Zeppelin and in 1919 he emigrated to Canada.

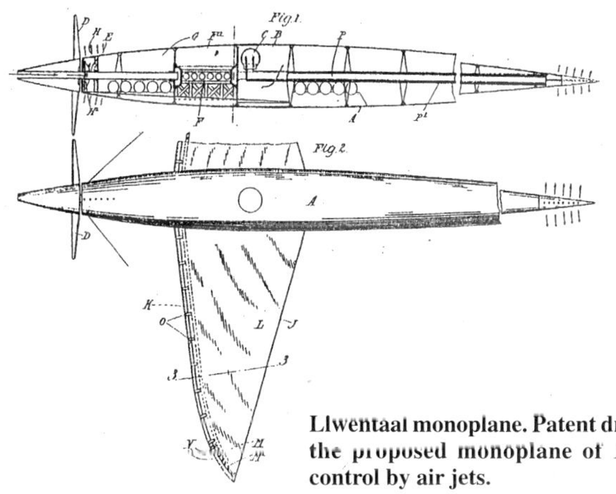

Liwentaal monoplane. Patent drawings of the proposed monoplane of 1911 with control by air jets.

LOBNITZ biplane (H.C. Lobnitz, Baring Rd., Cowes, Isle of Wight)

Nothing more has been traced on this machine reported to be under construction at Cowes in 1909, other than patent No.2111/1909, which describes a pendulum stability device for control of an elevator and ailerons. The patent does not describe an aircraft in any detail.

LOCKE glider (J.C. Locke, 47 Buxton Rd., Chingford, Essex)

This was built during 1910 and was tested at Barking on 26 July 1910. The machine was tailless with a swept back wing, braced to kingposts, above and below; the machine rested on skids.

The tailless Locke glider tested at Barking in 1910.

Data	Span	37ft
	Chord	7ft

LONDON & PROVINCIAL biplane (London & Provincial Aviation Co., Hendon)

Messrs. Warren and Smiles operated a flying school at Hendon and built a biplane, for tuition purposes, which was ready by September 1914. Further machines were to be completed later. Illustrations of the machine showed a typical Caudron biplane with no obvious differences from original French-built machines.

Power: 35hp Anzani three-cylinder Y-type air-cooled radial.

L & P (London & Provincial) biplane was of Caudron type built at Hendon in 1914 and later.

LONDON SCHOOL of AVIATION biplane

Under construction and shown in model form at the Stanley Show in November 1909.

LONDON SCHOOL of MOTORING monoplane

Reported in December 1909 but failed to materialize.

Long monoplane. JBD Long had this made by R Beney & Co. and tested it at Acton in 1911.

LONG monoplane (J.B.D. Long, London Aviation Ground, Acton)
This machine was built on Blériot lines, apart from its four wheel and skid type undercarriage and long tail-skid. It was constructed by Robert Beney & Co. of 7 Carlisle St., London W, using materials from Handley Page, and was tested early in 1911 at Acton.

LONG monoplane (Victor Long, The Firs, Arundel, Sussex)
This monoplane, built in 1910-1911, was similar in layout to a Blériot and incorporated a Blériot type undercarriage. However it differed in other respects, particularly in the tail unit, which consisted of a one piece elevator and tailplane, with a long triangular fin and square shaped rudder, mounted above the fuselage girder. The heavily cambered wings incorporated cutouts for small ailerons in the trailing edges.

A 20/24hp JAP vee four, side valve, air-cooled engine was fitted, originally with a Cochrane metal propeller with corrugated metal trailing portions, but this was replaced by a Chauvière wooden propeller at some stage. The original pilots controls consisted of two hand wheels, Antoinette style, mounted fore and aft on the top longerons, but these were replaced with a more usual control column.

No record of tests or flight exists. Long discontinued his aviation activities and offered a Blériot type monoplane with 'Farman type' landing chassis, without engine and propeller for £20 in *Flight* on 8 July 1911 and the JAP engine and Chauvière propeller for £70 on 5 August 1911, together with a Blériot chassis.

LOVETT & RANDALL man powered aeroplane
Other than the name nothing is known of this machine.

LOWY & SWINGLER monoplane (A.E. Lowy & Swingler, Birmingham)
These two were members of the Birmingham Aero Club who began testing their monoplane early in 1914, but with unsuccessful results. Lowy was previously a partner in building the Franklin monoplane.

LUMB monoplane (J. Lumb, Blackpool, Lancashire)
Lumb constructed this machine in 1909 and took it to the Blackpool Aviation Meeting that year, but it did not fly during the meeting. The engine was a 50hp JAP, of unknown type. No more was heard of this machine.

Data	Span	36ft
	Chord	6ft 6in

Lumb monoplane built at Blackpool was at the Meeting of 1909 but did not fly.

The completed Victor Long monoplane probably never flew and was disposed of in 1911.

Lyster-Smythe glider built in Ireland in 1909.

LYSTER-SMYTHE glider (Cecil and Richard Lyster-Smythe, Barbavilla, County Westmeath, Ireland)

The Lyster-Smythe brothers aged fourteen and twelve respectively built a primitive biplane glider, towards the end of 1909 and did all the work themselves, with the exception of the wheel axles. Although somewhat crudely made, the brothers were able to achieve a number of glides of thirty or forty yards. A rudder was fitted and a front elevator but with no obvious method of warping the wings. The undercarriage comprised of a pair of bicycle wheels.

Data	Span	20ft
	Chord	6ft 4in
	Height	6ft

M

McARDLE & GOULD monoplane (W.E. McArdle and George Gould, Gladstone Rd., Bournemouth and Bagshot Moor, near Beaulieu)

W.E. McArdle and J. Armstrong Drexel were the proprietors and chief flying instructors of the New Forest Aviation School near Beaulieu, which flourished in 1910 and early 1911 and which, at its height, had at least seven Blériot monoplanes in use; three of these were built by McArdle's garage and motor works at Bournemouth and fitted with Anzani engines.

McArdle and George Gould, who became chief engineer of the New Forest School, built a monoplane of Blériot type at Bournemouth in August 1909. Nothing was subsequently heard of it. George Gould was McArdle's foreman carpenter for the aero-construction.

In May 1910 McArdle was reported to be building three biplanes at Bournemouth. The Flying School closed down in early 1912 and some or all of their aeroplanes were sold to Major F.B. Fowler of the Eastbourne Aviation Company's Flying School.

McCURD multi-plane (W.A. McCurd & Co., Bellbrook, Ballard's Lane, North Finchley, London, N3)

This machine was being designed and constructed by this company in 1910, featuring wings of variable incidence. It was to be fitted with a 60hp engine, but was not completed. Patent No.8966/1907 refers. McCurd & Company were motor lorry body builders.

M'DONNELL monoplane (Reginald F. M'Donnell, Lower Mount Street, Dublin)

A tractor monoplane built in late 1909. No elevators were fitted as the main planes were designed to flex to give the same effect as elevators. An engine of less than 15hp was designed by M'Donnell.

Data	Span	13ft 6in
	Length	10ft

McDONNELL monoplane (McDonnell, Booterstown, near Dublin, Ireland)

This concern was reported to be making three monoplanes in 1909, of which one at least was constructed using steel tubing. The third machine was reported as having movable planes that would also be warped in combination with a box shaped lifting plane for steering. The frame of the fuselage was made of steel tubing. The powerful engine was of McDonnell's own design. No evidence that any of the monoplanes were finished. (It is possible that these machines relate to Reginald M'Donnell mentioned above)

McArdle and Gould monoplane was built at Bournemouth in 1909.

Macfie monoplane built and flown for a short time at Fambridge in 1909.

MACFIE monoplane (Robert Francis Macfie, Fambridge and Maplin Sands, Essex)

Macfie came from San Francisco to England in 1909, and began construction of the monoplane at Fambridge on 2 August 1909. Trials commenced on 16 September 1909 and flight was achieved in the period to 19 October 1909, after various modifications and repair of damage. Fambridge was unsuitable as a flying ground, and on 11 November 1909, the monoplane was taken to Maplin Sands, but was removed by the end of the month after damage, bad weather and a War Office order to move. Unable to find a suitable field in the Southeast and Midlands, Macfie took the aircraft to France, but was refused permission to fly at Pau, and Croix d'Hins was considered unsuitable. When the machine was received back in London on 2 February 1910 it was beyond repair.

The monoplane was a single-seat tractor with a triangular section fuselage consisting of a tapering ash girder, made with butt joints, reinforced with steel fittings, bound by tape, and glued. The whole structure was braced by wires and was mainly uncovered. The castering undercarriage was made of steel tubing and was sprung, as was the tail-skid.

The wing had a marked camber and small dihedral and incorporated a warping trailing edge portion. Two triangulated pylons above the fuselage provided anchorages for bracing wires and the warping control.

Power: 35hp JAP eight-cylinder air-cooled vee driving a 6ft 6in diameter propeller.

Data		
	Span	28ft 6in
	Chord	6ft 6in
	Height	5ft to center prop boss
	Area	164 sq ft
	Area elevators	24 sq ft
	Area rudder	4 sq ft
	Weight	680 lb

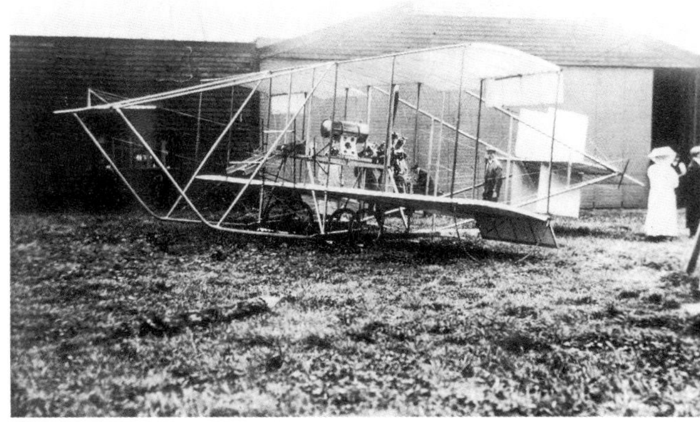

Macfie *Empress* biplane of 1910 flying at Brooklands.

MACFIE *Empress* biplane (Flying at Portholme, Hunts and Brooklands)

This machine was designed by Macfie, the parts were fabricated by W.H. Tothill and the aircraft was assembled at Portholme, between March and May 1910. First flight took place on 12 May 1910, but immediately afterwards, permission to use the ground was withdrawn and Macfie moved to Brooklands in June. First straights took place there on 18 June 1910 with insufficient power for proper flight. The machine was then altered to become more of a Farman type and loaned to J. Radley to take to Wolverhampton, but apparently was not used.

On its return in July, the aircraft reverted to its original form, and damage, including that caused by a fire, was repaired. In the week ending 14 August 1910, the aircraft reappeared at Brooklands with a 35hp Green engine in place, but lacking a suitable propeller, could not achieve flight. In September a 50hp Gnôme was fitted and with this engine straight flights were made. However, as control was too sensitive, the tail booms were extended by five feet, in which form the aircraft flew on 4 October 1910 more satisfactorily. Macfie flew in the Neill Cup Contest, but only accumulated 7min 15sec flying time on 6 October 1910, far less than the 2hr 59min 17sec of Gilmour, the winner. Nevertheless the biplane flew well and was used by Macfie to obtain his RAeC pilot's certificate No.49 in January 1911, and it was also flown by others including J. Valentine.

A new engine, the Empress rotary, was fitted and ground tested on 20 November 1910. The aircraft was being flown by O.C. Morison on 22 January 1911, when the engine cut out and he landed in the sewage farm. The aircraft was salvaged and some of the parts used to rebuild it, probably again fitted with the Gnôme. Straights were flown on 1 March 1911 and circuits on 8 March 1911.

The basic Macfie biplane was a typical pusher machine of the time, with front and rear elevators and fitted initially with a single rudder, which was replaced by two at some stage. The engine, pilot and passenger were mounted centrally on a platform, on struts between the wings, which carried single acting ailerons on both top and bottom planes. the rear booms were parallel in plan, but tapered to a point in elevation. The front lower booms were extensions of the undercarriage skids, which curved upwards to meet the straight top booms at the front elevator pivots. The skids were extended aft to support the tail and provided the mountings for the four main wheels on separate axles.

Macfie seems to have concentrated on his next aircraft and made little use of his biplane after it was rebuilt. At some time in 1911 he disposed of it to Herbert Spencer who rebuilt it on Farman lines and completed and flew the machine at Brooklands in April 1912.

No data is available other than information on the various power plants and the fact that the tail was extended by five feet at an early stage.

Power: 35hp JAP eight-cylinder air-cooled vee, originally in the monoplane.
35hp Green four-cylinder inline water-cooled from 14 August 1910.
50hp Gnôme seven-cylinder air-cooled rotary from 20 September 1910.
50hp Empress seven-cylinder air-cooled rotary from 20 November 1910. Gnôme fitted again later

MACFIE Circuit biplane

Macfie's entry for the Circuit of Britain of 1911 was built in three weeks and came out at Brooklands on 17 July 1911, but was damaged on its first attempt at taxiing. It seems that the center of gravity was too far forward, putting pressure on the skids, causing them to break and foul the propeller. Withdrawal from the contest resulted and no further attempt to rebuild the aircraft was made.

The intended 100hp ABC engine was not available in time and a 50hp Gnôme was fitted instead. The fuselage was an open rectangular section, wooden, braced girder tapering in plan and elevation, but finishing at a vertical post well behind the tail unit. The fuselage was mounted centrally on struts between the wings, enabling the pilot to see above the top wing. It may have been intended to cover the structure later for maximum performance to be attained. Long curved skids, similar to those used on the earlier biplane, were fitted, with a pair of castering wheels on diagonal spring struts.

The wings were of tapered form with rounded tips and were swept back at an angle of twenty degrees at the leading edge. The vertical pairs of interplane struts were closely spaced and allowed warping of a large portion of the trailing edge. The tail unit comprised three sets of divided rudders, with pairs of coupling rods top and bottom; the intermediate hinges being combined with the outboard pivots of the centrally mounted elevators.

Power: 50hp Gnôme seven-cylinder air-cooled rotary.

Data		
	Span	26ft
	Chord	3ft 6in tapering to 2ft 3in
	Area	130 sq ft
	Gap	4ft
	Weight allup	600 lb

MACKENSIE-HUGHES triplane (E.S.B. Mackensie-Hughes and A.W. Smith, Barking)

This machine, made by A.W. Smith at Barking, arrived at Brooklands in June 1910 and was housed in shed No.22. It carried the name *'Britannia'* in large letters on the fuselage sides, but was nicknamed 'The Staircase' by the Brooklands cynics. The machine failed to fly when first tested on 9 July 1910, or subsequently, being under-powered and too heavy. It was soon converted to become the Molesworth triplane (*q.v.*).

The mainplanes were staggered, with the middle one overlapping the trailing edge of the lowest by nine inches. The top plane was similarly placed above and behind the middle one. The center plane had a chord approximately double that of the

Mackenzie-Hughes triplane at Brooklands in 1910.

Macfie biplane built for the 1911 Circuit of Britain but sustained damage and was not rebuilt.

others, and it carried ailerons in addition, hinged to the trailing edge. Three pairs of interplane struts of N-form spaced each pair of wings, which were braced to vertical kingposts, above and below the fuselage. The tailplane was also a staggered triplane unit, with elevators on the center panel. The rudder was forward of the tailplane unit, and was hinged on a vertical strut, also carrying the tail wheel below the fuselage. Lateral control was by sideways movement of the back of the seat; rudder and elevator control was by separate levers.

The fuselage, which was fabric covered, was a rectangular section, wooden girder tapering to a point at both front and rear, with the engine housed in a compartment just forward of the pilot. The engine drove the overlapping twin propellers, ahead of the nose through chains and long shafts. The undercarriage was a widely spaced arrangement of struts and twin skids with two pairs of wheels, sprung mounted, on separate axles.

Power: 20hp JAP four-cylinder air-cooled vee driving twin metal propellers by chain and shafts at a reduction of 3 to 2, spaced at ninety degrees in overlapping planes.

Data	Span	23ft
	Area	250 sq ft
	Weight	600 lb

Mackenzie-Osborne multiplane built and tested near Lincoln in 1910.

MACKENZIE-OSBORNE multi-plane (Aubrey C. Mackenzie & Co., 21 Monks Road and Osborne Bros., Portland St., & Princess St., Lincoln)

A photograph of the partially assembled machine is the only illustration traced. Little is known of this machine, which was built at Princess St. and tested at Lincoln West Common in July 1910. The engine was a JAP two-cylinder and the span 21ft.

MACKWORTH biplane (Captain H.L. Mackworth RE, Shorncliffe, Kent)

This small machine of lightweight steel tube and bamboo construction, with silk covering, was built during 1909 by Capt. Mackworth and his brother, a Captain in the Royal Navy.

It weighed 250 lb. It was tested at St Martin's Plain Shorncliffe Camp in autumn 1909.

Power: Two FN engines producing a total of 14hp.

Data	Span	30ft
	Chord	5ft

McMULLEN aeroplane (Kenrie James McMullen, Brimpton, Berkshire)

The maker of Rational cars was reported to be making an aircraft in 1910-1911.

MacWILLIAMS glider (MacWilliams, 117 George St., Whitchurch, Glasgow)

This machine was described as of Valkyrie type, when it appeared in May 1911.

Data	Span	30ft
	Weight	70 lb

MADELER monoplane (A.W. Harrison, 42 Talbot St., Canton, Cardiff and Abergavenny)

This was reported to be a two-seater fitted with a 35/40hp two-cylinder water-cooled engine. No other information is available.

MADISON-BUTLER monoplane (Madison Co., Wade St., Littleover, Derby)

This single-seat tractor monoplane, known as the *'Silver Flyer'* was reported built during 1910 and was an adaptation of a Blériot monoplane. A separate report referred to a flight made at Clewes Farm in 1913-1914 by a machine built by the Madison Co. and flown by a Mr. Butler. The flight resulted in damage and the machine was abandoned.

Power: 30hp Anzani three-cylinder air-cooled, probably Y-type radial.

Data	Span	22ft
	Area	170 sq ft
	Length	24ft

MAILLARD glider (G.C. Maillard)

A Chanute type glider was reported made and crashed in early 1910. Span 20ft.

MANN monoplane (G.B. Mann, Wolverhampton)

Mann, a member of the Midlands Aero Club, made a small monoplane which was present at Dunstall Park on 3 September 1910. The machine was in company with the Seddon biplane and was similarly constructed of steel tube, supplied by the local firm of Accles and Pollock. Several unsuccessful attempts at flight were reported.

Power: 18/20 or 25hp JAP driving a 6 or 7ft diameter propeller.

Data Area 140 sq ft.

Mann monoplane at Dunstall Park, Wolverhampton in 1910.

Mann & Grimmer M1 with simplified undercarriage.

MANN & GRIMMER M.1 biplane (R. Frank Mann and Robert P. Grimmer, Arlington Rd., Surbiton, Surrey)
These partners had been involved since 1910 with models and gliders in the Arundel House School Aero Club. In August 1914 schemed the machine and in September they secured the support of W.H. Bonham-Carter to finance the project. Construction then began in a disused church building in Surbiton and by the end of January the machine was taken to Hendon.

Rowland Ding made a short straight flight on 19 February 1915 and a circuit of ten minutes the following day. Inevitably a number of things needed to be changed, and after further trials in March, the machine was returned to Surbiton for major work, including the fitting of a more powerful engine, new undercarriage and rudder and general strengthening.

The machine was returned to Hendon at the end of June and housed in a tent, as no spare shed was available. On 3 July 1915 the seventh flight was made and only quick action by Ding prevented a collision with a Caudron when landing. On 1 August 1915 Ding made his last flight, disenchanted no doubt after a propeller shaft failure and other setbacks. Sydney Pickles took over and flew the aircraft for a total of three hours in August, before being forced to give up. An ex-RFC pilot, A.E. Barrs took over on 2 October 1915, and carried out a number of satisfactory flights until 16 October 1915, when the gearbox seized, resulting in a broken propeller and chain. As a result of this and the deteriorating condition of the machine, due to the poor accommodation, it was moved back to Surbiton and dismantled, having flown only 18 hours. An improved successor, identified as M.2A, was half built by March 1916 and a government contract for it had been received, nevertheless the financier withdrew his support, and the incomplete machine was sold to Blériot Aeronautics at Brooklands. It was not developed.

The M.1 was unusual in layout, being similar to a tractor biplane with a front mounted engine, but original in that the engine was reversed to drive a pair of pusher propellers, through shafts and chains behind the wings. The drive shaft from the engine passed under the cockpit to a gearbox, from which chains were taken out through the fuselage sides to sprockets on the propeller shafts. These revolved in bearings hung on cables and light struts between the wings. This flexible method of mounting was not satisfactory, and was replaced by substantial struts at the time of the engine change. The interplane struts of the two bay wings were also strengthened and streamlined at the same time. The original undercarriage, with twin skids and multiple struts, was changed for a simpler one of vee type. A larger rudder, with balance areas above and below the fuselage, was also fitted.

Power: 100 and 125hp Anzani ten-cylinder air-cooled radial driving twin pusher propellers by shaft, gearbox and chains.

Mann & Grimmer M1 biplane was built at Surbiton in 1914 and was tested at Hendon throughout 1915. Shown here with original undercarriage.

MANN & OVERTONS monoplane (Mann & Overtons Ltd., 15 Commercial Rd., Pimlico, London SW1)
This was one of three variants of the Santos Dumont *Demoiselle* which appeared at the Aero Show at Olympia in March 1910, this one built by a professional firm of coach builders.

The structure of three pine longerons was braced by oval steel tubes and wires. The steel tube undercarriage was unsprung, but relied on the pneumatic tires and the flexibility of the curved axle to cushion landing shocks. The wing had a large cutout at the center and was set at a considerable angle of incidence.

Mann & Overtons exhibited this *Demoiselle* type monoplane at Olympia in March 1910.

Above & Center: Manning-Flanders MF1. This is the 1974 replica of the abandoned design, made from the original drawings.

Flight and *The Aero* disagreed on the method of lateral control, *The Aero* stating that this was catered for by the dihedral, whereas *Flight* stated that the left hand lever operated both rudder and warp; the right hand lever controlled the elevator. The pilot sat low down in the fuselage, below the center section. The engine was mounted on a cast aluminum bracket on the front of the single top longeron, with a streamlined oil and fuel tank behind.

Power: 30hp Anzani three-cylinder fantype semi-radial driving a 6ft diameter propeller.

Data		
	Span	18ft 4in (20ft *)
	Chord	7ft
	Length	20ft
	Area	120 sq ft (133 sq ft *)
	Area rudder	6 1/2 sq ft
	Area elevator	16 sq ft
	Weight	326 lb (340 lb*)
	Weight allup	536 lb
	Speed	45mph
	Price	£300

*alternative figures from *The Aero.*

MANN & WELLINGS monoplane (Brooklands, Hangar No.18)

This was described in *The Aero* 24 May 1910 (p.407) as 'a curious little machine with a fuselage of steel tubing, nicknamed the *Bedstead*'. No further information available.

MANNING-FLANDERS MF.1 monoplane (W.O. Manning and Howard Flanders)

This aircraft is one that was not completed in 1910-1911, when it was originally conceived. The basis of it was apparently the Flanders F.1, which was abandoned in June 1910-1911 (*q.v.*) and which Flanders passed on to Manning, at the time when Manning was employed by the Coventry Ordnance Works. Pressure of work prevented Manning completing the design and construction, the aircraft being broken up at Battersea in 1912. A Crossley aero engine, intended to be fitted, was also not produced.

The was a conventional style tractor monoplane of the period, using a triangular section fuselage, uncovered behind the pilot. A central skid undercarriage and a cabane above the fuselage provided anchorages to brace the wings, which probably employed warping control in the original design.

However a replica aircraft based on some of the original drawings was built in 1974 by Personal Plane Services of White Waltham. The replica MF.1 has flown successfully at intervals, including a Channel crossing in 1984, powered by a 65hp Continental flat four air-cooled engine.

MARLBORO aeroplane (Marlboro Aeroplane Works, London Aviation Ground, near Ealing)

It was reported that a machine, type unknown, with a 36hp Coventry Simplex six-cylinder engine, was built in 1911.

MARSH Bros., Glider (Marsh Bros., Holmewood, Derby)

Various reports in the Press during 1907-1909, but no details.

Martin-Handasyde monoplane No.1 under construction in 1908-1909.

Martin-Handasyde monoplane No.1. Modified version after completion at Welsh Harp.

MARTIN & HANDASYDE aircraft (Helmuth Paul Martin and George Harris Handasyde)

From the beginning of their association in 1908 the partners built monoplanes and, for certain, from the second aircraft onwards, these owed much in their design to the Antoinette. Both partners names were used originally and were actually painted under the wings of No.3, but for convenience the name was later contracted to the more manageable 'Martinsyde'.

Martin was a director of the carburetor makers, Trier and Martin, and although a statement in the aviation Press, *The Aero* 28 June 1910 (p.512), denied the aviation connection with that company, the stand at Olympia in March 1911 was listed in that name. Trier and Martin were based at Trinity Works, New Church Rd., Camberwell, London SE and, until the partners became established at Brooklands in 1911, some of the early work took place there. At Brooklands the No.2 aircraft was housed in the original shed used by Paulhan's aircraft, but the new sheds were being built and soon Shed No.12 in Swallow Lane was available and a second, No.29 adjacent to the Byfleet Banking in 1911. In 1914 two big hangars with curved roofs were built behind the Flying Village. In March 1915 large premises in Oriental Rd., Woking became the manufacturing center throughout the war period. The company went into receivership in November 1920.

It was just prior to the war in 1914 that the first Martinsyde biplane, a tractor type, was designed, which was the first type to be produced in quantity during the early part of the war.

MARTIN-HANDASYDE monoplane No.1

Little information on this aircraft remains, but photographs show the general configuration. The aircraft was a tractor monoplane built in 1908-1909 with a triangular section open fuselage structure with the wing mounted on the top longerons. In its original form the chassis consisted of a single trailing wheel, centrally mounted, and outriggers. The wing had extended areas towards the tips, presumably for warping control. A Beeston-Humber car engine, although a water-cooled type, was arranged to run without a radiator by use of an exposed condensing header tank.

At a later stage major changes were made. The undercarriage became a more usual cross axle type with possibly rubber cord springing in the vertical members and a single curved skid supported at the fuselage nose. The overhung trailing edge of the wing was removed. The tail surfaces were now of elongated triangular shape, similar to those of an Antoinette.

The photograph of the first version shows the machine housed in a shed, later the assembled aircraft was in the ballroom of the Old Welsh Harp Inn. It was tested in a field nearby at Edgware, where the propeller disintegrated and tore the engine from its mountings. Thereafter repairs took place and the machine was moved to Handley Page's ground at Barking Creek. Martin just succeeded in lifting the machine off the ground and the partners were sufficiently encouraged to enter a meeting at Halifax. The machine was damaged on the train and finally wrecked in a gale which blew down the tent in which it was housed. The engine and parts were salvaged.

Power: 29hp Beeston-Humber four-cylinder inline car engine (12-14hp rated) driving a three-bladed propeller.

Data

Span	22ft
Root chord	5ft 6in
Tip chord	3ft
Weight	580 lb

MARTIN-HANDASYDE monoplane No.2

The second monoplane was again built in the ballroom at Hendon and fitted with the original Humber engine. It was reported by *The Aero*, 24 May 1910 (p.407),.to be nearly ready for trials in shed No.12 at Brooklands.

The layout, similar to the earlier monoplane, was given greater wing area and the fuselage was strengthened with a three-ply wood covering throughout. The undercarriage was mounted on a single post, which also extended upwards, and together with kingposts at mid-span, provided anchorages for the bracing wires and pivots for the warping cables. The wheels, on separate half axles, were sprung by a laminated wooden spring. A tail-skid was supported by the rudder post, although the machine normally rested on the nose skid.

It is uncertain whether the machine actually flew with the Humber engine, but a 40hp side valve JAP had later been fitted and, on 15 June 1910, Martin lifted the machine off the ground. Tests continued until 30 July 1910 when it crashed heavily. It was not flown again until Martin took it out on 10 October 1910 and later that month was reported to have covered 300 yards. Graham Gilmour flew the aircraft on 2 November 1910, but disliked the Antoinette type separate hand wheel controls. He had these modified and retested the aircraft on 12 November 1910, when its identity was changed to No.3.

Power: 29hp Beeston-Humber four-cylinder inline car engine driving a Curva propeller.
40hp JAP eight-cylinder air-cooled vee.

Data

Span	32ft
Root chord	6ft
Tip chord	4ft

This view of the fuselage of No.2 shows the Antoinette-type controls and the side valve JAP engine in use in 1910.

MARTIN-HANDASYDE monoplane No.3

When the Antoinette type controls of No.2 were criticized by Graham Gilmour and changed to a central stick for the elevator, but with a wheel to control wing warping, the rudder bar was retained. A large metal shield was attached to the stick to protect the pilot from the large amount of oil emitted by the JAP engine, which had auxiliary exhaust ports drilled in the cylinder walls. An extended fin and a rudder with squarecut trailing edge were now fitted.

Gilmour expressed himself as very satisfied with the controls and stability of the aircraft and the general strength of its construction. He seems to have 'acquired' the aircraft, his name appearing on the rudder and 'No.3 Martin-Handasyde' under the wings.

Martin-Handasyde monoplane No.3 was the previous machine converted to a control stick with wheel for warping and was finally powered by an OHV JAP engine.

Martin-Handasyde monoplane No.2, built at Hendon and tested at Brooklands, was much like an Antoinette including the control system.

A wing tip was damaged when the machine was taken out by him on 8 November 1910, but it was flying well on the 12 November 1910. In the following weeks the engine was changed for the 35hp ohv JAP from the Blériot '*Big Bat*', and Gilmour flew with this for the first time on 11 December 1910. Sopwith tested the machine in the middle of February as a potential purchaser of the next machine already in the course of construction. Soon after this, in April or May, a 35hp Green was fitted for a short time, which involved the fitting of a sloping radiator behind the engine. By the 4 June 1911 a JAP engine had been reinstalled and pilot E.V.B. Fisher was still being blinded by the oil. The following day the machine was damaged by falling timber when Latham crashed on the roof of shed No.29, where the machine was housed. It is uncertain whether No.3 flew again, nevertheless it was still virtually intact in August 1914.

Power: 40hp JAP eight-cylinder side valve air-cooled vee driving a 7ft diameter propeller.
35hp JAP eight-cylinder ohv air-cooled vee.
35hp Green four-cylinder inline water-cooled.

Data		
	Span	32ft
	Chord	6ft 2in reducing to 5ft at tips
	Length	28ft
	Area	175 sq ft
	Area tailplane	19 sq ft
	Area elevators	6 sq ft
	Area rudder	8 sq ft
	Area fin	8 sq ft
	Weight	560 lb

MARTIN-HANDASYDE monoplane No.4B *Dragonfly*

Flight reported on 4 February 1911 (p.40), the advanced state of construction of a new two-seater with Gnôme engine, purchased by Tom Sopwith, while *The Aero* 8 March 1911 (p.193) referred to two machines in course of construction, the first to be fitted with a Gnôme.

Sopwith was out on a Martin-Handasyde aircraft several times in February culminating with one flight on 22 February 1911, when the machine tipped on its nose, but this was on No.3 and No.4B seems to have been sent to Olympia for the Aero Show opening on the 24 March 1911, unflown. After the show closed, the machine was assembled in time for Sopwith to fly it on the 17 April 1911, but a wing was damaged when landing. Sopwith was soon to leave for his American tour and, with insufficient time to repair the *Dragonfly* before he left, he was forced to look elsewhere for a suitable aircraft to take. The machine remained with the firm and never became Sopwith's property.

The Gnôme was of inadequate power and was replaced by an Antoinette provided by Hubert Latham, who became interested in the aircraft during his visit to Brooklands in May.

The new engine was installed and Radley tested the aircraft on 14 June 1911, followed by Morison on 19 July 1911 after the wing had been reconstructed; Gilmour, Fisher and Hamel also flew the aircraft in July but, on 23 July 1911, Morison crashed badly near the sewage farm and the machine was not rebuilt. There is no evidence that a second machine of the same type was built.

No.4B followed the general pattern of its predecessor, but was larger for passenger work. The undercarriage still consisted of hinged half axles connected to a sliding central fitting, but rubber cords in tension replaced the laminated wooden spring used earlier.

Power: 50hp Gnôme seven-cylinder air-cooled rotary.
65hp Antoinette eight-cylinder water-cooled vee.

Data		
	Span	37ft
	Chord	6ft (average)
	Length	33ft
	Area	240 sq ft
	Weight	800 lb
	Speed	60mph
	Price	£1,100
	Data for aircraft with Gnôme	

MARTIN-HANDASYDE monoplane (No.5)

The next monoplane to appear followed the same general pattern of No.4 with an Antoinette engine, necessitating the long condensing radiator along the fuselage. It was a natural evolution of the type and was designed from its inception to use the Antoinette engine, the production of which was unexpectedly terminated soon after.

Improvements had been made to the cowling of the engine and decking. The undercarriage was of a strengthened type, to that previously used, but still with hinged half axles connected to a sliding central fitting and sprung by four rubber cords in tension.

The machine still adopted a nose down attitude on the ground until loaded. The plywood shell of the rear fuselage

Martin-Handasyde monoplane No.4B *Dragonfly* at Olympia in March 1911 was fitted with a Gnôme engine later replaced by an Antoinette.

Martin-Handasyde monoplane No.5 built in 1911 crashed at Richmond with fatal results to Graham Gilmour in February 1912.

was cut away forming diamond shapes, visible through the fabric, a feature peculiar to this machine. A new shape of fin and rudder, with curved outlines was used.

The first flight was made by Tom Sopwith on 13 November 1911 and he piloted the machine on successive flights until 31 December 1911, when he flew to St Albans, damaging the propeller and undercarriage on landing. The machine was taken by road back to Hendon, where repairs were made and it was then flown back to Brooklands on 31 January 1912 by Graham Gilmour. New wings were fitted and Gilmour tested these on 13 February 1912. On 17 February 1912 he left Brooklands to fly over London. Following the Thames, in snow and freezing conditions, he was forced to land, but crashed fatally in the Old Deer Park at Richmond. The cause of the crash was not determined, although some onlookers reported that a wing broke off.

Note. This and subsequent monoplanes were not given type numbers at the time but are given numbers here for convenience only. The '*Aeroplane*' referred to this machine at the time as the '*Magnificent Martinsyde*'.

Power: 65hp Antoinette eight-cylinder water-cooled vee. No data available but probably similar to No.4. One aircraft only built.

MARTIN-HANDASYDE monoplanes (No.6) and variants 1912-1914

After the loss of Gilmour in No.5, there were no reports of flying by Martinsyde until 27 June 1912, when Gordon Bell flew a new monoplane with Antoinette engine. This was soon followed by a different type of monoplane, designed specifically for the Military Trials of August 1912, (No.7) and so for the first time two Martin and Handasyde aircraft were flying at Brooklands.

Some confusion can arise from the lack of clarity of the early reports and from the use of the description 'Military Monoplane' for (No.6) at present under consideration. This arose from a War Office contract reported to be for four aircraft, against which (No.6) was the first to be built. The need for a separate design probably arose when the conditions of use were made known later.

Martin-Handasyde Serial No.278 with Antoinette engine was supplied to the RFC in 1912.

These aircraft were being produced at the time of a number of disasters to monoplanes of various types, which resulted in the War Office ban on their use by the RFC and although it is possible that two Martin and Handasyde monoplanes were delivered to Farnborough, they were not used by the Service, one actually being used as a gunnery target at Shoeburyness.

The company was greatly committed to the monoplane design, notwithstanding the loss of Gilmour on (No.5) and the War Office ban, they continued to develop and fly variants of the type into early 1914. There were several setbacks along the way, including the fatal crash of Edward Petre in gusty weather at Marske on 24 December 1912, on the Antoinette powered aircraft completed the month before. The accident report stated that the wings collapsed, and suggested an investigation into the design was required. Although the firm disagreed with this finding a change to box type spars was made in a new machine with Laviator engine described in February, prior to its appearance at Olympia in March 1913 on the stand of T.B. André and Co., the successors of Trier and Martin as selling agents. This engine was unsatisfactory and was changed for an Austro-Daimler in April, and the aircraft was then flown by Gordon Bell on 8 May 1913 for the first time. In his hands it was remarkably maneuverable and achieved speeds up to 82mph, but Bell overdid the aerobatics on returning from Eastchurch to Brooklands on 13 June 1913 and the machine was destroyed and the passenger killed.

A further Austro-Daimler powered machine was flying at Brooklands in September 1913, in the hands of Barnwell, who

Harold Barnwell flew this Austro-Daimler powered Martin-Handasyde in the 1913 Aerial Derby.

competed in races at Brooklands in it and in the Aerial Derby, where he obtained second place at a speed of 72.5 mph over ninety-five miles.

The last monoplane to be completed flew in February 1914 and may have been intended to become a floatplane. This again had an Austro-Daimler engine and was flown by Barnwell initially, and later by R.R. Skene and V. Waterfall. The machine was at Shoreham at the outbreak of war and was commandeered by government officials, but apparently was not flown again.

No doubt these machines incorporated many improvements, and there are few features to distinguish one from the other, also contemporary information is sparse. Apart from the engine changes, there were less obvious alterations, such as the large-scale use of duralumin in parts of the structure, and changes to the undercarriage springing by enclosing the rubber springing inside the central tube. Serial No.278 was one RFC machine which had rubber cord springing on the diagonal arms, as on the Military Trials aircraft.

In some cases the extent of the plywood covering, or reinforcing, of the rear fuselage was varied. The 1913 Show machine seems to have fabric over the girder only in this area, whereas the 1913 Aerial Derby machine had plywood and a small area only covered by fabric.

The last two Austro-Daimler machines rested on their tailskids, when on the ground, indicating a change to the weight distribution. Particular attention was paid to the final machine, which had fairings added to the undercarriage and bracing posts, together with improved cowlings and decking. A fore and aft strut connected the lower ends of the kingposts at mid-span. Two pairs of new wings tried in June proved to be disappointing.

Power: 65hp Antoinette eight-cylinder water-cooled vee. Probably used in three aircraft.
80hp Laviator (Dansette-Gillet system) eight-cylinder water-cooled vee. Used in one aircraft only, replaced by an Austro-Daimler.
120hp Austro-Daimler six-cylinder inline water-cooled, probably used in 3 aircraft.

Data

Span	42ft 6in
Chord	7ft
Length	35ft 4in
Propeller	8ft 10in diameter Regy
Area	286 sq ft
Weight	1,100 lb
Weight allup	1,760 lb

Data is for 1913 Show machine with Laviator.

MARTIN-HANDASYDE monoplane (No.7) Military Trials type

This machine was one of two flying at Brooklands in mid-July 1912, but it left by road for Larkhill on 20 July 1912. Although it was apparently satisfactory at Brooklands, the aircraft was soon let down by its engine at Larkhill and, apart from completing the assembly test, made only one or two flights in windy conditions. In addition the aircraft was overweight.

Martin-Handasyde produced two machines for the Military Trials. This version was fitted with the unsatisfactory Chenu engine, the other had an Antoinette.

There were a few features to distinguish the Military Trials machine from the No.6 type. The Chenu engine incorporated a two to one reduction gear and drove a four-bladed propeller. The coolant radiator was shaped in the form of a saddle and fitted on top of the fuselage above the wing. The main undercarriage post, and the kingposts, were enclosed in streamlined fairings, however four rubber spring cords were now incorporated in both diagonal arms and these remained exposed. The object of the stronger springing being to satisfy the ploughed field case, specified in the rules of the contest. A propeller-driven pump to pressurize the main fuel tank was mounted on the undercarriage vertical post.

After the Trials the Chenu engine was replaced with an Antoinette and Petre continued flying the aircraft until it is believed to have been wrecked on 24 October 1912, after a forced landing.

Power: 75hp Chenu six-cylinder inline incorporating a 2 to 1 reduction gear.
65hp Antoinette eight-cylinder water-cooled vee.

Data	Span	42ft 6in (*Flight* 42ft)
	Length	38ft
	Area	310 sq ft
	Weight	1,250 lb
	Weight allup	1,800 lb
	Speed	75mph

MARTIN-HANDASYDE *Waterbus* monoplane

This machine was reported to be under construction at Brooklands in 1913. It was not completed as a seaplane, with proposed center float and balancing floats at the kingposts, but may have been completed as the last land-plane of (No.6) type in 1914. Ailerons were to be fitted instead of wing warping.

Power: 120hp Austro-Daimler eight-cylinder water-cooled vee.

The wing and rudder of the Transatlantic monoplane were actually completed at Brooklands.

MARTIN-HANDASYDE Transatlantic monoplane

The partners last monoplane was designed for the sole purpose of competing for the £10,000 Prize offered by the *Daily Mail* for an Atlantic crossing by air. The machine was sponsored by Canadian financier, E. Mackay Edgar, and was to be flown by Gustav Hamel. He was unfortunately lost in the Channel on 23 May 1914, when the machine was well advanced in construction.

Although superficially an enlarged version of earlier types, the Transatlantic monoplane incorporated a number of original features. The fuselage girder was still of triangular section with the center portion, some fourteen feet in length, enclosed by watertight bulkheads to ensure flotation should the need arise. The engine compartment, with frontal radiator, was cowled in the usual manner and the rear fuselage was fabric covered. A fuel tank of cylindrical shape, 9ft long by 3ft diameter, was housed in the front of the enclosed section, with the two crew members seated side by side behind. The top decking was continuous from the nose of the machine and faired out just behind the cockpit.

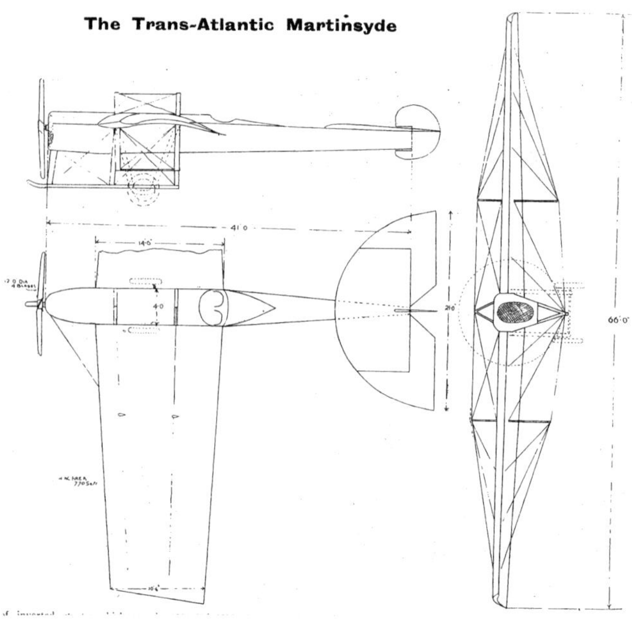

Artist's impression of the Transatlantic Martinsyde which was not completed.

The wings were built in four sections. The front and rear spars of the center portions extended above and below the top longerons respectively, to meet on the aircraft center line. The outer wings continued the tapered plan form and included the warping area of the trailing edge portion. Balanced tail surfaces were employed, the large combined tailplane and elevators forming a semicircle.

Struts forming a vee-shape, at both front and rear spar stations provided anchorages for the extensive bracing system and the warping control. The lower pylons, and a similar one below the nose, supported the central skid. The wheels were mounted on separate struts to points at the top longerons, and the whole undercarriage was arranged to be dropped after take off, leaving the central skid for the landing with balancing skids under the mid-span kingposts.

Construction of the machine continued with the possibility of another pilot taking on the task, but inevitably the project was abandoned with the outbreak of war.

Power: 215hp Sunbeam twelve-cylinder water-cooled vee driving a 12ft diameter Lang four-bladed propeller.

Data	Span	66ft
	Chord	14ft 6in tapering to 10ft 6in
	Length	46ft 4in
	Area	770 sq ft
	Weight	2,400lb
	Weight allup	4,800 lb
	Speed	80-85 mph

MARTIN-HANDASYDE pusher biplane

This aircraft was being built in the spring of 1914, and was entered for the Aerial Derby starting at Hendon on 6 June 1914 to be flown by Guy Blatherwick. The machine was not completed in time and construction seems to have been abandoned.

A pusher biplane was built by Martin-Handasyde for the 1914 Aerial Derby which was abandoned.

A silhouette by Geoffrey Watson, and a brief description, showed a typical pusher biplane of the time, with tail unit carried on booms converging together in plan. The wings were of unequal span, the top having considerable overhang. A 65hp Antoinette engine was mounted in the rear of the nacelle. Steel tubing was used in the construction of the booms and the undercarriage.

MARTIN-HANDASYDE monoplane

Flight 7 August 1914 (p.842) reported that drawings of a large new monoplane were being prepared. It was to follow the general lines of its predecessors, but the pilot was placed far back and the passenger forward with good vision over the leading edge of the wings. The aircraft was not proceeded with.

Martinsyde S1 single-seater scout biplane of 1914 with early-type undercarriage.

Martinsyde S1 with the later V-type undercarriage.

MARTIN-HANDASYDE S.1 biplane

The first biplane made by the firm, produced in any numbers, was a small single-seater scout, similar in layout to the Sopwith Tabloid and Bristol Scout. Delivery of eleven aircraft was made to the RFC by the end of 1914, and production continued well into 1915.

The machine was a single bay biplane with ailerons on all wings. The early examples had a robust undercarriage of four struts and twin skids, with small wheels at the tips of the skids. The aircraft rested on the rear of the skids when on the ground. Later the undercarriage was changed to a conventional vee type with tail-skid. A peculiar feature was the use of external joint plates for the strut connections. A Lewis gun could be mounted on the top central section, some aircraft being so fitted in the early war years.

Power: 80hp Gnôme seven-cylinder air-cooled rotary.

Data		
	Span	27ft 8in
	Chord	4ft 9in
	Gap	4ft 6in
	Length	21ft
	Area	280 sq ft
	Area tailplane	20 sq ft
	Area elevators	13 1/3 sq ft
	Area rudder	5 1/3 sq ft
	Area fin	2 1/2 sq ft
	Area ailerons	28 sq ft
	Speed	87mph

MASON monoplane (G.J. Mason, Hanwell, Middlesex)

Nothing further was heard of this machine that was reported to be under construction in 1909.

MAUND monoplane (A.V. Maund, Market St., Craven Arms, later Shifnal, Shropshire)

Maund was a cycle dealer who built a monoplane, which received a mention in 1910, but presumably failed to achieve successful flight, as nothing further was reported.

The photographs that exist show a monoplane with wings mounted at the top longerons, a triangular section fuselage,

Maund monoplane of 1910 had only a brief existence.

with a trailing fork undercarriage of welded steel tube, fitted with twin skids. The vee-twin engine, probably a JAP motorcycle engine, no doubt gave insufficient power for flight to be achieved.

MAXFIELD biplane (Alfred P. Maxfield, Gosta Green, Birmingham)

A motor engineer who experimented from 1902, Maxfield described in 1909, in a letter with sketches to *The Aero*, a biplane with an unusual central portion to the top and bottom wings. These were made in overlapping segments and able to flap to assist lift at takeoff and landing, and could assist control in flight. The machine had a biplane tail, and biplane front elevators carried on booms, and was propelled by a six-cylinder engine driving a pusher propeller. The design was superseded by a monoplane later that year.

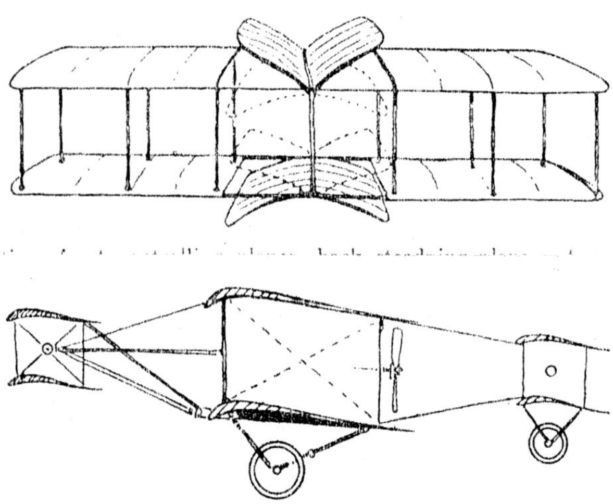

Maxfield biplane. A sketch of a project of 1909 showing unusual center section of top wing.

Maxfield monoplane which was flown at Castle Bromwich in September 1909 on show at the Midland Aero Club Show in Birmingham.

Maxfield monoplane (1/2 size) shown at Olympia in March 1910.

MAXFIELD monoplane

This was probably the first aeroplane to fly in the Midlands on 24 September 1909, confirmed again by witnesses on 30 September 1909, at Castle Bromwich Golf Links. The flights were at low height and of short duration.

The machine had a fuselage and undercarriage made of cycle tubing and was mounted on three cycle wheels, the front pair being steerable. A biplane front elevator and biplane tail with single rudder, provided control in pitch and yaw. The curved wings were single surfaced and were mounted on the top rails of the structure, and were braced to kingposts.

Power was provided by a 3hp V-twin Garrard-Maxfield motor cycle engine which was later to be changed to a 40hp. The machine was exhibited at the Royal Hotel, Birmingham at an exhibition of models by the Midland Aero Club on 13 October 1909. A half scale development was also shown at Olympia on the Royal Aero Club stand in March 1910.

MAXIM biplane (Mr. later Sir Hiram Maxim, Baldwyns Park, Bexley, Kent)

Maxim, who came to England from America in 1881, had become interested in the problem of flight by 1887. He carried out many experiments, including making a large whirling arm, tested a great variety of propellers and patented many of his ideas. These culminated in his design of a large biplane, powered by two steam engines and designed to run on rails, but prevented by upper rails from lifting to a good height. On 5 July 1895 it is reported that the machine lifted off the ground and was airborne for 600ft, until an axle broke and the machine burst through the upper rail and was badly damaged.

There had been earlier tests from 1893 onwards but this, the last trial, was carried out in the presence of members of the Aeronautical Society and others. Although Maxim began repairs and modifications, he was forced to abandon the machine

Maxim. The huge steam-powered biplane of 1894-1895 which lifted off the starting rails but became entangled with the upper restraining rails and was badly damaged at Baldwyn's Park, Kent in July 1895.

and to close down his operation at Baldwyns Park, which, in any case, did not allow sufficient length for his purpose.

He had spent about £17,000 on the machine and it was about fifteen years before he constructed another aircraft. The large hangar was removed and re-erected at Eynsford in Kent, where it was later used by Pilcher for his gliding experiments.

The machine, as tested, was a biplane built around a light welded steel tube structure, the base of which provided a platform on which was positioned the boiler with its fuel and water tanks, the controls and the crew. The lifting surfaces consisted of a large octagonal, double-surfaced, top center plane, to which parallel chord outer wings were attached. Projecting at this level, were balanced front and rear elevators. At the level of the platform a pair of single surfaced lower wings were braced by light struts and cables. The sides of the base were used to mount the four running wheels. The restraining wheels, on their separate axles, were attached to the tubular frame.

Since the whole of the project was experimental, the configuration of the machine changed during the trials and, at one stage, an additional wing was fitted above the center section. In the final version it was proposed to fit five pairs of outer wings.

The two wide chord propellers had two blades consisting of laminated pine with fabric covering, and were braced by wires to an extension of the driving shaft. The engines were mounted horizontally at midgap, behind a wing shaped fairing and drove the propellers direct. Naphtha was the fuel used to raise steam in the Thorneycroft Marine boiler.

The crew of up to four stood on the platform, with Maxim controlling the power at a station behind the boiler stand, with a crew member behind controlling the elevators. The other crew members were there to take readings or as passengers.

Power: Two 180hp Maxim two-cylinder compound steam engines driving 17ft 10in diameter propellers.

Data		
	Span	104ft
	Gap	20ft
	Area	3,875 sq ft (6,000 sq ft with 5 pairs of wings)
	Length	120ft
	Center section span	50ft
	Center section length	47ft
	Fuselage platform	40ft long by 8ft wide
	Weight allup	8,000 lb (3 crew & 600 lb water)
	Speed achieved	42mph

MAXIM biplane 1910

In 1904 Sir Hiram began the drawings of a new aircraft, but did not complete the design immediately; in 1908 work was recommenced. The machine was built at the works of the Wolseley Tool and Motor Car Co., a subsidiary of Vickers, Sons and Maxim Ltd. at Crayford in Kent and was sufficiently advanced in construction to be photographed and described by *Flight* in April 1910, after Sir Hiram had himself described the principle features of the machine in an earlier article. Despite the passage of time, Maxim designed a scaled down version of his 1894 machine, but powered by an internal combustion engine of his own design.

The machine was a pusher biplane with front and rear biplane elevators and rear rudder. The two outer wing bays were arched and could be warped for lateral control, operated by a foot bar. The rudder was interconnected to steerable tail wheels operated by a hand wheel or handlebar on the control column, which worked in a fore and aft manner for moving the elevators.

The center structure of the aircraft was parallel in plan and consisted of two steel tubular top longerons, which were continued fore and aft with aluminum tubular booms to carry the elevator units. The lower longerons and bracing tubes were of aluminum. The main wheels were carried in wooden forks, trailing from the bottom longerons and were sprung by pneumatic struts at the rear; curved skids of ash extended forward to limit nosing over.

Maxim biplane built at Crayford was an obsolete design when it appeared in 1910 and was abandoned without being tested.

The lower center section of the wing provided the mounting for the engine, which drove one propeller direct and two others of larger size by rope drives, tensioned by jockey pulleys. The starboard propeller revolved in the opposite direction to the others and, to reduce any asymmetric tendencies, the pitch of the port propeller and its speed of rotation, were lower than for the starboard. The wide chord propellers were thin and made of laminated pine and were braced back to the drive shafts by steel tapes, themselves set with pitch to give thrust. The outer propellers rotated on bearings around the main top longerons.

The pilot sat in a short nacelle ahead of the lower wing and two passengers were accommodated in exposed positions on adjacent seats with footrests on the sides of the nacelle. The lower members of the nacelle were extended to provide a mounting for the fuel and oil tank and these terminated in a small lifting surface. The radiators were carried horizontally below the top center section of the wings.

Maxim was seventy years old and weighed seventeen stone and was being pressured by the other directors of Vickers, Sons and Maxim Ltd., so it was no surprise when he announced in October his intention to rest and not to do anything further with the machine for the present In the following March he resigned as a director and the company was then renamed Vickers Ltd. The machine never did fly after all the effort expended.

Power: 80hp Maxim four-cylinder inline water-cooled driving one 5ft diameter propeller direct and two 11ft 4in diameter propellers by rope drive.

Data	Span	44ft
	Chord	6ft 6in
	Length	35ft 6in
	Gap	6ft 6in
	Area	572 sq ft
	Area front elevator	81 sq ft
	Area rear elevator	81 sq ft
	Area rudder	16 1/2 sq ft

MAYNARD glider

This was reported built early in 1910 at Birmingham.

MEGONE biplane (W.B. Megone, Hawkinge, Kent)

The Megone biplane was designed and built by a Dutchman with the assistance of Victor Hunt and others, between 1912 and 1914. The machine appeared in a second version after early unsuccessful trials.

The aircraft was a two-seater pusher biplane, housing the passenger behind the pilot on a lifting seat, giving a good view through celluloid panels in the lower wing for observation purposes. The inverted triangular section tail boom required that the propeller revolve around its top member. The engine was mounted low down in the front of the nacelle with a Hele-Shaw clutch, rear extension shaft and chain drive to the propeller. This arrangement necessitated taking the control cables

Megone biplane with shortspan lower wing in 1912-1913 at Hawkinge.

through the center of the propeller boss, and inside the top tube, to the tail. The tail surfaces consisted of normal tailplane and elevator with a large shield shaped rudder above and a small surface attached to the tail-skid below.

It was reported that various engines were used, including 35 and 60hp Green types, the radiators for cooling being hung on the sides of the nacelle.

The early version of the machine had a short span lower wing with pronounced anhedral apparently without external bracing. The upper wing was braced independently to an inverted vee-shaped pylon above the center section, and by cables to the bottom of the nacelle. After the failure of this version, the lower wing was increased in span and two pairs of interplane struts were introduced; the center pylon was then discarded. The top wing, originally of parallel chord, which embodied the warp, was extended rearwards towards the tips, providing a larger overhanging warping section. The top supports for the warp cables were tilted backwards to match.

The main wheels were mounted on a continuation of the lower longerons, which were curved upwards and originally fitted with small bumper wheels. Small wheels were fitted to the lower wing tips when they were extended.

The machine, in its first version, was ready for testing early in 1913 by Victor Hunt, who acted as test pilot for Megone,

The longspan version of the Megone biplane in 1913-1914.

but damage resulted. Repairs and experiments continued into 1914, and the machine was still in existence in September of that year. Without warning, Megone suddenly disappeared leaving the machine locked in its shed and creating rumors that he was a spy for Germany.

Power: 60hp Green four-cylinder inline water-cooled. A 35hp Green and other types may also have been tried. A 10ft diameter propeller was fitted.

Data		
	Wing area	440 sq ft
	Tailplane area	54 sq ft

MERLIN biplane (Merlin Aeroplane Co., Greyfriars Lane, Coventry)

This was built in 1909 and was exhibited in Upper St., Islington at the same time as the Stanley Show in November. It was described as combining the designs of Cody and Curtiss with a four wheel chassis of Farman type.

Power: 25hp of unspecified make.

Data		
	Span	32ft
	Chord	5ft 4 1/2in
	Gap	6ft
	Weight	670-785 lb

The Mersey monoplane which crashed with fatal results during the Military Trials of 1912.

MERSEY monoplane (Mersey Aeroplane Co., Freshfield, Liverpool, Lancashire)

R.C. Fenwick and S.T. Swaby began the design and construction of this machine when managers of Planes Ltd., and completed the aircraft at their premises. The aircraft flew for the first time late in 1911 with Fenwick, the designer, as pilot. The partners took over the development of the machine to prepare for entry in the Military Trials of August 1912, in the course of which it crashed, killing Fenwick.

The machine was a side-by-side two-seater with a nacelle between the wings, which were attached at the top longerons. The engine in the front drove a pusher propeller at half speed though an extension shaft, passing between the two seats. Steel tubing was used for the booms, wing spars and undercarriage structure. The wings were braced by wires to tall king posts and to the undercarriage, including to the wheel centers. The tail unit was carried on the pair of horizontal tubes, which, with single wire bracing, appeared quite inadequate.

The tail surfaces consisted of a single tailplane and elevator with balanced rudder above and a small fixed fin below, although originally a smaller rudder and no fin were fitted.

Power: 45hp Isaacson seven-cylinder radial air-cooled driving a pusher propeller by a 2 to 1 reduction gearing and shaft.

Data		
	Span	35ft
	Chord	7ft
	Area	220 sq ft
	Weight	750 lb
	Weight allup	1,150 lb
	Speed.	55mph

METZGAR & LENO monoplane (Bert and Ted Metzgar and Herbert Dan Leno, Old Town, Clapham, London, SW)

The construction of this machine, which was the design of Bert Metzgar, was started in 1907, but it was not until the summer of 1911 that trials at Shoreham took place. These resulted in short hops of no more than 50ft up to a height of 3ft or so, before the financial strain was too much.

The aircraft was a pusher monoplane, much like an ASL Valkyrie in layout, and used tube in its construction. There were twin booms and these carried the four wheels on twin axles, and these curved upwards to provide the mounting for the front elevator. The rear booms carried a rudder and tailplane, and lateral control was by single acting, broad chord ailerons.

The pilot was originally seated behind and below the engine, which drove, at reduced speed, the propeller by chain and shaft, above the pilot. Later the propeller was driven direct by the crankshaft and the engine moved back and the pilot forward. The engine was a four-cylinder water-cooled type, reported by *The Aero* as having been made by the partners, although elsewhere it was referred to as a 60hp Brasier. It had cast aluminum water jackets with cast iron liners and cylinder heads and may have been a modification of a car engine. The radiator was finally mounted horizontally above the aircraft together with the fuel tank

Data		
	Span	36ft

The Metzgar and Leno monoplane was built at Clapham and tested at Shoreham in 1912.

MIDLAND AERO CLUB glider

A small tail first glider made by a carpenter member of the club was mounted on bicycle wheels and was shown at a model meeting at Sutton Park on 13 November 1909. This may have been the glider tested there unsuccessfully by J.H. Else (*q.v.*).

MIDLAND GLYDERS (Midland Aeroplane Co. Ltd., Grey Friars Lane, Coventry)

In 1909 this firm was advertising its 'Man Lifting Glyders' from £10 each, also aeroplanes, either monoplane or biplane. There is no indication of any being built or sold.

MILLER helicopter (J. Miller, 62 Abbott Rd., Putney, London, SW)

This device, built in 1909, consisted of twin screws of spiral design spinning in opposite directions on a vertical shaft, driven by a 50hp motor. A narrow vertical surface extended for the full height outside of the rotating screws to provide directional control.

Data	Two coaxial rotor screws 10ft diameter	
	Area	909 sq ft
	Weight	1,190 lb

The Miller helicopter was built at Putney in 1909.

MILLER ornithopters (W. Miller MRCS, Hart St., Covent Garden, London)

Towards the end of the 18th century Miller devised a primitive ornithopter with two pairs of winglets operated by ropes. The winglets were hinged to plates worn on the chest and back of the flyer. The ropes were pushed down by alternate movement of his legs and imparted the necessary flapping motion. As with most ornithopters the optimism of the inventor was not matched by the performance of the apparatus.

Millers second ornithopter, the *Aerostat*, appeared in about 1840 and was described as based on the form of a West Indian crow, the wings were operated manually by levers. It had a fixed wing center section and a moveable tailplane aft of the pilot. Weight 500 lb.

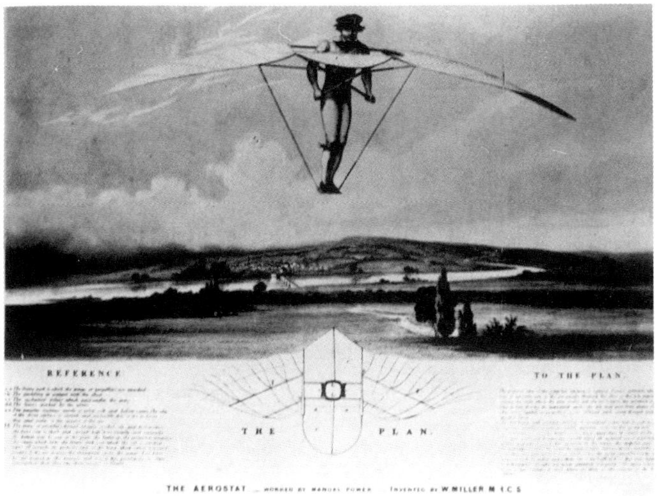

Miller ornithopter; a project of 1784.

MILLS-FULFORD monoplane. See Fulford.

MINES biplane (Edward Mines, Brightholme, W. Kirby, Cheshire)

This impractical machine appeared at the Doncaster Meeting of October 1909 but did not fly. It was variously nicknamed *'the meat safe'* or *'the coffee stall'*. It was for sale in February 1910.

The machine was a pusher biplane with bicycle front wheel and handlebars. There was no tail, but a large elevator flap was hinged on the front of the top wing; rudders were fitted to each outboard front interplane strut, and the lower wingtips could be pivoted for lateral control. The power was supplied by a

The Mines biplane was present at the Blackpool Meeting in October 1909 but did not achieve flight.

10hp JAP or Anzani vee-twin, air-cooled engine driving the propeller by chain and shaft. Mines is reported to have attempted to fly a glider at Llandudno in 1908

In March 1910 Mines registered a company Edward Mines Aviation Syndicate Ltd. to raise funds to purchase a Farman type all-British aeroplane from the Aerial Manufacturing Company of Great Britain and Ireland. Nothing further is known of this venture.

Data	Span	14ft
	Chord	6ft

The Mitchell brothers seaplane was taxied on the River Medway in 1914 but capsized before flight was possible.

MITCHELL hydro-biplane (Three Mitchell Bros., 32 Greenfield Rd., Gillingham, Kent)

The Mitchell brothers seaplane was tested on the Medway in the middle of 1914, but could only be taxied in an area where flying was restricted. On the last of these runs the machine nosed over in the water at Bennett's Wharf.

The aircraft was a typical tractor biplane, mounted on two main floats and a tail float. Ailerons were fitted on the interplane struts. Assembly took place behind the White Horse Inn in Pier Road, Gillingham. The war prevented any further activity.

The brothers are believed to have built a 25hp monoplane in 1912.

Power: 100hp Mors, probably a car engine.

Data	Span top	37ft
	Span bottom	34ft
	Chord	4ft 6in
	Gap	6ft
	Weight	1,050 lb

MITCHELL monoplane (Mitchell & Sons, Edinburgh)

No details are available of a machine reported to have been shown at the Scottish National Exhibition in May 1911.

MOLESWORTH-HUGHES triplane (H.B. Molesworth, Brooklands)

The Mackensie-Hughes triplane, *Brittannia*, was apparently taken over by Molesworth and redesigned in the autumn of 1910. In September it was reported as still unable to fly on 20hp after changes of propeller, both geared and direct drive being tried. In November a report suggested that a 20hp British Rotary engine was to be fitted, with a larger one of 50hp later. This engine did not materialize. On the 10 February 1911 the triplane was rolling, but although improved, it would still not lift off. The same applied on 1 March 1911, but then the undercarriage collapsed and the machine stood on its nose, when the machine was on the verge of flight. Nothing more was reported.

The Molesworth-Hughes triplane was modified from the Mackenzie-Hughes *Britannia* but failed to fly on test at Brooklands in 1911.

The modification from the Mackensie-Hughes triplane was primarily to replace the original engine driving twin propellers, with a more powerful 60hp ENV eight-cylinder, water-cooled vee, driving a single propeller direct. At the same time the complicated triplane tailplane was discarded and replaced with a single tailplane on the top longerons, with the rudder mounted behind.

MOON *Moonbeam* monoplane Nos.1 and 2 (Moonbeams Ltd., Royal Pier Gates, Town Quay, Southampton)

Eric Rowland Moon managed the family marine business and, in 1909, began the construction of his first aircraft. This was of *Demoiselle* type, with a light steel tubular frame and tapered wings with warping control. The machine had been undergoing trials for some weeks and had flown by early June 1910, when Moon wrote to *Flight*. It is believed that the trials took place from fields near Fawley.

The second version may have used the same fuselage structure, but had a more substantial undercarriage with twin skids.

Moon *Moonbeam No.1* monoplane of *Demoiselle* type at Eastleigh in 1910.

Moon *Moonbeam No.2* monoplane had a modified undercarriage with smaller wheels and skids.

Moore-Brabazon glider built by Short Bros. in 1907 was unsatisfactory on test.

A rigid wing with single-acting ailerons was used and the tail surfaces were changed to ones with straight edges.

The second machine flew more successfully than the first, from fields at Beaulieu, Ower and North Stonham, Eastleigh. Moon became a pilot in the RNAS after the outbreak of war.

Power: 20hp JAP four-cylinder air-cooled vee driving a 6ft diameter propeller.

Data	Span	24ft
	Weight	260 lb

MOORE ornithopter (Major Ross Franklin Moore, machine built by John Shaw of Hale St., Coventry. Wings made by Spencer & Sons, Highgate)

The machine was designed and patented by Moore, No.13311/1890, and was finished by 1892. The patent gives little detail, but shows a machine with oscillating bat-like wings, driven by an electric motor, which received current from a pair of ground power lines by means of trailing cables. The object of the machine was presumably for reconnaissance or surveying work.

The wings were mounted at the top of a frame, housing the mechanism and the load. The patent was not specific in stating the type of construction, but referred to the use of bamboo, steel or aluminum for the wing structure, covered with either gold beaters skin, silk or light fabric. The machine, which weighed 700 lb, never left the ground.

MOORE-BRABAZON biplane powered glider (J.T.C. Moore-Brabazon)

Moore-Brabazon's original intention was to make a biplane glider of 25ft span with cambered wings. However, in consultation with Short Bros. at Battersea, who made the machine for £25, the final version became a structure of ash and bamboo of 35ft span, with almost flat wings. This was the first heavier than air machine made by the Shorts, who had previously made only balloons.

The glider had a front elevator at the level of the lower wing. The undercarriage had skids and four wheels and could be dropped. The machine was built in late-1907, and taken to Brooklands, where it was housed adjacent to the first aircraft of A.V. Roe, in the motor racing paddock area. After a few unsatisfactory tests as a glider, Moore-Brabazon arranged for Howard Wright to reconstruct it as a powered aircraft.

Apart from installing a 16-18hp Buchet four-cylinder inline air-cooled engine driving a four-bladed aluminum pusher propeller, the undercarriage was made a fixed arrangement of four castering wheels. A fixed tailplane on booms was added, together with square shaped flaps, hinged vertically to the central interplane struts as balance controls.

In this form the undercarriage proved to be weak during runs along the Finishing Straight, so the machine was converted back to a glider with all control surfaces removed and a strengthened undercarriage fitted. It was taken to Chelmsford in Essex, where it was flown briefly as a kite. Moore-Brabazon then went to France, where he acquired a Voisin aircraft.

Moore-Brabazon biplane. The glider was modified and fitted with an engine by Howard Wright. The undercarriage failed on test on the Finishing Straight at Brooklands.

MORTIMER & VAUGHAN biplane (Mortimer & Vaughan, Edgware, Middlesex)

A photograph of a model of this machine, which was being built at the time, appeared in *Flight* 4 December 1909 (p.784). It was a three propeller machine, with semicircular wings, and named a 'Safety' biplane, but it crashed and burned on test in 1910. A second modified version is said to have been made in 1911. No reports of this can be traced.

Mortimer and Vaughan biplane. The fullsize version of this model was built but burned on test in 1910.

Moy Aerial Steamer which lifted, unmanned, off a circular track at Crystal Palace in 1875.

Moy ornithopter was tested in model form in 1901 but did not result in a practical aircraft.

MOY Aerial Steamer (Thomas Moy, Crystal Palace)

This tandem wing monoplane was the tangible result of inventions by Moy, and initially with Richard Edmund Still, patented in 1871 (No.3238) and 1874 (No.2808). The machine was powered by a 3hp steam engine and, although intended to carry a man, lifted itself only, when tried at the Crystal Palace in 1875. The tests were carried out on a circular track and a speed of 33mph was reached.

The machine was constructed of light tubing with a semi-circular front plane, behind which a pair of 12ft diameter propellers, of unusual design, revolved. These took the form of circular wheels, made of tubing, with six spokes on which eight 'aeroplanes' of reducing length were mounted; the attitude of these was adjustable to provide lift as well as thrust.

A biplane tail was used as an elevator. The patent also refers to pendulum control as well as manual control.

Moy was still involved in aeronautics at a much later date. Photographs of a model ornithopter on test, dated 1901, exist; however none of his ideas resulted in practical aircraft.

MOYA Balloonoplane No.3

Nothing is known of Nos.1 and 2, but No.3, a combination of a balloon, on end, above a small monoplane was exhibited at the

Moya Balloonoplane No.3 exhibited at Olympia in March 1910.

Aero Show at Olympia in March 1910. No contemporary reports of trials have been found, but the device would appear to be uncontrollable. Patent 30604/1909 refers.

MULLINER COACHWORKS Ltd. (1325 Long Acre, London WC; 7385 Bridge St., Northampton & Vardens Rd., Battersea)

The firm of Mulliners decided to enter the aircraft industry and to this end acquired a former roller skating rink in Battersea in December 1910, but reversed their decision at the end of 1911. The firm continued in being as coach builders for many years after.

In the time that they operated at Vardens Rd., and at their previous premises, the firm could claim the design of two original aircraft. In addition they manufactured aircraft to other people's designs, including building, or repairing, Blériot, Farman and other standard types. The Skinner monoplane was an example of a 'one off' machine, built for a customer; a biplane glider was reported as being made in November 1911 for an Irishman named Newton-Gordon.

MULLINER monoplane Model B

This was designed by Gordon Stewart and was exhibited at Olympia in March 1910 Although the machine was praised for the attention to detail in the design and quality of its construction, it failed to attract attention and soon disappeared from the scene.

The fuselage longerons were of H-section and lightened further by holes in the webs. The undercarriage was of light tubing with the wheels separately mounted in forks, which were sprung in a knee action manner. An unusual triple arm tail skid was fitted. The tail unit comprised a tailplane with single elevator and rudder above the fixed surface. Wing warping was controlled by a wheel on the control column.

Power: 35hp JAP eight-cylinder air-cooled vee driving a 6ft 3in diameter Spencer propeller

Mulliner monoplane Model B. One of the few machines bearing the firm's name was shown at Olympia in March 1910.

Data		
	Span	33ft
	Chord	6ft 6in
	Length	27ft
	Area tailplane	13 sq ft
	Area	220 sq ft
	Area elevator	14 sq ft
	Area rudder	4 sq ft
	Weight	420 lb
	Weight allup	605 lb
	Max speed	40 mph
	Price	£450-500

MULLINER Knyplane monoplane

E. Cecil Kny was the manager at Vardens Rd. and he designed the second Mulliner aircraft. Kny was of Danish extraction and also acquired the agency for Normale propellers.

This two-seater tractor monoplane had a boat shaped fuselage, covered with aluminum to behind the rear cockpit, with the rest fabric covered. A searchlight was fitted in front of the observer, when the machine was exhibited at Olympia in March 1911.

Unconventional features were the ability to control the incidence of the wings, which rotated about a tubular spar and also the ability to adjust the camber by separate controls. A fixed tailplane extended well along the fuselage; a cruciform elevator and rudder were used.

The undercarriage was of lever suspension type, with torsion springs and with a central sprung skid; a long tail skid was fitted.

Mulliner Knyplane designed by the firm's Danish manager was at Olympia in 1911. It was not seen subsequently.

The engine was buried inside the fuselage and was cooled by long external radiators extending well down the fuselage sides.

Power: 60hp ENV type F eight-cylinder water-cooled vee driving a Normale propeller through a clutch.

Data

Span	39ft
Length	36ft
Area	300 sq ft
Weight allup	1,250 lb

Mumford helicopter built at Denny Bros. shipyard achieved tethered flight in 1913.

MUMFORD helicopter SO.60, 1st and 2nd versions. (Edwin R. Mumford and J. Pollock Brown, Denny Bros., Leven Shipyard, Dumbarton, Strathclyde)

The partners were two senior men running the test tank at the shipyard, and the project was sponsored by Mr. Maurice Denny in 1905. A great deal of experimental work was carried out, and the first machine was virtually complete in 1906, but by 1909 was still in difficulties with the unsatisfactory Buchet engine, which was not achieving its claimed 25hp. This was replaced that year with a four-cylinder inline NEC, which also proved to be inadequate.

A second NEC, this time the new four-cylinder vee was fitted and first tests took place during September 1911. Thereafter this engine was used in both of the machines that were built.

The two versions were of the same basic layout, but varied mainly in improved detail construction, better materials and general points arising from the tests.

The frame was of rectangular planform, initially made of 11/2in o/d and 2in o/d by 22 swg aluminum tube, joined by steel castings and braced by wires. Two long wooden skids were fitted below the outside longitudinal members. The pilot sat behind the engine in the center of the craft with an elevator above him and a rudder at the rear of the frame.

Six 25ft diameter rotors, with two blades each, positioned at the four extremities and at either side at the center, were mounted on shafts tilted forward at ten degrees from the vertical, thus providing forward thrust. In October 1906 the blades were formed of bamboo rims with silk covering, but these suffered from water soakage causing reduction in strength and increase in weight. Steel rims were tried, but elm rims proved to be the most satisfactory and these survived from 1909 to the completion of the trials. In all positions the blades were braced back to their center shafts. The rotors were designed to revolve at 40 rpm.

A gearbox coupled through a clutch to the engine had two output shafts carrying sprockets, running at 3:1 reduction, and drove by chain two lengthwise side drive shafts, from which bevel gearing translated the drive to the rotor shafts. The radiator was made in four sections, which were mounted high up in the frame behind the pilot.

In 1912 the helicopter had made a number of tethered flights to about 10ft height, which culminated in one on 6 January 1913 when damage occurred, due to failure of a side drive shaft. The strain was telling on the aluminum tubes, and it was evident that a major reconstruction was necessary. A new machine was started in 1913 with a structure made of composite material of paper, wood and cloth, a foretaste of bonded materials of many years ahead.

Late in 1914 the new machine was fitted with floats, in which form it taxied out onto the water and flew 100 yards at 10ft height. The machine was towed back to the slipway, where it was destroyed that night by a gale, which ended this promising experiment.

Power: 25hp Buchet eight-cylinder air-cooled vee. The specified weight of 110 lb was exceeded and the power was not achieved.
25hp NEC four-cylinder inline two liter water-cooled two-stroke, weight 187 lb
40hp NEC four-cylinder 2.4 liter water-cooled two-stroke vee, weight 159 1/2 lb
A speed of 1,100 rpm was required to drive the mechanism and at this speed the first NEC gave less than 23 bhp; the second gave 37 bhp at 1,100 rpm.

Data

Max length	60ft
Max width	41ft
Weight	first version originally 886 lb increased to 1,577 lb prepared for flight.
Weight	second version 1,508 lb prepared for flight.

MUMFORD monoplane (W. Mumford, Ebrington St., Plymouth, Devon)

A single seater monoplane of Blériot type was reported to have been built by Mumford at Albert Rd., Devonport in 1910; this was the first monoplane made in the West Country. There were no further reports.

MUSGRAVE biplane (Major Musgrave, East Grinstead, Sussex)

This machine was under construction in 1911, but still only partly built when Musgrave died before war was declared in 1914. A.M. Low was originally involved with the design and later A.W. Judge. The machine was mainly made of bamboo, a material already obsolete.

Musgrave biplane was under construction from 1911-1914 by which time it was obsolete and Major Musgrave had died.

N

NALL biplane (W. Nall, 1 Ling St., Holt Rd., Liverpool)
No information is available on this machine reported in the Press other than that it was fitted with a device to give it automatic and lateral stability. It was for sale in April 1911.

NAPIER aeroplane (Lieut. Napier RN, Devonport)
An aeroplane of unknown type was reported under construction at the end of 1909.

NATURA ornithopter (N.R. Gordon)
This machine was reported in *The Car Illustrated* as being at the 1910 Aero Show, but was ignored by the aviation press. The illustration published was of a small flapping wing monoplane. The wings, which were heavily cambered and gull-shaped, were mounted at the top of a tubular structure. The operator was seated below and pedaled to provide the motion, through a system of levers. An extension of the chassis, which was mounted on three wheels, carried a fixed horizontal tail surface.

NEALE *Pup* monoplane (J.V. Neale, Aeronautical Society Experimental Ground, Dagenham, Essex)
A report in *The Aero,* 31 May 1910 (p.425), stated that Neale had been involved in aeronautics for many years and had built a successful airship twelve years before, in 1898, although this has not been confirmed. However, Neale brought his first monoplane to Dagenham in the middle of 1909, where he tested and improved it, prior to taking it to the Blackpool Meeting held 18 to 25 October 1909. Although the machine was assembled, the weather conditions made flying impractical and Neale complained that his machine had been damaged in a leaking shed. On returning from Blackpool Neale transferred to Brooklands shed No.6.

The *Pup* was a small tractor monoplane with a triangular section, wooden, braced fuselage girder, originally mounted on a transverse bamboo axle, with bicycle wheels and forks, which were also used at the tail. The undercarriage was improved before the machine reached Blackpool.

The wings were heavily cambered and included chordwise endplates; stiffeners at the outboard of the outer section provided hinge points for the wide chord, overhanging ailerons. The tail surfaces consisted of a biplane elevator and long trailing rudder behind. An inverted vee-shaped pylon, behind the engine, served to anchor the bracing wires and as a mounting for fuel and oil tanks. Spruce and bamboo were used in the construction of this rather crude machine.

Power:	9/12hp JAP twin-cylinder air-cooled vee driving a 6ft 6in diameter propeller at 3 to 1 reduction by a gearbox below the propeller.	
Data	Span	18ft
	Chord	5ft
	Weight allup	420 lb.

Natura ornithopter at the Aero Show at Olympia in March 1910 received little mention and was an obvious lost cause.

Neale *Pup* monoplane. Neale took this to the Blackpool Meeting of October 1909 but did not fly due to damage and weather conditions.

Neale monoplane at Brooklands in 1910. This was much changed from the Blackpool version and made a number of straight flights in December 1909 and later.

NEALE monoplane 1909-1910 (J.V. Neale, Brooklands Shed No.6)

At Brooklands Neale was reported in *The Aero* to have made a successful flight on 5 December 1909 on his 'Blériot' monoplane, when he covered the whole length of the aerodrome at a height of fifteen feet, having previously made several short flights at no great distance from the ground. This was almost certainly on the machine now described, which was considerably different from that taken to Blackpool.

The fuselage girder was rectangular in section with a Blériot-like undercarriage, with trailing axle and wheels and with a tail skid. The tail unit consisted of a long fixed tailplane extending well forward along the top of the fuselage, with triangular shaped divided elevators and rudder. The wings incorporated shortspan ailerons, inset into the wings. A coolant radiator stood above the fuselage by the pylon in front of the pilot.

Power: 24hp ENV type H four-cylinder water-cooled horizontally opposed.

NEALE VI monoplane

This was, it seems, a reconstruction of the previous machine that was being tested in May 1910 at Brooklands. Neale flew the machine on 21 May 1910 with only a 9hp JAP, but the fuselage broke on landing, after the failure of the undercarriage. It was later rebuilt with a more powerful engine. The Petre brothers, Howard Flanders and possibly W.O. Manning were involved with the rebuild.

Neale VI monoplane flown at Brooklands in 1910 was reconstructed from the earlier machine.

Neale VI a modified version at Brooklands.

The fuselage, undercarriage and wings were the same as those described before, except that the fuselage was fully covered with fabric. The tail surfaces were completely new; there was now a biplane tail with flexible trailing edge for control in pitch. The rudder was a flexible continuation of the fuselage. A long shallow fin was fitted ahead of the tail unit, on top of the fuselage, with a bicycle type tail wheel below.

Power: 9hp JAP twin-cylinder air-cooled vee driving a 7ft 8in diameter propeller
20hp JAP four-cylinder air-cooled vee driving a 7ft 10in diameter propeller
Reduction gearing of 6 to 1 and 3.25 to 1 were fitted to these engines with the propeller shaft passing through the gap between the cylinders.

Data		
	Span	27ft (30ft *)
	Chord	5ft
	Area	140 sq. ft
	Length	19ft 10in
	Biplane tail	9ft 6in by 2ft 6in
	Biplane tail area	50 sq. ft
	Weight	360 lb.
	Weight allup	490 lb. (530 lb.*)
	*rebuilt version	

NEALE VII biplane (J.V. Neale, Neale's Aeroplane Works, Baker St., Weybridge)

Neale had by now acquired a works in Weybridge, but continued to fly at Brooklands He brought out the biplane from his shed on 8 August 1910, taxied it for the first time and flew it soon after. Its origin and ownership became the subject of some

Neale VII biplane of 1910 with an original system of control. It could be flown with ailerons disconnected.

controversy, with a pupil pilot named Rippin, who damaged it at the end of September, after which only Neale seems to have handled the machine until flying ceased in November.

The machine was designed with a control system to avoid problems over the Wright patent for lateral control. Although different from the methods of control in general use, its originality was contested from as far afield as America and Australia in letters to the press.

This was Neale's last design, although in November, he was proposing the construction of a large dirigible, for which he was seeking funds. It is believed that the Neale biplane was dismantled and the parts used in a machine by another hopeful constructor.

The layout was of a conventional, equal span pusher biplane of the period, with front and rear elevators carried on booms, which were parallel in plan. A special feature was the omission of a rudder at the tail, this being replaced by rudders at each wing tip. These were of single acting type and were hinged on posts just behind the front outer interplane struts, where they not only provided directional control, but caused a 'screened area' over the top wing tip and aileron. Lift was thereby lost on one side, sufficient to make the ailerons redundant, and these were disconnected.

Power: 35hp Green four-cylinder inline water-cooled driving a 7ft 9in diameter propeller

Data		
	Span	34ft
	Chord	6ft 6in
	Gap	6ft
	Area	400 sq. ft
	Area tailplane	52 1/2 sq. ft
	Area rear elevator	26 sq. ft
	Area front elevator	24 sq. ft
	Area screen rudder	15 1/2 sq. ft each
	Length	41ft
	Weight	800 lb.
	Max speed	41mph

Newington (Ling) monoplane built at Hull and exhibited there in 1909. Failed to achieve flight on test in 1910.

NEWINGTON monoplane (Thomas L. Bell, St. George's St., Fenny Arches, Hull and E.M. Ling, 22 Parliament St., Hull)

Bell was the superintending engineer of the Newington Steam Trawling Co., the sponsors, and construction was the responsibility of his uncle, Mr. Jackson. E.M. Ling was presumably involved with design, being concerned with patents Nos.21956/1909 and 29614/1909. John Work was appointed president of a subsidiary Newington Monoplane Co.

The machine was exhibited at Hull Royal Institute in 1909 and was then taken, it is thought, to Hedon Racecourse for trials. Mechanical problems and damage from outside storage persisted up till the time when an attempt to fly was made, when the machine was finally wrecked.

The machine was a tractor monoplane with high wing positioned clear of the sides of the fuselage. This was built with a boat-like front section with vertical prow, the rear portion continuing as a diagonally braced open girder carrying fins above and below, with one piece rudder and a tailplane on top, with divided elevators. The undercarriage was sprung and consisted of a pair of motorcycle size wheels at the front, and a pair of bicycle type wheels, in separate forks, at the rear.

Power was apparently provided by a three-cylinder 40hp engine, mounted well back in the fuselage, driving the Handley Page propeller through a long shaft. The engine was water cooled by an array of pipes under the body, the underside of which, was partly open to reveal the presence of a water pump and other parts of the engine installation.

Data		
	Span	30ft 9in
	Length	26ft 3in

NEWTON GORDON glider

There is no information on this biplane glider reported as being made by Mulliners in November 1911 for an Irishman.

NICHOL glider (George Nichol, Holborn, London)
Reported to have flown at Farnborough in Kent, it used cycle tubing in its construction and weighed 160 lb.

NICHOLSON monoplane (Built by Holland & Holland Ltd., 479488 Oxford St., W)
This Blériot-like monoplane was built by a well-known firm of coach builders for Mr. B.H. Nicholson, and was shown on the stand of the Royal Aero Club at Olympia in March 1910. Its subsequent use is not recorded.

Some differences from the standard Blériot were apparent, such as the addition of a tall triangular shaped fin above the fuselage and a triangular shaped tailplane below the bottom longerons.

Power: 25hp NEC four-cylinder inline water-cooled two-stroke driving a 8ft diameter propeller

Data		
	Span	32ft 6in
	Chord	7ft
	Area	227 sq. ft
	Area elevators	18 sq. ft
	Area rudder	5 1/2sq ft
	Weight	800 lb.
	Length	32ft
	Price	£550

Nicholson monoplane built by coachbuilders Holland and Holland Ltd. was on the Royal Aero Club stand at Olympia in March 1910.

NOBLE glider
Reported built in May 1910, of monoplane type.

NOLAN monoplane (Bob and John Nolan, Whalley St., Accrington, Lancashire)
The brothers were interested in flight, and built kites and models. In 1904 a glider was made and tested on the hills near Accrington. In 1908 a machine with a motorcycle engine was tested but did not fly. A patent, No.4829/1910, was taken out.

NORMAN & M'KNIGHT monoplane (Norman & M'Knight, Glasgow)
This machine was shown in January 1910 at the exhibition of the Glasgow Model Aero Club in the Charing Cross Halls. Instead of having movable elevators at the rear, the monoplane had ailerons (elevons) under the wings to control pitch; these were operated by a steering wheel from the pilot's seat. The undercarriage comprised of two wheels and a tail-skid.

Power: Norman & M'Knight 12hp four-cylinder weighing 12lb

Data		
	Span	30ft
	Length	28ft

NORTH & MYERS glider (J.D. North and H.A. Myers)
This reference may be connected with the glider attributed to A.N. Myers of the Aeroplane Building and Flying Society (*q.v.*), of which North was also a member in 1910-1911.

NORTHAMPTON Polytechnic Institute monoplane
In a short article in *Flight* 7 October 1910 (p.866), Handley Page, who was a lecturer there, stated that 'a full size monoplane was under construction and a 30hp engine purchased'. There is no evidence of its completion as a flying aircraft, more likely it was retained as a student exercise. The airframe was almost certainly the Type C donated by Handley Page.

NORTHERN AIRCRAFT Co. Ltd. (Cockshot, Bowness-on-Windermere, Cumberland)
This company was formed in February 1914, with W.R. Ding as pilot and general manager, to operate commercial flights, initially using the Handley Page Type G biplane. Ding also applied to the Lakes Co. for the post of pilot, starting there in September 1914. By November he had arranged for the Northern Aircraft Co. to take over the whole of the Lakes company's activities.

As the successors to the Lakes Co., NAC inherited the monoplane that was nearing completion (*q.v.*), which then became known as the NAC monoplane. The first new aircraft initiated after the change was a pusher biplane, which did not materialize until early in 1915.

Although Ding has been credited with the design, the aircraft was mainly the work of C. Fleming Williams

NAC pusher biplane seaplanes PB.1 and PB.2
The machine that appeared after the outbreak of war was a two bay biplane with strut braced top wing extensions, and ailerons in the top wing only. The twin tail booms tapered in plan to a vertical rudder post, and joined the wings at the junctions of the rear spars and inboard interplane struts. A triangular fin was inset in the booms, and a rudder, with curved outline, was fitted. The rectangular tailplane, with tapered tips, was mounted on the top longerons and carried a divided elevator.

The nacelle was long and shallow and basically square in section. It carried a nose mounted radiator and had a raised decking to protect the pilot and passenger, aft of which the top was open. The Green engine was mounted aft with the cylinders protruding above the top of the nacelle, with fuel provided by a gravity tank under the top wing. The machine was mounted on a complex chassis of struts and cross members on twin floats.

The PB.1 was converted to the PB.2 by the change to an air-cooled rotary. This enabled the nose radiator to be dispensed with, with a consequent improvement to the shape of the nacelle and rearrangement of the rear portion for the new engine installation.

Northern Aircraft Co. PB1. The original machine in 1914 with water-cooled Green engine.

Northern Aircraft Co. PB2. This was PB1 reengined with a Gnôme rotary.

The first test flight was carried by J. Lankester Parker, who later was for many years the chief test pilot for Short Bros. Ding also flew the machine but little is known of its use in the early war years.

Power: **PB.1** 60hp Green four-cylinder inline water-cooled
PB.2 80hp Gnôme seven-cylinder air-cooled rotary

NORTHERN Polytechnic Club monoplane

In January 1914 the Flying Club of the Northern Polytechnic was reported to be building a monoplane with a 30hp engine, work on which was probably curtailed by the impending outbreak of war.

NORTH LONDON "*FLYING HOUSE*" multiplane (built at Alexandra Palace, Hornsey, London N)

It is difficult to believe that the *Flying House* here illustrated was seriously expected to fly and might just have been a fantasy in the mind of an eccentric inventor. However, the *Illustrated London News* of 10 November 1906 reported that it was built by a party of Frenchmen in the grounds of the Alexandra Palace (perhaps with the assistance of Auguste Gaudron, the balloon-maker, whose workshop was nearby). The machine had eight aluminium wings, each fifty-four feet long, and four propellers. The proposed powerplant is not known. The multiplane was designed to carry 100 passengers in some comfort. Needless to say there are no reports of it having flown and it was probably never completed.

This is almost certainly the flying machine described by "A.D." of the Alexandra Park Aviation Works, 77 Duke's Avenue, Muswell Hill, London N in his letter of 10 January 1905 to Patrick Alexander in which he offered to sell him patent rights to enable the construction to be finished. "A.D." stated that the machine was half-built and all the pieces of the mechanism were at hand. £2000 was required to complete the project.

The machine was 65ft long, 13ft wide and 19ft 6in high. It had eight wings, each 48ft 9in long and of 6ft 6in chord. The total wing span was 104ft. The engine was of about 100hp and drove four propellers. It was designed to carry 200 people at 60mph.

North London "*Flying House*" seen in unfinished state at Alexandra Palace in November 1906.

Whilst Patrick Alexander was very generous with his support of aeronautical projects, it would seem that on this occasion he decided that the project was without any merit. Nothing further was heard of the *Flying House*.

NOTTINGHAM monoplane (Messrs. Searby, Allen & Searby, Hartley Rd., Nottingham)

This rather fragile looking two-seater tractor monoplane, of conventional layout, was built by the above individuals, who wrote to *Flight* (25 November 1911 p.1031), outlining their lack of success in forming an aero club in the area.

Power: 30hp Alvaston two-cylinder horizontally opposed water-cooled

Nottingham monoplane built in 1911 by three friends soon faded into obscurity.

NUTTALL aeroplane (J.M. Nuttall)

Reported in the press on 11 December 1909 with a span of 24ft and chord of 3ft 6in.

NYBORG monoplane (T.G. Nyborg, 9 Chelsea Grove, Newcastle-on-Tyne, Northumberland)

Although trials of this machine were carried out at Bolden Racecourse, there is no evidence that flight was achieved. An entry in *Jane's* of 1909 gave no details, but photographs reveal the general configuration.

The machine consisted of a central structure of tubes, in which the engine, of unknown type, and the pilot were accommodated. Twin tractor propellers were driven by chains or belts. The narrow track undercarriage consisted of two bicycle wheels with a central skid and a smaller rear wheel mounted well forward.

Nyborg monoplane at Bolden racecourse, Newcastle on Tyne in 1910.

The gull-shaped wings were constructed with the inboard trailing sections divided into smaller flexible sections, apparently to simulate a bird's wing and these sections may have been controllable. A fixed horizontal surface, perhaps with a flexible trailing section for pitch control, was provided but there were no vertical tail surfaces.

The machine was quite low, probably no more than four feet high, but with a wing span approaching 40ft. Patent No.11961/1909 may be relevant.

The Northumberland and Durham Aero Club, of which Nyborg was a member, was unsuccessful in raising sufficient interest to run meetings in June and September 1910.

O

Ogilvie glider. A modified Wright-type bought from TWK Clarke.

Ogilvie biplane. A modified "Baby Wright" bought from Short Brothers.

OGILVIE & SEAWRIGHT glider (Alec Ogilvie and T.P. Seawright, Finchley, London)

A quadruplane glider was constructed in 1908 by the partners. It was fitted with a four wheeled undercarriage and was towed behind a car. It apparently came to grief when the tow rope broke, when the machine was at a height of thirty feet, piloted by Ogilvie's brother.

The following year Ogilvie bought a Wright type glider from T.W.K. Clarke and a Wright aircraft from Short Brothers.

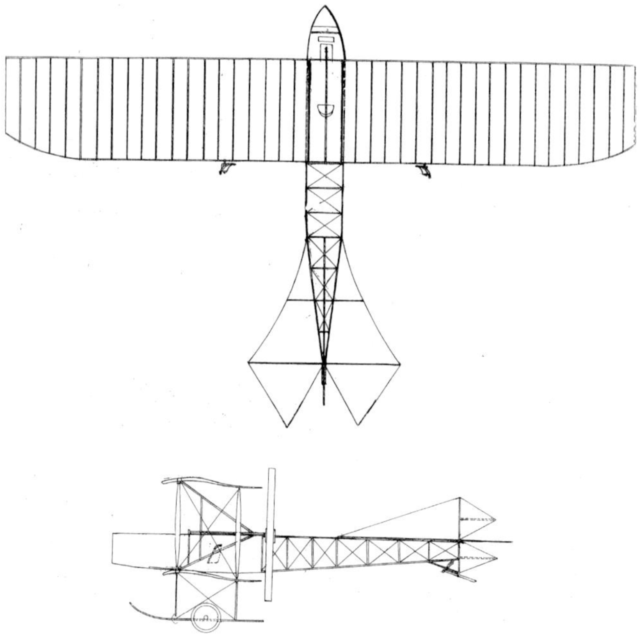

Line drawing of the Oliver biplane.

OGILVIE-HAWKINS triplane. See Hawkins-Ogilvie

OLIVER biplane (R.G. Oliver, Café Windmill, Cricklewood, London NW)

This biplane, designed in 1911, was similar in many respects to the early Avro biplanes except that it had twin pusher propellers mounted mid-gap behind the trailing edge of the wings. The propellers were driven by chains or belts from a water cooled engine "of Wright type" in the nose of the machine.

The biplane appears not to have been completed and its wings and other parts were offered for sale in *Flight* of 14 October 1911.

O'NEILL biplane (Miss Sheila O'Neill, 5 Salter's Hall Court, London EC)

A report of 30 November 1909 referred to a machine constructed at the London School of Aviation by Mr. Chandler and tested at Shellbeach in preparation for a cross-Channel flight. It was made of spruce and weighed 200 lb. Patent No.5908/1910 is relevant.

The Aviation World Directory gives details of an unorthodox monoplane invented by Miss O'Neill in which the wing was not attached rigidly to the fuselage but could be moved vertically and/or longitudinally to give control of height and direction. For transportation and housing the wing could be rotated to lie parallel with the fuselage. Nothing else is known.

OTTINO-WYLLIE Direct Lift device (G.P. Ottino and A. Wyllie)

This impractical device was exhibited at the Aero Show at Olympia in March 1910. It consisted of a series of tilting planes on an endless chain, intended to act as 'paddles'. No motor was fitted and there were no controls. Nothing further was heard of this machine. Patent No.9378 of 1909 refers.

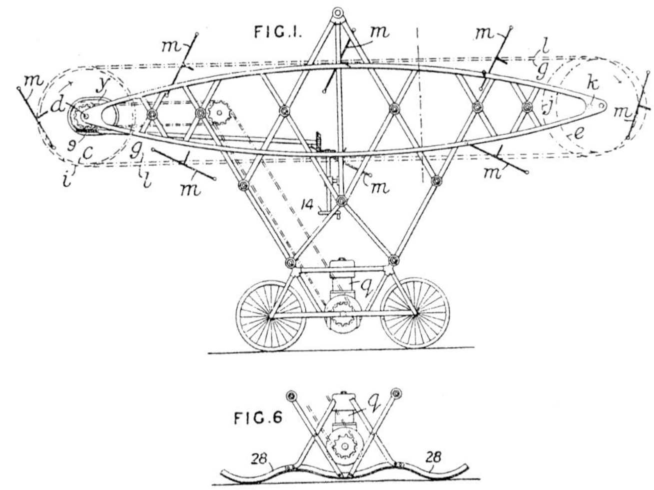

Ottino & Wyllie direct lift machine was exhibited at Olympia in March 1910.

OXFORD Ornithopter

This was reported built in November 1911. Area. 25sq.ft. Weight. 180lb.

P

PADDINGTON & DISTRICT Aero Club gliders (Sec. J. Hurlin, 2 Edbrooke Rd., with workshop at 14 Malvern Mews, London W and later at Windsor Place, Harrow Rd., from December 1910)

One of many model clubs in 1910-1911, which later joined with the Parkside Aero Club and used their ground at Sudbury. Their first full size machine, presumably a glider, was made available by J. Hurlin and Son in August 1910, for the members to rebuild. In December three machines were reported under construction by members at various locations, including one at Southend.

PADDINGTON Gliding Club glider (118 Elgin Rd., London W)

This club also used the Parkside ground at Sudbury and C.W. Tinson flew a glider there in April 1911

PALMER monoplane (L.N. Palmer & Co., 9a Trevelyan Rd., Merton Tram Depot, Tooting)

A machine with a 16hp JAP was reported being built in February 1910.

PARKER & WINN aircraft (Charles Parker and Rowland Winn, Holbeck, Leeds)

The partners built an aircraft in late 1909 with a 30hp Humber car engine. It crashed after takeoff from a field at Harewood. This may have been the machine offered for sale in *Flight* in October 1910 as 'Monoplane the first Yorkshire-built machine to fly; 40ft spread ready for engine'.

Parkes No.1 pedal-powered biplane in early 1909.

PARKES biplane No.1 (C.H. Parkes)

This first experimental machine, made by Parkes, was a two-bay biplane mounted on a bicycle and fitted with a biplane front elevator and tail rudder. The pedals drove a pusher propeller only, the chain to the rear wheel being disconnected. A speed of nine mph was reached on the level, but the machine would only lift off for a yard or two, when tested down a slope in 1909.

Fig. 2.—The improved biplane built by Mr. Parkes was equipped with a twin-cylinder air cooled engine of 4-h.p.; and with this machine he was able to accomplish "jumps" of from 10 to 40 ft.

Parkes No.2 motor-powered biplane in the summer of 1909.

PARKES biplane No.2

This was a motor powered pusher biplane, with large front divided elevator, capable of being operated together or independently. Separate single-acting ailerons were mounted behind the trailing edge of the top wing. The machine was mounted on a three-wheeled chassis with a 4hp vee-twin air-cooled engine behind the pilot, driving by chain a shaft, on which was mounted a pusher propeller. The allup weight was 350 lb. Jumps of 10-40 feet in length were made in the summer of 1909.

Parkes monoplane of 1910 was built by coachbuilders at Monmouth.

PARKES monoplane (C.H. Parkes & Parkes, Monmouth)

This machine was constructed for Parkes by T Preece & Co., coach builders of Monmouth.

It was a single-seat tractor monoplane with front and rear elevators and rear rudder, all mounted on an open girder structure on a three wheeled undercarriage. The high wing was heavily cambered and double surfaced, and was surmounted by a bracing pylon. Single acting wide chord ailerons were hinged behind the trailing edge of the wing and there were hoop shaped wingtip skids. The engine was positioned just ahead of the wing on a level with the top longerons, with the pilot seated behind the trailing edge.

The machine flew for the first time on 7 July 1910, but there were no reports of any further progress. A biplane, said to be under construction by Parkes brother, did not materialize.

Power: 20hp JAP vee-twin air-cooled with auxiliary exhaust ports, driving a variable pitch Chauvière propeller.

Data		
	Span	28ft 3in
	Area	200 sq. ft
	Length	28ft 3in
	Weight	500 lb.

PARSONS biplane (J.G. Parsons and Percy Maxwell Muller, Shed No.6, Brooklands)

P.M. Muller, who later became manager of Vickers at Brooklands in wartime, designed and built this biplane primarily for testing the Parsons Pendulum Paddle-Wheel Stabilizer. The machine was flown in May 1913, by John Alcock, later of Atlantic crossing fame, and also by Jack Humphries, without the stabilizer fitted; although using a heavy car engine weighing 400 lb. with radiators, the machine flew well reaching 1500 feet in twenty minutes. The stabilizers, which were mounted in the gap between each wingtip, were presumably tried and

Parsons biplane of 1913 at Brooklands. The heavy Aster engine was later replaced by a Gnôme rotary.

found wanting, as was the Aster engine, for it was replaced in August by a Gnôme.

Alcock was involved in a collision on the ground with the Champel biplane on 4 August 1913, in spite of which he started soon after in a race, but was soon forced to land. In September the machine was sold to a recent pupil of the Bristol School named Boger, who crashed it on 11 December 1913 at Ripley, when attempting to land there to breakfast at the Talbot Hotel. The remains were bought by Pemberton Billing and some parts were used in the construction of the Gaskell-Blackburn biplane.

Some features of the Caudron were embodied in the Parsons. The wing structure in particular employing closely spaced interplane struts and a similar system of warping of the top outer wing panels. Flexing of the tailplane for control, instead of a separate elevator, was also similar to the Caudron.

The fuselage was an ash and spruce girder, fabric covered in the area of the pilot's seat with the fabric tapering to a cone shape inside the structure, which was open back to the tail. The main undercarriage had a wide track and consisted of six ash struts and twin skids with a rubber cord sprung axle. The tail was supported by a cane, hoop-shaped tail skid. The engine was cooled by two radiators mounted behind, partially in the fuselage, behind which were fuel and oil tanks, followed by the pilot in his hammock type seat.

Power: 40hp Aster four-cylinder inline water-cooled driving a 8ft 2in diameter Normale propeller
70hp Gnôme seven-cylinder air-cooled rotary from July 1913

Parsons pendulum paddle-wheel stabilizer was tested on the biplane.

Data		
	Span top	39ft 6in
	Span bottom	25ft
	Chord	4ft 7in
	Area	295 sq. ft
	Area tailplane/elevator	40 sq. ft
	Area rudder	9 sq. ft
	Length	26ft 4in

PASHLEY biplane (Cecil Lawrence Pashley and Eric Clowes Pashley, Shoreham)

This pusher type two-seater was built by the Pashley Brothers in 1914 and it competed soon after its first trials, in a race for the Brighton Cup and £70, which Eric Pashley won on 11 July 1914.

The machine was basically of Farman type with ailerons on the top wing and a cumbersome four wheeled undercarriage, which it was planned to replace with a neater two wheeled vee type.

The engine was a 50hp Gnôme, also to be replaced by a 100hp Monosoupape Gnôme, but it is believed that these further developments were curtailed by the outbreak of war. The machine continued in use as a trainer at Shoreham.

Data		
	Weight	800 lb.
	Speed	61mph

Pashley biplane built by the Pashley brothers in 1914 and used by them at Shoreham. (Courtesy Nonie Cellier).

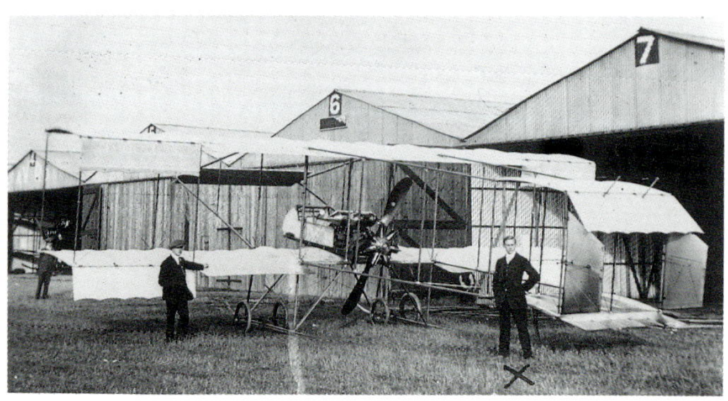

Passat ornithopter No.1 built at Wimbledon in 1908 and tested on the common.

PASSAT ornithopter No.1 (Monsieur J.B. Passat, Wimbledon, Surrey, assisted by Monsieur Martiniere)

Passat was a Frenchman living in England, who experimented with flapping wings from 1905. In 1908 he built his first machine in his garden at Wimbledon, and took it on to Wimbledon Common for its first and only trial, when it traveled at speed for some distance, being airborne for at least twenty yards, before being wrecked against a tree. He was present at Olympia in March 1909 demonstrating his flapping wing principle to visitors at the Aero Show.

The machine consisted of a welded tubular frame, faired to streamlined shape and mounted on three wheels. Two pairs of wings were fitted in tandem; the front pair serving as elevators and for lateral control. The rear pair flapped to provide propulsion, driven by the motor, which also drove the rear wheel. A tailplane, fin and rudder were fitted, together with a small fixed surface at the nose. The wings could be folded to enable the machine to be driven along the road.

Power: 4 1/2hp Werner two-cylinder air-cooled motor cycle engine

Data	Span	24ft
	Length	20ft
	Weight	300 lb.

PASSAT ornithopter No.2 *Seagull*

A second machine was built in 1909-1910 on generally similar lines to No.1, but strengthened and with a tractor propeller. Passat took this machine to Brooklands and tested it in the summer, without success. It was further reported to have flown between 100 and 150 yards after taking off at less than 15 mph.

Power: 8hp George of unknown type

Passat ornithopter No.2 *Seagull* was reported to have flown at Wimbledon in 1912.

Passat monoplane *Seagull* was tried at Brooklands in 1911.

PASSAT monoplane *Seagull*

Although lacking confirmation, it seems likely that Passat converted the ornithopter into a monoplane with variable incidence wings, as illustrated in *Flight* 3 February 1912 (p.109). Reportedly fitted with a 15hp Anzani engine, it suffered from engine problems, but may have been the machine with which flights of 100-150 yards were achieved on Wimbledon Common in 1912.

Passat became a naturalized Englishman and, much later, made further attempts with flapping wings at Surbiton, on a machine he termed a 'Helithopter.'

Paterson biplane No.1 built at the Liverpool Motor House where he was a director and tested at Freshfield sands in 1910.

PATERSON biplane No.1 (Compton Cecil Paterson., Freshfield Sands, Lancashire)

Paterson was a director of the Liverpool Motor House Ltd., and his biplane was built there in 1909-1910, making its first flight from the beach at Freshfield on 14 May 1910. The second flight on 23 June 1910 resulted in damage, which was repaired by the end of July. After this the aircraft flew well and was used by Paterson to obtain his aviators certificate No.38 issued on 6 December 1910. He used the machine for passenger carrying and acquired a considerable amount of flight experience. In November it was reported to have flown over 160 miles and that a second machine, with a more powerful engine, was being built. This was for Gerald Higginbotham and was completed in January 1911, the two Paterson machines being housed in adjacent hangars at Freshfield.

The Paterson biplane was basically of Curtiss type, constructed of ash, spruce and bamboo with a tubular steel undercarriage. Paterson arranged the machine for easy transport, by designing the outer wings for quick removal and assembly.

The machine had a biplane front elevator with a small triangular fin on top, the single rear elevator carried the rudder, which was cut away at the center to allow it to swing clear of the elevator. The ailerons were originally mounted in midgap on the interplane struts, but were moved to the trailing edges of the top wings; they were of single acting type and were enlarged with a part circular trailing edge shape. For early flights wheels were fitted at the wing tips.

Paterson biplane No.1. The version with Gnôme engine, modified ailerons and other improvements late in 1910.

Power: 1st machine 25-30hp Anzani three-cylinder air-cooled fan type semi-radial
2nd machine 50hp Gnôme seven-cylinder air-cooled rotary

Data		
	Span	34ft 4in
	Weight with 6 gal fuel for 90 miles	600 lb.
	Length	31ft
	Speed	41mph
	Price	£625

Paterson biplane No.2 was built at Cricklewood by Lawtons Motor Body Works in 1911 and soon dispatched to South Africa.

PATERSON biplane No.2

In the spring of 1911 Paterson left Liverpool to become an instructor for the Grahame-White School at Hendon, remaining there till late in the year, during which time he designed a new aircraft, which he planned to take to South Africa. This time the machine was constructed by Lawton's Motor Body Works, Cricklewood London NW. Paterson flew the machine for the first time from a field near the Welsh Harp on 18 October 1911, with some damage occurring, when he was forced to land in a restricted space with a blocked fuel pipe.

The machine was soon repaired, but it seems unlikely that it reached Brooklands when ready, as reported to be Paterson's intention. Time was short before the machine was to be packed for shipment on the SS *Garth Castle* on 11 November 1911. Paterson's team in the African Aviation Syndicate consisted of his assistant pilot, E.F. Driver, manager Guy Livingston and several mechanics. Apart from having the Blériot agency for Africa, the object was to interest the government and public in aviation, and to set up a flying school.

The party arrived at Cape Town in early December and gave demonstrations straight away, but Paterson crashed on 26 December 1911, when the fabric on the tail split open and he suffered injuries. The early problem overcome, Paterson and Driver flew considerable distances to give demonstrations,

Paterson biplane No.2 was also flown as a seaplane.

using the biplane, Driver's Blériot, and a Farman built earlier by John Weston at Brandfoort in the Orange Free State.

In March 1912 Driver crashed the Blériot badly, although it was later rebuilt, but by September the African Aviation Syndicate was in liquidation. However, Paterson acquired the assets and persuaded local business men to finance a new scheme, the Paterson Aviation Syndicate, for which he obtained a contract from the government to train ten military pilots, beginning in September 1913. For this he required a new instructor and E.W. Cheeseman late of the Grahame-White school, came out, but he crashed and was killed on 13 October 1913. At that time there were two Paterson biplanes in use at Alexandersfontein, near Kimberley; the second biplane, built by Mr. H. Carpenter was described as 'of Henry Farman type with Paterson section'.

The machine built in England was a typical pusher of the time, with front elevator and fixed tailplane carried on booms, parallel in plan and tapering in side view. The rudder made of aluminum sheet was cut away at the center to clear the tailplane. The wings were double surfaced and of equal span and carried single acting ailerons on all four. A nacelle enclosed the two occupants, seated in tandem, with fuel and oil tanks behind. The propeller was mounted in front of the rotary engine.

In the early part of 1913, Paterson fitted floats to one of his machines, which also had top wing extensions and this was the first seaplane in South Africa. On both the seaplane, and when operating on wheels, the nacelle was removed to reduce weight to aid the performance in the heat of South Africa.

There are shadowy references to a further Paterson aeroplane having been built in October 1913 by Lawtons Ltd., 25 Hope St., Liverpool, but this cannot be confirmed.

Paterson returned to England in 1914 and formed a partnership with Capt. E. Bass to build the Bass-Paterson flying boat, constructed by S.E. Saunders Ltd. at Cowes (*q.v.*).

Paterson is well remembered for his work in training the first pilots of the SAAF and a memorial hangar was built in 1981 on the original site, where this took place, housing a replica of Paterson's second aircraft. A road near Kimberley Airport was also named after him in 1987.

Power: 50hp Gnôme seven-cylinder air-cooled rotary

Data		
	Span	32ft without extensions
	Chord	5ft 6in
	Gap	5ft 3in
	Area	329 sq. ft
	Area tailplane	24 sq. ft
	Area elevator	31 1/2 sq. ft
	Area rudder	8 1/2 sq. ft
	Length	32ft 9in
	Weight	750 lb.
	Height	10ft 4in

PAUL glider (Paul., Sutton, Birmingham)

A biplane glider with a span of 24ft of which nothing more is known was reported.

PAYZE glider (Arthur Payze, Marsh Mills, Henley-on-Thames, Oxfordshire)

This was described as of Chanute type and of about 20ft wing span.

PEEBLES biplane (Peebles Motor Co., Edinburgh)

This machine was a replica of a Caudron built for a Mr. Lawrence Bell in June 1914.

PEMBERTON BILLING aircraft (Noel Pemberton Billing)

Pemberton Billing's early activities in aviation began in 1903-1904 with a manlifting kite or glider (PB.0). He then built three variations of a small monoplane, one using a rotary motor of his own design (PB.1). On the second of these he lifted off the ground in 1909, but crashed and was injured.

In 1909 Billing was responsible for the abortive attempt to create an aerodrome, with factory and support facilities, at Fambridge in Essex. This was a failure owing to the nature of the ground and its remoteness. Later, apart from editing a magazine, he turned his activities back to boats, for which he had facilities at Woolston, Southampton from 1912. Late in 1913 he again turned to aeronautical, work with the design of a small flying boat PB.1 (PB.7) using the name Supermarine for the first time.

Following the lack of success of this and other types, and little official interest, he gradually gave up his interest in the company, the responsibility for which was taken over by his works manager, Hubert Scott Paine, as the Supermarine Aviation Works Ltd. in 1916.

The identification of Pemberton Billing types is confused by the renumbering that took place retrospectively; the bracketed number is the later identity.

PEMBERTON BILLING glider (PB.0)

This was a triangular kite on which PB tried to glide from the roof of his house at East Grinstead in 1904.

Pemberton Billing PB0 glider was his first attempt at flight.

PEMBERTON BILLING monoplane (PB.1 and 3) (Woolston & Park Royal)

The machine was a small pusher, based on a tubular chassis on a tricycle undercarriage, tested unsuccessfully at Woolston in 1909.

The wings were extended aft at their inboard ends, terminating in flaps for lateral and directional control. On the second version the pilot was enclosed in a nacelle. It was the second version that lifted off briefly on several occasions, finishing with a jump of 60ft and a crash which injured PB.

Power: PB.1 Pemberton Billing valveless rotary
PB.3 JAP two-cylinder air-cooled

Pemberton Billing PB5 monoplane of 1909 built at Woolston was evolved from the earlier PB1 and PB3.

PEMBERTON BILLING monoplane (PB.5)

This was similar to PB.3 with the addition of a front elevator. It was not flown, but was sold to pay debts, while PB was recovering from his injuries.

Power: 25-40hp NEC four-cylinder inline water-cooled two-stroke.

PEMBERTON BILLING flying boat Supermarine PB.1 (PB.7 and 9)

This small single-seater flying boat was displayed at the Olympia Aero Show in March 1914 and was tested later with little success. The machine was broken up later in the year after an altered layout had been tried.

The PB.1 had a well shaped, circular section hull with an added lower section, with flared sides, ending in a step. The pilot was seated behind the wings, with the engine in a nacelle on struts ahead of the wings, mounted at a steep thrust angle. The wings were of unequal span and were made in one piece, without dihedral. Sprung, streamlined, circular section floats were mounted below the outboard interplane struts. Ailerons and conventional tail surfaces were fitted. An instrument named a 'Supermeter', to register the height above the water on alighting, was a special feature. A grapnel in the nose, controlled from the cockpit, was another original fitment.

After unsuccessful attempts to takeoff, the engine and pilot's position were interchanged. The engine was buried in the hull and drove twin pusher propellers by chain drives. The pilot was positioned ahead of the wings to balance the aircraft. In this form the machine was tested in May 1914 by Howard Pixton and was later identified as PB.9.

Power: 50hp Gnôme seven-cylinder air-cooled rotary driving a 8ft diameter three-bladed PB Trinity propeller, replaced by twin chain driven outboard propellers

Data		
	Span top	30ft
	Span bottom	28ft
	Chord	5ft 7 1/2in
	Gap	5ft 3in
	Area	293 sq. ft
	Area elevator	20 sq. ft
	Area rudder	12 sq. ft
	Area tailplane	31 sq. ft
	Weight	750 lb.
	Price	£525
	Weight allup	970 lb.

Flying boat PB1 reidentified later as PB7 was shown at Olympia in March 1914.

Flying boat PB1 reidentified as PB9 after modification to twin-pusher layout.

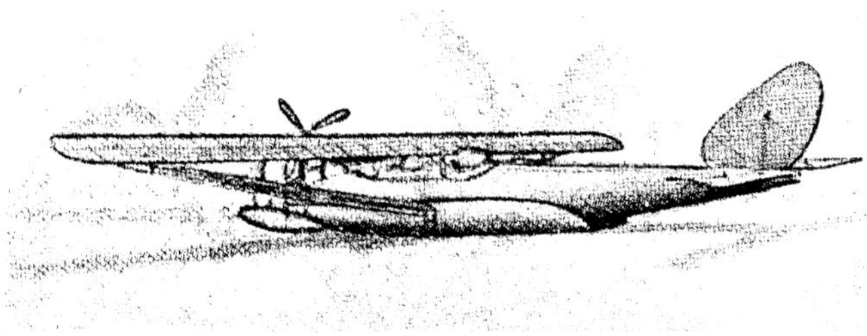

Flying boat PB2 later PB11, a 1914 project for a two-seater that was not built.

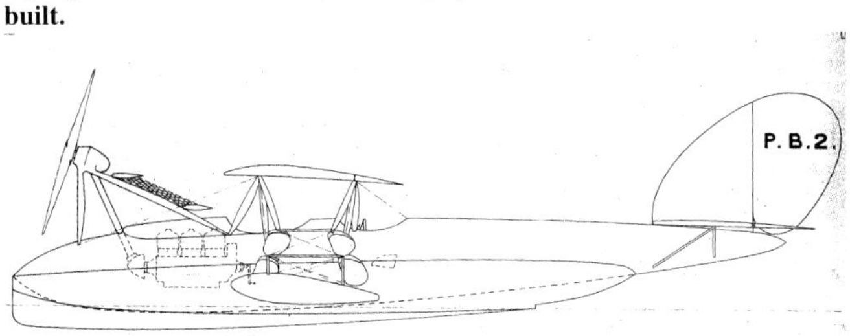

PEMBERTON BILLING flying boat Supermarine PB.2 (PB.11)

This project, which remained unbuilt, was to have the engine installed in the hull, driving the propeller through bevel gearboxes and shafts, at a high thrust angle. The hull design was an enlarged version of PB.1, with the lower hull portion extended further aft. The pilot sat in the rear cockpit; the front cockpit, under the wing, was larger and could probably hold more than one passenger. The monoplane wing was mounted above the hull on struts, and was braced to the hull by narrow chord aerofoils. The balancing floats, mounted under the aerofoil struts, were sprung by a scissors-like system of struts, passing on either side of the aerofoil struts, to attach to the main wing.

Power: 120hp Austro-Daimler six-cylinder inline water-cooled driving a 9ft diameter three-bladed PB Trinity propeller

Data Weight allup 2,000 lb.
Price £1575

PEMBERTON BILLING flying boat Supermarine PB.3

This was a further enlargement of PB.2 which remained unbuilt. This time wing warping was employed. The machine was to have two engines driving a propeller through bevel gearboxes and shafts as before. In his elaborate publicity material, produced for issue at the Aero Show, PB refers to 'alternate drive', which is presumably for a water propeller, and perhaps the first intended 'slip wing' machine.

Power: Two 90hp Austro-Daimler six-cylinder inline water-cooled driving a 9ft 6in diameter three-bladed PB Trinity propeller

Data Weight allup 3,000 lb.
Price £2310

PEMBERTON BILLING flying lifeboat, Supermarine PB.5 (PB.21) and PB.7 (PB.19)

The first version of this 'slip wing' design was not built, but two of the PB.7 type were ordered by the German Navy and construction of these was commenced in May 1914. The machine was illustrated and details given in advertisements dating from 29 May 1914.

The design consisted of a biplane with engine in the hull, driving twin pusher propellers by chain. The hull was designed as a motor boat, with water propeller and rudder, to which the rear portion of the hull, and complete wing cellule, were attached as one unit. One of the German machines was basically complete by the end of July, but further work was discontinued at the outbreak of war. The boat portion is believed to have been finished and used later on the Solent.

Flying lifeboats PB5 and 7 later renamed PB21 and 19 respectively with slipwing allowing the hull to proceed separately. One was basically complete in July 1914.

Power: PB.7 200hp Salmson (Canton-Unné) fourteen-cylinder water-cooled radial or 225hp Sunbeam twelve-cylinder water-cooled vee

Data

Span	57ft 6in
Length	34ft
Endurance	4hr
Weight allup	3,000 lb.
Speed	45-70 mph

The boat portion had a speed of 35 knots and a range of 200 miles maximum

Scout biplane PB9 later PB13 was built in seven days in August 1914, was tested at Netley and Brooklands and taken over by the RNAS.

PEMBERTON BILLING biplane PB.9 (PB.13)

This scout type biplane was constructed in August 1914 at the time of the outbreak of war. It was produced in a hurry, on an initiative by Billing, and became known as '*the Seven Day Bus*', from its construction between the 3rd and 11th of August. On the 12th, it was taken to a field at Netley, where after repair of damage caused while taxiing, it was flown by Victor Mahl of Sopwith's. Thereafter it was transferred to Brooklands, where it was flown by Jack Alcock, who refused to fly it again after two flights. Billing then attempted to take off himself, but damaged the undercarriage before becoming airborne.

The machine was taken over by the RNAS in early 1915 and became serial No.1267 at Hendon and Chingford, until struck off charge in June. Billing acquired the discarded PB.9 and used it to further his unsuccessful election campaign of early 1916.

The machine was comparable to other scout biplanes of the period, such as the Sopwith Tabloid and Martinsyde S.1, but it had nothing to commend it for general use, so was not ordered in quantity. A proposed enlarged version with staggered wings (PB.17) was not built.

One feature, which was unusual, was the construction of the complete wing assembly in one unit, which was passed over the tail to its correct position, where it was then clamped to the longerons. This feature may have stemmed from an original set of wings, bought by Billing from Radley, made for the Radley-England pusher biplane, which was not completed.

The engine was the Gnôme originally used in the Farman, on which Billing had learnt to fly in 1913, and was also used in the PB.1 flying boat, which had been discarded by this time.

Power: 50hp Gnôme seven-cylinder air-cooled rotary driving a 8ft diameter propeller

Data

Span	26ft
Chord	4ft 6in
Gap	4ft 6in
Area	205 sq. ft
Area tailplane	13 sq. ft
Area elevators	11 sq. ft
Area fin	2 sq. ft
Area rudder	6 sq. ft
Weight	560 lb.
Weight allup	750 lb.
Length	20 ft
Speed range	40-75 mph
Endurance	3hr

PEMBERTON BILLING biplane PB.11 (PB.15)

Serial No.1374, which the RNAS acquired in February 1915, was a pusher 'Gun Bus' type with a Gnôme engine. This was probably no more than the Farman III, which Billing had purchased in order to learn to fly in twenty-four hours on 17 September 1913, to win a wager with Handley Page, although this machine did not have an enclosed nacelle.

Pemberton Billing Boxkite biplane PB11 (PB15) became RNAS 1374.

PERCIVAL *Parseval* I biplane (N.S. Percival, Hangar No.2, Brooklands)

This machine was a further reconstruction of the biplane built by Eardley Billing, which was itself a reconstruction of a Voisin. During early trials it was reported to be difficult to fly, attributed to the replacement of the ailerons with warping control, and the fact that the fore and aft balance was incorrect. The latter defect was corrected by moving the fuselage by four inches in relation to the wings.

The machine was flown by Percival from the 16 August 1911 and, despite odd damage occurring, was successfully used for instructional work until April, when more extensive damage took place. In August the machine reappeared, after repair, fitted with the new Adams-Dorman engine for flight testing. This engine failed to attract interest and by October Percival

Percival *Parseval I* biplane a reconstruction of the Billing and a Voisin of 1911-1912.

had transferred his interest to a Caudron and later to the Dunne tailless aircraft.

Power: 40hp ENV type D eight-cylinder water-cooled vee
60-80hp Adams-Dorman eight-cylinder water-cooled vee from August 1912

PERRY, BEADLE biplane 1913 type T.1 (Perry, Beadle & Co., Aviation Works, Gould Rd., Twickenham, Middlesex)

The first result of collaboration by E.W. Copland Perry, a qualified pilot, and F.P. Hyde Beadle, a technician, who had met at the RAF Farnborough in 19112, was a small single-seater biplane built at Beaulieu in 1913. The machine apparently flew well at Beaulieu, making at least one flight to Cowes. It was described by C.G. Grey as being on 'Avro cum BE lines' but no description appeared.

However in 1914, after Perry, Beadle & Co. was formed, a catalogue was issued with a photograph showing an aircraft conforming to this description. The associated data refers to a developed version of this machine, Type T.2, with a six-cylinder radial Anzani, but this is probably applicable, in part, to the 1913 type.

The illustration shows an equal span biplane with a three-cylinder Y type engine and fuselage mounted with a gap between it and the lower wing. A twin skid undercarriage with wheels sprung on the skids was shown. Parallel ailerons were fitted to the top wing only.

Power: 25hp Anzani three-cylinder Y type air-cooled semi-radial

Data		
	Span	25ft
	Gap	c.5ft 9in
	Chord	4ft 6in
	Area	c.225 sq. ft

Perry Beadle T1 biplane built at Beaulieu in 1913.

PERRY, BEADLE biplane 1914 type T.2

The biplane, which arrived at Brooklands in May 1914, revealed differences from the earlier version. The fuselage now rested on the lower wing, the gap having been reduced. There was a large cutout in the trailing edge of the lower wing, below the fuselage. The top wing had been extended, and there was taper on the trailing edge of the ailerons, and bracing struts to support the extensions had been added. The undercarriage was of simple vee type with cross axle and no skids. The engine was a six-cylinder radial.

Copland Perry flew the aircraft on 26 June 1914 and later M.F. Glew and A. Dukinfield Jones also. It was requisitioned by the RNAS later in the year, becoming serial No.1322, and was wrecked at Hendon in March 1915.

Power: 45hp Anzani six-cylinder air-cooled radial

Data		
	Span top	c.35ft
	Span bottom	25ft
	Chord	4ft 6in
	Gap	4ft 9in
	Area	c.270 sq. ft
	Climb rate	400 ft per min
	Speed range	45-65 mph
	Endurance	2hr
	Price	£450

Perry Beadle T2 biplane modified from the T1 at Brooklands in 1914.

PERRY, BEADLE biplane type P.6

This project was to meet a military specification for 'Reconnaissance Type B', for which the following information appeared in the 1914 catalogue.

The aircraft was to be a pusher biplane, with a two wheeled undercarriage and skids and double acting ailerons. Accommodation was provided for a crew of three.

Power: 100-110hp Anzani ten-cylinder air-cooled radial

Data		
	Span	38ft
	Chord	6ft
	Area	450 sq. ft
	Area tailplane	58 sq. ft
	Area elevators	30 sq. ft
	Area rudders	18 sq. ft
	Weight	1,000 lb.
	Speed range	34-63mph
	Climb rate	380ft per min
	Endurance	4hr

Perry Beadle B3 flying boat in the works of SE Saunders Ltd. at Cowes where it was constructed in 1914.

PERRY, BEADLE flying boat type B.3

This biplane flying boat appeared for the first time at Olympia, incomplete, in March 1914, where it received praise for the quality of its workmanship. Much of this was the work of S.E. Saunders Ltd. of Cowes, who had designed the hull shape, and built it using the patented 'Consuta' system of copper wire sewn plywood. The lower wing was surfaced with 'Consuta' also and was made buoyant. The fin and tailplane were integral with the hull. The use of the lower wings as sponsons was covered by patent No.4634/1914.

The engine was mounted in the nose under a detachable cowling and drove, by means of an extension shaft and chains, the twin tractor propellers. These were mounted in streamlined housings and were supported between the planes by the intermediate interplane struts and bracing wires. Behind this, a single open cockpit, in the rounded top decking, housed the passenger in front and the pilot behind. The radiator was fitted high up across the inboard interplane struts.

The top wing carried inset ailerons in the overhanging portion. The other control surfaces consisted of a divided elevator and a rudder, the lower portion of which was immersed when the machine was at rest, to serve as a water rudder. The immersed tail surfaces were also covered with 'Consuta' ply.

Perry Beadle B3 flying boat on test on Lake Windermere in 1915.

After the Aero Show the machine was assembled in Saunders' Columbine Works and, soon after, the visiting impressment officer recorded it as a potential machine for Service use, by which time the ENV engine had been replaced by a 90hp Curtiss. The extent of the testing carried out at Cowes is not clear, but in August the machine was moved to the Eastbourne Aviation Co. Perry had been killed in an air crash on 16 August 1914 in France and the company was being closed down. Later the aircraft was sold to the Lakes Flying Co. and tests were carried out on Lake Windermere until July 1915, without flight being achieved. The aircraft was then broken up.

Power: 60hp ENV type F eight-cylinder water-cooled vee driving two Integral propellers through an extension shaft and chains, one crossed for opposite rotation. 90hp Curtiss type OX eight-cylinder water-cooled vee substituted in mid-1914

Data		
	Span top	35ft
	Span bottom	23ft 3in
	Chord top	6ft
	Chord bottom	4ft
	Gap	6ft 8 1/2in
	Area	290 sq. ft
	Area rudder	11 sq. ft
	Area elevators	6 sq. ft
	Area tailplane	25 sq. ft
	Weight	1,100 lb.
	Speed	72 mph
	Endurance	3 1/2hr
	Weight allup	1,600 lb.

PETRE monoplane (Henry A. Petre and Edward Petre, Brooklands, Hangar No.11)

The brothers designed and constructed this machine in 1909-1910 and it was shown in skeleton form on the stand of Leo Ripault Ltd. at Olympia in March 1910. The aircraft was taken to Brooklands for completion and testing and was housed in hangar No.11.

On the 26 July 1910 Henry Petre was airborne 23ft. above the ground, but, after flying about 100 yards, a gust of wind caused the machine to crash on the port wing. It was not rebuilt and the brothers went on to fly other people's aircraft. Edward was killed on a Martinsyde on 24 December 1912, but Henry went to Australia, that same month, to train the first military pilots at Point Cook and continued to serve in the Australian Flying Corps and its successor service, the RAAF.

The layout of the monoplane was unusual in that the propeller was behind the tail, driven by a long shaft from the engine mounted in the central position, with the pilot positioned ahead of the wing in the nose. The wing was built in two halves and could be rotated about the main spar to vary the angle of incidence in flight.

The Petre brothers' monoplane built by them in 1909-1910 at Brooklands.

The fuselage, which was split at the center to assist transport, was built with four longerons, vertical members and diagonal cross bracing, mainly of laminated ash. The top longerons were set wider apart than the lower pair and tapered towards the tail. The aircraft rested almost horizontally on the ground by virtue of the tall tail wheel mounting, a consequence of the incidence changing feature.

The main undercarriage consisted of two A-frames with twin skids and sprung cross axle. The control surfaces consisted of ailerons with diagonal hinge lines, their outline extending behind the main trailing edge shape. The all moving tail surface was mounted above the fuselage on a structure braced to the tail wheel support and provided lift as well as control in pitch. A small rudder was positioned below the fuselage. The two radiators for cooling the NEC engine were located on the fuselage sides alongside the cockpit.

Power: 25-35hp NEC four-cylinder inline water-cooled two-stroke driving a 7ft diameter Handley Page propeller through a 15ft long 2in diameter shaft.

Data		
	Span	30ft
	Chord	7ft
	Area	195 sq. ft
	Area tailplane	36 sq. ft
	Area rudder	7 sq. ft
	Length	29ft
	Weight	540 lb.
	Weight allup	740 lb.
	Speed	30mph
	Price	£800

PETTER flying bicycle (Percival Petter, Yeovil, Somerset)
Percy Petter, a member of the family that established the Westland Aircraft Works, was acquainted with Frederick J. Stringfellow, son of the famous John Stringfellow. He became interested in aeronautics towards the end of the nineteenth century, as a result of seeing the surviving Stringfellow models, and commenced experimenting on his own account. In 1894, at the age of twenty-one, he made a flying bicycle and a test rig of a helicopter mechanism, consisting of four boxkite surfaces on long rotating arms, driven by a Petter oil engine. The demands of the engine business caused the aeronautical work to be discontinued until 1915, when, in conjunction with his brother Ernest, aircraft construction was begun to support the war effort.

PHILLIPS helicopter (Councilor Henry Fentum Phillips, 4 Chertsey Street, Guildford)
Phillips was an electrical engineer who designed and made a number of propellers which were flown on model aircraft. No details are available of his full size machine built in 1909-1910 and tested at the Guildford Sewage Works at Bellfields.

PHILLIPS convertiplane (Arthur Phillips, Market Drayton, Shropshire)
See British Matchless.

PHILLIPS monoplane (Phillips & Co., Newcastle-on-Tyne)
Trials of this machine 'designed by a young foreign gentleman' were reported as being carried out at Gosforth in September 1909.

PHILLIPS helicopters and multi-planes (Horatio Frederick Phillips, Norwood, Harrow, Southminster, Streatham etc.)
Phillips, born 1845, was an inventor who began experimental aeronautical work with models, including making experiments with various forms of propulsion, aiming at a lightweight engine for flight purposes. His particular contribution was his work on aerofoil sections, from which the term 'Phillips Entry' is still used as a description and which were the subject of Patent Nos.13768/1884 and 13311/1891. His experimental equipment included a large steam powered Whirling Arm and a Circular Track as part of his costly experiments.

He aimed initially to develop direct lift types, i.e. helicopters, but lacking success with these, turned to gliders and models. His experience with the use of flexible fabric covering led him to decide to use solid 'sustainers', i.e. aerofoils, and he used examples of these, ranging in chord from 1.5 to 24 inches in the course of his trials. The nature of these narrow sections led to the design of 'Multi-planes' with many individual aerofoils in supporting frames. Several were built before a limited degree of success was achieved. His Patent No.20435/1890 covered machines with two or more frames.

Phillips carried out aeronautical experiments from about 1864 until 1910. Some of his early work was facilitated by his position as works manager, from 1890 to 1902, of the factory at Harrow of gun makers Cogswell and Harrison Ltd. This position, and their mutual interests, brought him into contact with Hiram Maxim. Support was also provided in 1893-1894 by the Comte de Lambert, but this was withdrawn owing to problems arising as a result of Phillips drinking habits. After retirement from business in 1902, Phillips devoted his time to experimental aeronautics and later to work as a motor and general mechanical engineer, with premises at West Barnham, Sussex. He died in 1926.

PHILLIPS helicopters and monoplane

In about 1870, a direct lift machine was constructed at Battersea, south London, in one of the railway arches. This was powered by a steam engine, of Phillips' own design, and drove contra-rotating steel rotors, later replaced by 20ft diameter wooden rotors. No success was achieved during tests in Battersea Fields. In 1903, with his son A.H. Phillips, he was again testing direct lift machines at the Crystal Palace, but again without success. One of these had 33ft diameter contra-rotating rotors driven by a 15hp petrol engine.

At this time Phillips also built a petrol engined monoplane on floats, which was launched on the Thames from Messengers Island, Surbiton with his son in control. The first trial was stopped when the port wing lifted 56ft and the machine, banking steeply, was at risk as it rounded the bend in the river towards Hampton Court. The trials were discontinued after the propeller broke and were not resumed.

In 1904, at Furzle Down Farm, Mitcham a machine with a more powerful motor driving two 20ft diameter rotors, set at nine degrees from the vertical, on the ends of a girder, was also tried. The tilt was to create some forward motion to aid take-off, but as with previous machines, this was a complete failure.

Phillips returned to direct lift machines in the autumn of 1909, when a new machine gave promising results and encouraged him to again pursue the principle, which he regarded as the practical flying machine of the future, particularly in England with its small fields and open spaces.

PHILLIPS multi-planes

Phillips pursued his aeronautical work with models after his early disappointments with rotary wings. Using models weighing up to 20 lb., which were shot from a catapult, he then constructed at Norwood a circular concrete track of 500ft circumference, 160ft diameter approximately, and tethered his machine to the center. First he used a full size type of box-kite with fabric covering, but soon decided on the use of more rigid aerofoils. This work led to the patents of 1884 and 1891 and the favoring of solid narrow aerofoils and the multi-plane type aircraft. To assess the best form for the blades, he proposed to use a steam driven whirling arm machine traversing a circumference of 323ft, 103ft diameter, to test them.

PHILLIPS multi-plane No.1 1893

Resulting from the work on the whirling arm and other tests, Phillips constructed a large unmanned multi-plane, with fifty aerofoils in a frame, mounted on a body on three wheels, with the single wheel at the front and was propelled by a compound steam engine driving a two-bladed tractor propeller. Phillips was sponsored by the Comte de Lambert at this time, which resulted in this engine being sold by the Count to the Musée de l'Air in 1936.

Phillips No.1 multiplane was steam-powered and was tested on a circular track at Harrow in 1893.

The machine was tested on a circular track at Harrow, 628ft. in circumference, 200ft diameter, anchored to a central post, where it lifted itself and a 56 lb. load, a total of 416 lb., about 2ft in 1893.

Power: 6hp Phillips compound steam engine driving a two-bladed tractor propeller

Data		
	Span	22ft
	Chord of slats	1 1/2 in
	Area	136 sq. ft
	Length	20ft
	Height	9ft 6in
	Weight	360 lb.

PHILLIPS multi-plane No.2 1903

There was a period of work on direct lift, before the next machine was built on the lines of the unmanned 1893 multi-plane. The new machine was constructed at Southminster in Essex and was tested in about September 1903.

In addition to a thirty-two slat wing, a thirteen slat front mounted elevator and five slat rudder, were fitted. The contra-rotating propellers were probably driven by a petrol motor. There are reports of a flight reaching 2-3 feet in height for 20-30 yards, at St. Lawrence Hall, near Southminster. A witness of this, Mr. J.J. Cant of Tillingham, confirmed this in November 1972, but there is no other evidence. The only known photographs came from Phillips' nephew.

Phillips No.2 multiplane was tested at Southminster in Essex in 1903 and is reported to have lifted off briefly.

PHILLIPS multi-plane No.3 1904

Another multi-plane was tested at Furzle Down Farm, Mitcham and was found to be longitudinally unstable, although it lifted at a speed of 50ft per second, 34mph. The total weight was 600lb.

Phillips No.3 multiplane. This petrol-engined machine was able to take off and fly a short distance at Mitcham in 1904.

The machine, of which a photograph exists and some details, was a tractor driven by a 22hp petrol engine. The pilot sat behind the lifting surfaces, of which there were twenty. The machine had a cruciform tail unit with elevator and a rudder and was mounted on three bicycle wheels.

PHILLIPS multi-plane No.4 1906

This version was tested on fields at Mitcham, later to become the golf course. No details are known, but it was probably a revision of No.3.

PHILLIPS multi-plane No.5 1907

This was a most unusual multi-plane, built with 200 slats, 1_ inches wide and 1/8" thick and with two inch spaces between them, carried in four separate frames. It was a tractor with, presumably, the previous engine driving a wide bladed propeller. There were no tail surfaces and the pilot sat in the center between two of the frames. The body of the machine consisted of a triangular girder with lattice bracing of steel ribbon. The machine rested on two twenty eight inch bicycle wheels in tandem with two smaller wheels at the wingtips.

It seems possible that Phillips lifted off the ground with this machine at Streatham in the middle of 1907. The airborne distance of about 500ft in a 30mph wind was calculated later. Although Phillips never claimed this as a flight, it could well have been the first sustained flight in the country. Further tests anchored to a pole by a 200ft wire were also made. Multi-planes, and No.5 in particular, were more in the nature of test equipment and not suitable for development into successful flying machines.

Power: 22hp petrol engine, eight-cylinder air-cooled make unknown, driving a 7ft diameter propeller

Data		
	Span	20ft
	Length	15ft
	Area	140 sq ft
	Height	10ft
	Weight	500 lb.

Phillips No.5 multiplane was airborne at Streatham for a 500ft straight flight and for circular flights anchored to a pole in 1907.

Pickering & Willoughby glider built by two youths at Knutsford in 1911.

PICKERING & WILLOUGHBY glider (Charles Leigh Pickering, and Norman Dean Willoughby, Knutsford, Cheshire)

This was constructed in the summer of 1910 by two friends, aged sixteen and fifteen respectively, and was a tail-first type, based on the Valkyrie monoplane. It was substantially built and designed for easy transport; it had cost £150 and been built in a fortnight. Span was 26ft, Length 20ft.

PICKERSGILL monoplanes (James Thompson Pickersgill, Park Lane, Longlee, Keighley, Yorkshire)

Patent No.307/1909 was applied for in January 1909, for a machine with a near circular wing, below which hung the tubular steel airframe, balanced so that, in the event of engine failure, the machine would descend as a parachute. The patent shows a twin-engined pusher with coupled drive by chains, with a front biplane elevator and twin rear rudders, mounted on a four wheeled undercarriage.

The local press reported that Pickersgill made three machines, one of which he assembled in 1909 in a marquee on a farm near Harden, with the intention of entering this for the Blackpool Meeting, to be held from 18-23 October 1909, but he was unable to complete the aircraft in time. He appears to have abandoned the circular wing by this time, for the only illustration found, is of a high wing, parallel chord monoplane with single acting ailerons, hinged at the trailing edges, but based on an airframe similar to that in the patent. The front and rear control surfaces were different and the power was provided by a single JAP vee-twin engine driving the propeller direct. However, it would seem that the machine was originally fitted with a chain drive to the propeller and this broke on test on 16 December 1909. This prevented any attempt to fly at a meeting arranged by the Northern Aero Syndicate (J.W. House a Blériot owner) at Halifax, which was to be held over the Christmas holiday. The machine was, however, exhibited at Halifax Zoo unflown instead.

On 4 and 5 January 1910 the machine was being towed back to Pickersgill's home at night by horses, when it slid off the bogie, which had shed a wheel. The effort was too much and it was decided to store the machine in the yard of the 'T'Gawmless Inn' at Cullingworth, to be collected the following day. However a gale destroyed the machine in the night and it never did fly.

Another version, probably a reconstruction, was fitted with two JAP vee-twin engines. The two engines drove separate propellers from countershafts, but were coupled together to give 16hp. A Mr. Taylor of the Central Motor Co. of Keighley supervised the preparation of the power.

Two 6ft 8in diameter propellers are known to have survived until at least 1971. It is unclear which power came first, but no doubt, it was one based on the scheme shown in the patent sketches. Patents 307/1909 and 21411/1910, which appear not to have been accepted, are relevant.

PIFFARD Biplane No.1 (Harold Hume Piffard, Bedford Park, Chiswick, London W4)

The first Piffard biplane was largely manufactured in the studio at his home, with the aid of two carpenters, and was taken to a shed rented from Thorneycrofts at Back Common Rd., Turnham Green, Ealing for final assembly and engine testing. In September 1909 a field was rented at North Ealing adjacent to Hanger Lane Farm, in the area which soon became Acton Aerodrome. After various problems, the machine was coaxed off the ground to a height of twelve feet for about 100 yards on 2 December 1909. That night a gale destroyed the marquee and the aircraft.

The machine was a pusher biplane with long curved cane skids, reinforced with steel, similar to a Sommer in appearance. The wings were double surfaced, with fabric back to the main spar on the lower surface. The engine was originally spring

The Pickersgill monoplane with parallel chord wing built in Yorkshire in 1909-1910.

Piffard biplane of 1909 was built at Chiswick and assembled and tested at Ealing.

loaded, on a sliding bed, in a fore and aft direction and could move forward as power was applied. This was to bring the center of gravity forward and reduce the angle of incidence and risk of a stall; the system could also be locked by the pilot, but was abandoned at the suggestion of W.O. Manning.

A front mounted elevator and rudder were fitted, together with a fixed stabilizer above and further forward. A rear non-lifting stabilizer was also fitted. Ailerons on the front outboard interplane struts were controlled by a rocking seat backrest. Vertical fabric panels were fitted to the intermediate interplane struts.

The undercarriage consisted of a sprung nose wheel and two rear wheels, the latter arranged to lift above the level of the skids for landing. A pair of radiators faced the airflow on either side of the pilot.

Power: 35hp ENV type D eight-cylinder water-cooled vee

Data		
	Span	34ft 6in
	Chord	6ft 6in
	Area	510 sq. ft

Piffard biplane No.2 used the engine and parts of No.1 and achieved a number of straight flights at Shoreham in 1911.

PIFFARD biplane No.2

Construction of a second biplane was begun immediately, utilizing parts and the engine of the first. It was ready for testing at Shoreham by 3 May 1910, but, after a number of short straights at low level, the machine was badly smashed and Piffard injured on 21 May 1910.

The machine was rebuilt by 23 June 1910, but bad weather prevented tests until 10 July 1910, after which further short flights were made, with damage occurring and repairs necessary. Flights of up to half a mile were made. On 4 October 1910, after a successful flight, including a turn, a gust caused the machine to crash. The tail booms broke behind the wings and extensive damage was done although the engine was unharmed. Piffard had already decided to make a machine of a new design.

The layout of the second machine was generally as No.1 although some important changes had been made. The lower fabric covering of the wing, back to the spar, was previously attached to the ribs on both sides, but on this machine it was released on the lower surface to allow it to take the shape created by the pressure of the air. The ailerons, initially on the interplane struts, were removed and single acting ailerons were hinged to the trailing edges of all four wings. The rudder was transferred to the tail, and the front fixed stabilizer removed. The main wheels were brought forward and a steerable tail wheel, in addition to cane skids, was fitted. The tail wheel was frequently broken and was soon discarded.

Power: 35hp ENV type D eight-cylinder water-cooled driving a 7ft diameter Weiss propeller

Data		
	Span	34ft
	Chord	6ft
	Area	408 sq. ft
	Area elevator	56 sq. ft
	Area rudder	10 sq. ft
	Area stabilizer	63 sq. ft
	Length	31ft
	Weight allup	850 lb.

PIFFARD hydrobiplane No.3

In January 1911, in an article in *The Aero*, Piffard indicated that a successor was in hand, to the machine wrecked in the previous October. This time his intention was to fly from the water. The information on his seaplanes is not available in much detail, but it would appear that he produced a pusher biplane fitted with his original engine, making use of tube in its construction.

The flotation gear consisted of a single large central float, possibly also intended to provide lift, with a pair of nose floats and a small tail float. Air bags for the minor floats were originally used. The machine was a sesquiplane, with the lower wings extending only out as far as the main booms, whereas

Piffard hydro-biplane No.3 tested without success at Shoreham in 1911.

the top wings had considerable overhang, braced by struts back to the ends of the lower spars.

The front elevator was carried on booms and was on a level with the lower wing. There was a fixed tailplane and rectangular rudder and single acting ailerons on the top wing, with the outer ends cut back diagonally.

The machine was housed at Bungalow Town, Shoreham. On 8 August 1911 the hydroplane was taken down to the sea for the first time but it capsized and had to be beached. There were further trials with no satisfactory results and the machine was reconstructed later that year.

Piggott No.1 biplane built at Ongar, Essex from 1910 was eventually taken to Hendon for testing in 1912 but was a failure.

Piffard hydro-biplane No.4 was also tested at Shoreham in 1911 but was wrecked before flying.

PIFFARD hydrobiplane No.4

The rear portion of the fuselage and the center section of No.3 were used again, to form the basis of No.4, later in 1911. The wings were of two bays, with extensions on the top wings tilted upwards and braced with light struts to the lower spars. Two separate single acting ailerons on each top wing hung on the trailing edges of the main wing and the extensions. All of Piffard's wing surfaces were made with spars at the leading and trailing edges with little or no camber to the ribs. The buoyancy was now provided by two rather short main floats with a tail float as before. The main floats had, in addition, curved hydroskis attached to their undersides, which were braced back to the main structure of the aircraft.

The aircraft taxied, but would not unstick from the water; Piffard achieved no success with his seaplanes and this was his last aviation activity. The aircraft was wrecked in a gale on the beach at Shoreham at the end of 1911. The ENV engine, propeller, fuel tank and radiator were retained and now form an exhibit at the Science Museum.

PIGGOTT No.1 biplane (Piggott Bros. & Co. Ltd., 220-224 Bishopsgate St, London EC. Construction at Stanford Rivers, Ongar, Essex)

This unusual machine was designed by S.C. Parr and was reported to be built on the theories of F.W. Lanchester. The aircraft was a side-by-side two-seater pusher biplane, constructed of light aluminum tubing joined by sockets. The front and rear booms, which were single tubes only at the top, were braced by wires, the front one carrying the elevator, the rear one a fixed tailplane with a shield shaped rudder, between the top and bottom members. The rear booms were spread vertically as much as possible; the top member attaching to the bracing pylon above the wing, the bottom to the lowest point on the understructure. Two large stabilizing fins were mounted on the top wings.

The wings, ailerons and the front and rear surfaces were elliptical in shape, the ailerons being mounted on the front outboard interplane struts. The main undercarriage, a pair of wheels in tandem, consisted of a pair of fabricated radius arms cantilevered from points on the main frame, with suspension by helical springs between the arms and the main frame. The skids between the wheels became effective when the wheels lifted on landing. Outboard wheels, on similar radius arms, were positioned below the last pair of interplane struts.

The power plant was rather complex, incorporating a multiple disc clutch, which was disengaged for starting. The outboard end carried a multiple disc sprocket for the wide chain taking the drive, at about 2 to 1 reduction, to a differential gearbox on, the front of the propeller shaft. The output of this gearbox was through inner and outer drive shafts, each mounting a propeller of different diameters, each with four blades. The smaller propeller was mounted about 4ft 6in behind the larger.

The differential gear was designed to balance the loading on the two propellers as they counter-rotated.

In May 1910 the machine was far from completion, but it was eventually taken to Hendon for testing in 1912. It was a failure and almost certainly never left the ground.

Power: 80hp Vivinus four-cylinder inline water-cooled driving contra-rotating four-bladed metal propellers 13ft 6in and 8ft 4in diameter with adjustable pitch, through clutch, reduction chain drive and differential gear box.

Data		
	Span	60ft
	Max chord	5ft
	Area	568 sq. ft
	Area elevator	33 sq. ft
	Area tailplane	44 sq. ft
	Area rudder	20 sq. ft
	Length	34 sq. ft
	Weight allup	1,150 lb.
	Speed	48 mph

PIGGOTT monoplane

This machine was exhibited at the Olympia Aero Show in March 1911, with a fully enclosed cockpit. However, before it was tested at Hendon by S.C. Parr, it was modified to provide an open pilot's cockpit. During the tests the machine left the ground, but the undercarriage was damaged on landing. Although the damage was repaired, there were no further reports and the machine was abandoned.

As originally built, the Piggott monoplane had a light body totally enclosing the engine, pilot and passengers, who gained access to the interior cabin through a door in the underside; the only vision was through portholes. The engine was set back in the fuselage and required an extension shaft to carry the propeller, which had a streamlined boss to blend into the lines of the body. The radiator was shaped to lie on top of the front fuselage. A twin skid Farman-type undercarriage, with two pairs of rubber sprung wheels and a large tail wheel were fitted. A rather small rudder, elongated tailplanes and a one piece elevator comprised the tail unit. Lateral control was by warping, and all control cables were duplicated.

Power: 80hp Vivinus four-cylinder inline water-cooled.

Data		
	Span	30ft 6in
	Weight	850 lb.
	Length	24ft 6in
	Weight allup	1,400 lb.
	Max speed	75 mph

Piggott monoplane shown at Olympia in March 1911 was abandoned after damage incurred while on test at Hendon.

Piggott biplane was an unsuitable machine for military use although entered for the trials of 1912.

PIGGOTT biplane

The third, and final, Piggott aircraft was a small two-seater biplane entered for the Military Trials of August 1912 (No.29), for which purpose it was unsuited, being under powered, and consequently unable to perform any of the tests.

The design was of conventional tractor layout, but featured three bay wings, with a narrow gap, and single interplane struts. The crew sat exposed above the level of the top wing and were vulnerable if the machine should overturn. The tailplane extended well forward along the fuselage sides and a crescent shaped rudder extended, equally above and below the fuselage.

Power: 25-35hp Anzani three-cylinder Y-type air-cooled radial.

Data	Span	25ft 6in
	Gap	2ft 6in
	Area	100 sq. ft
	Weight	300 lb.
	Weight allup	700 lb.
	Length	17ft 6in
	Max speed	55 mph

PILCHER gliders (Percy Sinclair Pilcher, Glasgow and London)

Pilcher became interested in flight whilst at Glasgow University. He began by making and testing models from 1893 and made his first full size glider in 1895. He corresponded with Lilienthal and visited Germany that year, to witness and discuss matters at first hand, at a time when his machine was well advanced in construction. In the space of a year or so Pilcher had made his fourth glider, at which time he moved south to London, to work for the Maxim, Nordenfeldt Gun Co. as Hiram Maxim's assistant; this was around March 1896.

This arrangement terminated, when Maxim was no longer managing his own company and Pilcher then became joint managing director of Wilson and Pilcher Ltd., an engineering company registered in November 1897, with workshop premises in Clerkenwell. The latest of his gliders, the *Hawk*, was brought down with him for further development and trials were carried out at Eynsford in Kent. This glider was still in use in September 1899 when he agreed to demonstrate it at Stanford Park, near Rugby, which resulted in an accident on 30 September 1899, from which he died two days later.

Also at Stanford Park, for exhibition only, was a triplane glider, to which Pilcher intended to fit a small 'oil engine', that had already been on test, but required further development.

PILCHER *Bat* glider

This first glider was made early in 1895 and was tested from slopes at Cardross by the River Clyde from June of that year.

The machine was a monoplane hang glider with considerable dihedral, controlled by weight shift only. A triangular front frame, and a single rod to carry the fin, comprised the center structure of the machine. A tailplane was fitted later. The wings were made separately, the main spars being lashed to the front frame and extensively braced by wires. Later the excessive dihedral was reduced, by curving the wings downwards, and introducing greater camber.

Pilcher *Bat* glider. The first version of 1895 built and tested near Glasgow.

Pilcher *Bat*. The second version was fitted with a tailplane.

Data	Area	150 sq. ft
	Weight	45 lb.
	Weight allup	190 lb.

PILCHER *Beetle* glider

Pilcher's second machine was ready for use in about June 1895, but not for very long, as he found the machine ungainly. It had been assembled in a barn at Auchensail, north of Cardross, where he was able to spend more time at practice.

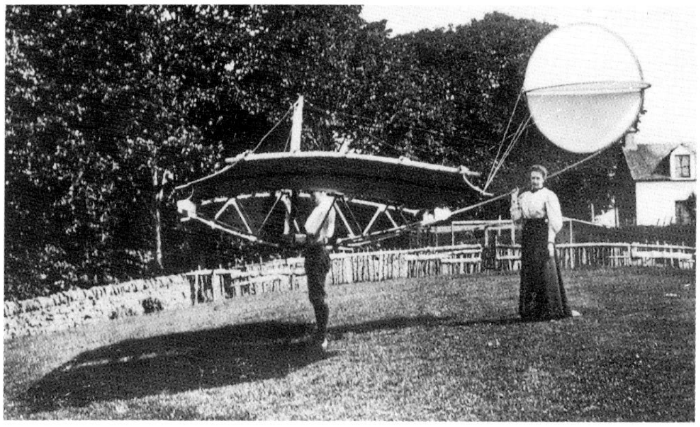

The *Beetle*. Pilcher's second glider was unsatisfactory and was soon abandoned in 1895.

He had built the *Beetle* with the intention of fitting an engine, after the initial gliding trials, and consequently it was stronger and heavier than would otherwise be the case. The body consisted of two parallel Warren girders, between which the shoulders could pass to enable the pilot to rest his arms in sleeves on the bottom longerons. The wing, of angular shape, was built round five spars and was braced to kingposts above, and to the girders below. A circular fin and tailplane, in the form of a cross, were fitted.

The *Beetle* was discarded in favor of the modified *Bat*, which was put back into use again in September.

Data	Area	170 sq. ft
	Weight	80 lb.
	Weight allup	225 lb.

Pilcher *Hawk* the fourth glider was begun late in 1895 and was flown successfully at Eynsford from 1896 until Pilcher was killed when flying it at Stanford Park, Rugby in September 1899.

Pilcher constructed the *Gull*, his third glider, in Scotland but moved to Eynsford in Kent in 1896 where it flew successfully.

PILCHER *Gull* glider

The third glider was constructed late in 1895, but was not really proved until the following year, although there is evidence of it being complete, but damaged, in October 1895. The light wing loading made the *Gull* difficult to handle in gusty conditions in Scotland, but with better conditions in the summer of 1896, after the move to Eynsford, the *Gull* was flown. It suffered some damage at times and was still awaiting repair in January 1897.

The body consisted of a tapered, wire braced girder, from which a frame extended below, and which carried, at its outboard ends, the two vertical kingposts, later connected by a crossbar. The cruciform tail, as used on the *Beetle*, was employed. The wing structure consisted of a number of radial ribs, similar to those of the *Bat*, which could be swung round the kingposts to fold back for transport.

Data	Area	300 sq. ft
	Weight	55 lb.
	Weight allup	200 lb.

PILCHER *Hawk* glider

Work began on the fourth glider towards the end of 1895, which was to be of smaller wing area than the *Gull*, and thus more manageable. The move to London involved Pilcher with other activities, which included the design and construction of a large exhibit, to be shown at an International Exhibition at the Imperial Institute, South Kensington. This was a lifting platform for launching gliders, and the *Hawk* was also displayed alongside it, under the names of Maxim and Pilcher, 32 Victoria St., London SW. The exhibition lasted four months so it was towards the end of 1896, before the *Hawk* was available to test. By the end of the year Pilcher had flown distances up to 100 yards and at least 150 yards had been achieved at a public demonstration in July 1897 at Eynsford.

The Pilcher gliders had been housed in Maxim's large shed on property owned by the reorganized Vickers company, who now required the removal of the shed by the end of the year, with the consequent loss of gliding facilities. Thereafter the *Hawk*, which was not flown at all in 1898, was used as an exhibit at lectures, until the final demonstration at which Pilcher was killed.

The center structure consisted of two curved bamboo members, joined at front and rear by cast metal fittings, the latter also providing a mounting for a hinged bamboo member to carry the tailplane, which was designed to pivot upwards only; a fin was added later.

A spar, clamped to the center structure, terminated in the uprights, which served as bracing kingposts and pivots for the axles of a pair of sprung wheels, a new feature. The wings, as before, were built with radial ribs designed to fold back fanwise; these were of bamboo and heavily cambered. The pilot stood in the center with his arms in sleeves, resting on the bamboo members.

Pilcher's intention to power the *Hawk*, as with his earlier gliders, was covered by patent No.9144 /1896, but this was never realized.

The *Hawk* still exists as an exhibit at the Royal Scottish Museum. At least six reproductions have been made for films and for display purposes, and include examples at Stanford Hall and the Science Museum in London.

PILCHER multi-plane

By 1897 Pilcher was corresponding with Chanute, a protagonist of multiple wings, which may have influenced him to construct a multi-plane as his fifth machine. On 30 September 1899, at Stanford Park, in addition to the *Hawk*, there was an untried glider described as a 'three decked machine' at the time. There is some uncertainty whether it was in fact a triplane or perhaps an incomplete quadruplane, as one of Pilcher's surviving drawings shows such a machine. T.W.K. Clarke produced a drawing of a triplane in 1910, from the remains of the machine and the balance of evidence favors a triplane.

Since his early experiences with the *Beetle*, Pilcher had contemplated the use of a power unit using steam, carbonic acid, oil or other material to extend the range of his flights. The patent No.9144 for a small pusher driven machine was based on the *Hawk*. A small 4hp petrol motor was designed by his partner, Walter G. Wilson in 1898-1899 and may have been of twin-cylinder horizontally opposed type. The installation of this engine in the triplane, driving a pusher propeller on a long shaft above the pilot's head, is probably the configuration of the machine that Pilcher was envisaging.

The Pilcher multiplane was not completed but this artists impression is based on the surviving drawings.

The Pilcher Triplane, probably named *Duck*, would have conformed to the drawing shown, which is based on the surviving data.

Data	Span	20-24ft
	Area	150-170 sq. ft
	Weight	50-60 lb. as a pure glider
	Weight allup	256-266 lb as a powered machine

Planes biplane was modified from the Handley Page Type B (HP2) to a single pusher propeller at Freshfield in 1910.

PLANES biplane (Planes Ltd., Shed No.1, Freshfield, Lancashire)

This machine was made originally by Handley Page at Barking to the basic ideas of W.P. Thompson, a patent agent of Liverpool. The main principle was to concentrate the weight below the wings, to provide 'pendulum stability'. Thompson's original type of tubular construction was heavy and was discarded in favor of wood and fabric. Much of the work at Barking was carried out by R.C. Fenwick, an assistant of Thompson's.

The machine was tested on the rough ground at Barking in October 1909 and damage to the undercarriage resulted. Immediately afterwards, the shed in which it was housed collapsed in a gale, causing further damage. Thereafter the machine was rebuilt at Barking by Fenwick and, although previously known as the Handley Page type B, it now became the Planes biplane. It was delivered by rail to Freshfield in about September 1910, without being tested further. Fenwick taught himself to fly on it, receiving Aviators Certificate No.39 on 29 November 1910. Within a week it was badly damaged on takeoff, in a crash caused by the downwash from a Farman flown by Compton Paterson, flying low overhead. However the machine was repaired and continued in use.

This large biplane was converted from the chain driven, twin pusher type, of the Handley Page type B, to a single pusher propeller. The tail booms were raked downwards to carry a biplane elevator and twin rudders, below the level of the bottom wing. The undercarriage consisted of a pair of main landing wheels and skids, positioned on either side of the pilot, with a pair of nose wheels added to deal with any tendency for the nose to dig in. In addition there were small wheels on long supports at the wingtips and a tail skid. Ailerons replaced warping of the upper wing tips.

Power: 60hp Green four-cylinder inline water-cooled

PLATTS glider (R. Platts, Birmingham Aero Club and 168 Hampton Road, Birmingham)

This machine was flown from a tow, as both a monoplane and a biplane, at various times in 1911 at Billesley Farm, King's Heath. It was reported to have reached a height of 33ft and flown 100 yards. A biplane glider, made of bamboo, was advertised for sale in 1912. A powered monoplane, to be completed in December 1911, was also reported.

Data	Area	227 sq. ft
	Weight	70 lb.

POCHIN biplane (A. Whitewell Pochin, Freshfield and Southport, Lancashire)

This small biplane is reported to have flown short distances on three occasions on 28 October 1911 at Freshfield, although earlier in September *Flight* had reported that Pochin was building a monoplane. The following month he received instruction from Higginbotham at Freshfield. He took over the hangar at Southport from John Gaunt in August 1911, and again the following year, when he shared a small biplane with a Mr. Arthur Rimmer.

POLLOCK monoplane (George William and Stewart Pollock, Robertson St., Glasgow)

This was a monoplane of Blériot type built in 1910; the framework was of ash covered with North British Rubber Co. No.6 fabric.

Power: 60hp Pollock four-cylinder air-cooled driving a 6ft 9in diameter propeller

Data	Span	40ft
	Chord	7ft 6in
	Rudder	4ft 6in x 4ft 6in
	Elevator	9ft 6in x 4ft 6in
	Length	24ft 6in

POLYTECHNIC Aero Club & Flying Society (Regent St., London W)

This society was active from 1910 with B. Graham Wood in charge of design and construction. In 1912 the society was busy with camps at Easter, Whitsun and Summer at Amberley in Sussex, where a shed and facilities were made available by José Weiss. In *Flight* 5 December 1912 (p.906) the Secretary advised that the following gliders were available for students:

Weiss glider *'Olive'*.

Weiss glider *'Joker'*.

Vaughan II glider. A Wright type of 33ft span.

Polytechnic built portable monoplane of 44ft span.

Vaughan and Wood both wrote articles for *The Aero,* in which their gliders were described and illustrated. An early effort by Wood was the Wood and Maas glider *(q.v.)*.

During 1910, a grant of £200 to buy materials, was made by the London County Council, for a modified Blériot type tractor monoplane to be made to the society's designs. The engine was to be designed by A.G. Clark, at one time a designer with the Green Engine Co., and to be made at the Polytechnic. The machine may have evolved as the No.4 glider listed by the Secretary.

Porte *Demoiselle*-type monoplane built at Portsmouth in 1910.

PORTE monoplane (Lt. J.C. Porte RN, Haslar, Portsmouth, Hampshire)

After the biplane glider, made in conjunction with Lt. Pirie, Porte made a *Demoiselle* type monoplane. This followed the general lines of Santos Dumont's design, with the pilot seated at the level of the lower longerons in the triangular section fuselage. There were some original features, such as the extra pair of wheels fitted on a long axle, below the pilot's position, to aid balance during the taxiing trials, and also a forward projecting skid.

A biplane, which Porte was reported to be making with a 40hp Green engine and using metal construction, failed to materialize.

Power: 35hp Dutheil-Chalmers two-cylinder horizontally opposed water-cooled

Data	Span	22ft
	Length	27ft

PORTE & PIRIE glider (Lts. J.C. Porte and W.B. Pirie RN, Haslar, Portsmouth)

Two reported attempts to fly the glider at Portsdown Hill from a trolley on a track both resulted in damage. Porte continued with his *Demoiselle* monoplane, and later became a director of the British Deperdussin Co. A project to fly the Atlantic in a Curtiss flying boat was stopped by the outbreak of war. In wartime he was the leading figure on flying boats at Felixstowe. The glider had equal span wings, notable for the heavy stagger, the trailing edge of the top wing being vertically in line with the leading edge of the bottom wing. Four pairs of N-shaped interplane struts spaced the gap, with ailerons mounted on each outboard pair. There were also rudders on each vertical strut at the wingtips. An elevator was carried on the tail booms and the machine rested on two skids on its launching trolley.

Porte & Pirie glider was tested on Portsdown Hill in 1909 without success.

The two officers sat side by side in the center section, Porte controlling the elevator and ailerons, possibly also serving as elevators, by means of a lever, while Pirie operated the rudders directly by a cable.

Data Span 26ft

PORTER helicopter (J. Robertson Porter, 9 Gray's Inn Square, London EC)

Porter exhibited his direct lift machine at Cordingley's Motor Car Exhibition at the Agricultural Hall, Islington in March 1908 and again at the Sports and Travel Exhibition at Olympia in July 1909.

The design was covered by patent No.21261/1908 and consisted of two ducted fans in a frame, driven through shafts, gearboxes and a chain drive from the motor mounted on the base. Each fan had four separate stages and was enclosed in a cowl to direct the air pressure downwards. A pusher propeller was mounted on an extension of the horizontal main shaft; a rudder was provided at the rear and was controllable from the operator's station at the front of the machine.

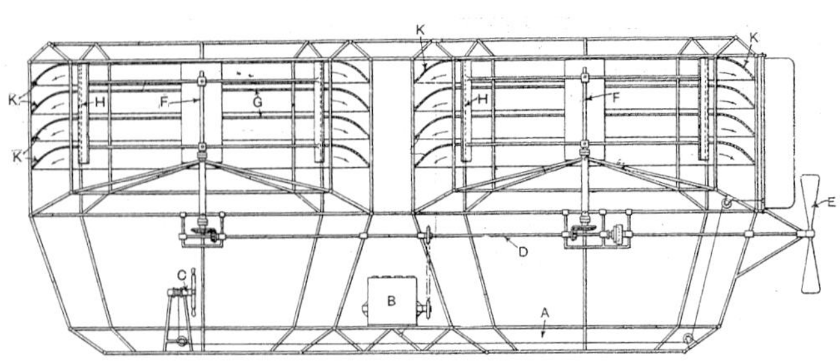

Porter helicopter of 1908 which was an early form of hovercraft.

A similar but smaller machine with a single two-stage ducted fan was also proposed.

Power: 6hp JAP two-cylinder air-cooled

Data		
	Length	16ft
	Width.	8ft
	Height	14ft
	Weight	200 lb.

PORTER '*Gyropachute*' (J. Robertson Porter, 9 Gray's Inn Square, London EC)

This vertical lift machine was exhibited at the Aero Show at Olympia in March 1913 by Messrs. Dagnall, Mallinson and Porter. It consisted of a frame of steel tube, over which was formed a bell shaped covering with a flexible skirt. Inside this was a rotor driven from below, by a Gnôme rotary engine, with the fifteen gallon and two gallon fuel and oil tanks revolving

Porter Gyropachute was an unusual type exhibited at Olympia in March 1913 of which no more was heard.

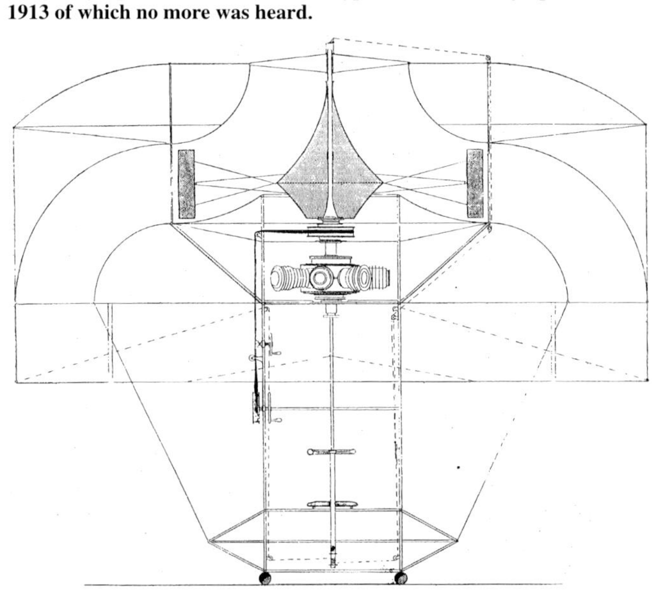

with the engine. A smaller parachute surface at the bottom encompassed the pilot's station, at the base of the machine. The control of horizontal movement was obtained by manipulating sections of the skirt to release some of the air pressure.

Power: 50hp Gnôme seven-cylinder air-cooled rotary

Data		
	Diameter	14ft
	Height	11ft
	Area	400 sq. ft
	Weight	350 lb.

The Portway monoplane of 1910 was a *Demoiselle* variant built at Bromley.

PORTWAY monoplane (A.P. Portway & Portway, Bromley, Kent)

A modified form of the Santos Dumont *Demoiselle* was made by the Portway brothers in early 1910. It was flown at West Wickham, although flying was limited by the nature of the ground and damage to the wheels. The fuselage was made of bamboo with cast aluminum sockets at the joints.

Power: 35hp Lascelles four-cylinder air-cooled fantype semi-radial driving a 7ft diameter propeller

Data		
	Span	25ft
	Area	150 sq. ft
	Chord	6ft 6in
	Weight	350 lb.

POWER monoplane (A. Grattan Power, Acton)

Power designed a monoplane and arranged for it to be built by Lascelles and Company and fitted with a 35hp Lascelles engine. When tested at the London Aviation Ground it refused to fly; Power then purchased a Lascelles 'Ornis' monoplane as exhibited at the 1910 Olympia Show with the notice 'flight guaranteed'. This monoplane was also a failure and Power sued Robert Beney, the director of the Lascelles Company for the return of his money. He won his case.

POYNTER monoplane (E.J. Poynter, Twickenham, Middlesex and Brooklands)

Poynter was an artist and model maker, who had a full size monoplane of his own design constructed by Howard Wright at Battersea. It was completed and taken to Brooklands for assembly early in June 1910. It was apparently never flown, as Poynter had a car accident and, despite being well made by a reputable constructor, the monoplane found no further use.

The machine had a large tapered monoplane wing braced with kingposts, in the style of the Antoinette. There was a heavy

Poynter had his monoplane built by Howard Wright in 1910. It was taken to Brooklands but never flown owing to an accident to the owner.

twin skid main undercarriage, with two pairs of sprung wheels. The large cruciform tail, with radial ribs, was mounted on a universal pivot, similar to that of the *Demoiselle*, and required a very tall tailwheel strut to provide adequate ground clearance. The engine was mounted high on the top longerons and was controlled by two foot pedals, one for the throttle and the other for a valve lifter for starting.

Power: 60hp Green four-cylinder inline water-cooled

Data		
	Span	45ft
	Chord	9ft tapering to 6ft
	Area	300 sq. ft
	Area elevators	60 sq. ft
	Area rudder	30 sq. ft
	Length	28ft

PREMIER monoplane (Premier Aeroplane Manufacturing Co., 144 Offord Rd., Islington)

A monoplane, built on Blériot lines, was exhibited at the Stanley Cycle Show in an unfinished condition in November 1909. It was designed by G.J.M. Fitzjohn, using ash as the main material for its construction. There were no reports of completion or of trials.

Data		
	Span	25ft 6in
	Length	26ft 6in
	Weight	469 lb.

PREMIER biplane.

A biplane of 30ft span and 6ft chord was also to be built, but this failed to materialize. In its advertisements, the company offered to provide skilled labor to build aircraft at customer's own premises.

PRIDE monoplane (Christopher Pride, 4 Guinea Lane, Fishponds, Bristol)

This machine was built during 1909, and on 18 July 1909 flew for a distance of 976 yards before hitting a tree and being wrecked. It was a tandem monoplane powered by a 50hp motor with propellers front and rear.

Data		
	Span	32ft 8in
	Weight allup	684 lb.

PROSSER & BONEHILL glider (E. Prosser and A.M. Bonehill)
This Chanute type glider was reported built in 1910, but was wrecked on 26 August 1911.

Data	Span	26ft
	Chord	5ft

PRUDEN cycloplane (E.C. Pruden, Elm, Near Cambridge)
Described as being built on the 'principles of a cycloplane and useless as a flying machine', it became the subject of a court case after it was burnt.

It was a pedal-powered machine, mounted on a launching trolley, from which it was to detach when reaching flying speed. The machine failed when tested on land at Needham Hall, Elm, where a shed had been built to house it.

PUPIN motoplane. (Emile Pupin and Marc Denaus, 114 Wardour St., London W)
Early in 1911, a machine of unusual layout arrived at Hendon. It was a monoplane with an open girder fuselage mounted on a two wheel undercarriage with two skids, and a tail wheel on a tall structure. The engine, of unidentified type, was mounted in the center and drove, through chains and shafts, a pair of propellers mounted alongside the rear fuselage. A long rectangular fin was mounted above the forward fuselage with a rudder at the tail.

Three pairs of equal span wings were fitted, which could be varied in incidence and moved differentially in conjunction with the rudder. Patent No 27 476/1909 is relevant.

On test on 10 February 1911, three of the wings were damaged and that was the last heard of the motoplane.

R

RADLEY (& MOORHOUSE) monoplane (James Radley, St. John's St., Huntingdon and William B. Rhodes Moorhouse)
James Radley was one of the leading British pilots of a Blériot in 1910. After he returned from a successful tour of America at the end of the year, he turned to design and construction and gradually gave up flying. He was joined in this work by W.B. Rhodes Moorhouse, who took on the flying and instructional work, and later was to win the VC in the RFC in 1915.

The first aircraft built was a single-seater with gull wing, which flew at Portholme, Huntingdon for the first time 27 July 1911. The machine was of conventional tractor layout with the fuselage faired to a circular section. The engine was set back in the nose, with the propeller mounted on an extension shaft supported on a front bearing. A four wheeled main undercarriage with two skids, and a tall tail skid, supported the machine on the ground, this providing clearance for the bottom rudder. The tail unit consisted of a long triangular shaped fin, with

Radley monoplane with gullwing was built at Huntingdon in 1911 but was wrecked in January 1912 by the inexperienced purchaser.

Pupin monoplane, with three pairs of wings, seen at Hendon in 1911.

tailplanes extending along the center of the fuselage sides, together with a one piece elevator and top rudder.

The inboard section of the wing was set with a small dihedral angle but curved downwards from about mid-span; the tips were extended rearwards and could be warped for lateral control.

The machine was later referred to by the names of both Radley and Moorhouse, and was bought by a newly qualified pilot from Brooklands, named Tom Garne. He wrote it off in a crash at Huntingdon on 19 January 1912.

Power: 50hp Gnôme seven-cylinder air-cooled rotary

Radley & Moorhouse monoplane was evolved from a Blériot into a single-seater for racing.

RADLEY & MOORHOUSE monoplane, Blériot type (Portholme, Hunts)

The Blériot monoplane was a machine that was copied, with variations, by a number of small establishments, when embarking on construction work. The Radley and Moorhouse concern at Portholme was no exception. A two-seater was flying from about October 1911 and was used by Moorhouse to gain his Aviators Certificate No.147 on 7 October 1911. The following month he was testing a new R & M two-seater. This was the machine illustrated on the 14 December 1911 in *The Aeroplane*.

A machine flown by Moorhouse into third place in the Aerial Derby on 8 June 1912 had changes made for racing. It may have been the same machine, converted into a single-seater, with the addition of a streamlined fairing around the pilot's cockpit on the fuselage top. The wheels had been given fabric covers.

In April 1912 Moorhouse had left Huntingdon and established himself at Brooklands with the single-seater machine and later in the year undertook to fly a French Breguet in the Military Trials. On the flight from France on 4 August 1912, with two passengers, the machine crashed in Kent and took no part in the trials.

Power: 50hp Gnôme seven-cylinder air-cooled rotary

Radley-England Waterplane No.1. This twinhull flying boat was built at Portholme in 1912-1913.

RADLEY ENGLAND waterplane No.1 (James Radley and Eric Gordon England, Portholme & Shoreham)

This big six seater, twinhull flying boat was built in 1912-1913 at Portholme. Although not eligible for competing in the *Daily Mail* Circuit of Britain Contest to be held in August 1913, it was to form the basis of a suitable machine with a British made engine.

The initial flight trials were carried out with a temporary four wheel undercarriage and tail wheels on the floats, at Portholme and flotation tests also on the local river. The machine then was moved to Shoreham, where it was tested on the

River Adur. It was soon moved to Brighton and housed in Volk's hangar, but it sank close inshore in early June and was moved back to Shoreham.

It was possible that some salvaged items were used in the second machine, but since there were so many differences, this seems unlikely.

The wings contained three bays outboard of the tail booms, with a slight overhang of the top wing; within the tail booms the center section bays were smaller. The wings were unstaggered and carried ailerons on the top only.

The aircraft was mounted on each float with three struts only and bracing wires. Each float had a front and rear watertight compartment, between which was an open cockpit for three persons, the pilot seated in front on the starboard side, the passengers being seated side by side behind. The floats were rather like punts with flat undersides, without steps, and with curved top deckings.

The power plant was the most unusual feature of the machine, and consisted of three Gnôme rotary engines on a common axis, which drove a separate propeller shaft above them. A U-shaped cowling surrounded the front engine to act as an oil collector. The fuel and oil were carried in a cylindrical tank above the propeller shaft, the fuel and oil being pumped to each engine, which each had their own throttle control.

Power: Three 50hp Gnôme seven-cylinder air-cooled rotaries driving a 9ft.10in. diameter four-bladed propeller, by triple chain at a ratio of 29 to 19.

Data		
	Span top	45ft 4in (47ft 11in *The Aero*)
	Span bottom	44ft 6in
	Chord	5ft 8in
	Area	505 sq. ft
	Area tailplane	25 sq. ft
	Area elevators	32 sq. ft
	Area rudders	15 sq. ft
	Hulls length	15ft
	Hulls beam	3ft
	Length	29ft 3in
	Weight	1,400 lb.
	Endurance	1 1/2 hr

Radley-England Waterplane No.2 was built to compete in the Seaplane Circuit of Britain of 1913 but engine trouble on test caused withdrawal of the machine.

RADLEY ENGLAND waterplane No.2

The second version of the Waterplane, launched on 6 August 1913, was so different and appeared so soon after the damage to the first, that it seems to have been a new machine built at Portholme, where the capacity for construction existed. It had a British-made engine and greater endurance, to make it suitable for the Circuit Contest, apart from strengthening and other major changes.

The wings had greater span and area and were now mounted on four splayed struts, closer together, and were connected by a pair of struts, which ran from each bow to the engine bearers. The earlier punt-like floats had been replaced by stronger, clinker-built boat hulls made by the South Coast Yacht Agency, and now accommodated four crew in separate cockpits, with the pilot in the starboard front position as before.

The engine bearers carried the radiator at the front, with a large combined fuel and oil tank holding eighty-two gallons of petrol and eight gallons of oil, between the radiator and the engine.

Serious trouble was found with the engine immediately the trials began and the machine was withdrawn from the Circuit Contest straight away. There were no further reports and Gordon England left to take up other work and later to sue Radley for arrears of salary and expenses. The machine was bought by Pemberton Billing but was not used.

Power: 150hp Sunbeam. eight-cylinder water-cooled vee, geared to drive a 9ft 6in diameter Lang four-bladed propeller at half engine speed

Data		
	Span top	51ft 7 1/2in
	Span bottom	45ft 10in
	Chord	5ft 10in
	Area	560 sq. ft
	Area tailplane	24 sq. ft
	Area elevator	35 sq. ft
	Area rudders	18 sq. ft
	Hulls length	18ft
	Hulls beam	2ft 6in
	Length	29ft 9in

RANDALL aircraft (David Randall, Coach builders, E. Finchley, London N)

This manpowered aircraft was constructed in 1913-1914 and was broken up after the outbreak of war without being tested.

The aircraft was described as 'like an inverted pagoda' with a top plane eight feet square and a lower plane three feet square. The pilot sat between the wings and was provided with pedals to drive a pusher propeller, and handlebars to control a rudder. The machine was fitted with skids for landing, but was to be launched from a four-wheeled trolley on which it rested.

RANKIN KENNEDY biplane

In his book '*Flying Machines*" published in 1907, the author, a well known writer on engineering subjects, sketched and described a machine based on the Wright Flyer, but with a 'mov-

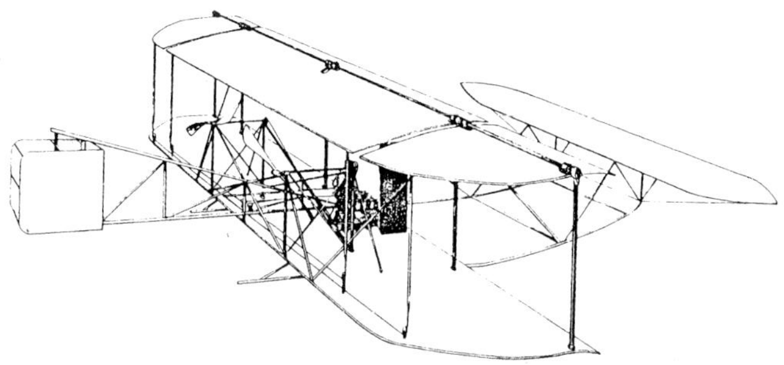

Rankin Kennedy schemes for a stable biplane 1907 and a twin-rotor helicopter were not built.

able upper plane' intended to provide stability in flight. As an alternative form of control to that of the Wright Flyer, it would avoid infringement of the Wright Bros. patents. The machine remained a project and was not built.

With the exception of the twin rudders, the control of the machine was all embodied in the top wing, which was in three parts. The main center portion was adjustable for incidence to control lift; the outer sections worked differentially, to provide correction in the rolling plane. The prime object was to maintain the horizontal attitude of the body of the machine and to avoid displacing this, by over correction of the controls.

Power: 25hp driving two 18ft diameter pusher propellers.

Data	Span	38ft
	Chord	6ft
	Gap	6ft 6in

ROYAL AIRCRAFT FACTORY (RAF) Farnborough, Hampshire)

This military establishment began life in 1904-1905 on the Farnborough site, after the transfer of the Balloon Factory and School of Ballooning. The early work carried out included the building and flying of airships and kites, which developed into gliders and, from 1906, into powered flying machines. The earliest machines made at Farnborough were those for which Cody and Dunne were primarily responsible, but their work came to an end in March 1909, when all heavier-than-air work was stopped as an economy measure.

The Factory was reorganized as a civilian establishment from October 1909, with authority to carry out experimental work and to repair and maintain the aircraft operated by the Balloon School. It was necessary to acquire aircraft to carry out experimental work, among the first being the machine constructed by Geoffrey de Havilland, who was himself engaged at the same time as an engineer and pilot. The Factory was retitled HM Aircraft Factory in April 1910, and again in April 1912, when it became officially the Royal Aircraft Factory.

The building of a few aircraft for experimental work soon developed into quantity manufacture of a variety of aircraft types and engines. This competition caused considerable disquiet in the aircraft industry, where many firms were suffering from lack of orders. This contentious situation was finally resolved in 1917 by a major reorganization and a return to the original objective, of concentrating on research and experimental work for the benefit of the industry and the flying services. Contracts for various RAF designs were placed with industry from 1912 and made a major contribution to the growth of the industry for wartime production.

Aeroplanes built at Farnborough from 1911 to 1914.
(Extract from a record made in 1919 at the RAE)

Type.	1911	1912	1913	1914
BE.1		1		
BE.2		1	5	
BE.3 & 4		2		
BE.5 & 6		2		
BE.7			1	
BE.8		2	1*	
BS.1			1	
FE.2	1			
FE.3			1	
FE.6				1
HRE.2			1	
RE.1			2	
RE.3			1	
RE.5				22
SE.1	1			
SE.2			1	
SE.4				1
Totals	**2**	**8**	**14**	**24**

BE.1, 2, 5 and 6 biplanes (Blériot Experimental)

BE.1 was nominally the 'reconstruction' of a Voisin, although probably only the engine and little else was used in the new machine. Designed by Geoffrey de Havilland, he flew it for the first time on 4 December 1911.

BE.2 was virtually an identical aircraft, but with an air-cooled engine, perhaps from the Breguet, which was the basis for its 'reconstruction'. This machine flew for the first time on 1 February 1912 and again was flown by de Havilland.

BE.5 began life on 27 June 1912 with a water-cooled ENV engine from a Howard Wright, which was soon changed to an air-cooled Renault, which became the usual engine for other aircraft of a similar type.

BE.6 first flew on 6 September 1912 and may have been fitted briefly with an ENV before a Renault was installed. Later this machine was used for testing an oleo undercarriage with a single skid.

BE.1, 5 and 6 were similar machines to BE.2 and there were five more made at Farnborough in 1913, of the type loosely referred to as BE.2 type aircraft.

These machines were conventional two bay tractor biplanes, with the pilot in the rear seat of the single cockpit, although later a decking between the crew was fitted together

RAF BE.1 December 1911 saw the first flight of this Geoffrey de Havilland design of which many variants were built for the RFC and RNAS.

RAF BE.2 was also fitted with unequal span wings but had a Renault air-cooled engine.

with one behind the engine. The latter took the place of the vertical radiator, when the air-cooled engine was fitted. The exhaust pipes of the BE.2 were originally taken inside the fuselage instead of underneath. Differences in the length of the undercarriage skids, and the substitution of equal span wings, were other variations introduced over the period of the aircraft 's life. Changes to the rigging of the wings were made during testing, but these were originally unstaggered and without dihedral and with warping control.

The BE.2 was fitted with twin main and tail floats as an amphibian and as a pure seaplane, in which latter form it apparently lifted off when tested on the Fleet Pond.

Power: 60hp Wolseley eight-cylinder water-cooled vee BE.1
60hp Renault eight-cylinder air-cooled vee BE.1 and 2 60hp ENV type F eight-cylinder water-cooled vee BE.5 and 6
70hp Renault eight-cylinder air-cooled vee BE.2, 6 and others

Data for BE.1. Some variations for other machines.

Span top	38ft 7in
Span bottom	34ft 11 1/2in
Chord	5ft 6in
Gap	6ft
Area	374 sq. ft
Area tailplane	52 sq. ft (Originally smaller on BE.1)
Area elevators	25 sq. ft
Area rudder	12 sq. ft
Length	29ft 6 1/2in
Weight allup	1,700 lb.
Height	10ft 2in
BE.1 with Wolesley	Speed range 42-59mph
	Climb 155ft per min. to 600ft
BE.2 with 70hp	Speed range 40-70mph
Renault	Climb 305ft per min. to 1,000ft

BE.2a and BE.2b biplanes

The BE.2a was evolved from the previous variants of the BE.2 in the middle of 1912, into a machine suitable for use by the Military Wing, and orders were placed with a number of firms. Approaching one hundred BE.2a aircraft were built.

The type inherited the unequal span wing initially, although the majority were fitted with wings of equal span and a new section, rigged with reduced incidence. The aircraft incorpo-

RAF BE.2a. The first machine built by Armstrong Whitworth & Co. in 1913-1914. One of many contractors which built BE types.

RAF BE.2b had improved crew protection and control surfaces but initially used warping for lateral control.

rated the front and rear deckings added to previous machines. A semicircular tail plane was standardized which had reduced area.

Individual machines were used at Farnborough to test various new installations in connection with research into stability, such as triangular fins above the center section, increased width at the top of the interplane struts, and alternative tailplanes.

A BE.2a No.218, converted to a single-seater, with an extra fuel tank in the front cockpit, flew 560 miles nonstop in 7hr 20min on 22 November 1913, flown by Capt. Longcroft. On the 13 December 1913, Capt. Salmond established a British height record, by reaching an altitude of 13,140 feet.

The BE.2b gave the crew better protection, by changes to the top decking and embodied revised control runs, but was basically the same aircraft as the BE.2a. The BE.2b was developed in early 1914 and was ordered in considerable numbers from contractors, the last deliveries being made to the RFC, the prime user, in February 1917. The development of the basic BE.2 series was continued in 1915 and later.

The earliest manufacturers of the BE.2a were Vickers, Sir W.G. Armstrong-Whitworth, British & Colonial (Bristol), Coventry Ordnance Works (COW), S.E. Saunders, Handley Page and Hewlett & Blondeau. Some of these firms continued to make the BE.2b and later versions.

Power: 70hp Renault eight-cylinder air-cooled vee driving a 8ft 10in diameter four-bladed propeller

Data		
	Span	36ft 11in
	(Early aircraft, top 37ft 7in, bottom 34ft 10 1/2in)	
	Chord	5ft 6in
	Gap	6ft
	Area	376 sq. ft (Early aircraft 357 sq. ft)
	Area elevators	25 sq. ft
	Area rudder	12 sq. ft
	Area tailplane	34 sq. ft
	Length	28ft 4in (29ft 6 1/2in; long skids)
	Height	10ft 2in
	Weight	1,274 lb.
	Weight allup	1,600 lb.
	Max speed	70 mph
	Climb to 3,000ft	9 min
	Endurance	3hr

BE.2c biplane

The BE.2c was the result of a good deal of development work at Farnborough on stability, carried out on examples of its predecessors. The ability of the aircraft to fly, with the minimum of control movement by the pilot, was regarded as a great attribute for a reconnaissance aircraft, the intended role of the BE.2c. The lack of maneuverability was later to become a great disadvantage, when air fighting developed.

The most apparent change from the BE.2b was the use of wing stagger, which was obtained by moving the lower wing back, to compensate for the loss of lift, occasioned by the use of a smaller non-lifting tailplane. The ailerons replaced wing warping on all four wings and a fin was fitted. The twin skid undercarriage of the BE.2b was used on early aircraft, but was later replaced by a vee type with cross axle on most aircraft. The long exhausts, fitted under the fuselage, were shortened later and some machines had exhausts taken up over the top wing. Many changes and operational additions came much later, among which was the introduction of the RAF.1a engine, made by the Factory.

The BE.2c was in production by a number of contractors over a long period and was in service until the end of the war.

Power: 70hp Renault eight-cylinder air-cooled vee
90hp RAF.1a eight-cylinder air-cooled vee

RAF BE.2c prototype flew in May 1914. Staggered wings and a tail fin made the aircraft extremely stable.

Data	Span	37ft
	Chord	5ft 6in
	Gap	6ft 3in
	Area	354 sq. ft *
	Area tailplane	36 sq. ft
	Area elevators	27 sq. ft
	Area rudder	12 sq. ft
	Area fin	4 sq. ft
	Length	27ft 3in
	Height	11ft 1 1/2in

*Alternative sources quote 371 & 396 sq. ft

Data RAF.1a

	Weight	1,370 lb.
	Weight allup	2,142 lb.
	Max speed	86 mph at sea level
		72 mph at 6,500ft
	Climb to 6,500ft	20min
	Climb to 3,000ft	6 min
	Ceiling	10,000ft
	Endurance	3 1/4hr

BE.3 and BE.4 biplanes

These two machines, designed by John Kenworthy, were structurally identical even though the BE.3 was allegedly the 're-construction' of a Paulhan biplane, No.F.2 and BE.4 of a Bristol Box-kite No.F.9.

The design was of a two-seat tractor biplane, similar to the BE.1 and 2, but with staggered wings and a rotary engine. The fuselage, which was wide and deep to accommodate the rotary engine, had one cockpit for the two crew in tandem and was mounted on short struts above the lower wing. The tailplane and elevator were in one piece, with a balanced rudder above and no fin. The wings were of equal span and control was by warping.

BE.3. was flown by G. de Havilland for the first time on 3 May 1912 and was passed to No.3 Squadron RFC on 13 May 1912 becoming No.203. It continued in use to the end of 1914.

BE.4 was also flown by de Havilland for the first time on 24 June 1912 and was handed over to the RFC on 8 May 1912, becoming No.204. The original 50hp Gnôme was replaced with a 70hp and again with an 80hp engine. The aircraft was involved in a fatal crash on 11 March 1914, when the rudder post broke, resulting in an ugly new design with a fin being introduced. This was fitted to No.203 and the three other machines of the type, which were also built by contractors, namely Nos.303, 417 and 418.

RAF BE.3 was the first of a small number of rotary-powered aircraft supplied to the RFC in 1912-1913.

RAF BE.4 was structurally identical to the BE.3.

Power: 50, 70 & 80hp Gnôme seven-cylinder air-cooled rotaries driving two-bladed or occasionally four-bladed propellers.

Data	Span	39ft 6in
	Chord	5ft 6in
	Length	27ft 3in

BE.7 biplane

Only one of these high-powered versions of the BE.3/4 was built, and it began life with a first flight of twenty minutes flown by de Havilland on 28 February 1912. The machine became Serial No.408 later 438 at the Central Flying School, where it was in use to at least the end of 1913. On 2 May 1913 it established a British altitude record of 8,400ft. with pilot and two passengers aboard.

The changes made to the BE.3/4 design were seemingly only those necessary to accommodate the two-row 140hp Gnôme fourteen-cylinder rotary, primarily the reduction of the front fuselage length by one bay, to adjust the position of the center of gravity.

BE.8 biplane

Kenworthy was responsible for the development of the BE.3/4 type into the BE.8, a similar rotary-powered tractor biplane, which was first flown in August 1913. Two prototypes were built at the RAF, and were probably the machines delivered to the CFS, Nos.423-424 in November 1913. Subsequent machines were manufactured by Vickers, Bristols and the Coventry Ordnance Works, to a total of twenty four.

The main changes from the BE.3/4 included mounting the lower wings directly to the fuselage bottom longerons, and the fitting of a BE.2 type rudder. The two prototypes had one large cockpit for the two occupants, but production aircraft incorporated a central decking, a triangular fin and a four-bladed propeller.

RAF BE.8 developed from the BE.3/4. The prototype originally had an undivided cockpit and no fin.

Although some aircraft went to France with squadrons, the BE.8 saw little operational use and was mainly used for training. A further development, the BE.8a, appeared in 1915 with alternative engines, ailerons replacing warping, larger fins and other improvements adopted from the BE.2c.

BS.1 (Blériot Scout) Biplane

Geoffrey de Havilland, assisted by H.P. Folland and S.J. Waters, was responsible for the design of the BS.1. The aircraft was a scaled down BE.3 with single bay wings, and was the first ever known as a 'Scout', which later evolved into the traditional single-seater fighter.

The circular engine cowling was cut away at the bottom and also allowed space for a 140hp two-row rotary, although the single-row Gnôme only was fitted. The forward part of the fuselage was constructed on four longerons to a point aft of the cockpit, from where it continued to the tail as a wooden monocoque.

Lateral control was by wing warping; the tail unit consisted of a semicircular tailplane with one piece elevator. The small balanced rudder was due to be replaced by a larger unit, at de Havilland's suggestion, but before this could take place the machine spun into the ground. The machine was wrecked and de Havilland was severely injured. The aircraft was rebuilt in a new form as the SE.2.

Power: 100hp Gnôme fourteen-cylinder air-cooled two-row rotary

RAF BS.1 scout biplane flew and crashed in May 1913 injuring Geoffrey de Havilland.

Data		
	Span	27ft 6in
	Length	20ft 6in
	Weight allup	1,232 lb.
	Height	8ft 10in
	Speed range	51-92mph
	Climb	800 ft per min

BS.2 biplane.

This was an aircraft which was proposed, but not built. It was to be larger than the BS.1, and to have a welded steel tubular fuselage structure.

FE.1 was built privately by de Havilland and sold to the Factory in January 1911. It became the basis for the FE.2 range used extensively in wartime.

FE.1 (Farman Experimental) biplane

This aircraft was not built at Farnborough, but was the second aircraft designed and built by Geoffrey de Havilland, which became Government property in January 1911, when he joined the Factory. It was designated FE.1, and survived at Farnborough until 15 August 1911, when it was crashed by Lt. T. Ridge and became the subject for a 'reconstruction' as FE.2.

In its short life at Farnborough, it was improved and used for experiments with new tailplane, elevator and wing extensions, including flights with the front elevator removed. An altitude of 920 feet was reached and passengers were carried on a number of occasions which included tuition for Lt. Ridge, who was the Assistant Superintendent of the Factory.

Power: 45hp Iris four-cylinder horizontally opposed water-cooled

Data		
	Span	33ft
	Length	40ft
	Area	340 sq. ft
	Weight allup	1,100 lb.

FE.2 1911 and 1913 versions

Two distinct versions of this aircraft were built, the first of which made its first attempt at flight on 16 August 1911, which was aborted due to engine failure.

1911 version. This machine was a typical pusher biplane with a nacelle for the crew with single acting ailerons on all four wings, which consisted of two bays outboard of the tail boom struts.

The tailbooms comprised three bays and originally carried a biplane tail with single rudder between; the lower tailplane

FE.2 the 1911 version was a reconstruction of the FE.1.

was soon removed and the rudder area increased by making it taller. The two wheel and twin skid undercarriage was mounted below the wings.

In April 1912 the FE.2 was tested on Fleet Pond as a seaplane, with large central float fitted under the undercarriage skids, together with a tail float and small wingtip floats. A new and more powerful engine was fitted, enabling a passenger to be carried successfully.

After reconversion to wheel undercarriage, limited tests were carried out with a Maxim gun fitted to the front cockpit.

Power: 50 and 70hp Gnôme seven-cylinder air-cooled rotary with four-bladed propeller

Data		
	Span	33ft
	Weight allup	1,200 lb.
	Area	340 sq. ft
	Max speed	48 mph
	Length	28ft
	Climb with passenger	140 ft per min

RAF FE.2 seaplane was tested on Fleet Pond with a 70hp Gnôme engine.

1913 version. The second version of the machine was quite different from the first, although some parts of the earlier machine may have been used in its construction. It does not appear on the RAE list as a new machine in 1913.

The center section, enclosed within the tailboom struts, was broad; the outer wings, reported to be of BE.2a origin, were of single bay type with a large unsupported outer section with rounded tips and warping control. A deeper nacelle than its predecessor's, was mounted on a similar two wheel and twin skid undercarriage, but with mounting struts forward of the wings.

RAF FE.2 the 1913 version had a Renault engine and new wings among other changes.

The tail booms tapered inwards more acutely and the top members sloped down more steeply. The tailplane, with divided elevators, was mounted on the top members and braced with struts to the lower members. A large unbalanced rudder was fitted but no fin. This FE.2 was wrecked on 2 October 1914, when being flown by R. Kemp, the passenger E.T. Haynes being killed in the crash.

Power: 70hp Renault eight-cylinder air-cooled vee with four-bladed propeller

Data		
	Span	42ft
	Area	425 sq. ft
	Length	30ft
	Weight allup	1,865 lb.
	Max speed	67 mph
	Climb	330ft per min initially
	Ceiling	5,500ft

FE.2a biplane

The FE.2a was a new type of two-seater pusher biplane, the design of which began in the middle of 1914, the first aircraft flying on 26 January 1915, piloted by Frank Goodden.

The aircraft was considerably bigger and heavier than the earlier FE.2 types, with a very deep nacelle, reduced at the

RAF FE.2a was a much enlarged aircraft developed in 1914. The prototype flew with a Green engine but the small number of production aircraft used the Beardmore.

front to seat the observer in an exposed position, with a wide area of view. The machine was fitted with BE.2c wings, but without stagger, for the two bays outboard of the wide center section. The center section trailing edge was hinged to form a flap intended to be used as an airbrake, but this was discarded on later production aircraft. The undercarriage had oleo legs and a nose wheel which required a complex arrangement of supporting struts.

The tailplane was mounted on the top of the tail booms, with a triangular fin above, as a kingpost for bracing wires. The rearmost struts in the tail booms were wider in chord to provide additional fin area ahead of the large aerodynamically balanced rudder.

Small side air intakes provided cooling air to the radiator for the cowled Green engine, which was originally fitted, but which was soon replaced by a Beardmore engine of lighter weight.

Twelve machines were ordered straight off the drawing board and, although delay occurred in delivery due to slowness of engine deliveries, the aircraft were dispatched by October 1915, and all saw service in France. The FE.2a. was later developed through the FE.2b to FE.2h, and was a successful aircraft in the early part of the war and was used in various roles until the Armistice.

Power: 100hp Green six-cylinder inline water-cooled driving a two-bladed propeller
120hp Beardmore (Austro-Daimler) six-cylinder inline water-cooled

Data		
	Span	47ft 10in plain ailerons
		50ft 1in balanced ailerons
	Chord	5ft 6in
	Gap	6ft 3 1/2in
	Area	494 sq. ft
	Length	32ft 3 1/2in
	Weight	1,993 lb.
	Weight allup	2,680 lb.
	Max speed	75 mph at sea level (Green)
		80 mph at sea level (Beardmore)
	Climb to 3,000ft	8 1/4min

RAF FE.3 originally AE.1 (Armed Experimental) built in 1913 was soon abandoned.

FE.3 biplane (AE.1 Armed Experimental)

This machine was a two-seater pusher designed, under the leadership of S.J. Waters, to carry a one-pounder COW gun. The testing, by de Havilland and Kemp, revealed the inadequate rigidity of the tail and flying was abandoned. The effect of the recoil of the gun was assessed with the machine suspended from a hangar roof.

The water-cooled engine was mounted at the front of the nacelle and drove a four-bladed propeller through a shaft and enclosed chain. A circular air intake in the nose provided air to a radiator in front of the crew, the air escaping through louvered panels in the sides.

The tail unit was mounted on a single tubular boom, with a support for the extension shaft behind the propeller, the tail itself being braced by wires to the wings and undercarriage Triangular fins, above and below the boom, formed a post for the tall rudder; a large semicircular tailplane and divided elevators completed the tail unit.

The wings were of unequal span, with the top overhang braced by struts. Lateral control was by warping.

Tests of the machine, which was built in 1913, were not continued, but a successor of the same general configuration followed in 1914.

Power: 100hp Chenu six-cylinder inline water-cooled.

Data		
	Span	40ft
	Chord	5ft 9in
	Gap	6ft 4in
	Area	436 1/2 sq. ft
	Length	29ft 31n
	Height	11ft 3in
	Weight	1,400 lb.
	Weight allup	2,080 lb.
	Speed	75mph
	Rate of climb	350ft per min to 1,000ft
	Ceiling	5,000ft

FE.6 biplane

The FE.6 was designed and built in 1914 as a successor to the FE.3, to carry on the work of investigating the requirements for the carriage of the one pounder COW gun. Frank Goodden flew the aircraft on 14 November 1914 for the first time, but the flight resulted in damage which, although not severe, was not repaired, indicating problems of a more fundamental nature.

The aircraft was a heavier and larger version of the FE.3 and was of the same basic layout. The tail boom was more substantial and was not braced externally, and had space to run the tail surface controls internally. The boom would, of necessity, have been firmly anchored back to the structure of the aircraft. The gun was mounted above the lip of the front cockpit to fire forward, but was probably never actually fired.

Power: 120hp Austro-Daimler six-cylinder inline water-cooled

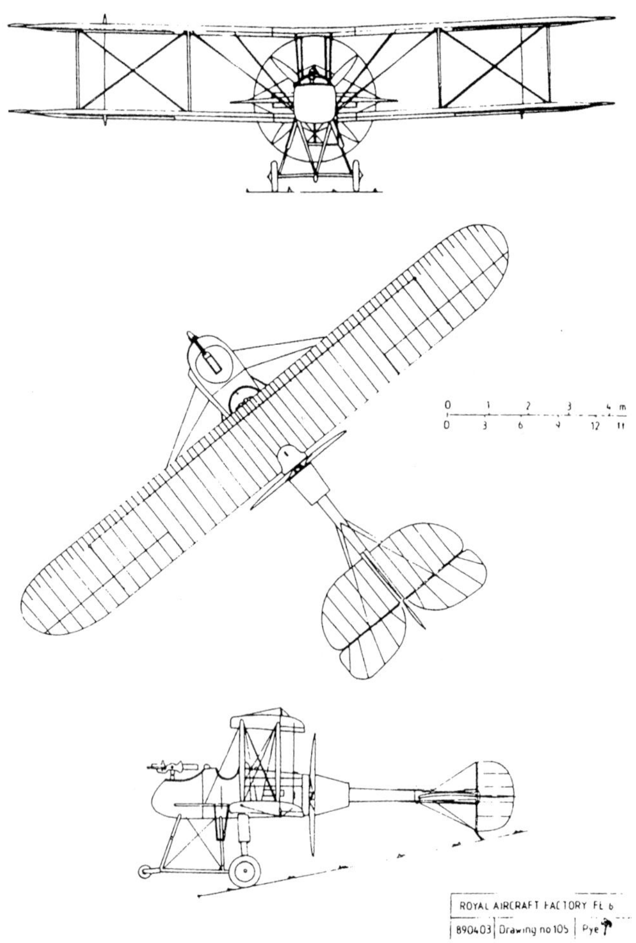

RAF FE.6 with the same one-pounder COW gun as FE.3 was built in 1914 but was also fundamentally unsound.

Data		
	Span	49ft 4in
	Chord	6ft
	Area	542 sq. ft
	Length	29ft 6in
	Height	15ft
	Weight	2,000 lb.
	Weight allup	2,630 lb.

RE.1 biplane (Reconnaissance Experimental)

The first of two prototypes, serial No.607 was flown in May 1913, the second No.608 following in September 1913. Both were used for test flying at Farnborough, mostly by E.T. Busk, in connection with improvements to aircraft stability, although the second machine was transferred to the RFC in March 1914.

The RE.1 was a single bay biplane, which in its original form, had staggered wings with warping control. The fuselage incorporated some steel tube in addition to wood in its structure and was mounted on a two wheeled undercarriage by four struts, a neater arrangement than that of the BE.2, from which the aircraft was derived. The tail unit was conventional, with tailplane, fin, rudder and divided elevators. However, changes were made to a number of features during development. For a time on No.608, four small triangular fins were fitted above the top wings. Wings of greater span with ailerons were fitted, the stagger reduced and the dihedral varied. The fin was reduced and a larger aerodynamically balanced rudder fitted, together with a new tailplane with straight leading edge, square tips and new elevators to match. Provision for armor plate was made in the basic design and may have been fitted.

Power: 70hp Renault eight-cylinder air-cooled vee driving a four-bladed propeller

Data		
	Span	34ft later 36ft
	Chord	5ft 6in
	Area	316 sq. ft (later 328 sq. ft)
	Weight	1,000 lb.
	Weight allup	1,580 lb.
	Speed range	48-78 mph
	Initial climb	600 ft per min

RE.2 and HRE.2 biplane (Hydro Reconnaissance Experimental)

Only one of this type was built, incorporating features of the RE.1 and some from the BE aircraft. The machine was a two bay biplane with warping and staggered wings, a tall aerodynamically balanced rudder and one piece elevator and no fin. The undercarriage was derived from the RE.1 as, no doubt, was much of the fuselage design.

RAF RE.1. Only two of these machines were built in 1913.

RAF RE.2. One of these was tested as a landplane in 1913.

RAF HRE.2. Tested as a seaplane in 1913 on Fleet Pond. It was unable to take off and reverted to a modified form of the land version for the RNAS.

The RE.2 was flown first as a land plane on 1 July 1913, and was then converted to the HRE.2 hydroplane. The twin main floats had steps and the tail float mounted a water rudder. A triangular fin and tall rudder were fitted.

The first attempts to take off from Fleet Pond were unsuccessful. A 100hp engine was then fitted and wings with ailerons, but despite the increased power, the aircraft would not rise and finally hit the bank, wrecking the floats and causing other damage, when it overturned. After repair, the machine reverted to a land undercarriage and was passed to the RNAS as No.17 from August 1914, until it was wrecked on 10 January 1915.

Power: 70hp Renault eight-cylinder air-cooled vee
100hp Renault twelve-cylinder air-cooled vee

Data		
	Span	45ft 3 1/2in
	Chord	6ft
	Gap	6ft 3 1/2in
	Area	498 sq. ft
	Length	32ft 3in
	Height	12ft 2in
	Max speed	60 mph (70hp)
		75 mph (100hp)
	Climb to 3,000ft	5 min
	Endurance	4 1/2hr

RE.3 biplane

This aircraft was the same as the RE.2 in its final form, but was fitted with an Austro-Daimler engine. It first flew in November 1913, but was wrecked at Farnborough, as a result of engine failure, when being flown by E.T. Busk on 27 September 1914.

RAF RE3. This was the same as the RE.2 in its final form but had an Austro-Daimler engine.

Power: 120hp Austro-Daimler six-cylinder inline water-cooled

Data		
	Span	45ft 3 1/2in
	Chord	6ft
	Gap	6ft 3 1/2in
	Area	498 sq. ft
	Length	32ft 3in
	Height	12ft 2in

RE.4 biplane

This aircraft, intended to operate from small fields, was not built. It would have been developed from the RE.2/3, with unequal span wings and fitted with a 120hp Austro-Daimler engine or the large Gnôme fourteen-cylinder two-row rotary as used in the SE.4.

RAF RE.5 twenty-four of which were built in 1914-1915 with short or long span wings.

RE.5 biplane

The basic RE.5 was a large two-seater with the pilot seated in the rear cockpit. The engine was closely cowled and ducted cooling air to a radiator mounted behind the engine; the cooling air exhausted through louvered panels on either side of the fuselage. A tall rudder, with curved trailing edge, and triangular fin were fitted, together with a conventional tailplane and divided elevators. Much use was made of steel tubing in the fuselage structure and tail unit.

This further development of the RE.2/3 first appeared with equal span wings, and with ailerons fitted to both top and bottom. Later, on four aircraft, the top wings were increased in span, braced to the lower wing by diagonal struts, and the ailerons were moved outboard. One of these aircraft, No.380, flown by Norman Spratt, reached an altitude of 18,900 feet on 14 May 1914. Certain of the extended span aircraft had increased fuel tankage and were flown as single-seaters.

A total of 24 RE.5 aircraft were delivered between January 1914 and early 1915, and most served with the RFC. Experiments, to reduce their long landing run, were made with both a 'plough brake' on the undercarriage and airbrakes on the fuselage sides. One aircraft served with the RNAS, and others were retained at Farnborough for trials, including the dropping of cast iron weights of up to 10 cwt., the testing of the RAF.4 and Sunbeam engines, also the RAF type of undercarriage oleo legs.

Power: 120hp Austro-Daimler or Beardmore six-cylinder inline water-cooled driving either two or four-bladed propellers.

Data		
	Span	45ft 3 1/2in or 57ft 1/2in
	Chord	6ft
	Area	498 sq. ft or 569 sq. ft
	Max speed	78 mph
	Climb	400ft per min

RE.6 biplane

This was a design for a three-seater biplane powered by a 250hp engine and fitted with a float undercarriage. The project was discontinued at an early design stage.

SE.1 biplane (Santos Experimental)

The 'reconstruction' of a Blériot XII monoplane was the basis for a 'canard' biplane identified as SE.1, due to its similarity in layout to the biplane, Type 14bis, of Santos Dumont, in which he made the first flight of a powered heavier-than-air machine in Europe.

This was the first original design from the Aircraft Factory after the reorganization, which followed the discontinuing of the work of Cody and Dunne under Col. Capper, the aircraft making its appearance in June 1911. After the repair of minor damage, which occurred when taxiing on 7 June 1911, the aircraft was flown the following day for about a mile. Geoffrey de Havilland was the pilot and largely responsible for the design, although Mervyn O'Gorman, the Superintendent and F.M. Green, the Chief Engineer of the Factory, exerted their influence on various aspects. Many changes were made in the course of testing the machine in June and July to improve the handling, particularly the tendency to sideslip.

The aeroplane that emerged from the Factory was rather different to earlier concepts outlined by O'Gorman in March 1911, and other sources in the technical press. It had a two bay wing structure, the top wing slightly greater in span than the bottom, with lateral control by warping sections in the top wings. Control in pitch was by a divided elevator mounted at the nose of the nacelle. The machine was mounted on a twin skid undercarriage with two wheels, on a rubber cord sprung axle, and with small tail skids at the ends of the main skids.

Power was provided by the ENV engine from the Blériot, which drove a pusher propeller at two-thirds engine speed, through a chain and shaft. The engine was cooled by radiators mounted on either side of the nacelle, alongside the pilot's position.

The separate wire braced tail booms carried rudders of approximately the height of the wing gap. The boom lengths were shortened early in the trials and were planned to be lengthened again later. A supplementary nose mounted rudder was fitted at one stage and the fabric covering of the nacelle was temporarily removed without improvement in the yawing plane. Also the foreplane was later changed to provide a fixed portion with separate elevators.

The position of the wheels was varied and this change was accompanied by the deletion of the small tail skids, which were replaced by a tall sprung nose skid and additional vee-struts connecting the skids to the lower longerons and wing spars.

Despite these and possibly other alterations, de Havilland still regarded the machine as difficult to handle. However, on 18 August 1911, Lt. Ridge the Assistant Superintendent an inexperienced pilot, insisted on being allowed to fly the machine, which side slipped into the ground. Ridge succumbed from his injuries and the SE.1 was destroyed.

RAF SE.1 first flown in June 1911 was destroyed in a fatal crash which killed Lt. Ridge.

Power: 60hp ENV type F eight-cylinder water-cooled driving a two-bladed propeller by chain and shaft at 2/3rds engine speed

Data		
	Span top	38ft
	Span bottom	36ft
	Chord	6ft
	Gap	6ft
	Area	382 sq. ft
	Length	29ft short booms
	Height	11ft 6in
	Foreplane span	18ft 4in
	Foreplane chord	4ft
	Foreplane area	64 sq. ft
	Weight	1,220 lb.
	Weight allup	1,640 lb.
	Max speed	55mph

RAF SE.2 was a reconstruction of the crashed BS.1 which flew in October 1913.

SE.2 biplane (Scouting Experimental)

The BS.1, which crashed on 27 March 1913, had by that time been reidentified as the SE.2. However, approval for its reconstruction was given, within a cost of £900, and so the opportunity was taken to carry out a number of design changes. Among these was the fitting of an 80hp Gnôme engine in a new circular cowling and a completely new tail unit. The latter consisted of small triangular fin surfaces, above and below the fuselage datum, with a high aspect ratio rudder, having a reinforced lower edge member, serving as a tail skid. The new vertical surfaces necessitated the fitting of a divided tailplane and elevators.

In this form the SE.2 was flown by de Havilland in October 1913 and was retained by the Factory until instructions to release it to the Military Wing caused it to be handed over on 17 January 1914, becoming No.609, before being returned to Farnborough in April for further modification. This rebuild included the replacement of the rear fuselage monocoque with a conventional wooden girder of four longerons, with formers and stringers fairing the outside shape. The tail unit was changed to provide a larger fin and rudder and there were new tailplane and elevators. The engine was enclosed in a new cowling, tapering steeply from its largest diameter down towards the propeller boss, and rearwards to align with the fuselage shape. A small spinner was fitted. All external bracing was made of streamlined section wires known as Rafwires, the first use of these, which were adopted extensively.

RAF SE.2a was transferred to the RFC in January 1914 and served in France armed with two rifles until March 1915.

Frank Goodden flew the rebuilt SE.2 on 3 October 1914 and it was soon serving with No.3 Squadron in France armed with two rifles. It was returned to England in March 1915 and no further details of its use are known.

Power: 80hp Gnôme seven-cylinder air-cooled rotary

Data		
	Span	27ft.6 1/4in
	Chord	3ft 9 1/2in
	Gap	4ft 7in
	Area	188 sq. ft
	Length	20ft 10in
	Height	9ft 3in
	Weight	720 lb.
	Weight allup	1,132 lb. (1st form)
		1,246 lb. (2nd form)
	Max speed	91mph (1st form)
		96mph (2nd form)
	Endurance	3hr

SE.3 biplane

This was a project for a development of the SE.2, with the more powerful 100hp Gnôme nine-cylinder engine, for which some preliminary design work was carried out by HP Folland. It was to be an unstaggered biplane with single I-shaped interplane struts and with ailerons usable as either flaps or airbrakes.

SE.4 biplane

The SE.4 was a single-seat tractor biplane designed by H.P. Folland, which was of similar style to the projected SE.3, but with more power.

The fuselage was of girder type, with formers and stringers to fair the shape to circular section. A triangular dorsal fin, and a smaller ventral fin, were fitted together with a conventional rudder. The tailplane and divided elevators had the gaps sealed with an elastic material to reduce drag. The wings were like those of the unbuilt SE.3 with I-shaped interplane struts

RAF SE.4 was tested from June 1914 with both 100 and 160 hp Gnôme engines, but the undercarriage caused landing problems.

RAF SE.4a prototype with large spinner.

and full span ailerons capable of use as flaps, when raised, to reduce drag for high speed flight. The undercarriage was unusual, consisting of a transverse leaf spring mounted to an inverted tripod of struts. Owing to excessive rolling during taxiing trials it was soon altered to a conventional type with rubber cord sprung axle. The 160hp Gnôme proved to be unreliable and was changed for a 100hp Gnôme Monosoupape, with the consequent reduction of performance.

The SE.4 was ready for testing on 17 June 1914 by Norman Spratt and was passed to the RFC in August, but was badly damaged in a landing accident on the 12th of the month and was not rebuilt. Work on a second machine was discontinued.

Power: 160hp Gnôme fourteen-cylinder two-row air-cooled rotary
100hp Gnôme Monosoupape nine-cylinder air-cooled rotary

Data		
	Span	27ft.6in.
	Chord	3ft.9 1/2in
	Gap	5ft 1in
	Area	188 sq. ft
	Length	21ft 4in
	Height	9ft tripod u\c, 9ft 10in V u\c
	Max speed	135mph (160hp)
		92mph (100hp)
	Min speed	52mph
	Climb	1,600 ft /min (160hp)
	Endurance	1hr

SE.4a biplane

The design of the SE.4a was begun late in 1914 by H.P. Folland but the aircraft did not fly until 25 June 1915, when Frank Goodden made the first flight. Three more of the type soon followed, but these were a modified version which had flat sided fuselages with only short fairings behind the open fronted, circular engine cowling, whereas the first machine had the circular fuselage shape obtained with formers and stringers. The large spinner of the prototype was open at the center and a cooling fan was fitted to supplement the airflow through the narrow opening around the periphery. This was not used on the subsequent aircraft.

The SE.4a was a new design with few of the SE.4 components used in its construction, perhaps only the tailplane and elevators. The full span ailerons used as flaps or to reduce drag when raised, were again fitted, but in a new staggered wing cellule with conventional interplane struts. There was no center section and the top wings met at the centerline at an inverted V-cabane.

The prototype, No.5609, probably remained at Farnborough, but the other three Nos.5610-5612, were used by the RFC with Lewis guns fitted to the top center section, the first British aircraft with forward firing guns.

Power: 80hp Gnôme seven-cylinder air-cooled rotary

Data		
	Span	27ft 6in
	Chord	4ft 2in
	Gap	4ft 9in
	Length	20ft 10 1/2in
	Height	9ft 5in

RAF SE.4a production version, three of which served in the RFC with Lewis guns on the center section.

RAS monoplane was built at Gosport in 1910 but was not developed.

Rathen biplane intended to compete in the *Daily Mail* flight from Glasgow to Edinburgh was not completed in 1911.

R.A.S. Monoplane (R.A.S. Aeroplane Co., Gosport, Hampshire)

This tractor monoplane appeared in October 1910, but reports disappeared from the Press soon after and it was presumably unsuccessful on test.

The fuselage was an open braced girder of triangular section, the two top longerons and the struts being of circular section hickory wood; the bottom longeron was of rectangular section forming a keel which extended from below the engine to the tail. The fuselage rested on an undercarriage structure made mainly of steel tubing with two hickory cross members. A combination of air pressure with compression and tension springs provided the suspension of the fork mounted wheels. The tail wheel steered in unison with the rudder and was sprung by air and compression spring.

The tail unit comprised a one piece tailplane, adjustable for incidence, and triangular shaped elevator. Both the fin and the rudder were triangular in shape, with a small rudder below the tailplane on an extension of the rudder post. A further fin was fitted below the pilot's seat.

The wing was built with a main spar which, in combination with the top and bottom bracing, contained the compression loads without loading the fuselage. The full span rear spar could be rocked bodily to provide lateral warping control activated by a system of cables and pulleys to a wheel on the horizontal control shaft; the elevator was operated by the fore and aft movement of this wheel. Foot pedals controlled the rudder and tail wheel.

Power:	35hp Lascelles four-cylinder air-cooled fantype semi-radial driving a 7ft diameter propeller	
Data	Span	32ft.4in
	Chord	7ft tapering to 6ft 6in
	Area	210 sq. ft
	Length	28ft 6in
	Weight allup	700 lb.

RATHEN biplane (George Morrice, John Scott and George Dean, Rathen near Fraserburgh, Scotland)

This machine was built by three friends in 1911 with the object of competing for the *Daily Mail* £1,000 prize then on offer for a flight between Glasgow and Edinburgh. Morrice and Scott, originally from Buchan, returned from America and set up an engineering works at Rathen in conjunction with Dean, who was from England.

The aircraft structure was the work of Morrice and Scott and the engine was Dean's handiwork. It was to be flown by a pilot named Nutter from Nelson in Lancashire. Dean was apparently aiming to make a machine capable of flying the Atlantic, hoping to win the much larger prize on offer. However the aircraft was not completed, work being discontinued when Dean, who had a marital problem, disappeared.

The type of machine can be seen from the only available photograph, which shows a single-seater biplane with an elegant fuselage of circular frames and stringers. The wings had straight leading and curved trailing edges, blending into rounded wing tips. Among the parts are six-cylinder barrels and two propellers with corrugated blades, the latter probably bought from Cochrane or Beney.

Power:	60hp Dean probably water-cooled	
Data	Span	22ft
	Length	20ft
	Speed	68mph

RETFORD glider

A biplane glider made in 1911 of which there are no details
.

RIDLEY gliders (Cyril Burfield Ridley, Arundel House School Aero Club, Surbiton, Surrey)

A Chanute type biplane glider was built by Cyril Ridley of Thames Ditton in 1910, while he was a pupil at Arundel House School. The machine was flown successfully at Sandown Park on several occasions and also at the Crystal Palace. It was made

Cyril Ridley from Thames Ditton built this glider in 1910 while still at Arundel House School. He flew it at Sandown Park and the Crystal Palace.

of cheap materials and did not last long. For 1911 Ridley proposed to make a monoplane glider and wrote to *The Aero* asking for comments on his suggested design. After leaving school Ridley worked for Sopwiths in the Rink, before the War, but by April 1916 had learned to fly at the Hall School before joining the RNAS on 25 June 1916.

Data	**Glider No.1**	**Glider No.2**
Span	18ft	24ft
Chord	4ft 6in	6ft
Gap	4ft 6in	
Area	sq. ft	sq. ft
Length	14ft	16ft
Weight	140 lb.	

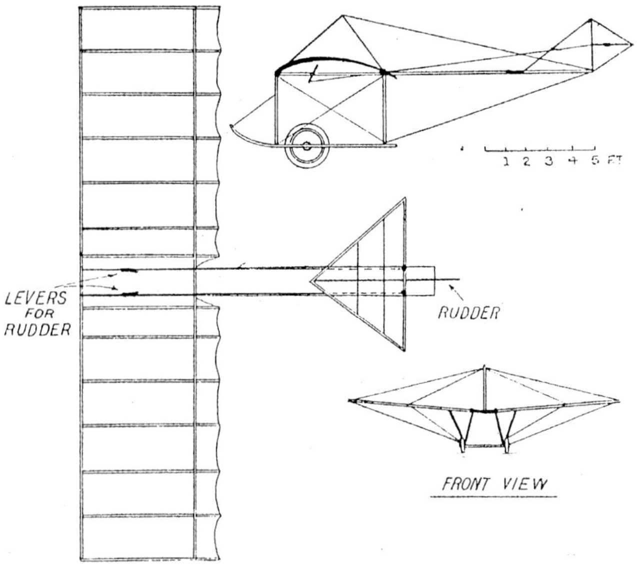

Ridley monoplane glider proposed for construction in 1911.

ROBINSON gliders (G.J. Robinson, Salisbury, Wilts.)
Robinson is reported to have made two different gliders, details as follows:

Data	**Cody Type**	**Chanute Type**
Span	23ft	20ft
Chord	5ft	4ft
Gap		3ft 6in
Weight	80 lb.	35 lb.

ROBINSON glider (P.G. Robinson)
The only information available is that the span was 20ft.

RODGERS glider (Edward E.U. Rodgers, 2 Colchester Villas, Elmgrove Rd., Weybridge, Surrey)
As a lad living in Weybridge, Rodgers saw the building of Brooklands from its inception and was able to enter the area because his father, a motor engineer, made parts for some of the constructors, including Frenchman, Monsieur Bellamy, the earliest aviator at the Track. Young Edward delivered the parts to the customers and could use this as a pretext to gain entry on many occasions, which inspired a wish to become involved in aviation himself.

After one aborted project, Rodgers at about fourteen years of age, made a glider which he and a friend towed fifteen miles to Newlands Corner behind their bicycles. They made three short glides down the slopes before the machine was wrecked and abandoned.

The glider was based on a *Demoiselle* monoplane and was described as having a ladder-like fuselage 10ft long, on which the pilot sat below the wing leading edge. The wing was 18ft in span and made in halves joined on the centerline and could be warped. A normal elevator and rudder were fitted. The two wheels had flat rims of steel sheet, 3in wide rolled to 1ft diameter and were unsprung. The intention had been to install a 79hp Indian vee-twin motor cycle engine later, but this did not take place.

ROLLS RPG powered glider
C.S. Rolls acquired his first glider from Shorts in July 1909, which was based on an early Wright type. (Short-Wright glider S.9). Soon after Rolls placed on order with Shorts for a second glider (S.12), which seemingly was not delivered in complete

Rolls powered glider (RPG) of 1909-1910 built by Shorts at Eastchurch and developed by C.S. Rolls just prior to his death.

form, but was developed by stages into a powered aircraft (RPG.IV). The extent of Rolls' influence on the design is not clear, for features of Shorts construction at the time were used in a basic layout resembling a Wright twin pusher aircraft, fitted with a 35hp Green four-cylinder inline water-cooled engine.

The aircraft was virtually complete at Eastchurch in May 1910 and was for sale, but may not have flown before Rolls died at Bournemouth on the 12th of July. It was almost certainly regarded as an obsolete type and, if Rolls had lived, his own designs would probably have been based on Farman or Sommer types, for the indications were that Rolls had intentions of becoming a constructor himself.

The fate of the RPG is not known but it was probably dismantled soon after Rolls' death. The unsuccessful Short No.3 (Constructors No.S.24), probably contributed it's engine and parts of the airframe in the construction of RPG IV.

ROOTS helicopter (J.D. Roots)

This machine was fitted with a 40hp Antoinette engine and had a 24ft diameter rotor. No other details are available.

ROSS monoplane (Ross, Hilgay Park Rd., Chelmsford, Essex)

This was reported as being a Blériot type monoplane with a span of 25ft, and fitted with an unidentified air-cooled engine of 30-38hp. Advertised for sale in *Flight* on 28 January 1911.

ROSSINI aeroplane (Clarence Rossini., Oswestry, Cheshire)

A manpowered type reported in 1908. No further information.

ROWELL monoplane (R. Rowell & Co., Springwell House, Bath Rd., Cheltenham, Gloucestershire)

This firm of cycle agents built a Blériot type monoplane covered with Pegamoid in 1910-1912 It was offered for sale in June 1912.

Power: 30/35hp Lascelles four-cylinder air-cooled semi-radial.

Rowell monoplane built in 1910-1912 at Cheltenham.

Data Span 36ft
Length 28ft

RUCKER biplane (Martin Diedrick Rucker, 48c Gillingham St., Pimlico, London SW)

A biplane based on the Voisin was being constructed in 1910, but may not have been completed. Rucker leased shed No.37 at Brooklands and during 1911 was involved with the unsuccessful Edwards Rhomboidal.

Power: 50hp Simms six-cylinder water-cooled vee

Ruffy-Baumann Caudron-type biplane built in 1914 for use at the school at Hendon.

RUFFY biplanes (Ruffy School of Flying. Works at 34 Kendall Mews, George St., Portman Square, London W. Flying operations at Hendon)

The Ruffy School of Flying was a training establishment that built Caudron biplanes in 1914 and used them for instructional purposes at Hendon. Both single and two seater machines with 45hp Anzani six-cylinder radial and 50 and 60hp Gnôme seven-cylinder rotary engines were built. The company became the Ruffy-Baumann School of Flying in early 1915, to which the James brothers *(q.v.)*, moved their Caudron type built in Wales in late 1914.

RYLEY glider (L.G. Ryley, Coventry Aero Club)

This canard type biplane, named *'Dragonfly I'*, was made in 1914 and, on 21 August, was tested by towing, carrying operators of up to ten stone in weight. The war soon stopped these activities.

The machine was constructed mainly of bamboo, with steel joint fittings, all glued and bound with twine; ash was used for the skids and deal for the interplane struts and internal compression ribs in the wings

Data		
	Span	24ft
	Chord	4ft
	Gap	4ft 6in
	Area	184 sq. ft
	Area elevator	27 sq. ft
	Area rudder	4 1/2 sq. ft
	Length	17ft 6in

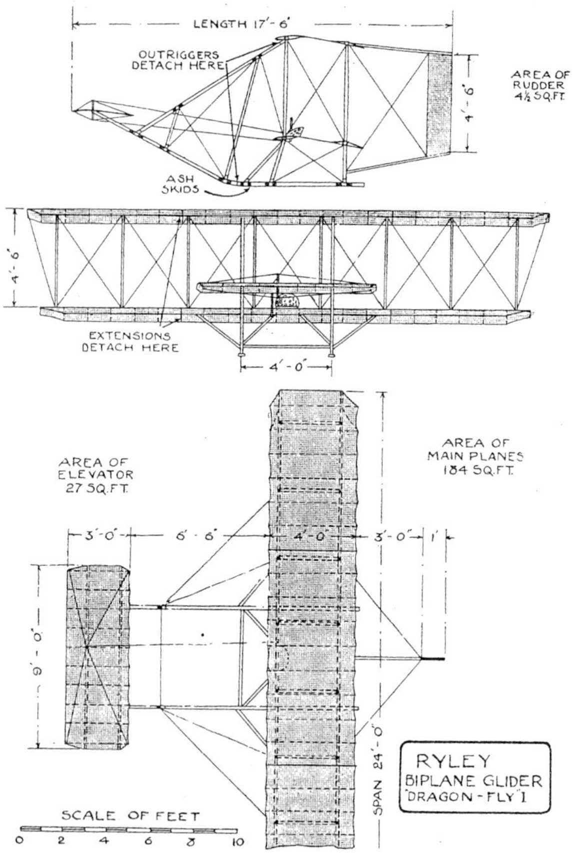

Ryley glider *'Dragonfly 1'*. Its use was curtailed by the outbreak of war in 1914.

S

SALMET waterplane (Henri Salmet, Brooklands)

The first report of this machine being constructed by the French Blériot mechanic and pilot was on 5 May 1913. It was being built at Filey for entry in the *Daily Mail* Seaplane Circuit of Britain Contest, but when mentioned on 26 June 1913, it was at Brooklands ready for testing on wheels. The general arrangement was similar to a Curtiss type F, which had flown recently as a four seater and is best described by extracts from the report in *The Aeroplane:*

"The hull of the boat is of peculiar design, having a bottom coming to an edge, without actually forming a keel. What would in an ordinary boat form the counter, is in this machine tapered far aft to form a fairly long fuselage on which is mounted the tail, the elevator and a large balanced rudder, which is placed above the tail. At first sight it strikes one that with so large a rudder placed so high, there would be an enormous twisting strain on the comparatively small fuselage and that the staying of the tail is also on the light side, but doubtless the correctness of the view will be proved in the test flights, before the machine is taken on the water. Underneath the fuselage and behind the deepest part of the boat proper a water rudder is fitted, acting in conjunction with the air rudder.

Above the boat are placed the mainplanes which appear to be almost identical with those of the Avro biplanes. Mounted over the deck of the boat is a 100hp ABC engine which drives a tractor screw by chain gearing, the propeller shaft being situated in the neighborhood of the upper plane, the propeller tip only clearing the foredeck, thanks to a little trough built therein, so that, at the lowest point, the tip of the propeller is actually inside the deck. The pilot is seated in what would be the stern of the boat and control is by means of a wheel mounted on a tubular rectangular arch, somewhat in the manner originated in Deperdussin machines. On each side of the boat are fitted landing wheels on a tubular chassis, pivoting somewhat in the manner of the Blériot."

In August, *Aeronautics* described the tests of the machine as "not very successful" and promised further details when "the balance, etc. is obtained correctly". This was the last report and presumably the machine was abandoned.

It is believed that the engine was the first 100hp ABC made by Armstrong-Whitworth at Newcastle-on-Tyne, which was illustrated on its mounting in *Flight* of 28 June 1913.

SALTMARSH biplanes (A. Saltmarsh, Sunnyside, Staines, Middlesex)

An advertisement appeared in *Flight* in 1912 for four unfinished biplanes, one costing over £100, and a quantity of ash timber and parts. A fair offer would be accepted. It is not known if Saltmarsh was the actual maker or perhaps just a dealer in secondhand items.

SANDERS biplane No.1 (Capt. Haydn Arnold Sanders MD, London Aeroplane and Navigation Co., 23 Blenheim Park Rd., Croydon, Surrey)

Capt. Sanders was assisted by his brother, Kempton Sanders and F.L. Rawson, a consultant.

The machine was built at Kessingland, near Lowestoft, and was tested at Benacre Denes, making its first flight in October 1909, but was wrecked on hitting a telephone wire on 13 February 1910. In its short life the Sanders biplane had achieved a flight of 2 1/2 miles, but had only flown at heights up to twenty-five feet.

The biplane was constructed mainly of steel tubing and was a canard design with a fixed boxkite biplane unit on the front booms. There was a single rudder behind the elevators,

Sanders biplane No.1 was built and tested near Kessingland in 1909-1910. It flew successfully until it hit telephone wires.

Sanders biplane No.2. The second version with ENV engine and single pusher propeller.

which were hinged to the top and bottom wing trailing edges. The wings had tips which curved downwards outboard of the X-shaped interplane struts, the centers of which served as mountings for the aileron hinges.

The pilot was seated ahead of the engine which drove twin pusher propellers on long shafts by chains. The engine, although reliable, was heavy at 289 lb., for the power developed. The undercarriage was a tubular structure with two main wheels and twin wooden skids and a single tail wheel. Certain features were covered by patent No.4624/1909.

Power: 30hp Brooke four-cylinder inline water-cooled 'marine' engine driving two 8ft diameter propellers by chains.

Data Span 50ft

SANDERS teacher.

After the crash of No.1, Sanders moved to Beccles where a ground trainer was constructed with control surfaces and structure representative of those to be used for No.2.

SANDERS biplane No.2 Types 1 and 2 (Sanders Aeroplane Co., Beccles, Suffolk)

The second Sanders biplane was made in two versions and had quite a different fuselage from No.1, although contemporary reports referred to 'reconstruction' taking place. The only report of flight refers to three over Beccles Common, where the machine was constructed. These took place on 30 April 1911, when a height of forty feet was achieved and a speed of 40mph. This was soon after the Type 2 aircraft had been exhibited at Olympia in March and probably refers to that machine. The Type 1 was illustrated in March and must have been tested earlier in the year, but no evidence of flight can be traced. In advertisements in June 1911 the machine was referred to as the 'Aerial Cruiser' with a 'Sanders Patent Aerial Gun'.

Sanders biplane No.2 of 1911. This is the first version with Alvaston engine and twin pusher propellers.

The two versions were basically identical, except for the power, which on the Type 1, had twin pusher propellers, whereas the Type 2 had a single larger propeller. This change necessitated widening the gap between the tail booms to provide clearance, which then tapered to the tail in plan, instead of being parallel as on No.1.

The main girder took the form of an extended undercarriage structure and formed a mounting for the front biplane elevator, which incorporated two small fixed fins. The undercarriage consisted of a nose-wheel and a pair of main wheels on a rubber cord sprung axle, which after takeoff could be retracted to allow the machine to land on the skids to reduce the landing run. The top wing had the down turned tips, as used previously, but the lower was of reduced span, cut squarely at the outboard interplane struts with wingtip skids below. Outboard of the center section were three pairs of interplane struts, the two rear outers being used to mount the ailerons. On the twin pusher version, further pairs of struts were necessary to mount the bearings for the propeller shafts. The outer wing bays, with the ailerons, could be folded forward to reduce the span for storage. Elevator flaps were again fitted to the top and bottom center sections on the first version but were not present on the machine at Olympia. Three narrow chord rudders were fitted at the tail. The pilot again sat ahead of the engine which was flanked on either side by coolant radiators, with a fuel and oil tank above.

The first machine had an Alvaston engine with the twin propellers driven by chains. The second machine had an ENV, driving a single propeller by chain on a shaft mounted above. A report that Sanders' aircraft was converted to a waterplane and was in Northumberland in 1913 has not been confirmed.

Power: No.1 50hp Alvaston four-cylinder horizontally opposed water-cooled driving two 8ft 6in diameter propellers by chains and shafts.

No.2 60hp ENV type F eight-cylinder water-cooled vee driving a 9ft diameter Sanders propeller by chain and shaft at a ratio of 1.4 to 1.

Data for Type 1.

Span	40ft
Chord	6ft
Area	472 3/4 sq. ft (inc. tailplanes)
Length	31ft 6in
Front elevator span	12ft
Front elevator chord	3ft
Front elevator area	72 sq. ft
Rudders height	6ft
Rudders chord	1ft
Rudders area	18 sq. ft
Ailerons span	7ft
Ailerons chord	2ft
Ailerons area	4 sq. ft each
Weight	1,100 lb. (Type 1) 1,000 lb. (Type 2)

SAUL tandem biplane No.1 (G.P. Deverall Saul)

Also described as a quadruplane, the machine was a tandem biplane, built by Handley Page at 36 William St., Woolwich, London SE. It was completed in May 1909 and Handley Page tested it on a trolley towed by a car to aid the low powered motor, at his new flying ground and premises at Dagenham. The machine apparently lifted off briefly on 25 May 1909 with H.P. aboard, and Saul was sufficiently encouraged to order a two-seater version. The machine was also flown as a kite. The Saul biplane was exhibited at the Sport and Travel Exhibition at Olympia which opened on 6 July 1909.

The structure was built of wood, mainly hickory, and rested on two long skids. The front wings had rigid curved leading edges and flexible trailing portions; the rear wings were of greater span and of similar construction. The pairs of wings were heavily staggered in opposite directions, the whole arrangement being patented to cover a claim for automatic stability. There were no other control surfaces or systems. A small motor drove a pusher propeller.

Power: 8hp air-cooled petrol motor of unknown make.

Data

Span	20ft
Chord	3 ft
Gap	3ft
Area	216 sq. ft
Length	20ft
Weight	162 lb.
Weight allup	320 lb.
Speed	30mph
Price	£250

Saul tandem biplane No.1 built at Woolwich by Handley Page and flown briefly by him in 1909.

Saul tandem biplane No.2 outside the Barking works of Handley Page where it was built in 1909.

SAUL tandem biplane No.2

The two-seater, of similar design, was under construction in mid-1909 by Handley Page at Dagenham, also a similar machine for the Bridgwater Motor Co., Handley Page's agent in the West of England. There were no further reports of these machines. Price £550.

Power: 20-25hp air-cooled petrol motor of unknown make.

Saunderson monoplane No.1 *'Mayfly'* was taken to the Blackpool Meeting in October 1909 but did not fly.

SAUNDERSON monoplane *Mayfly* (H.P. Saunderson & Mills, Elstow, Bedfordshire)

The Saunderson monoplane arrived at the Blackpool Meeting on 18 October 1909. Although it was reported to have flown previously, when tested on Dunstable Downs, it was unable to lift off at Blackpool.

The machine was a single-seat tractor with wings mounted high on either side of a fuselage of tubular construction. Power was provided by a three-cylinder air-cooled fantype radial also made by Saunderson.

Saunderson monoplane No.2 *'Verifly'* was tested at St Albans in 1910.

SAUNDERSON monoplane *Verifly*

The second monoplane built by Saunderson in 1910, was tested on the Earl of Verulam's estate, Gorhambury Park, St. Alban's. It probably used the same engine as the first machine. No record of flight can be traced.

The fuselage of the aircraft was very shallow, so the seat and the pilot were exposed to the slipstream. The box girder was tapered and fully enclosed, probably with plywood covering. Two triangular shaped fins were fitted, above and below the rear fuselage, the lower protected by a separate skid. The rudder is not shown in surviving photographs, but was apparently fitted to the top fin only. The tailplane consisted of triangular shaped halves extending well along the fuselage sides. The elevator was in one piece and was also triangular. The undercarriage and wings were similar to those of the earlier machine and may have been transferred from it. The name '*Verifly*' was carried on both sides of the fuselage with a decorated surround.

Schmarz ornithopter built in Liverpool in 1909-1910 with the intention of flying across the River Mersey.

SCHMARZ ornithopter '*Airship*' (Rudolf Schmarz, 14 Dorset St., W. Derby together with C.W. Sandler, Sutton Lodge, Great Sutton, Cheshire)

Known as the '*Airship*', the machine was in fact heavier than air with a rotor and flapping wings. It was built by Archibald Ford & Co. at Liverpool in 1909-1910, with the intention of flying across the Mersey. Patent No 2165/1909 is relevant.

Data	Height	12ft approx.
	Rotor diameter	20ft

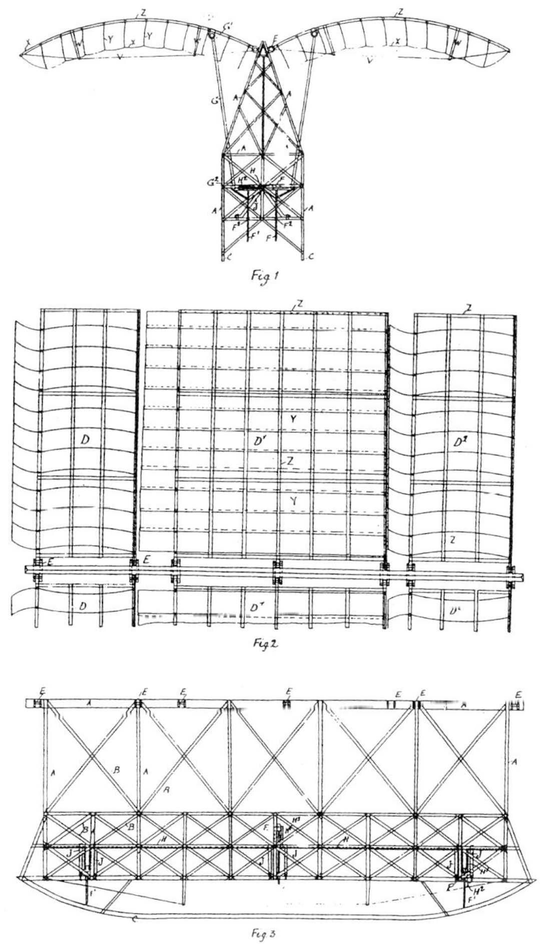

Scott ornithopter. This is the patent drawing No.24247 of 1907.

SCOTT ornithopter (Bateman D. Scott and Wilfrid E. Scott, Anslea House, Parliament St., Gloucester)

In 1907 the brothers covered the design of an ornithopter by patent No.24247/1907, but it has not been established whether construction of the machine took place; it would seem to have been quite impractical.

It had three pairs of flapping wings of gull like shape, the front and rear each being of half the area of the center plane and these operated at equal intervals, counter to one another. Each plane was covered with aluminum membranes, which closed on the down stroke.

In 1910-1911 the brothers were involved with the design of the Webb-Peet rotary engine, an example of which was exhibited by Weston, Hurlin and Co. at Olympia in 1911.

SCOTTISH AVIATION Co., *Caledonia* and *Dart* monoplanes. (185 Hope St., Glasgow)

This company published a prospectus on 14 December 1911 stating, among other things, the acquisition of Barrhead Aero-

Scottish Aviation Co. *Caledonia* monoplane flown at Barrhead near Glasgow in 1911.

drome and an agreement to build Avro-type aircraft. A.V. Roe was himself on the list of directors. The managing director, Frederick Norman, reported in *The Aero* on 9 October 1911, that an Avro-type aircraft was nearing completion and that a Scottish-built Farman type pusher biplane, already flying, was to be re-engined with a 50hp Alvaston, built at Glasgow, to make it eligible to compete for the £1,000 prize on offer for a flight from Edinburgh to Glasgow by an all-Scottish machine.

The first aircraft, of which illustrations have been traced, is the *Caledonia* monoplane, which was flying at Barrhead from November 1911 piloted by R.W. Philpott, the company pilot. Norman was the designer of this single-seat tractor monoplane, with triangular section fuselage and twin skid, four wheeled undercarriage. It was powered by a 35hp JAP eight-cylinder vee air-cooled engine.

A Blériot with a modified undercarriage and skids, also fitted with a 35hp JAP engine, appeared for the first time on 3 February 1912, and became known as the SAC *Dart*. The two monoplanes and the Avro and Farman type biplanes were destroyed in a fire at Barrhead on 13 April 1912. This set back the prospects for growth of Scottish Aviation and no further reports of flying at Barrhead or of the company followed this disaster.

SAC *Dart* was a much modified Blériot monoplane flown at Barrhead in 1911 and destroyed by fire with the other SAC machines. The pilot is the comedian Harry Tate.

Seddon tandem biplane with a structure of tubular rings at Wolverhampton in 1910. The machine was too heavy and did not fly.

SEDDON biplane (Lt. J.W. Seddon RN, and A.G. Hackett, Wolverhampton)

This tandem biplane, jointly designed by Seddon and Hackett, was possibly the world's largest at the time of its appearance in the late summer of 1910, at Dunstall Park flying ground, Wolverhampton. It was built by Accles and Pollock Ltd., tube manufacturers of Oldbury, Worcestershire and employed mainly circular members rolled from tube in its construction. About 2,000 feet of tubing was used.

The machine was a twin-engined tractor with a front biplane elevator, with two rudders between each pair of main wings, and a fin aft of the front wings. Separate fuel tanks and radiators were provided for each engine, these being mounted on either side of the crew enclosure, intended to carry six people amidships.

Seddon carried out an initial run on 7 November 1910, but testing was halted when a wheel collapsed. The machine was extremely heavy and did not fly, and was later scrapped.

Power: Two 65hp NEC four-cylinder inline water-cooled two-stroke engines driving Beedle propellers by chains.

Data	Area	1,000 sq. ft
	Weight	2,600 lb.

SHARPE biplane (J.P. Sharpe)

A full size version of an electrically powered model was reported to be under construction in November 1909.

A photograph of the model in flight showed an equal span pusher biplane with wings tapering in chord, the widest section being at the tips. Control was by front elevator and rear mounted rudder and dart shaped tailplane. The 'butterfly' shape was provisionally patented. No further reports have been traced.

SHAW glider (James Shaw, Whalley New Rd., Blackburn, Lancashire)

A monoplane glider of Blériot/Antoinette type was reported constructed by Shaw in 1910.

Data	Span	13ft 7in
	Length	21ft

SHEFFIELD monoplane (H. Sheffield, 86 Herries St., Queen's Park, London W.10)

Nothing more was heard of this machine reported to be under construction in 1910-1911.

Sheffield and District Aero Club glider flown under tow in 1910.

SHEFFIELD & DISTRICT Aero Club gliders

A tailless biplane glider of 42ft span was illustrated in flight under tow at Tinsley on 16 July 1910. In May 1911 it was presented to the Conisbrough and District Aeroplane Society. A member of the society, S.H. Dewsnap, constructed a full-sized biplane glider with tail. This was reported in towed flight on 13 May 1913, with two small children aboard, when it reached a height of 20ft.

SHORT Bros. aircraft

The two younger brothers, Eustace and Oswald began constructing balloons in 1900 at Hove, but moved to Battersea railway arches in 1903. In November 1908 the partnership, which now included the eldest brother Horace, was established to build heavier-than-air aircraft. Their first products were an unsuccessful biplane for F.K. McClean, and a more successful machine for J.T.C. Moore-Brabazon. These were followed by a license agreement with the Wright brothers to build six Wright Flyers, the demand coming from would-be British aviators, headed by C.S. Rolls, who also ordered a glider of Wright type on which to gain his first experience of flight.

This work established Short Bros. as the first aircraft manufacturers in the world and necessitated an increase in the work force. A new factory was established at Leysdown, also known as Shellbeach, on the Isle of Sheppey, in March 1909 for aircraft manufacture, leaving Battersea to continue for a period beyond 1914 on balloon and airship work. The following year the factory was reestablished a few miles away at Eastchurch on ground providing a more suitable airfield. The increased seaplane work in 1913, necessitated a more convenient site and an enlarged factory, which was built on the banks of the River Medway at Rochester. Shorts became the major contractor for seaplanes for the RNAS in the Great War and specialized mainly in marine aircraft until the demand for these ceased. The company remains in being, but has been located in Belfast for many years resulting from major rearrangements of the industry following the Second World War.

The early products of the company were recorded in numerical sequence in an order book; the number, prefixed by the letter S, indicated the sequence in which work was initiated, but was not related to the first flight or delivery date. The book included orders for aircraft of types other than Short's designs, also major spares orders were entered, creating breaks in the continuity of the constructors number sequence.

The particulars of the aircraft follow in groups of related type and are not in strict numerical order. The naval serial numbers and Admiralty type numbers are also quoted to facilitate identification; the latter became the more usual means of identification in the later part of the period covered.

Left: Sheffield Model Aero Club glider built by SH Dewsnap.

Short S.1. The brothers' first aircraft was built at Battersea and shown at Olympia part finished in March 1909. The machine failed to fly when on test at Shellbeach.

SHORT AIRCRAFT.

Constructors Number	Type	Details of Order	Date/ Service	Service Number
S.1.	Pusher biplane	FK McClean No.1	January 1909	
S.2.	Pusher biplane	JTC MooreBrabazon	April 1909	
S.3.	Short-Wright No.1	CS Rolls No.2 McClean No.2	March 1909	
S.4.	Short-Wright No.2	A Ogilvie	March 1909	
S.5.	Short-Wright No.3	FK McClean	March 1909	
S.6.	Short-Wright No.4	M Egerton A Ogilvie	March 1909	
S.7.	Short-Wright No.5	CS Grace	March 1909	
S.8.	Short-Wright No.6	E Pitman CS Rolls – A Ogilvie	March 1909	
S.9.	Short-Wright Glider	CS Rolls No.1	May 1909	
S.10.	Huntington-Dunne Triplane		April 1909	
S.11.	Dunne D5 Biplane	Blair-Atholl Syndicate	June 1909	
S.12.	Biplane glider	CS Rolls No.2, not completed.	August 1909	
S.13.	Biplane	M Egerton twin propellers, not completed	December 1909	
S.14.	Spares for Ogilvie Wright biplane		December 1909	
S.15.	Monoplane	GC Colmore. 60hp Green, not completed	December 1909	
S.16.	Spares 2 Wright propellers	CS Grace	January 1910	
S.17.	Spares 2 propellers	JTC Moore-Brabazon	January 1910	
S.18.	Biplane as S.13	GC Colmore, not completed see S.26	January 1910	
S.19.	Spares	A Ogilvie	January 1910	
S.20.	Biplane	as S.13, not completed	January 1910	
S.21.	Spares & launching tower & rail	CS Rolls	February 1910	
S.22.	Propellers & other spares	CS Grace	February 1910	
S.23.	Propeller for S.13	M Egerton	February 1910	
S.24.	Small biplane	CS Rolls No.3. 35hp. Green converted to powered glider	March 1910	
S.25.	Repairs for CS Grace			
S.26.	Pusher biplane	FK McClean No.3 used by GC Colmore		
S.27.	Pusher biplane	CS Grace No.2 FK McClean		
	Rebuilt as Tandem Twin	McClean No.11	RNAS	4/B4
S.28.	Pusher biplane	Moore-Brabazon No.6		66
	McClean No.5	RNAS		
S.29.	Pusher biplane	CS Grace No.3, lost in Channel	December 1910	
S.30.	Pusher biplane	Not built		
S.31.	Pusher biplane	Not built		
S.32.	School biplane	CS Grace McClean No.8 rebuilt as his No.14	RNAS	904
S.33.	School biplane	McClean No.13A floatplane No.13B		
S.34.	Biplane	McClean No.6	RNAS	1/B1/T1/1
S.35.	Pusher biplane	M Egerton, rebuilt as Triple Twin S47 11	December 1911	
S.36.	Tractor biplane	McClean No.12		
S.37.	Monoplane.	Rebuild of Birdling McClean No.9		
S.38.	Pusher biplane	Delivered 23 May 1911	RNAS	2/B2/T2/2
S.39.	Triple Twin biplane	McClean No.10		
	Rebuilt as S38 type.	Contract CP31105/13/X2087	RNAS	B3/T3/3
S.40.	Biplane	FK McClean, not completed, rebuilt as S58		
S.41.	Tractor biplane hydro		RNAS	H1/10
S.42.	Monoplane	FK McClean	RNAS	M2/8
S.43.	School biplane	as S32, for War Office	RNAS	H5/401
S.44.	School biplane	as S32, for War Office	RNAS	402
S.45.	Tractor biplane		RNAS	T5/5
S.46.	Monoplane.	Twin engines 'Double Dirty'	RNAS	H2/12

S.47.	Triple Twin tractor biplane	'The Field Kitchen'	RNAS	T4/4
S.48.	No record			
S.49.	Tractor biplane	as S45, initially for War Office	RNAS	423/1268
S.50.	Tractor biplane	as S45, initially for War Office	RNAS	424/1279
S.51.	Tractor biplane	as S45	RFC	413
S.52.	Tractor monoplane hydro	two 140hp Gnôme, not completed	RNAS	H3/14
S.53.	Tractor biplane	for FK McClean, two 70hp Gnôme, not built		
S.54.	Tractor biplane hydro folder		RNAS	H6/19
S.55.	S38 type nacelle biplane		RNAS	28
S.56.	Improved S41 biplane	Contract CP43037/13	RNAS	H8/20
S.57.	Improved S41 biplane	Contract CP43037/13	RNAS	H9/21
S.58.	Nacelle side-by-side biplane	rebuilt. Contract CP65176/14	RNAS	904
S.59.	S38 type nacelle biplane	M Egerton, impressed	RNAS	190 (?)
S.60.	Tractor biplane hydro		RNAS	42
S.61.	S38 type nacelle biplane		RNAS	34
S.62.	S38 type nacelle biplane		RFC	446
S.63.	S38 type nacelle biplane	War Office not completed		
S.64.	Tractor folder seaplane	Contract CP35829/13	RNAS	81
S.65.	Tractor folder seaplane	Contract CP35829/13	RNAS	82
S.66.	S38 type nacelle biplane	Contract CP36191/13	RNAS	62
S.67.	S38 type nacelle biplane	Contract CP36191/13	RNAS	63
S.68.	Seaplane for Round Britain Contest	McClean No.15		
S.69.	Tractor seaplane	Admiralty Type 74. Contract CP38479/13	RNAS	74
S.70.	Tractor seaplane	Admiralty Type 74. Contract CP38479/13	RNAS	75
S.71.	Tractor seaplane	Admiralty Type 74. Contract CP38479/13	RNAS	76
S.72.	Tractor seaplane	Admiralty Type 74. Contract CP38479/13	RNAS	77
S.73.	Tractor seaplane	Contract CP38479/13	RNAS	78
S.74.	Tractor seaplane	Contract CP38479/13	RNAS	79
S.75.	S38 type nacelle biplane	Contract CP36191/13	RNAS	65
S.76.	S38 type nacelle biplane	Contract CP36191/13	RNAS	64
S.77.	S38 type nacelle biplane	Contract CP36191/13, S28 rebuilt	RNAS	66
S.78.	S38 type nacelle biplane	Contract CP31105/13X, S39 rebuilt.	RNAS	3
S.79.	Pusher biplane hydro	Contract CP38479/13	RNAS	80
S.80.	Nile pusher biplane hydro	FK McClean No.16		
	Rebuilt for Admiralty	Contract CP65176/14	RNAS	905
S.81.	Pusher biplane seaplane	Gunbus. Contract CP01764/13	RNAS	126
S.82.	Tractor folder seaplane	Contract CP53783/13	RNAS	119
S.83.	Tractor folder seaplane	Contract CP53783/13	RNAS	120
S.84.	Tractor folder seaplane	Contract CP53783/13	RNAS	121
S.85.	Tractor folder seaplane	Contract CP53783/13	RNAS	122
S.86.	S38 type nacelle biplane	rebuilt	RNAS	1
S.87.	Tractor folder seaplane	Admiralty Type 135. Contract CP30774/14	RNAS	136
S.88.	Tractor folder seaplane	Admiralty Type 135. Contract CP30774/14	RNAS	135
S.89.	S38 type nacelle biplane	Contract CP38688/14	RNAS	152
S.90 - S.95	Tractor biplane seaplane	Admiralty Type 166. Contract CP37385/14	RNAS	161-166
S.96 - S.100.	Tractor biplane seaplane	Russian Navy, not built		
S.101 - S.105.	Tractor biplane seaplane	Gun machine, not built		
S.106 - S.107.	Tractor biplane seaplane	Admiralty Type 184. Contract CP46634/14	RNAS	184-185
S.108 - S.115.	Tractor biplane seaplane	Admiralty Type 74. Contract CP02129/14	RNAS	811-818
S.116 - S.118.	Tractor biplane seaplane	Admiralty Type 830. Contract CP57105/14	RNAS	828-830
S.119 - S.121.	Tractor biplane seaplane	Admiralty Type 830. Contract CP57105/14	RNAS	819-821
S.122 - S.127.	Tractor biplane seaplane	Admiralty Type 827. Contract CP57105/14	RNAS	822-827
S.128.	Tractor biplane seaplane	Admiralty Type 74. Contract CP46634/14	RNAS	183
S.129 - S.138.	Tractor biplane seaplane	Admiralty Type 184. Contract CP57105/14	RNAS	841-850

SHORT biplane S1

An order for an aeroplane from F.K. McClean, early in 1909, was the start of aircraft manufacture by the Short brothers. No.1 was designed by Horace Short and was built at Battersea, in quick time, to be exhibited as an uncovered structure at the Olympia Aero Show of March 1909. After the Show it was taken to Shellbeach for completion and covering. The first trials were in September with the Nordenfeldt car engine, later used in Short-Wright No.3 with which it failed to leave the starting rail. A Wright engine, made by Barriquand et Marre in France, arrived in October, but after McClean made several attempts to take off, damage occurred and the machine was discarded. Although *The Aero* later reported that 'the machine had the ability to fly' there is no evidence to support this statement.

Short No.1 was a three bay biplane with front biplane elevator incorporating a fixed fin. The wings were double surfaced and heavily cambered, with the chords of the outer bays considerably increased in width to provide a flexible surface for lateral control. Four rudders were pivoted between these flexible panels for directional control.

The pilot's position was ahead of the lower wing front spar and the engine, which drove twin pusher propellers by chains and shafts. A narrow radiator filled the gap between the wings on the starboard side of the engine. The chassis was a long twin skid structure without wheels.

Power: 30hp Nordenfeldt four-cylinder inline water-cooled car engine weighing 600 lb.
30hp Wright four-cylinder inline water-cooled made by Barriquand et Marre
in France driving two 10ft diameter propellers by uncrossed chains and shafts.

Data

Span	40ft
Length	24ft 7in
Area	576 sq. ft
Weight allup	1,200 lb.

SHORT biplane S2

Horace Short designed his second machine to Moore-Brabazon's order of April 1909, for a machine on which to attempt to win the *Daily Mail Prize* of £1,000, for a circuit of one mile on a British aircraft flown by a British pilot. The aircraft was built at Shellbeach and was ready to receive the engine in September. For the initial trials a Belgian Vivinus engine was fitted, until the British-made Green became available in early October. However two flights were made with the heavy Vivinus, including one of nearly a mile on 27 September 1909. After repair of damage and fitment of the Green, Moore-Brabazon flew for a mile to win the prize on 30 October 1909. S2 was then flown successfully from Shellbeach to the new Short works at Eastchurch, a distance of four and a half miles, on 7 January 1910 and on 1 March 1910 flew nineteen miles in thirty-one minutes before the crankshaft broke. This flight won him the British Empire Michelin Trophy of 1910. The aircraft was exhibited on the Royal Aero Club stand at Olympia in March 1910.

Short S.2 on which Moore-Brabazon won the *Daily Mail* prize for a circular flight of one mile in October 1909.

Short S.2 the final version with extended tail.

Short S2 was a twin pusher biplane with three-bay wings outboard of the twin girder chassis. The wings were mainly parallel in chord, of higher aspect ratio than on S1 and with tips curving to a point, where a strut was provided to mount the 'balancers' for lateral control. These each consisted of two surfaces of different areas on an arm with a central pivot and these worked differentially controlled by foot pedals.

The main girder structure was carried upwards at the front to mount the biplane elevator with rudder mounted behind. A fixed tail was carried on wire braced top and bottom booms. In the early part of 1910 the tail was modified; twin booms tapering in side elevation, were fitted with a much extended fin and fixed tail surfaces.

The pilot sat on the leading edge of the lower wing on the starboard side of the centrally mounted engine with a space to port for a passenger; a pig was carried on one occasion for a wager to prove that pigs could fly! The single narrow radiator was later replaced by two of greater width and lower height on the lower wings in line with the chassis members.

Power: 32 1/2hp (?) Vivinus four-cylinder inline water-cooled reported 'to weigh 200 lb. more and give 15hp less than the Green engine'.
50/60hp Green four-cylinder inline water-cooled driving twin pusher propellers by uncrossed chains and shafts.

Data

Span	48ft 4in
Length	32ft
Area	450 sq. ft
Weight allup	1,485 lb.
Speed	45mph

Short-Wright biplane S.5 was the third of these built for F.K. McLean.

SHORT-WRIGHT biplanes S3-S8

Six of these biplanes were built at Shellbeach under license from the Wright brothers, to the same design as the Wright Flyer used by the Wrights for demonstrations in the U.S.A. and in France. Some changes however were made to the last two which were built.

Four aircraft were completed but waiting for engines by July 1909. The third, for McClean, was tested with a Nordenfeldt car engine, but could not maintain flight after launching. The car engine was replaced with one of Wright type built by Barriquand et Marre. No.1 was tested in October by C.S. Rolls, with a Wright engine made in France by the Léon Bollée company of Le Mans. Some damage was sustained during the initial tests, but this was soon repaired and flight was achieved on 1 November 1909.

The six machines were delivered to private owners as follows:

S3 No.1	C.S. Rolls	1 November 1909
S4 No.2	A. Ogilvie	3 September 1909
S5 No.3	F.K. McClean	20 November 1909 (McClean No.2)
S6 No.4	M. Egerton	21 December 1909
S7 No.5	C. Grace	14 February 1910
S8 No.6	C.S. Rolls	14 February 1910 to A. Ogilvie later

The Wright Model B biplane was of equal span with warping of the two outer bays. The twin skid chassis was extended forward to mount a biplane elevator with 'half-moon' fins between. The twin rudders were mounted on a single width, wire braced tail boom on the first to fourth aircraft; on the fifth and sixth the tail boom was of double width. The water-cooled engine was mounted on the right of the lower center section and drove twin pusher propellers by chain and shafts mounted outboard between the wings. The pilot and passenger sat to the left of the engine, with the pilot in the center. The machine was launched from a rail and landed on the skids.

Ogilvie later modified his second machine at Camber Sands to delete the front elevator and its supporting structure, as on the later version known as the 'headless' Wright. The single elevator was placed behind the rudders and blinker type fins were added to the front skids. The machine was operated on wheels and continued in use until August 1914. It is reported to have flown with three passengers and to have been tried unsuccessfully as a floatplane.

Power: 30hp Wright four-cylinder inline water-cooled driving twin pusher propellers by chains and shafts.
French made by Léon Bollée or Barriquand et Marre.
50hp NEC four cylinder vee water-cooled two-stroke fitted to No.6 September 1910.

Data		
	Span	41ft
	Area	515 sq. ft
	Length	29ft
	Weight	885 lb.
	Weight allup	1,200 lb.
	Speed	50 mph

SHORT-WRIGHT glider S9 (CSR.1)

This was a copy of the Wright glider of 1902-1903, the structure of which was made at Battersea for C.S. Rolls in the spring of 1909, and was then taken to Leysdown for covering. The machine was completed by the end of July and Rolls commenced testing immediately, using a rail on a slope near Leysdown. A number of glides were made terminating on 10 October 1909 when, having acquired sufficient experience of handling the glider, Rolls offered it for sale in March 1910. A

Short-Wright biplane S.6 acquired by Ogilvie after Rolls died. It was much altered by him at Camber Sands and was flown successfully until 1914.

Short-Wright glider S.9 as delivered to Rolls in July 1909 for practice prior to completion of his powered Short-Wright.

Short-Wright glider S.9 at Leysdon in August 1909 after rear rudder and 'halfmoons' had been fitted.

second order for a glider, S12, was apparently not completed but may have covered work for a later powered glider.

The glider had equal span two bay wings with warping control, a front biplane elevator and twin rear rudders. It differed from the original Wright in that the pilot sat upright controlling the machine with a lever on the right hand side operating the elevator and one on the left operating both the warp and the rudders. Fins, known as 'half-moons', were added between the biplane elevators during the time that the glider was in use.

Data	Span	32ft 10in
	Area	325 sq. ft
	Length	18ft

SHORT biplane. S13 (Refers also to S18 and S20)

A biplane for the Hon. Maurice Egerton, of which there is no evidence of completion, was on order from December 1909. A further entry, S23, in March 1910 covered the supply of a propeller for this machine. It was to be powered by a Wolseley engine driving two propellers and to have a detachable skid. S18 and S20 were to be similar machines for C.G. Colmore and another.

SHORT biplane S24

This machine, usually referred to as Short No.3 in the press, was ordered by C.S. Rolls (his CSR3), and was the last Short machine built at Battersea, before the move to Eastchurch. It appeared at Olympia in March 1910 but failed to fly, when tested briefly by Rolls, soon after the Aero Show closed.

The machine was a smaller and lighter version of S3, but with a single pusher propeller, driven directly by the engine, which necessitated twin tail booms to clear the propeller. The rail and derrick system for launching was abandoned in favor of a four wheeled undercarriage, which could be retracted under spring tension, to permit landing on the skids. The engine was mounted on the centerline in the top half of the wing gap with the main fuel tank below, flanked by twin radiators. A gravity fuel tank and oil tank was mounted above the engine, for which a starting handle was provided. The pilot sat ahead of the wing leading edge between the top longerons.

The tail surfaces were fixed, and consisted of a long rectangular fin mounted at the center of the tailplane, which was of high aspect ratio and was adjustable for incidence on the ground only. The front biplane elevator was without camber and, with the rudder mounted separately behind, was similar to S2. The balancers, mounted on a single strut at the extreme tips of the wings, were single panels of spring tensioned fabric and were controlled by the right-hand lever; foot pedals controlled the rudder, and a separate lever on the left side controlled the front elevator. The main wooden girder structure was similar to that of S2, with flat strip bracing twisted at the point of crossover.

The lack of success in the short time that Rolls had available, before leaving for the Nice Meeting, caused him to discontinue trials, and to dismantle the aircraft, parts of it being used later in the machine known as the Rolls Powered Glider (RPG) costed against Rolls' order for S12.

Power: 35hp Green four-cylinder inline water-cooled driving a Short 7ft 6in diameter propeller direct.

Data	Span	35ft 2in over balancers
		31ft 8in to wingtips
	Chord	5ft 4in
	Gap	4ft 4in
	Area	282 sq. ft
	Area tailplane	21 1/2 sq. ft
	Area elevator	55 sq. ft
	Area rudder	10 sq. ft
	Area balancers	12 sq. ft each
	Length	31ft.
	Weight	615 lb.
	Weight allup	857 lb.
	Speed	45-50 mph
	Price	£650

SHORT biplanes, pusher types S26-S29

The Short brothers discontinued the manufacture of Wright type aircraft and changed to those similar to the Farman/Sommer types. The change coincided with the move to Eastchurch and identification by construction numbers beginning at S26. The first production was a batch of four aircraft S26-S29, which

Short S.26-S.29 and 34 were fitted with various engines. This is C.S. Grace's S.27 with French ENV engine without wing extensions.

Short S.27 with Gnôme engine and extended wings.

Short S.27 (Grace No.2) with 60hp ENV F.

were usually referred to as the S27 type, this being the machine which was most successful in the hands of Cecil Grace. The first of the type appeared in June 1910.

The S27 type was a typical pusher biplane of the period with either equal span wings or with top wing extensions. The undercarriage was a simple two wheel, two skid arrangement and the tail was supported by hoops of rattan cane under the booms, although a sprung tail skid was soon fitted. Single acting ailerons were originally fitted to all four wings but Grace had these removed from the lower wings, which suffered from flexibility and which were stiffened at the same time.

The front elevator was split in two and originally extended outboard of the front booms but these extensions were later removed. The rudder was originally one rectangular surface below the tailplane but, when this proved to be inadequate, a similar rudder was mounted above the tailplane. A section of the tailplane trailing edge was hinged and controllable by the pilot as a trimming control.

Water-cooled Green and ENV engines were used originally but later the air-cooled Gnôme rotary was fitted. The four aircraft of the type were as follows:

S26. 35/40hp Green: For G.C. Colmore, originally under No.S18, not completed. The machine crashed but was rebuilt and passed to F.K. McClean (McClean No.3), who loaned it to the Navy for pilot instruction. It survived until April 1911, but was sluggish on the controls and known as 'The Dud'. (Naval biplane No.1)

S27. 60hp ENV French-made type F Series 1 No.4: C.S. Grace No.2. Flown by him at Wolverhampton, Bournemouth, Lanark and elsewhere from June 1910. After Grace was lost in December 1910 the aircraft was sold to McClean and converted to become the Tandem Twin with two 50hp Gnôme engines. (McClean No.7 became No.11 after conversion)

S28. 35/40hp Green also 50hp Gnôme: For J.T.C. Moore-Brabazon, but transferred to F.K. McClean after Brabazon gave up flying in July, and allocated to the Navy for pilot instruction. (Naval Biplane No.2) Converted for longer range as S38 from May 1911. (Moore-Brabazon No.6. McClean No.5)

S29. 60hp ENV British-made type F: Originally a spare machine but became C.S. Grace No.3 as an all-British machine, to compete in the Baron de Forest Contest. This was the machine in which Grace was lost on 22 December 1910 in the Channel, when returning to England for a second attempt for the Prize

Data	Span	34ft 2in (46ft 5in with extensions)
	Chord	6ft 9in
	Gap	6ft 4in
	Area	480 sq. ft (517 sq. ft with extensions)
	Length	40ft 6in
	Weight	1,000 lb.
	Weight allup	1,400 lb.
	Speed	40 mph

Short S.29 was similar with a British ENV.

Short S.28. McClean had a Green engine fitted but trouble with this prevented his attempt on the Baron de Forest prize. A Gnôme was refitted later and a nacelle as on S.35 was added.

Short S.27 Tandem-Twin. A major conversion of the original S.27 was acquired by McClean and loaned by him to the Admiralty.

S27. Tandem Twin biplane

Although a major reconstruction was carried out to S27 to fit a second engine, this machine received no separate construction number. The aircraft was built to investigate the effect of rotating airscrews on stability and control and was the subject of patent No.22675/1911 by Horace Short. The first flight took place on 29 October 1911 by McClean, who acquired this as McClean No.11, as well as many of the early Short aircraft. The Tandem Twin was loaned to the Navy for instructional use from December 1911, but was crashed on 11 March 1912 by Lt. Samson. It was identified as Serial No.4 in the first Admiralty series.

The machine was flown both with and without extensions, and the main structure was unchanged apart from strengthening of the undercarriage. A nacelle was added to house the crew side by side, between the two engines, to which access was gained through an aperture in the floor, the draught through which inspired the nickname '*Vacuum Cleaner*'. Two extra rudders were added on top of the original cambered tailplane.

Data		
	Span	34ft 2in (46ft with extensions)
	Chord	6ft 9in
	Gap	6ft 4in
	Area	480 sq. ft (571 sq. ft with extensions)
	Weight	1,800 lb.
	Weight allup	2,100 lb.
	Speed	55 mph

S32. Tractor biplane

This machine was built for Cecil Grace towards the end of 1910, originally as a tractor version of the S27 with a 60hp ENV type F water-cooled engine. Reports appeared that Grace was out at Eastchurch with the aircraft on 12 November 1910, and again after alterations with a mechanic on 19 November 1910. On 27 November 1910 a pair of calipers lying around on the ground, flew into the propeller and damaged it during engine running. No further report on this machine appeared and Grace left for Swingate on 11 December 1910, for his Baron de Forest attempt from which he did not return. No illustration of this machine has been traced, but it was almost certainly similar to the Gnôme engined S36.

S32. Pusher biplane, dual-control type (Refers also to S33, S43 and S44)

S32 was acquired by McClean, and was rebuilt for him as a dual control side by side pusher, which he flew for the first time on 29 July 1911, before lending it to the Navy (McClean No.8).

S33 was a similar machine, which he lent for a short time to the Army for training Territorials. Two more of the type, S43 and S44 were built for the Central Flying School at Upavon becoming Nos.401-402 in the service. These machines were of the same basic S27 type with wing extensions and a 70hp Gnôme rotary engine. An additional rudder was fitted above the tailplane. Originally, No.13A in McClean's fleet, S33 was converted to a floatplane for a flight along the Thames on 10 July 1912, then identified as No.13B.

In early 1913 S32 was again rebuilt for McClean to a new standard similar to the S38 type, then becoming McClean No.14. The machine was provided with a nacelle with tandem seating for two, with a structure at the prow carrying the front elevator. The tail booms were parallel in elevation, but tapered slightly in plan. A rectangular tailplane and elevator were mounted on the top boom and twin rudders pivoted between the booms. A 70hp Gnôme was fitted. The wings were extended by a further bay each side to a span of 70ft 6in, with dihedral and taper on the leading edge, outboard of the second pair of interplane struts. Kingposts supported the upper wing extensions.

S32 was intended to be a seaplane to take to Egypt to explore the River Nile, but in this form it was unsatisfactory, and was replaced by S80. S32 was passed to the Admiralty and later rebuilt as S38 with 52ft span wings and duplicated landing wheels. It became serial No.904 and was used for training purposes at Hendon.

Power: 70hp Gnôme seven-cylinder air-cooled rotary

Data			After Rebuild	as S38
	Span	46ft 5in	70ft 6in	52ft
	Area	517 sq. ft	772 1/2 sq. ft	580 sq. ft
	Length	40ft 6in		
	Weight	1,150 lb.		
	Weight allup	1,540 lb.		
	Speed	45 mph	38 mph	

Short S.33. A dual control seaplane version of the S.27 type. S.43-S.44 were similar but landplanes.

S33. Pusher biplane, twin float seaplane

This was a further variation of the S27/S32 type, bought by McClean and briefly loaned to the Army, but later converted for use as a seaplane (McClean No.13B). On 31 May 1912 McClean flew the seaplane from the River Swale at Harty Ferry and, on 16 July 1912, flew under all the bridges over the River Thames, including passing under the top span of Tower Bridge, finally alighting on the river between Charing Cross and Westminster Bridges, to attend a reception given by Lord Northcliffe at Carmelite House. Police permission to fly back was refused, the machine was then taxied down river but was damaged at Shadwell Basin and dismantled.

Short S.34. An instructional machine for the RNAS.(B.1, T.1 and No.1).

S34. Pusher biplane

This was an 'improved S27' fitted with a 50hp Gnôme and a nacelle for the crew of two in tandem. It was bought by McClean (his No.7) and loaned for instructional use as Naval Biplane No.1. It was later sold to the Admiralty for £500, becoming successively B1, T1 and serial No.1. A similar machine, S38, was bought by the Admiralty to replace S26, and became T2 and serial No.2. Increased tankage was fitted later to both machines for duration tests.

S35. Pusher biplane (1911 type)

This machine, also referred to as of 'improved S27 type', differed in certain respects it. The most noticeable feature was the nacelle to protect the crew. It was fitted with top wing extensions, single acting ailerons on the top wing only and a single sprung tail skid.

Short S.35 built for the Hon. Maurice Egerton in 1911.

An additional rudder was fitted above the tailplane as on S32. The aircraft was built to the order of the Hon. Maurice Egerton, and was flown by him on 9 March 1911.

Power: 50hp Gnôme seven-cylinder air-cooled rotary driving 8ft 6in diameter propeller.

Data		
	Span top	46ft
	Span bottom	32ft
	Chord	6ft 9in
	Gap	6ft 4in
	Area	517 sq. ft
	Area tailplane	65 sq. ft
	Area front elevator	30 1/2 sq. ft
	Area rear elevator	26 sq. ft
	Area rudders	31 sq. ft
	Height	11ft 6in tail down
	Weight	1,100 lb.
	Weight allup	1,540 lb.
	Speed	40 mph

Short S.38 after many improvements became the prototype of a standard biplane for the RNAS which was produced by Short and other manufacturers in wartime. This is S.62 which became RFC 446.

S38. Pusher biplane (Refers also to S55, 58, 59, 61-62, 66-67, 75-78, 86 and 89)

S38 began life basically as an S27/32 with top wing extensions and a 50hp Gnôme and was initially used for training by the Navy. Later it was used for experimental work as an amphibian, fitted with three streamlined air bags and a 70hp Gnôme engine. The aircraft was flown off the platform deck of HMS *Africa* by Samson on 10 January 1912, the first British shipboard takeoff and the first of a number of experiments, until it was wrecked on 9 July 1912.

Short S.38 amphibian made the first takeoff from a British warship.

Short S.38. This is constructors No.S.75 (RNAS No.65) a typical example.

The aircraft was rebuilt with the inevitable modifications and appeared in its new form as the prototype of a new type for the Navy, known as the S38 type. This was followed by further revisions, which became standardized in a production run of a dozen aircraft. Certain features of the S38 type appeared in other Short types. After the first batch built by Shorts, production was carried out by contractors Pemberton Billing, who built twelve, Nos.158-091 and White and Thompson, later the Norman Thompson Flight Co., who also built twelve, Nos.314-348 and 843-439, the last of which was delivered in June 1916.

The type featured a long tandem two-seater nacelle which, on the production aircraft, carried the smaller front elevator on a structure on its nose, to replace the separate front booms of the prototype. The wing gap was decreased and the span increased; the extensions were braced by wires and kingposts and had tapered leading edges and dihedral. The tailplane and rear elevator were placed on the top booms and twin rudders, with rounded profiles, were pivoted below. The chassis was fitted with twin main wheels on shortened skids from mid 1913. The first production was flown on 4 November 1912 by Lt. W. Parke. One aircraft, S62 became No.446 at the Central Flying School but the majority were acquired by the Navy.

Power: 70hp Gnôme seven-cylinder air-cooled rotary. The prototype was fitted with 50hp Gnôme which was used on some other aircraft, as was the 80hp later.

Data

Span	52ft
Area	500 sq. ft
Length	35ft 6in
Weight	1,050 lb.
Weight allup	1,500 lb.
Speed	58mph
Endurance	5hr

Production.

Shorts Construction No.	**RN Serial No.**	
S38	2 (Was T2)	Prototype 50 & 70hp Gnôme
S78	3	70hp Gnôme
S55	28	50 & 70hp Gnôme
S61	34	80hp Gnôme
S66	62	50 & 70hp Gnôme
S67	63	
S75	65	
S76	64	
S77	66	80hp Gnôme. Maxim gun fitted No front elevator
S89	152	80hp Gnôme Side by side seating
S58	904	
S62	446	War Office aircraft for CFS
S59	-	Shorts/M Egerton aircraft
S86	1	Reconstruction of McClean No.6 Rebuilt by December 1913 with 70hp. Gnôme. Float undercarriage fitted March 1914 used for trials of GRW wheel attachment, later reverting to land undercarriage. Deleted February 1916.

S39. Triple-Twin tractor/pusher biplane (Refers also to S35 rebuild)

This aircraft was built to test Horace Short's ideas to improve control at low speeds, by designing for slipstream over all control surfaces. The twin engines were also a safety measure, since the aircraft could fly successfully on one. McClean flew

Short S.39 Triple-Twin a development to improve control by providing greater coverage of the control surfaces by the slipstream.

the S39 solo for the first time on 18 September 1911 and also with Samson as a passenger the same day. (McClean No.10).

The aircraft was a strengthened form of earlier Short biplanes, with a central nacelle and side-by-side seating. It was necessary to place the front booms closer together, to provide clearance for the outboard propeller tips; the front elevator consequently extended beyond the booms. The tail booms were parallel in plan and with slight taper towards the tail in elevation. The tailplane, with rear elevator, was mounted on the top booms, with three rectangular rudders trailing behind their last uprights. A strengthened undercarriage with diagonal bracing struts and twin tail skids were fitted.

The wings were originally without extensions, which were added later, and were fitted with single acting ailerons both top and bottom, including to the extensions when these were fitted. The front engine drove the outboard counter rotating tractor airscrews by chains. The rear engine was mounted with the propeller in front on an overhung mounting. Two cylindrical fuel tanks were originally mounted below the top center section, but a further pair were added in December 1911, in similar positions on the next pair of interplane struts at the time when the extensions were first fitted. The extensions were removed in October 1912 and replaced by wings of 50 feet span, these also being fitted with extensions later when the span increased to 64 feet.

The Admiralty purchased S39 in June 1912; identified as T3, it became No.3 later. The machine was entirely rebuilt under Short's No.S78, as a single engined, tandem seated pusher with either 70 or 80hp Gnôme. From June 1913 it was used by the Navy, including wartime service in France, until finally discarded in April 1916.

Short S.39 with extended top wing.

Short S.39 rebuilt to resemble S.38 type and became RNAS 3.

A second Triple Twin was constructed from S35 for the Hon. Maurice Egerton and was flown from April 1912 and may have been rebuilt itself by early 1914.

Power: Two 50hp Gnôme seven-cylinder air-cooled rotaries driving two 8ft 6in diameter tractor propellers by chain and one 8ft 6in diameter pusher propeller direct.

Data		
	Span	34ft later 50ft and 64ft
	Chord	6ft 9in
	Gap	6ft 4in
	Area	435 sq. ft later 650 sq. ft and 745 sq. ft
	Area tailplane	65 sq. ft
	Area front elevator	29 sq. ft
	Area rear elevator	29 sq. ft
	Area rudders	45 sq. ft
	Length	45ft
	Height	12ft
	Weight	1,800 lb.
	Weight allup	2,100 lb.
	Speed	55mph

S40. Pusher biplane

Ordered by F.K. McClean, but apparently not completed and rebuilt as S58.

S79. Pusher biplane

This machine was ordered separately on Contract CP38479/13 and became serial No.80 in naval service. It was apparently a two-seater side by side training machine and probably similar in many respects to S81, in this machine's original or modified form. A large rectangular fin was mounted with most of its area above the tailplane.

Power: 100hp Gnôme Monosoupape nine-cylinder air-cooled rotary.

S80. Pusher biplane, the Nile seaplane

In 1913 F.K. McClean planned an aerial exploration of the River Nile, for which S32 was originally intended, but this aircraft proved to be inadequate for the purpose and a new seaplane was built. This was S80 (McClean No.16), which flew for the first time as a seaplane on 2 October 1913, piloted by Gordon Bell and although basically a four seater, it was soon flown with five aboard. The aircraft arrived in Egypt on 27 December 1913 and was flown from Alexandria to Khartoum by stages, suffering some damage and repairs en route. It was dismantled at Khartoum and returned to England in March 1914. At Eastchurch major repairs and modifications were carried out, including conversion to a two-seater side-by-side layout.

As originally built, S80 was a typical Short pusher biplane with a nacelle and with a front elevator on an outrigger on the nose. There were the usual tailplane and elevator carried on top of the tail booms, which were parallel in plan and elevation. Twin balanced rudders were fitted. The two bay wings carried top wing extensions braced by struts and wires with single acting ailerons on the top wings only. The main floats, parallel in width, tapered to a curved stern and were each mounted by three struts. The tail was supported by two air bags of circular section below the tail booms.

During the rebuild, on return to Eastchurch, the fuel tanks were lowered into the compartment previously housing the rear seats, and the front elevator was removed. In this form McClean presented the machine to the Navy in August 1914, when it was identified as No.905, although never marked as such. In October the original 160hp Gnôme was replaced by one of 100hp. A large rectangular fin was mounted with most of the area above the tailplane. The tail booms were tapered inwards towards the tail bringing the rudders, which were of a new shape, closer together.

Power: 160hp Gnôme fourteen-cylinder two-row air-cooled rotary. Replaced later by 100hp Gnôme Monosoupape nine-cylinder air-cooled rotary with a four-bladed propeller.

Data		
	Span	67ft
	Area	540ft
	Length	33ft 9in
	Weight	2,200 lb.
	Weight allup	3,600 lb.
	Speed	60 mph

Short S.81 with one and a half pounder Vickers gun. It was RNAS No.126.

S81. Pusher biplane. Gun carrying seaplane

A machine generally similar to S80 was constructed for the Admiralty for aerial gunnery experiments, and was first flown at Calshot in May/June 1914. Tests of the Vickers one and a half pounder, Lewis machine gun and Davis six pounder guns were carried out at various times, also tests of a dynamo and searchlight, before the machine, No.126, was discarded in October 1915.

Variations from S80 included rubber sprung floats and wingtip air bags. This machine was the last pusher design to be produced by Shorts.

Power: 160hp Gnôme fourteen-cylinder two-row air-cooled rotary with a four-bladed propeller.

Data As for S80

S42. Tractor monoplane

Shorts produced mainly biplanes, but in 1912 the firm did produce a monoplane after working on a *Birdling* acquired by McClean from the defunct Universal Aviation Co., his No.9. The Short S42 was flown by Samson on 24 February 1912 for the first time, and was acquired by the Navy and identified as M2 later No.8. It was flown as both a land and seaplane and was finally discarded in June 1914.

Short S.80 the Nile seaplane for McClean. No.16 in his fleet of aircraft later taken into the RNAS as No.905.

Short S.42 tractor monoplane similar to the Bleriót XI. It became RNAS No.8.

The machine was similar in general design to a Blériot, as was the *Birdling*, but with an altered undercarriage and engine mounting. The tail surfaces differed by the use of a cambered tailplane, with full width elevator and square shaped rudder with a balance extension. The rear fuselage was uncovered initially, but fabric was applied later. Lateral control was by wing warping.

Power: 50hp Gnôme seven-cylinder air-cooled rotary.

Data		
	Span	29ft 3in
	Area	165 sq. ft
	Length	25ft.
	Speed	55 mph
	Endurance	5hr

S46. Tractor/pusher monoplane., '*The Double Dirty*'

This machine was delivered to the Navy at Eastchurch, receiving serial No.12, and was flown by Samson on 23 September 1912 for the first time. It derived its nickname from the oil emitted by the two rotary engines, in close proximity to the crew members seated side by side between them. It survived some damage, but was not popular and was deleted in August 1913.

No detailed information for the machine is available, but it seems to have been fitted with a short nacelle, similar to that of the Tandem Twin, and widely spaced tail booms to clear the rear propeller.

Power: Two 70hp Gnôme seven-cylinder air-cooled rotaries.

S52. Tractor monoplane, twin-engined hydro

This was a project for the Navy (serial No.14), which was to be built at Eastchurch in 1912. It was to be a four seater and powered by two 140hp fourteen-cylinder two-row rotary engines.

S36. Tractor biplane

This machine was built for McClean to a design evolved from the original S32 for Cecil Grace, but fitted with a rotary engine. It was flown by McClean for the first time on 10 January 1912 (McClean No.12).

The fuselage was a wooden braced girder, originally uncovered behind the two cockpits; the engine and tank bays were closely cowled with aluminum panels, and plywood covering behind in the cockpit area. The fuselage was mounted centrally in the wing gap and rested directly on the twin skid undercarriage, which was carried on four struts ahead of the wings. The two bay wings had separate pairs of interplane struts adjacent to the fuselage, on which connections were made to attachment fittings, protruding from the fuselage sides at the top and bottom longerons. The wings were of S27 type with top extensions and single acting ailerons on the top wing only; there was a gap in the lower center section. The tail unit consisted of a tailplane with divided elevator, on the top longerons, and an unbalanced rudder. The placing of the fuselage in the gap was aimed at avoiding a change of trim with the non-lifting tailplane. Later the fuselage was lowered slightly and the rear portion was covered with fabric.

Short S.36. The basis for a succession of tractor biplanes was built for McClean in 1911 and followed the abandoned S.32 for the late Cecil Grace.

Power: 70hp Gnôme seven-cylinder air-cooled rotary driving a 9ft diameter propeller.

Data		
	Span	46ft 5in
	Area	515 sq. ft
	Length	35ft 6in
	Weight	850 lb.
	Weight allup	1,300 lb.
	Speed	60 mph

Short S.41 land or seaplane was first flown in April 1912 (RNAS H.1 and No.10).

S41. Tractor biplane, land and seaplane and Improved type S41 (Nos.S56 and S57)

S41 was a slightly larger and more powerful version of S36 for the Navy for use as either a land or seaplane. It was first flown by Samson as a landplane on 2 April 1912. The aircraft was soon converted to a seaplane by a complete change of fuselage. It was fitted with twin pontoon type floats and streamlined air bags under the tail and the inboard interplane struts. A board across the floats protected the airscrew from spray. The crew were seated in tandem in the fuselage, which was completely covered, and which was mounted lower in the gap than on S36. The two-row engine was provided with a front support and a lengthened cowl with rounded top.

In mid-1913 the center section gaps were covered and, in the following September, folding wings of increased span were fitted. The machine, originally H1, was reidentified as No.10 and a 140hp fourteen-cylinder two-row Gnôme was fitted in early 1914. A further reconstruction with Salmson engine took place in 1915.

Short S.41 improved versions S.56 and 57 (RNAS Nos.21 and 20).

Short S.51 went to CFS as No.424 in 1912. S.45 and S.50 were similar machines.

The Improved type S41, two of which were ordered for the Navy at the end of 1912, were basically seaplanes, although S57 (No.20) was also flown with the GRW amphibious gear and S56 (No.21) was later used as a landplane.

Gordon Bell tested and delivered the machines to the Isle of Grain in April 1913. Both were used for trials of various equipment and for operations early in the war. S57 survived until 14 February 1916 and S56 until 7 February 1915.

The improvements embodied in the Improved type, included a new wing section, wire and kingpost bracing for the wing extensions and double acting ailerons. The wing floats were narrower and mounted further outboard at the lower wing tips.

Power: 100hp Gnôme fourteen-cylinder two-row air-cooled rotary.

Data		
	Span	50ft
	Area	450 sq. ft
	Length	36ft 6in (landplane)
		39ft (seaplane)
	Weight	1,100 lb.
	Weight allup	1,600 lb.
	Speed	60 mph
	Endurance	5hr

S45. Tractor biplane (Refers also to S49, S51)

A second biplane for the Navy, similar to S36 41 was flown by Lt. Spenser Grey on 23 May 1912 as a landplane, identified as T5, later No.5. When used as a seaplane it was fitted with a central pontoon float, and twin streamlined airbags under the inboard interplane struts only. An airbag was fitted close-up under the tail later. S45 capsized on 4 October 1912 and was deleted.

Short S.45 (RNAS Nos.T.5 and 5) as a central float seaplane.

S45 (No.5) had been fitted with new wings in the autumn of 1912, in which the top wing extensions were braced by wires and kingposts instead of the diagonal struts originally used. Double acting ailerons were also fitted. The engine cowling was modified to a square section matching the fuselage shape and the top cowling line was continued as decking to the rear of the two cockpits.

In this form the machine was the pattern for three landplanes (S49 51) ordered by the War Office for use at the CFS (serial Nos.413, 423 4). No.413 was soon written off in December 1912. The airframes of 423 and 424 were transferred to the Admiralty in August 1914 and reidentified as 1268 and 1279. These were Shorts last landplanes built prewar.

Power: 70hp Gnôme seven-cylinder air-cooled rotary

Data		
	Span	42ft
	Area	450 sq. ft
	Length	35ft 6in
	Weight	1,080 lb.
	Weight allup	1,500 lb.
	Speed	60 mph

S47. Triple tractor biplane, '*The Field Kitchen*'

This development based on earlier tractor biplanes was a twin-engined machine with both engines contained in an extended fuselage under a long cowling. The rear engine was turned round and drove, by chains, the twin propellers mounted on the inboard interplane struts. This engine rotated in the reverse direction to the normally mounted front engine with directly driven propeller. The rear engine was inadequately cooled and the severe heat generated caused the use of the nickname. The crew of two were seated side-by-side.

McClean tested S47 for the first flight on 24 July 1912, and it was soon taken over by the Navy, becoming T4 later No.4. It was used for experiments with wireless transmission but was not repaired after damage on 22 January 1913.

Short S.47. The Triple-Tractor (RNAS Nos.T.4 and 4) was first flown by McClean in July 1912.

Power: Two 50hp Gnôme seven-cylinder air-cooled rotaries. The front engine drove a propeller direct, the rear rotating in the opposite direction drove twin propellers by chains.

Data

Span	48ft
Length	41ft
Area	500 sq. ft
Speed	60 mph

Short S.60 (RNAS No.42) after conversion to land chassis was built in 1913 and written off in France in September 1914.

S53. Twin tractor biplane

This was to be a machine for McClean with two 70hp Gnôme engines which was not built.

S54. Tractor seaplane

This machine was probably similar in most respects to S41, S56 and S57, but fitted with the more powerful two-row engine and funnel type exhaust, which was used on S64, S65 and later type machines. The top wing had broad chord, single acting ailerons with taper on the inboard sections. The aircraft was probably delivered in November 1913 and was wrecked off Felixstowe on the 14 December 1914.

Power: 140 or 160hp Gnôme fourteen-cylinder two-row rotaries.

S60. Tractor biplane, seaplane or landplane

This machine was built as a private venture, and was exhibited at the 1913 Aero Show at Olympia in February, on the stand of Percy Grace, brother of the late Cecil Grace, who acted as selling agent for Shorts. It was acquired by the Admiralty soon after the Show, being identified as No.42. In service the floats proved to be easily damaged in rough water; the machine was then converted to a landplane and was taken to France in August 1914, where it crashed into a tree, when being flown by Samson on 28 September 1914, and was written off.

The design included a number of new features, including lightweight main floats made of heavy fabric on wooden frames. These were connected to the fuselage by steel struts, the interplane struts being also of steel tubing. The pilot sat in the rear position; the front cockpit could accommodate two side-by-side and included an engine starting handle which a passenger could operate. The top wing extensions were braced by struts and single acting ailerons were fitted to the top wing only; the lower wing had rounded tips. The engine was carried on front and rear mountings and a board, to protect it from spray, was fitted below but was soon discarded. The wing tip floats were cylindrical canvas bags on steel frames, with horizontal tails. The tail float had parallel sides and small skids for protection, and was connected to the fuselage by five struts, the rearmost providing a pivot for a water rudder. No fin was fitted initially but one was added later.

Power: 80hp Gnôme seven-cylinder air-cooled rotary driving a 8ft 6in diameter propeller.

Short S.60 (RNAS No.42) in seaplane form in 1913.

Data	Span top	48ft
	Span bottom	30ft.
	Chord	5ft
	Gap	5ft
	Area	390 sq. ft
	Area tailplane	9 sq. ft
	Weight	1,200 lb.
	Weight allup	1,990 lb.
	Speed	65 mph

Short S.68 built for McClean to compete in the 1913 Seaplane Circuit of Britain. (Courtesy John Blake).

S68. Tractor biplane for 1913 Seaplane Circuit of Britain

This was McClean's aircraft No.15, built specially to compete for the *Daily Mail* £500 prize, the contest to start on the 16 August 1913. Vibration problems could not be resolved in time and the entry, No.4, was withdrawn. The machine proved to be slow and was soon abandoned.

The type was an individual design, but included features from other Short types and resembled an enlarged version of the S60, apart from the installation of the inline engine and its radiators. The aircraft was built initially with top wing overhang and single acting ailerons, but later the lower wings, which had rounded tips, were increased to a span equal to that of the top wings. The exhaust pipes for the six-cylinders were carried vertically up to a collector box exhausting over the top wing. Large radiators were mounted fore and aft on the center section struts and these were a source of leakage.

It is possible that this aircraft was impressed into the Navy as No.182 with Gnôme engine.

Power: 100hp Green six-cylinder inline water-cooled.

Data	Span	61ft
	Length	40ft
	Area	600 sq. ft later increased to 660 sq. ft

Tractor Seaplanes 1913-1914

In the period prior to the outbreak of war and to the end of 1914, Shorts produced several different designs of seaplane for the Navy, including the first folding wing machines for stowage aboard ships. These followed the same general configuration, but varied in size, weight and power. The basic layout was of a twinfloat seaplane with tail and wingtip floats. The wings were of unequal span and single acting ailerons were used. The tail unit was conventional and included a large fin. The crew were seated in tandem with the pilot in the rear cockpit. Descriptions of these follow.

S64 65. Tractor seaplanes. Admiralty Folder Type 74 (Nos.81-82)

These were the first aircraft with folding wings for the Navy and were completed in July and October 1913. The wings had two bays with struts supporting the top wing extensions. A funnel carried the exhaust over the top wing.

Power: 160hp Gnôme fourteen-cylinder two-row air-cooled rotary driving a 9ft 6in diameter propeller

Data	Span top	56ft
	Span bottom	40ft
	Chord	6ft
	Gap	6ft.
	Area	550 sq. ft
	Length	39ft
	Weight	2,000 lb.
	Weight allup	3,040 lb.
	Max. speed	78 mph
	Endurance	5hr

Short S.68 was later fitted with an extended lower wing.

Short S.64 (RNAS No.81) The first machine with folding wings known as Admiralty Type 74. Improvements were built into later batches.

Short S.69 Admiralty Type 74 (RNAS No.74).

Short S.115 (RNAS No.818) was one of a batch of Type C folder seaplanes based on S.69 Admiralty Type 74.

S69-72, S108-115 and S128 Tractor seaplanes, Admiralty Type 74 (Nos.74-77, 811-818 and 183)

These aircraft were built in two batches, four being delivered in January and February 1914 and all were completed by the end of the year. The three bay wings were non-folding with struts bracing the top wing extensions. The second batch, referred to as the 'Improved Type 74' were the first aircraft to be built at Rochester. A single aircraft, S128, was ordered on Contract No.57105/14 and this was delivered in November 1914 from Eastchurch.

Power: 100hp Gnôme ten-cylinder two row, or nine-cylinder Monosoupape air-cooled rotaries.

Data

Span	57ft
Area	580 sq. ft
Length	39ft
Weight	2,100 lb.
Weight allup	2,700 lb.
Max speed	65 mph
Endurance	5hr

S73-74. Tractor seaplanes Admiralty Type 74 (Nos.78, 79)

These aircraft were similar to S69-72 but with the more powerful engine with funnel-type exhaust and were delivered in February and March 1914.

Power: 160hp Gnôme fourteen-cylinder two-row air-cooled rotary.

Short S.83 (RNAS No.120) A batch of four built in 1913-1914 were identified as Admiralty Type 81 and were three bay versions of the Type 74 Folder.

S82-85. Tractor seaplanes Admiralty Folder Type 81 (Nos.119 122)

These aircraft had three bay wings of increased span, but were generally similar to S64-65.

No.121 was the first British aircraft to successfully drop a torpedo.

Power: 160hp Gnôme fourteen-cylinder two-row air-cooled rotary driving a 9ft 6in diameter propeller

Data

Span top	67ft
Span bottom	52ft.
Chord	6ft
Gap	6ft
Area	690 sq. ft
Length	42ft
Weight	3,050 lb.
Weight allup	3,500 lb.
Max speed	78 mph
Endurance	5hr

S87. Tractor seaplane Admiralty Folder Type 135 (No.136)

S87 and S88 were ordered in February 1914 with water-cooled Salmson engines, of two different types. The first to be delivered had the more powerful engine, which was cooled by a large block radiator mounted on top of the fuselage, ahead of the wings. A small cowl was fitted over the top of the engine. The wing extensions were braced by struts, and single acting ailerons with tapered inboard sections were used. An elongated fin and a tail float with small water rudder and the usual type of airbags were fitted.

Power: 200hp Salmson (Canton-Unné) fourteen-cylinder two-row water-cooled radial.

Short S.87 (RNAS No.136) was one of two machines built in 1914 with Sunbeam engines.

Data		
	Span top	54ft 6in (56ft ?)
	Span bottom	40ft
	Chord	6ft
	Gap	6ft
	Area	570 sq. ft (550 sq. ft?)
	Length	40ft (39ft ?)
	Height	12ft 6in
	Weight	3,000 lb.
	Weight allup	3,700 lb.
	Max speed	72 mph (78 mph?)
	Endurance	4hr

Short S.88 (RNAS No.135), fitted with a lower powered engine, was smaller but generally similar.

S88. Tractor seaplane Admiralty Type 135 (No.135)

S88 was delivered on 9 September 1914 and allocated for shipboard use on HMS *Riviera*, and was used in the Cuxhaven raid of 24/25 December 1914, from which it did not return. In appearance it resembled S87, but was smaller, lighter and with a lower performance.

Power: 135hp Salmson (Canton-Unné) nine-cylinder single-row water-cooled radial.

Data		
	Span top	52ft
	Span bottom	36ft
	Chord	6ft
	Gap	6ft
	Area	530 sq. ft
	Length	37 ft
	Height	12ft 6in
	Weight	2,700 lb.
	Weight allup	3,400 lb.
	Max speed	65 mph
	Endurance	4 1/2 hr

S90-95. Tractor seaplanes Type C, Admiralty Type 166 (Nos.161-166)

This batch of aircraft derived from S87 (No.136), but incorporated a number of changes. The two bay folding wings of unequal span had kingpost and wire bracing of the top wing extensions. The ailerons were shorter and in one piece, but still with a tapered chord. A larger elongated fin and balanced rudder were fitted. The spacing tubes between the floats were arched to provide for the carriage of a torpedo.

The aircraft were ordered in January 1914 and were built at the new Rochester works but were not delivered until March to August 1915. A further batch of twenty Type 166 aircraft was built by Westlands in 1916, without the arched cross struts, as carrying a torpedo had proved to be impractical.

Power: 200hp Salmson (Canton-Unné) fourteen-cylinder two-row water-cooled radial.

Data		
	Span	57ft 3in
	Area	575 sq. ft
	Length	40ft 7in
	Weight	3,500 lb.
	Weight allup	4,580 lb.
	Max speed	65 mph
	Endurance	4hr

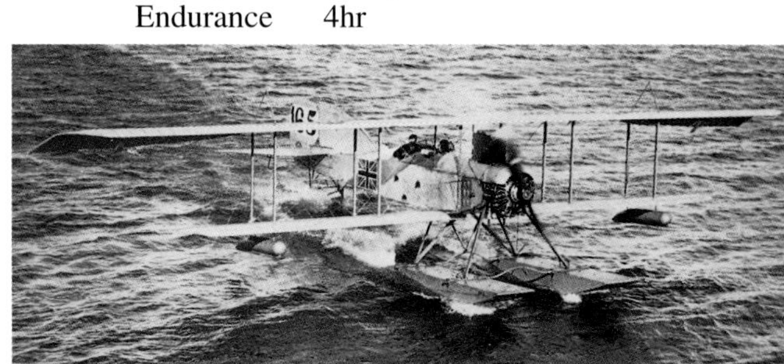

Short S.94 (RNAS No.165) one of a batch built in 1914-1915 known as Admiralty Type 166.

Short S.106 (RNAS No.184) also known as Admiralty Type 184 was a most important naval aircraft. Designed in 1914 it was built in considerable numbers in wartime.

S106-107 and S129138 Tractor seaplanes Admiralty Type 184 (Nos.184-185 and 841-850)

This type was a most important naval aircraft, which was put into large-scale production at several contractor's works. Shorts received an initial contract CP46634/14 in July for two prototypes S106-107, which became serial Nos.1845 in service. The company also received contract CP57105/14 for ten aircraft S129-138, which became serial Nos.841-850, before the prototypes had flown. The parent company made 117 of the total of about 650 aircraft that were built by ten manufacturers. The twelve aircraft on these first contracts were fitted with the 225hp Sunbeam instead of the 200hp Salmson originally selected, and a number of different engines were fitted later. Various airframe changes were also made to the Type 184, which remained in production throughout the war years and continued in service with the RAF until 1920.

Among the intended uses of the type was the carriage of a fourteen inch torpedo, which together with wireless, two crew and other equipment proved to be too much for practical use, particularly in tropical conditions. To overcome this, in 1916, S212 (No.8070) was modified by fitting extended span upper wings and reduced span lower wings, without ailerons. This was known as Type B, but only a further ten aircraft Nos.4085-9094, built by Mann, Egerton and Co., were so modified.

The Short was a developed and enlarged version of S68, the 1913 Circuit of Britain seaplane, made suitable for operations from carrier ships. The fuselage was a wire braced girder structure of four spruce longerons, spindled for lightness, with high tensile steel fittings at the joints. Apertures were provided in the rounded fairings on the upper longerons for the crew, consisting of the pilot in front and the observer/gunner behind. Below the cockpits the fuselage sides were plywood covered with inserted footsteps; rear of this, the fuselage was fabric covered. A radially slotted nose cowl and aluminum panels enclosed the engine installation back to the leading edge of the lower wing, and a large box shaped radiator was mounted above the engine.

The wings comprised three bays and were equal in span. The upper wing consisted of a center section of the same width as the fuselage, followed by a section tapered on the trailing edge to a point outboard of the second bay, where the chord then became parallel. Ailerons of wide span were fitted to both top and bottom wings, which operated downwards only. The ailerons on No.184 were originally arranged to droop, but this rendered the aircraft difficult to control on the water, and rubber cords were added to pull both to the horizontal. The lower wing was parallel and of narrower chord. Cylindrical airbags hung below the outboard interplane struts. The whole wing assembly could be folded and locked and employed ingenious means for controlling the complicated bracing system to keep it clear of the tail unit, and to provide for single-handed operation by the pilot from the cockpit. The wing structure was a conventional two spar wooden structure with closely spaced nose ribs; the interspar and trailing edge ribs were more widely spaced and the wired trailing edge produced a scalloped effect to the fabric.

The tail surfaces were similar in shape to those of the Types 827 and 830 which were being developed at about the same time. The long curved fin was shaped to match the aerodynamically balanced rudder and was extended to provide a top pivot point for the rudder. The tailplane was in one piece attached to the top longerons and a divided elevator gave clearance for the rudder. Stranded bracing cables, as used elsewhere in the aircraft, stabilized the tail unit.

The chassis for the main floats was made of faired steel tubes and consisted of a pair of single struts at the front and a pair of vee-struts at the rear, attached to cross tubes and braced by cables. The floats were attached to the cross tubes by rubber cords and were slotted to allow vertical movement. The tail float was also sprung by rubber cords and had a water rudder actuated by the main rudder control. The floats were parallel box structures. The torpedo was to be carried under the cross tubes of the chassis, for which purpose they were originally curved at the center. When this role was discontinued, straight cross tubes were fitted and the prime role became bombing, for which purpose a centerline rack was mounted below the fuselage, which could carry up to 500 lb. weight.

Power: 225hp Sunbeam twelve-cylinder water-cooled vee.

Data		
	Span	63ft 6in
	Span folded	6ft 4 3/4in
	Chord top	6ft 6in
	Chord bottom	5ft
	Gap	5ft 6in
	Area	688 sq. ft
	Main floats	16ft long by 2ft 10in beam
	Length	40ft 7 1/2in
	Length folded	44ft 2in
	Height	13ft 6in
	Weight	3,500/3,700 lb.
	Weight allup	5,100/5,500 lb.
	Speed	75 mph
	Endurance	2 3/4 5hr
	Service ceiling	5,500-9,000ft

Short Admiralty Type 827 in East Africa.

S116-118, S119-121 and S151-162 Tractor seaplanes Admiralty Type 830 (Nos.828-830, 819-821 and 1335-1346)

S122-127 Tractor seaplanes Admiralty Type 827 (Nos.822-827)
The development of the smaller version of the Type 135 (S88) resulted in contracts for two batches of three and four aircraft incorporating improvements, which became known as the Admiralty Type 830. Delivery of the first batch was made between November 1914 and January 1915 and the second batch between December 1914 and January 1915. A further batch of twelve ordered in 1915 were delivered in that year. S151-162 (Nos.1335-1346)

Aircraft Nos.828-830 had the 135hp Salmson engine, as fitted in the original machine S88, but Nos.819-821 had the 150hp Sunbeam, later known as the Nubian. A further batch of six aircraft, which were not delivered until 1915, were designated the Admiralty Type 827. In other respects the two versions were similar, but differed from the original Type 135 (S88) by the use of kingpost and wire bracing to the top wing extensions.

In addition to the aircraft built by Shorts at Rochester, Parnalls, Faireys, Sunbeam and Brush produced the 827 in quantity in wartime, and the type continued in service throughout the period of hostilities; quantities of some nineteen of Type 830, and 107 Type 827 aircraft were ordered. Certain of the later aircraft had wings with stiff, as opposed to wired, trailing edges and parallel chord, double acting ailerons.

Type 827.

Power: 150hp Sunbeam eight-cylinder water-cooled vee, later named Nubian.

Data		
	Span	53ft 11in
	Area	506 sq. ft
	Length	35ft 3in
	Weight	2,700 lb.
	Weight allup	3,400 lb.
	Max speed	62 mph
	Endurance	3 1/2hr

Type 830.

Power: 135hp Salmson (Canton-Unné) nine-cylinder single-row water-cooled radial.

Data		
	Span	53ft 11in
	Area	506 sq. ft
	Length	35ft 3in
	Weight	2,624 lb.
	Weight allup	3,324 lb.
	Max speed	70 mph
	Endurance	3 1/2hr

Abandoned Short Types of 1914

No doubt a number of types were projected of which there is no remaining evidence. However in June 1914 a prototype Short Type B two-seater tractor biplane, Admiralty serial No.178 with 200hp Salmson or Le Rhone engine, to be built at Eastchurch, was ordered. This was to be followed by nine production aircraft ordered in July, Nos.190-198. None of these was produced and actual manufacture had probably not commenced before cancellation, since no construction numbers were allocated.

SHORTER monoplane (Shorter, Coventry Aero Club)

This was a cycleplane type with tractor propeller driven by pedals and belts. The early results achieved were described as

Shorter monoplane cycleplane made by a member of Coventry Aero Club which hopped down slopes in early 1914.

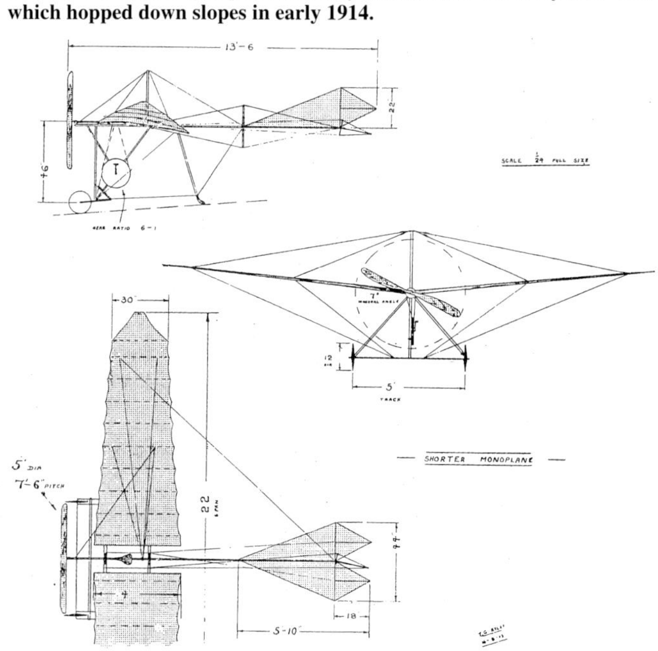

'encouraging'. Some hops down a slope were made during further trials in January 1914.

Data	Span	22ft
	Chord	4ft tapering to 2ft 6in
	Length	13ft 6in

SIM glider (A. Sim, Sundridge Park, Kent)
This amateur built two bay biplane glider was made originally with only a boxkite tail with three fins on booms, but trials suggested that the tail was too heavy and a front elevator was then added. The wings were built with the spars at the extreme edges of the chord, and were single surfaced with no camber. The operator stood in the gap in the lower center section with the machine resting on bamboo skids at the center and wing tips. Bamboo was also used for the structure.

The machine was flown as a kite and lifted off with one person aboard.

Data	Span	21ft
	Area	253 sq. ft
	Chord	5ft
	Gap	4ft 9in
	Area of tailplane	42 1/2 sq. ft
	Length less elevator	17ft 4in

Sim glider made at Sundridge in Kent in 1909-1910. Although it flew with a man aboard it was flown as a kite to adjust the balance.

SIMMS cycloplane (Frederick R. Simms, London)
Yet another flying bicycle. The motor was geared to a propeller at the end of a shaft at the back of the bicycle's frame. A series of wings were supported above the bicycle by rods and ties which could be moved to alter the angle of the wings.

Simms monoplane built in 1908-1909 at Chipping Norton.

Simms biplane cycleplane of 1910 was wrecked as it started to lift.

SIMMS gliders (Herbert Rutter Simms, Chipping Norton, Oxfordshire)
In 1908-1909 Simms constructed a high wing monoplane described as a 'Bedstead type', presumably for being mounted on four wheels. The rear wheels were arranged to drive a tractor propeller, the intention being to roll down a slope and to gain assistance from the rotating propeller. The machine was made of bamboo and the heavily cambered wing and tail were covered with linen.

In 1910 Simms made a biplane cycleplane, with staggered wings and triangular front elevator flaps and a high mounted tail. Outriggers, fitted with wheels, were fitted to stabilize the machine, but it was wrecked on the first trial when, apparently, it was beginning to lift. Simms became an apprentice with A.V. Roe at Manchester in June 1910, and learnt to fly at Shoreham, where he tested and instructed on Avro aircraft. In 1913 he moved to Sopwiths and then went to Greece in connection with the establishing of a Naval Air Service. On his return in 1914 he was commissioned in the RNAS, but drowned after being shot down on operations in May 1916.

SIPPÉ monoplane (Sidney V. Sippé, Arthur H. Sippé and James Jensen, Beckenham, Kent)
A small single-seater tractor monoplane was built by the brothers and a friend in 1910 and was tested at Addington in Surrey towards the end of that year.

Sippé monoplane was constructed by the brothers and a friend at Beckenham in 1910 and was tested at Addington.

The pilot sat below the wing in the fuselage structure, which was of welded steel tube and incorporated the undercarriage, engine mounting and bracing structure. The bamboo tail boom was supported by a tail wheel and carried conventional tailplane, elevators and balanced rudder. A 20hp Henderson four-cylinder air-cooled motorcycle engine was probably the power

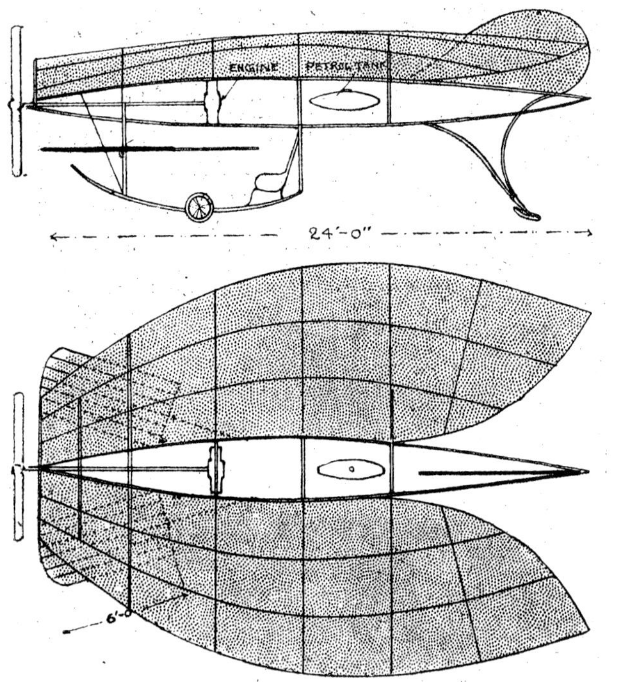

Sirie monoplane. An interesting design based on the leaf of an oriental tree.

SIRIE monoplane (M. Leslie-Miller, 2 Hamilton Terrace, East Partick, Scotland)

This inventor designed an unusual monoplane and named it after a tropical tree, the leaves of which resembled the wings of his machine. The curved wings were heavily swept back alongside the fuselage and had considerable dihedral. The rotary or radial engine was mounted inside the fuselage about one third the way back from the nose. There was a large rudder at the rear. The main wheels were mounted on twin skids, upon which the pilot was seated. There was a large tail skid. Two elevating planes were fixed at the at the front of the machine and were operated by rods. They could move in four different ways for both lateral and vertical control.

A scale model was made and numerous successful flights were carried out with it by the inventor but it seems unlikely that a full-size machine was ever completed.

Data	Span	18ft
	Length	24ft
	Elevators	6ft long

Skinner monoplane built by Mulliners at Clapham Junction and tested at Brooklands in 1911.

SKINNER monoplane

Few details are known of this machine constructed by Mulliner's Coachworks Ltd. at Varden's Rd., Clapham Junction, SW for Mr. Skinner. It was complete and ready for testing at Brooklands in August 1911, and was reported to have been bought by Jack Humphreys in September.

The machine was a single-seater tractor with triangular section fuselage girder, which was uncovered at the front, to just behind the pilot's seat. A substantial twin skid undercarriage extended well forward, to protect the propeller, and a long tail skid was fitted. An elongated fin and tailplane extended well forward on the top of the fuselage, and behind was mounted a cruciform tail unit, which was moveable on a single pivot. The wing had a well cambered section and marked dihedral, with lateral control by warping.

Power: 60hp ENV type F eight-cylinder water-cooled vee.

SKUDDER monoplane (G.F. Skudder & Co., 107 Tooley St., London SE1)

This monoplane was built in 1909-1910 and was described as of 'Blériot type', but with a forward elevator and skids. The engine was a four-cylinder Darracq. Nothing further is known.

SMITH aeroplane (J.W. Smith, Shoreham, Sussex)

No information is available on this 'unique' machine at Shoreham in 1913.

SMITH multi-plane (Samuel Smith, Daubhill, Bolton, Lancashire)

For many years the identity of this machine was confused, being thought to be a multi-plane built by Horatio Phillips, until clarified in 1984 by John Bagley of the Science Museum.

The machine was designed and built between 1910 and 1912 and was assembled at Closes Farm, Smethurst Lane, Daubhill with the help of the farmer F.C. Wallis. It was tested but it is doubtful if it lifted off and was broken up soon after.

The vertical frame contained about one hundred narrow slats, below which was the fuselage, whose bottom longerons were extended forward to form skids and also served to support the tail. The tail unit appeared to consist of a conventional rudder, tailplane and elevators. Below the vertical frame, wide chord surfaces probably served as ailerons. The engine was mounted high on a structure with the thrust line approximately at the center of the multi-plane frame. The pilot's position was an enclosure in the fuselage. Patent No.17362/1910 was taken out to cover a gyroscopic control device, which was to be employed.

Power: 60hp Edwards six-cylinder inline water-cooled two-stroke.

Data	Span	36ft
	Chord	7ft
	Area	252 sq. ft

Smith multiplane built at Daubhill, Lancashire in 1910-1912 used Phillips-style aerofoils. It is doubtful if it flew and was soon abandoned.

SMITH aircraft (Osborn Smith)

An aeroplane and a glider were reported in January 1910. No details have been found.

SMITH glider (W.W. Smith)

Smith wrote a letter giving details of a modified glider that he proposed to build, seeking assurances on some of its features. It was to be launched from a Wright type derrick and rail, was to be built of bamboo and could have an elevator in front, or a tail, or both. It is uncertain whether the machine was built.

SMITH and MATEYKA biplane (G.P. Bragg-Smith and S.F. Mateyka)

A biplane, similar to one of the models built by G.P. Bragg-Smith, was reported being built in 1910. Both partners were model makers in 1909-1910, when Smith proposed to build a full size version. The machine was probably based on his prize winning model shown at Olympia in March 1911. No evidence of its completion can be traced. See also Bragg-Smith.

Data	Span	31ft 4in
	Chord	6ft 6in
	Length	54ft 3in

SMITHERS See Langley-Smithers monoplane

SMYTH biplane (Arthur Smyth, Eastbourne, Sussex)

This machine was reported being built in 1909, but there is no evidence of completion.

Sonoda biplane was built at Barking by Handley Page. It was at Hendon in July 1912 but was wrecked in September.

SONODA biplane (T. Sonoda, Hendon & Japan)

Handley Page built this biplane at Barking and it appeared at Hendon for the first time on the 7 July 1912. It was displayed at the Naval and Military Aviation Day at Hendon on the 28 September 1912 but crashed with damage to the undercarriage and propeller two days later, when being flown by C.W. Meredith. There were no further reports of this machine.

The two bay biplane resembled a French Zodiac machine and the design incorporated original control features by Sonoda. The unequal span wings of parallel chord were staggered and the top overhang was braced by struts. Single acting ailerons were fitted to the top wing only, which mounted a gravity fuel tank above the center section.

The fuselage was carried on struts above the lower wing and was mounted directly on the undercarriage, which was of split axle type with a center skid and long compression struts connected at the top longerons. Protective hoops were attached to the lower wing at the tips.

The tail unit consisted of a long triangular dorsal fin and top and bottom rudders; there was no bottom fin, but the lower rudder was protected by a tall tail skid. The tailplane, mounted on the centerline of the fuselage, was similar in length and shape to the fin and carried a one piece elevator. Radiators were mounted on either side of the cockpit area.

The aircraft was painted pale blue with the designer's name in large letters, tapering in depth, on the fuselage sides. The Japanese flag appeared on the top rudder and the legend that the machine had been constructed by Handley Page Ltd. of 72 Victoria St., London SW.

Power: 60hp Green four-cylinder inline water-cooled.

SOPWITH AVIATION Co. (Canbury Park Rd., Kingston-on-Thames & Brooklands, Surrey)

The Sopwith name is so well known from the wartime successes of aircraft such as the Pup and the Camel, that only a brief introduction is required. Tom Sopwith who had previously flown as a balloonist, began flying at Brooklands on Howard Wright aircraft in October 1910. He achieved an immediate success by winning the £4,000 Baron de Forest Prize for the longest flight from Britain to the Continent, in December of that year. He followed this with a most successful tour of the flying meetings in America, commencing in May 1911, where he won a considerable amount of prize money. He returned to England in October and decided to set up a flying training establishment.

The Sopwith School of Flying began operations from 1 February 1912 with four aeroplanes, Howard Wright and Blériot monoplanes, a Howard Wright biplane and a Burgess-Wright biplane, bought in America. This latter machine was important in that it was the basis for a reconstruction, which became known as the Sopwith-Wright biplane, accepted as an all-British aircraft, and the first to bear the Sopwith name.

This aircraft, and a few that followed, were built at Brooklands in Sheds 21 and 31, but larger premises were acquired for manufacturing, in the form of a disused roller skating rink at Canbury Park Rd., Kingston, which was used from January 1913 onwards. In 1914, with the war approaching, a new building was constructed further along the road, which was progressively expanded to become Sopwith's wartime base and later, the main center of the successor Hawker company, who continued to use it as late as 1962. In all his early work Sopwith was ably supported by Fred Sigrist, a practical engineer, who contributed much to the early designs. From 1912 Sopwith was creating an organization and he gave up flying. The company was reconstituted in March 1914 as a limited company, registered as 'The Sopwith Aviation Co. Ltd.' An employee and former pupil of the school, H.G. Hawker, became chief test pilot from October 1912, and in addition had a great influence on the design of the early aircraft.

The early types were evolved by Sopwith, Sigrist and Hawker with R.J. Ashfield as a draughtsman detailing their ideas on drawings, and carrying out stress calculations, from late in 1912. He became the leader of a small drawing office in the rink, but later Herbert Smith, working separately for a time, eventually took on the role of chief designer in wartime.

The Sopwith Order Book has been used as the basic record of the work initiated from January 1913, when the first entry was made. The various types are listed in the order in which the first instruction to proceed, were recorded. Some anomalies arise with previously published information, but due weight has been given to the Order Book information.

Extract from Sopwith Order Book.

Office Order No	Date	Type	Engine	Qty	RFC/RNAS Serial nos.
		*Sopwith-Wright biplane	35hp Green 40hp ABC	1	
		*Hybrid biplane	or 70hp Gnôme	1	27
		*Bat Boat No.0	50hp Gnôme	1	
1	30 January 13	Tractor biplane D1	80hp Gnôme	1	33
4	20 February 13	Bat Boat BB1	90hp Austro-Daimler	1	
4a	1 April 13	Bat Boat BB2	90hp Austro-Daimler	1	38
5	1 April 13	Bat Boat BB3	100hp Green	1	118
			90hp Austro-Daimler		
6	24 February 13	Tractor biplanes D2, 3 and 4	80hp Gnôme	3	See O/No 23
7	16 April 13	Tractor hydrobiplane HT1, 2 and 3.	100hp Anzani	3	59-60
8	26 April 13	Hydro biplane ST.	2 120hp. Austro Daimler	1	61
10	20 May 13	Tractor hydro ADH	90hp Austro-Daimler	3	
11	20 May 13	Tractor biplanes D (part O/No 22)	80hp Gnôme	3	
13	27 May 13	Gun carrying pusher biplane GPH	200hp Salmson	1	93

14	2 June 13	Hydro biplane DM	hp Green	1	
21	5 August 13	Rebuild No.27	70hp Gnôme	1	27
22	8 August 13	Tractor biplanes D2	80hp Gnôme	9	See text
23	9 August 13	Tractor biplanes D3 & D4	80hp Gnôme	2	103-104
26	10 September 13	Rebuild DM aircraft			
28	6 September 13	Repair Bat Boat No.38			
38	14 October 13	School seaplanes SPAz	100hp Anzani	3	For Greece
39	25 October 13	Pusher seaplanes. SPAz	100hp Anzani	2	123-124
40	25 October 13	Two seater Tractor StB	80hp. Gnôme	1	
43	17 November 13	Bat Boat	200hp Salmson	1	For Germany
44	17 November 13	Bat Boat. BBS	200hp Salmson	1	127 then Greece
46	21 November 13	Mock Torpedo machine TT	100hp Green	1	
53	9 December 13	Recondition No1 or No.6 D2			
55	9 December 13	Single seat Tractor biplane SS	80hp Gnôme	1	Canceled
58	24 December 13	Land Tractor biplane Ds	100hp Gnôme	4	149
59	9 January 14	Tractor Scout biplanes SS	80hp Gnôme	9	See text
68	26 January 14	Special Tractor seaplane	200hp Salmson	1	170
70	2 February 14	Hydro Tractor HS Schneider Cup machine	100hp Gnôme Monosoupape	1	
73	11 February 14	Tractor seaplanes	120hp Austro-Daimler 200hp Salmson	1	137 138
74	12 February 14	Seaplane DM	100hp Green	1	151
76	18 February 14	Tractor biplane Type D3 Pusher seaplane Type SPGN	100hp Gnôme Monosoupape	1 5	896 897-901
82	5 March 14	Bat Boat DM 2 seater 1914 type	225hp Sunbeam	1	879
85	9 March 14	Military biplane 2 seater 1914 type D3	100hp Gnôme Monosoupape	1	880
87	14 March 14	Tractor Scout biplanes SS	100hp Gnôme Monosoupape	3	604, 611, 654
93	2 April 14	Hydro Tractor HS (repaired O/No 70)	100hp Gnôme Monosoupape		
98	25 April 14	Schneider Cup m/c	100hp Gnôme Monosoupape	1	
103	7 May 14	Tractor Scout	80hp Gnôme	1	
104	7 May 14	Tractor biplane (Warp control)	80hp Gnôme	1	1214
105	7 May 14	Tractor biplane (Warp control)	80hp Gnôme	1	1215
107	8 May 14	Tractor Scout	Less 80hp Gnôme	1	For Italy
108	8 May 14	Tractor Scout	Less 80hp Gnôme	1	For Blériot
110	14 May 14	Seaplane Admiralty Type C	200hp Salmson	3	157-159
110	14 May 14	Tractor biplane Type D	100hp Gnôme	1	160
117	2 June 14	Tractor Scout	80hp Gnôme	1	
122	8 June 14	Tractor Scout	Less 80hp Gnôme	1	For Russia
132	22 June 14	Stunt biplane StB repair	80hp Gnôme	1	
134	29 June 14	Gun carrying biplane Type GPH	200hp Salmson	1	93 (rebuilt)
136	3 July 14	Bat Boat Type BBS	200hp Salmson	1	For Greece
138	8 July 14	Land Tractor biplane Ds (Churchill)	100hp Gnôme Monosoupape	1	149 rebuilt
151	28 July 14	Gun carrying Pusher 2 seater	100hp Gnôme Monosoupape	1	
155	1 August 14	Hydro biplane Type ST (rebuilt)	2 120hp Austro-Daimler	1	61
160	14 August 14	Land Scouting Pushers	(110hp Sunbeam) (150hp Sunbeam)	6	801-806
161	11 August 14	Tractor biplane 2 seater Type D2	80hp Gnôme	1	906
165	21 August 14	Torpedo carrying seaplanes Admiralty Type 860	225hp Sunbeam	10	851-860
167	29 August 14	Tractor seaplanes. Admiralty Type 807	100hp Gnôme Monosoupape	4	807-81
170	3 September 14	Tractor biplane StB Stunt Bus	80hp Gnôme	1	169
181	26 September 14	Scout biplanes Tabloid SS3	80hp Gnôme	12	1201-1212
182	28 September 14	Tractor biplanes Type D5 "Spinning Jenny"	80hp Gnôme	24	1051-1074
183	28 September 14	Tractor seaplanes. Admiralty Type 807	100hp Gnôme Monosoupape	8	919-926
188	7 October 14	Scout Biplane	80hp Gnôme	1	1213
192	13 October 14	Torpedo carrying seaplanes Admiralty Type 860	225hp Sunbeam	12	927-938
206	7 December 14	Two seater Tractor. "The Sigrist Bus"	80hp Gnôme	1	
207	7 December 14	'T' (torpedo) seaplanes Admiralty Type 860	225hp Sunbeam	24	1280-1299 1347-1350

Sopwith-Wright biplane. The reconstruction of the American Burgess-Wright bought in 1911.

SOPWITH-WRIGHT biplane

This machine was a reconstruction of the Burgess-Wright biplane bought in America during Tom Sopwith's tour in 1911. The work, for which Fred Sigrist was largely responsible, was carried out at Brooklands. *Flight* reported that nothing of the original aircraft remained after reconstruction, although the basic configuration was much the same.

The aircraft was a twin pusher propeller type, with two bay wings, with extra interplane struts to mount the propeller shafts. The front spars were at the leading edges of the wings, with flexible overhangs aft of the rear spars for warping control. Closely spaced tail booms, parallel in both views, carried twin rudders of narrow chord, mounted on common pivots within the booms. The tailplane had tips tapering to a point and the rear portion was flexible to act as an elevator. Two square shaped panels hung below the top wing on the front intermediate interplane struts.

The engine, on the lower center section, was offset to port with the pilot and passenger in a nacelle to starboard, with a large fuel tank behind. The drive to the propeller shafts was by chains, the starboard being crossed to provide contra-rotation. A radiator was mounted fore and aft on the starboard inboard interplane struts and an oil tank above the engine.

The main structure incorporated twin skids with triangular 'blinkers' as additional surface area for directional stability, and the skids extended aft to support the tail as well as the extended tail booms. Two pairs of wheels were mounted and rubber sprung on the skids. The machine departed from the Wright system of interconnected warp and rudder control, having the Farman system of separate control lever and rudder bar, with which Sopwith had experience. The Farman control had been fitted to the original Burgess-Wright by Burgess at Sopwith's request.

The Sopwith-Wright flew for the first time with a Green engine on Thursday 2 May 1912, with further flights the following day and on Sunday 5 May 1912, when Raynham competed in the Cross Country Handicap and achieved third place. On Friday 17 May 1912 it was reported that the first flight was made with the ABC engine. The aircraft was flown in other competitions and Raynham won the Shell Speed Trophy at Hendon on 8 June 1912, but much of it's life was devoted to instructional work at the Sopwith School of Flying. The most significant success of the machine took place on 24 October 1912, when Harry Hawker flew for 8hr 23min, for which he won the 1912 British Empire Michelin Trophy No.1 and a prize of £500. This performance was a new British duration record.

Power: 35hp Green four-cylinder inline water-cooled.
40hp ABC four-cylinder inline water-cooled.
Both engines drove 8ft 6in diameter Bristol pusher propellers by chain, the starboard crossed to provide contra-rotation.

Data		
	Span	38ft 9in
	Chord	6ft 3in
	Gap	5ft 3in
	Area	475 sq. ft
	Area tailplane/elevator	35 sq. ft
	Area rudders	15 sq. ft
	Length	29ft 6in

SOPWITH Tractor biplane (The '*Hybrid*')

This was the first original Sopwith machine and was built entirely at Brooklands during the time that the flying school was in being. Sopwith tested it for the first time on Saturday 4 July 1912. On 12 July 1912, when flown by Gordon Bell with R.L. Charteris as a passenger, it crashed by the sewage farm without injury to the occupants, but the resulting damage was considerable. The rebuilt aircraft was flown by Sopwith on his return from motorboat racing in America on 18 September 1912. On 8 October 1912 Sopwith flew the machine to Farnborough, presumably for official approval, for it was delivered to the Navy at Eastchurch by Hawker accompanied by Kauper on 22-23 November 1912, the flight taking two days as they were forced to land due to fog on the first day.

In its first form the '*Hybrid*' had an uncovered rear fuselage and a four wheeled undercarriage. The wings, with four pairs of interplane struts, were derived from the Wright, hence the name, and had warping control. The fuselage was positioned well above the lower wing, to which it was joined by

Sopwith biplane known as the *Hybrid* because of its Wright-type wings with a new tractor fuselage. Flown at Brooklands in July 1912 in its first form.

Sopwith *Hybrid* with modified fuselage and two wheel undercarriage.

struts enclosed by a large ungainly fairing. The machine was supported on the rear portions of the main skids when at rest.

The tail unit consisted of a split rudder, each portion of equal size, and a one piece tailplane with warpable trailing edge serving as an elevator. The engine was covered by a curved top cowl and flat side panels, air entering from below to cool the engine, which was set back from the nose, and drove the propeller by an extension shaft. The pilot occupied the rear cockpit with room in front for two passengers side by side. During the rebuild the fuselage was lowered onto the lower center section and was entirely covered in. The undercarriage was altered to a cross axle with only two wheels.

The '*Hybrid*' became serial No.27 with the Navy and remained in use until the Spring of 1913, when contract CP45759/13X11122 was placed with the firm to rebuild it on the lines of the latest tractor biplane, Type D.

Power: 70hp Gnôme seven-cylinder air-cooled rotary driving a 9ft 6in diameter Chauvière propeller

Data		
	Span	44ft
	Chord	6ft
	Gap	6ft 4in
	Area	520 sq. ft
	Area rudder	12 sq. ft
	Area tailplane/elevator	40 sq. ft
	Length	26ft 4in
	Weight	950 lb.
	Speed	55 mph
	Endurance	3 1/2 hr

SOPWITH Bat Boat hydro-biplane

It was inevitable, with his marine interests, that Tom Sopwith should consider operating aircraft from water. His first experience was to fly a Donnet-Levêque flying boat from the Seine near Paris, when it was reported that he flew over and under the bridge at Juvisy on 2 June 1912. This machine was a small side by side flying boat, with retractable land undercarriage and a pusher propeller, driven by a 50 or 80hp Gnôme engine.

This was, no doubt, practice for his later attempt to fly a similar machine of his own construction, seen at Brooklands and briefly described in *The Aeroplane* of 1 August 1912 (p.110), of which no further reports were published. However a vague report in *The Aeroplane* of 5 September 1912 (p.244),

Sopwith hydro-biplane of the type later called Bat Boats. This shows the machine outside SE Saunders works at East Cowes in 1912. It had been built at Brooklands using a Saunders-built Consuta sewn hull but failed on test.

referred to an *Evening News* headline 'Waterplane Turns Turtle'. Could this have been the end of the first Sopwith Bat Boat?

The hull of the machine was built by S.E. Saunders Ltd. of Cowes, using two layers of spruce with a waterproof fabric interlayer, sewn together with copper wire by the 'Consuta' method, patented by Saunders. The hull had a flat top and sides and at least one step on the planing bottom, which curved down from the broad prow, but was otherwise flat, as it tapered in width to a point aft. The pilot was seated behind the wings with the passenger in front between the lower wings. The wings were attached to the top of the hull and were of unequal span, with a single bay and with warping for lateral control. Cylindrical stabilizing floats with pointed nose and tail were hung under the lower wing tips. The Gnôme engine, with fuel and oil tanks behind it, was placed on the front of the center section struts and drove a tractor propeller. This position resulted in a high thrust line and undoubtedly contributed to the machine's demise.

Note. The photographs of this machine only came to light in 1991 in Sir Thomas's personal possessions and show the machine at the Saunders works at East Cowes. In retrospect, the machine would appear to have had poor water handling characteristics and whatever happened on test was sufficient to discontinue further work and no mention was made of it subsequently.

Power: 70hp (probably) Gnôme seven-cylinder air-cooled rotary.

Data		
	Span top	30ft approx.
	Weight of hull	130 lb.
	Span bottom	22ft approx.

SOPWITH Tractor biplane Type D (Originally known as the Three-Seater)

The first entry on 30 January 1913 in Sopwith's Order Book was for a three seater biplane developed from the '*Hybrid*'. This was constructed with great urgency to be exhibited at the Olympia Aero Show opening on 14 February 1913. The machine was built in the Rink and was taken by road to Brooklands one week before the show opened. After brief tests by Sopwith and Hawker it was dismantled and dispatched to Olympia. After the show closed it was flown at Brooklands and then delivered to Hendon by a Naval pilot on 1 March 1913. The machine appears to have been accepted, although it was soon returned to the works for modification, nevertheless it was paid for on 20 March 1913, for a sum of £1,085, and became serial No.33. The aircraft saw service in Belgium and France, but was finally deleted on 14 October 1914.

Three more of the type were put in hand and two of these were covered by Admiralty Contract CP45488/13X11298, becoming Nos.103-104, when delivered in September 1913. They were retained in use until September and January 1915 respectively. The third machine was retained by the firm until impressed in August 1914 as No.906. In its short service life it operated in France and Belgium and was used to attack airship sheds at Dusseldorf, before deletion in October 1914.

The War Office bought nine of the type for use by the Military Wing, which were ordered on 8 August 1913 and were delivered between November 1913 and January 1914. These were numbered 243, 246-247, 300, 315, 324-325 and 333. No.243 was tested to destruction at Farnborough and saw no active service.

The original Sopwith Tractor was built as a three seater, but it lost this identity when ordered by the Services, who had a requirement for two seats only. The machine had two bay wings with stagger; warping control was fitted initially to Nos.33 and 906, although ailerons were fitted to these later and to all the subsequent aircraft. The area and the gap were smaller than on the 'Hybrid', at the time when the fuselage had been lowered to the level of the bottom wing on that machine.

The undercarriage was simplified, the skids being shortened and small bumper wheels added. The first machine rested on the trailing portion of the main skids, although a double tail skid was fitted; later machines rested on a modified tail skid, probably due to the revision of the undercarriage and weight redistribution. The original balanced rudder with curved outline and area mainly above the fuselage, was superseded by one with a flattened shape, with the area equally distributed above and below.

No.104 was used by Hawker on 31 May 1913 to establish a British altitude record of 11,450 ft.

Power: 80hp Gnôme seven-cylinder air-cooled rotary driving a 8ft 6in diameter Chauvière propeller.

Data		
	Span	40ft
	Chord	6ft 1in
	Gap	5ft 3in
	Area	365 sq. ft
	Area tailplane/elevator	39sq ft
	Area rudder	10 1/2sq ft
	Length	29ft 6in
	Weight	1,060 lb.
	Weight allup	1,810 lb. (3 seater)
	Speed range	40.6-73.5 mph
	Rate of climb	2 1/3 min to 1,000ft
	Endurance	2 1/2 hr

Note. In addition to the newly-built aircraft listed, No.27, the '*Hybrid*' was returned to the works for rebuilding. Entry No.21 in the Sopwith Order Book dated 5 August 1913, gives a price of £685 for the work. The aircraft left the Rink on 28 October 1913, and was delivered from Brooklands to Eastchurch by Harold Barnwell on 6 November 1913.

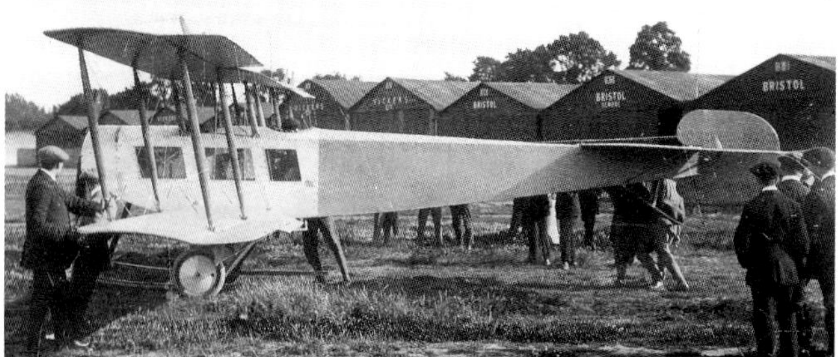

Sopwith tractor biplane Type D originally referred to as the 'Three-seater'. The prototype flew in February 1913 and became No.33 with the RNAS.

Sopwith Type D, in service with the Military Wing in 1913-1915, was used only as a two-seater.

SOPWITH Bat Boats (Type.1) Nos.BB1, 2 and 3

The earlier attempt to make a flying boat was relegated to the past and soon forgotten. This next design evolved in the latter part of 1912 and appeared as the first machine on the floor of the Rink at Kingston in January 1913. Work on it proceeded rapidly, with the object of showing it at the Aero Show, opening at Olympia on 12 February 1913. This was achieved, and the Bat Boat shared the stand with the new three-seater Tractor. The machine was much admired for its workmanship, which included the attractive Saunders-built hull, designed by Sydney 'Joe' Porter. The Admiralty selected it as one of the number of machines, which were ordered at the Show. Contract CP32098/13X2399 was issued as cover on 20 February 1913 at an agreed price of £1,500.

After the show closed on 22 February 1913, the Bat Boat was taken to Kingston and then on to Cowes, where it was housed in the Folly Sheds at Whippingham, and Sopwith and Hawker began testing. During March a number of attempts, by both, to lift the machine off the water were made without success. Sopwith eventually reached a height of a few feet but the Bat Boat dropped back onto the water, severely damaging the hull. The machine was left out overnight and further damage was caused by rough weather, the extent of which was such that Sopwith immediately instructed that a replacement be put in hand to satisfy the Admiralty contract. This was ordered on the works on 1 April 1913 and known as BB2. It was made together with a further machine, known as BB3, an amphibious version intended to compete for the Mortimer Singer Prize. This latter machine was seemingly a rebuild, but how much of BB1 was salvaged is not clear.

Although all three machines were of the same basic design, a number of variations took place during the course of their development. The type number is used here to distinguish these machines from the larger machine which came later; it was not in use at the time.

SOPWITH Bat Boat (Type 1) No.BB1

The hull was twenty-one feet long with a vee entry and a single thirty-four inch deep step at twelve feet from the stempost. The crew were seated side-by-side with the pilot on the port side. A hull of 'Consuta' construction of cedar ply was made by S.E. Saunders Ltd. for this first machine. The ply was applied in strips fifty-six inches wide lengthwise, caravel fashion, with a waterproof fabric layer between the plies.

The two bay wings were of equal span and parallel chord, mounted on short struts above the hull, immediately behind the cockpit. Cutouts in the trailing edges provided clearance for the propeller. The bottom center section was open and consisted of the two main spars only. The original bracing wires to the forward part of the hull were soon replaced by substantial struts to the engine mounting structure. Wing warping control was employed.

Twin strut and wire braced booms, of three bays, extended aft from the inboard interplane struts to a single sternpost. No

Sopwith Bat Boat (Type 1) BB.1 nearly complete in the Rink at Kingston January 1913.

fin was fitted and the single parallel chord rudder with curved top and bottom ends protruded above, and below, the booms. A parallel chord tailplane with divided elevators was mounted on the top booms. An additional elevator was fitted above the nose of the hull. The wing tip floats were of cylindrical form with conical ends and were made of light gauge copper sheet. Bicycle valves were provided to permit inflation for the removal of dents occurring in use.

The coolant radiators were of spiral tube manufacture and were mounted on either side of the engine between the wings. A handle for hand starting from the cockpit operated on the front end of the crankshaft. Fuel was carried in a cylindrical tank on the port side below the top center section.

Power: 90hp Austro-Daimler six-cylinder inline water-cooled driving a two-bladed pusher Levasseur propeller.

Data		
	Span	41ft
	Chord	5ft 6in
	Gap	5ft 6in
	Area	400 sq. ft
	Length	30ft 4in
	Span tailplane	9ft
	Span front elevator	8ft
	Hull length	21ft
	Hull beam	4ft
	Dihedral	2deg
	Incidence	4deg
	Weight of hull	180 lb.
	Weight allup	1,650 lb.
	Max speed	60-65 mph
	Endurance	2hr

The demise of Bat Boat BB.1 on the River Medina at Whippingham in March 1913.

Sopwith Bat Boat BB.2 as first built in mid 1913.

Sopwith Bat Boat BB.3 the amphibious version with Green engine in which Hawker won the Mortimer Singer prize in 1913.

SOPWITH Bat Boat (Type 1) No.BB2

This machine was assembled at the Salterns Yard of Hamble River, Luke & Co. and was delivered to the Navy at Calshot on 8 June 1913 to meet Contract CP32098/13, placed as a consequence of the 1913 Olympia Show. The aircraft was given Serial No.38 and served until September 1914.

After the first machine was damaged at Cowes in March, a number of improvements were made to the second Bat Boat. The twin strut and wire braced booms now had four bays and tapered to a single sternpost. The rudder was of a new type, aerodynamically balanced by extensions forward of the hinge line at both top and bottom and operated between a divided elevator. No fin was fitted and the front mounted elevator was discarded. The wings incorporated ailerons and there was a small increase in area and dihedral angle. The machine was rebuilt after severe damage, caused by heavy seas, while moored overnight at Brighton on 23 August 1913. A triangular fin was fitted together with an oval shaped rudder during the rebuild.

The hull was similar to that used on the previous machine except that it was built by Sopwiths at Kingston, using cedar ply skins and was not fabricated by the Consuta method.

Power: 90hp Austro-Daimler six-cylinder inline water-cooled.

Data Similar to BB1 except for following:

Length	32ft
Dihedral	3deg
Area	428 sq. ft

SOPWITH Bat Boat (Type 1) No.BB3

The machine numbered BB3 in the Order Book was to be used by Hawker to contest the Mortimer Singer Prize, in which he was successful on 9 July 1913. It is not clear which parts of BB1 were used in this aircraft. The hull may have been repaired and reused. However, apart from this the differences from, and damage to the first machine, were such that probably mainly minor items were used.

An undercarriage was necessary for the contest, the main element being a tubular member across the hull mounted in bearings on each side. The wheels were carried in forks on the outboard ends of the tube and were designed to be raised by a lever in the cockpit and to fall under their own weight. The wheels were fitted with tires, two feet in diameter and four inches wide, and were unsprung. The hull was supported at the rear by a tail skid, when on the ground.

The increased length tail booms were fitted, but were opened out to double stern posts to carry twin rudders, of a new shape, and providing more clearance of the propeller tips. A new tailplane, with curved leading edge, carried a one piece elevator. No fins were originally fitted, but small triangular fins were added when the machine was in Naval service in about May 1914.

To comply with the rules of the contest, a British-made Green engine was used, which was removed later, as was the undercarriage, when the machine was transferred to the Navy in February 1914. As No.118 it served until February 1915.

Sopwith Bat Boat BB.2 with later type tail in 1914. Became No.38 with the RNAS.

BB.3 later served as No.118 with the RNAS fitted with twin fins and a headlight and with the wheels removed.

Power: 100hp Green six-cylinder inline water-cooled driving a 9ft diameter pusher propeller.
90hp Austro-Daimler six-cylinder inline water-cooled. A report that a 120hp version of this type was fitted at first was probably incorrect.

Data Basically as for BB2 although an increase in the allup weight of the amphibious version to 1,950/2,000 lb. could be expected. The undercarriage track was 4ft 8in.

SOPWITH Bat Boat (1915 Project)

The small Bat Boat was used extensively by the naval personnel at Calshot, and during the various trials it was inevitable that improvements and other applications were suggested. Flight Commander J.L. Travers, an experienced pilot with design experience prewar at Shorts, carried out much of the flying. He proposed a redesigned version based on No.118, powered by a 100hp Gnôme engine and capable of carrying a machine gun or 150 lb. of bombs. To comply with this proposal, R.J. Ashfield produced a general arrangement drawing No.839, dated 27 January 1915, with tandem seating and a gun mounted forward of the front cockpit. Travers was interested also in a gun mounted on the upper center section, firing above and behind, and suggested that the pilot and passenger positions should be reversed. His request to carry out trials with a gun fitted to No.118 was not approved within the Admiralty and neither was the redesigned Bat Boat, which remained only a project. Details are in Air 1.349.15/227/4. PRO report dated 12 January 1915.

SOPWITH Type HT (Hydro Tractor) biplane

Instructions were given for work to proceed on three of these machines for the Admiralty against Contract CP36209/13/5287 in mid-April 1913, for a total price of £4,795. Delivery of serial Nos.58, 59 and 60 took place in the following June and July. No.59 was eventually converted to a landplane in service.

The design was unrelated to the earlier tractor landplane, except for similarly shaped tail surfaces. The fuselage was a conventional wooden girder structure with a fabricated sheet steel front engine mounting. The wings spanned two and a half bays, the bays adjacent to the fuselage being smaller, and there were strut braced top wing extensions. The top wing center section was completely cut away, and the wings enclosed with shaped endplates. Double acting ailerons were fitted to both top and bottom wings.

Sopwith Type HT (Hydro Tractor) biplane. Three were delivered in 1913. No.59 was used finally as a landplane.

The two main floats were widely spaced to avoid the need for wing tip floats and a cylindrical tail float, with tapered ends, supported the tail. The floats were sprung by four leaf springs and were attached, by the chassis, to the spars at the intersections with the inboard interplane struts. The exhaust, for the radial engine, took the form of two half rings placed out in front and exhausting downwards.

Power: 100hp Anzani ten-cylinder two-row air-cooled radial.

Data

Span	56ft 4in
Chord	5ft 2 1/2in
Floats	15ft long spaced at 10ft 3in at the centerline

SOPWITH hydrobiplane Type ST

At the end of April 1913, instructions were issued to the works to build one aircraft to Contract CP37570/13X3726, for delivery to the Isle of Grain by 11 October 1913, at a cost of £4,000. The machine was partly built at the works of Hamble River, Luke & Co. and was finally assembled at the Sopwith premises at Woolston. The nacelle was recorded as 'delivered' in February 1914, but the aircraft was canceled before final delivery could be made. The engines and wireless equipment were removed and returned to Farnborough and Calshot respectively.

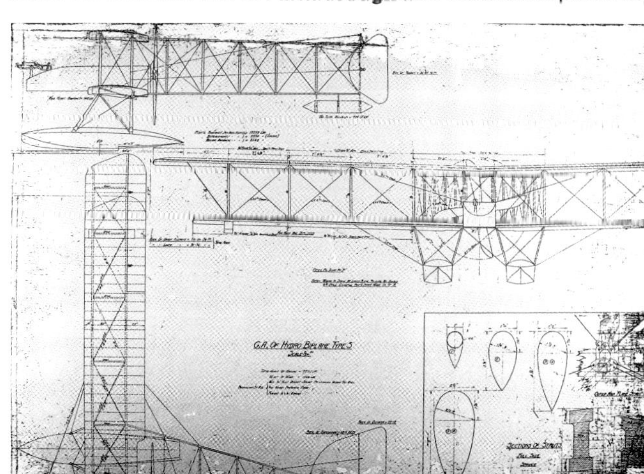

Above: Sopwith Type HT (Hydro Tractor) biplane was known as the Cromarty.
Right: Sopwith Type ST hydro-biplane. This large twin-engined pusher was built in late 1913 as No.61 for the RNAS but was canceled before delivery. It was intended to be a carrier of the Vickers 1 1/2 pounder gun.

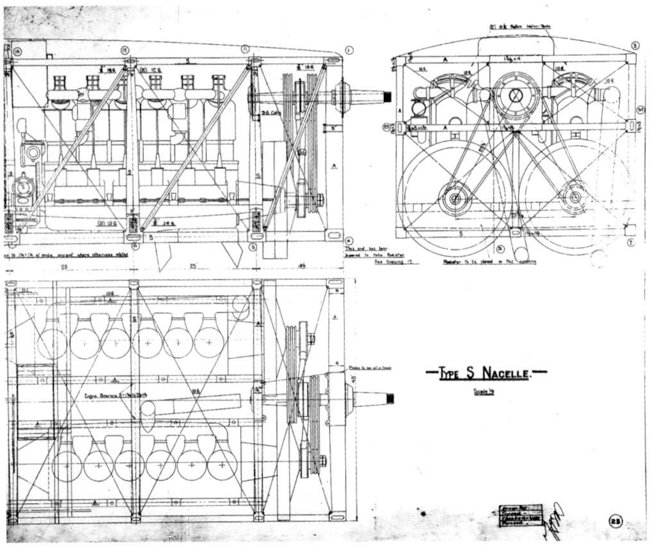

This drawing of the nacelle shows the coupled engine arrangement of the Type ST.

The airframe was held at Woolston for structural testing by loading with sand and was reported to be still there on 7 August 1914. The aircraft was reported to be 'still on order' in the list of 'HM Aircraft, Built, Building or under Repair.' for July 1914. This large aircraft was identified by Serial No.61 and was to carry the pilot and wireless operator side by side, with a gunner in front.

The layout was of a large three-bay pusher biplane with a further center section bay within the width of the tail booms. The top wing, made in halves and joined on the centerline, was of greater span than the bottom, and had extensions supported by struts; the tips were tapered and rounded off. The tail booms were parallel in elevation and tapered in plan to the tail. The control surfaces consisted of ailerons on all four wings, a large rudder with curved leading edge forming a balance area, and a semicircular tailplane with elevators, mounted on the top booms.

The two engines were mounted side-by-side in the rear of the nacelle, driving a single pusher propeller on a separate shaft, by triple link chain drives. Separate radiators and header tanks were provided for each engine at the extreme rear of the nacelle. The pilot and wireless operator were seated side by side just ahead of the wing. The nacelle tapered to a knife-edge at the front with the decking sloping down; above this a pillar mounting was provided for the Vickers 1 1/2 pounder gun, which was to be the subject of the trials.

The large twin floats, with single steps, were mounted on separate chassis under the center section bays; a cylindrical tail float, with tapered ends, supported the tail.

Power: Two 120hp Austro-Daimler six-cylinder inline water-cooled driving by chains and shaft a 12ft diameter four-bladed pusher propeller.

Data		
	Span top	80ft
	Span bottom	70ft
	Chord	7ft 6in
	Gap	6ft 9in
	Area tailplane	65 1/2 sq. ft
	Area elevators	37 1/5 sq. ft
	Area rudder	30 1/2 sq. ft
	Length	44ft 6in
	Height	13ft 10in
	Main floats	20ft long: 3ft 9in beam: 2ft.9in deep
	Weight	5,230 lb.
	Tail float	8ft 9in long: 1ft 6in diameter
	Speed	55mph.
	Endurance	5hr

SOPWITH Tractor hydro-biplane Type ADH

A batch of three of these machines with 90hp Austro-Daimler engines was authorized in advance of contract, on 20 May 1913. The type was not built but was presumably of similar basic design to the Type D, since the cost of work carried out was transferred over to a later War Office contract for more of this latter type on 8 August 1913.

SOPWITH Gun carrying hydro-biplane Type GPH

This large pusher seaplane, Serial No.93 was ordered on Contract CP01717/13X7344, and instructions to the works to proceed were issued on 27 May 1913, with delivery required to the Isle of Grain by 13 August 1913. Initially a 120hp Austro-Daimler was specified, for a total aircraft cost of £2,650 increased to £3,050, when the engine type was changed to a 200hp Salmson at the end of the year.

The aircraft was delivered on 4 February 1914 and had passed its tests by 29 February 1914, but the penalty for the delay, reduced the price by one third. Delivery was actually made to Calshot, where trials of various guns, up to a Vickers 12 pounder, were carried out.

No.93 was returned to the Sopwith Works for rebuild on 29 June 1914. The triangular tail frame was replaced by a rectangular one. The existing tail booms and tailplane were reconstructed by using as much as possible of the original parts. Two rudders were fitted in place of the existing one. The work was completed by mid-July and No.93 was returned to Calshot.

The aircraft was deleted in August 1914. No details or photographs are available but the type formed a basis for other Sopwith pusher types.

Power: 200hp Salmson (Canton-Unné) fourteen-cylinder water-cooled radial.

Data		
	Speed	60mph
	Endurance	4hr

SOPWITH Tractor hydro-biplane Type DM

Sopwith instructed the works on 2 June 1913 to proceed with the manufacture of a twin float seaplane to compete for the £5,000 prize offered by *The Daily Mail,* for the Seaplane Cir-

Sopwith Type DM hydro biplane built to contest the *Daily Mail* Seaplane Circuit of Britain in 1913.

The Sopwith Type DM was bought by the Admiralty to use both as a land and seaplane but was deleted in August 1914.

cuit of Britain, to commence from Southampton Water on 16 August 1913. In the event the Sopwith machine was the only competitor to reach the starting line. However after flying the 240 miles to Yarmouth, accompanied by Kauper, Hawker was forced to give up due to sickness, caused by fumes from a broken exhaust, and sunstroke. A second attempt was made on 25 August 1913, in which they crash landed in the sea near Dublin, after covering 1,043 miles. On the way they suffered a broken oil pipe, and Kauper broke his arm in the crash. Hawker was awarded £1,000 by *The Daily Mail,* in recognition of a remarkable flight.

The aircraft was recovered and rebuilt with an extra fuel tank, to compete in the Michelin Contest later that year. However the machine did not compete, and in February 1914 the Admiralty purchased the aircraft for £1,275 under Contract CP31415/14X2430, for use as both a seaplane and landplane. As serial No.151 it was delivered to Calshot in April, but was soon damaged in a forced landing and was deleted on 19 August 1914. The Navy considered the machine quite unsuitable for observation purposes due, no doubt, to the restricted view from the front cockpit.

The machine was a conventional biplane, with wings of three-and-a-half bays, the smaller bays adjacent to the fuselage. The single step floats were attached at the first interplane struts, and were stabilized by struts to the engine mounting and by cables. the small streamlined tail float was replaced by one of box type, of greater volume.

The control surfaces were conventional with a balanced rudder, but no fin, and there were ailerons on all four wings. The top wing center section was cut away for ease of exit by the crew, and the inboard ends of the wings were fitted with end plates. Radiators were mounted on either side of the fuselage, alongside the front passenger's cockpit.

Sopwith Type DM landplane version.

Power: 100hp Green six-cylinder inline water-cooled driving a 8ft 6in diameter Chauvière propeller.

Data		
	Span	49ft 6in (50ft.)*
	Chord	5ft 3/4in (5ft 11/2in)*
	Gap	5ft 4 1/2in
	Area	500 sq. ft
	Area tailplane	120 sq. ft
	Area elevators	26 sq. ft
	Area rudder	12 sq. ft
	Length	31ft (31ft 6in)*
	Floats	10ft long by 2ft beam (14ft long by 2ft 4 1/2 in beam)*
	Weight	1,600 lb.
	Weight allup	2,400 lb.
	Speed	60-65 mph

*Data from *Flight a*lternatives from *The Aeroplane*

SOPWITH school sea or land pusher biplane Type SPAz & SPGn

Admiral Kerr, who had been seconded to the Greek Navy, ordered three dual controlled seaplanes for training purposes by letter dated 13 October 1913, on behalf of the Greek government, and these were immediately put into work. The total price was £6,205 less one per cent for the fund for widows, with delivery to be made by sea in April, May and July 1914.

The British Admiralty followed with an order for two similar machines Nos.123-124 on Contract CP54578/13X15783, instructions for which were issued to the works on 25 October 1913, with delivery to the Isle of Grain finally made in March 1914 at a total cost of £2,605 as eventually negotiated.

These aircraft were assembled at Hamble and Woolston and test flown by Howard Pixton, with official acceptance, including that for the Greek machines, at Calshot. A further batch of six aircraft was ordered by the Greek government and instructions for these were issued to the works on 18 February 1914. Five were completed as pushers but one was built as a tractor seaplane similar to the 'Circuit of Britain' seaplane Type D.3 (*q.v.*). At the outbreak of war these aircraft known as Type SPGn were commandeered by the British Admiralty, and were delivered between July and October 1914 as serial Nos.897-

Sopwith Type SPAz. with 100hp Anzani. The type was first ordered by the Greek government.

901. They differed from the first batch in being fitted with Gnôme Monosoupape engines and there was a requirement for the dual controls to be made removable, to enable a Lewis gun to be fitted for an offensive role. Cover was on Contract CP50330/14X13443 for a total of £18,356.

The design was of a typical pusher biplane of the period, with three bay wings and smaller center section bays. The tail booms were attached at the intersection of the outer center section struts, with the wing spars. The tail booms tapered in plan to the rudder post and the upright struts were staggered, and divided the booms into four bays. The crew were housed in tandem cockpits in the nacelle forward of the wings. The machine was mounted on two main floats on inverted vee struts, below the outer center section struts; the floats were also stabilized by struts to the lower longerons and by cables, and were sprung by leaf springs at the four main attachments. A small cylindrical tail float was fitted. The RNAS operated some machines as landplanes with four wheel, twin skid undercarriages.

Sopwith Type SPGn Gun Bus. The British Admiralty ordered machines similar to the Greek pusher biplanes but with Gnôme engines.

The control surfaces consisted of ailerons on top and bottom wings, a semicircular tailplane, which was soon enlarged to a rectangular shape to improve stability, and a shield shaped rudder with a small balance area; the rudder was mounted between the booms and was also subject to enlargement after early tests. The engine starting gear was also subject to breakage on the early machines.

Power: 100hp Anzani ten-cylinder air-cooled two-row radial driving a 8ft 6in diameter propeller fitted to the first three Greek and two British machines Nos.123-124. 100hp Gnôme Monosoupape nine-cylinder air-cooled rotary fitted to 5 Greek machines commandeered by British Admiralty as Nos.897-901.

Data	Seaplane.	Landplane.
Sopwith drg. no.	119	199
Span	50ft	50ft
Chord	5ft 1 1/2in	5ft 1 1/2in
Gap	5ft.4 3/16in	5ft 4 3/16in
Area	480 sq. ft	480 sq. ft
Length	30ft 2 3/4in	30ft 3/8in
Height	12ft	10ft 9in
Weight	1,485 lb.	1,185 lb.
Weight allup	2,200 lb.	1,900 lb.
Tail float	4ft 4in long x 1ft dia.	-
Main floats	14ft long 2ft 5in beam 1ft 11in deep	-
Tailplane span	11ft	11ft
Tailplane area	35 sq. ft	35 sq. ft
	increased to 55 sq. ft	
Elevators span	12ft 3in	12ft 3in
Elevators area	22 sq. ft	22 sq. ft
Rudder area	9 sq. ft	9 sq. ft
	increased to 12 sq. ft	

SOPWITH two-seater tractor Type StB (Stunt Bus)

A brief entry in the Order Book at the end of October 1913, against the name of Hawker, authorized the manufacture of the machine which became known as the Tabloid. A separate contract was made to sell the machine to Hawker for £275 for a tour of Australia, with various conditions relating to its use and eventual return, if unsold.

This machine was the type that evolved into the Schneider and the Baby and provided the basis for later small scout bi-

Sopwith Type StB *Stunt Bus*. The first of the type later known as *Tabloids* was a two-seater and was taken to Australia by Hawker.

planes. Hawker had obtained Sopwith's approval to build this aircraft and, together with Sigrist, decided on the configuration.

The machine was flown at Brooklands by Hawker for the first time towards the end of November and, at the weekend of 29-30 November, it was flown to Farnborough for appraisal and to Hendon for display, before returning to Brooklands. The performance figures recorded at Farnborough, particularly the speed range, were outstanding at that time.

Hawker made the first flight in Australia on 27 January 1914 at Melbourne and carried out demonstrations and passenger flights until late April, when the aircraft was dismantled and shipped back to England. By late June the machine was back at the works for overhaul and Hawker flew it again to Farnborough on 13 July 1914. It was then taken briefly on charge by the War Office in August as No.604, but was passed on to the Navy as No.169 soon after. Sopwiths received instructions from Sqn. Comdr. Spenser Grey to convert it to a single seater with an armored seat and, in this form, it was delivered to Eastchurch in September.

The aircraft was a neat little single bay biplane with warping wings and a small amount of dihedral and slight stagger. The wings were built in halves, the lower portions joined to the sides of the fuselage at the longerons, the upper halves to a small center section mounted on four vertical struts.

The fuselage was a conventional braced wooden girder, fully fabric covered. The tail unit comprised a tailplane with curved leading edge, divided elevator and balanced rudder but no fixed fin. The undercarriage consisted of four struts and incorporated twin skids; the axle was split and pivoted at the center, and was sprung by rubber cords. The crew were seated side by side in the narrow fuselage, only 3ft 3in wide, with the pilot on the port side having a wheel for the wing warping, on the control column, and a rudder bar.

At the time looping was becoming a popular stunt, but Hawker had refrained from carrying out the maneuver in Australia. However on his return he practiced on other Sopwith machines, before performing on the Australian machine. For this purpose, the rear fuselage was left uncovered and during the overhaul a vee-strut type undercarriage was fitted, the wheel covers were omitted and extra dihedral was given to the lower wing.

Power: 80hp Gnôme seven-cylinder air-cooled rotary driving a 8ft 2in diameter propeller

Data		
	Span	25ft 6in
	Chord	5ft
	Gap	4ft 3in
	Area	240 sq. ft
	Area tailplane	22 sq. ft
	Area elevator	12 sq. ft
	Area rudder	7 sq. ft
	Length	20ft
	Weight	670 lb.
	Weight	1,060 lb. (with pilot & 3 1/2hr fuel)
	Speed range	36.9 92mph
	Climb to 1,200ft	1min fully loaded

SOPWITH Bat Boats (Type 2) BBS

The Order Book shows, on 17 November 1913, two separate orders for Bat Boats of an enlarged version with the more powerful Salmson (Canton-Unné) engine. The first order was placed by Capt. von Pustau, acting as agent for the German Navy, and the second by the British Admiralty on Contract CP57461/13X18411.

The first appearance of the larger Bat Boat was in March 1914 at the Aero Show at Olympia, when the machine for Germany appeared on the Sopwith stand. After the show it was assembled at Woolston and was flying there in May. Although one source reported that the machine was flown away by Herr Hillman, the firm received in March, an order for packing cases to transport the machine to Germany, which is almost certainly the way in which it was delivered. The aircraft received serial No.44 in the German Naval Air Arm and was based at Kiel-Holtenau for training purposes.

The machine for the Admiralty was flying over Southampton Water in April and early May, when it was taken to Calshot by Howard Pixton. The aircraft was identified as No.127, but it was not well received by the Navy and was passed on to Vice-Admiral Kerr for £3,000 for use by the Greek Naval Air Corps.

Sopwith Bat Boat (Type 2) BBS. Two of this type were built with Salmson engines. The first became RNAS No.127.

One of the Salmson engined Bat Boats was delivered to the German Naval Air Service just before war was declared.

Sopwith Bat Boat (Type 2) BBS. A third machine with Sunbeam engine and other changes was built to compete in the abandoned Seaplane Circuit of Britain of 1914.

A third machine of the type, but with a Sunbeam engine was ordered on 5 March 1914, to be flown by Pixton as No.3 in *The Daily Mail* Seaplane Circuit of Britain, to be held in August, but which was abandoned at the outbreak of war. The aircraft was delivered to Calshot on 9 August 1914 by Pixton and it was taken over by the Navy as No.879 for £2,800. It remained mainly unserviceable, due to engine trouble and damage, until November, when it was suggested that two 100hp Gnôme Monosoupape engines should be fitted. The engine change did not take place and No.879 was dismantled in the sheds at the Salterns at Hamble Spit in April, although the machine was still on Admiralty lists in September 1915.

The Circuit Bat Boat (Type 2) BBS was impressed in August 1914 as RNAS No.879.

Bat Boat No.127 after going to the Greek Naval Air Corps.

SOPWITH Bat Boat (Type 2) German and Greek aircraft

This enlarged version had double the power of the earlier Bat Boats, but the basic design was generally similar. The hull was covered in mahogany, it was 20ft long and was broader and deeper. A new feature were the external ducts, which fed air to the single step to break the suction on take off. The crew were seated side-by-side with the pilot to starboard.

The three-bay wings were of unequal span, and greater in both span and chord than the earlier Bat Boats. The lower wings were mounted direct to the gunwale with three degrees of dihedral, and with 1ft 6in of stagger. Box shaped wing tip floats were mounted on four short struts. The tail booms contained four bays and tapered to a single vertical sternpost. There was no fin and a single oval shaped balanced rudder was fitted. The tailplane with divided elevators was mounted on the top booms and braced to the bottom booms by cables. Sketches in *Flight,* made prior to the Aero Show, showed three bays and a parallel chord unbalanced rudder, with clearance for a one piece elevator, but these features were not incorporated in the machine as exhibited.

The engine was mounted high up on the rear center section struts on cross mountings, with a radiator in front mounted across the two forward struts. A small gravity tank was fed by a wind pump from the main fuel tank in the rear of the hull, giving a total fuel capacity of seventy gallons. The engine was started by compressed air supplied by a system housed in the front of the hull. The machine supplied as No.127 to the RN was fitted with a wireless telegraphy set, ahead of the passenger, and this was powered by a generator driven by a single-cylinder Motosacoche engine.

Power: 200hp Salmson (Canton-Unné) 2M7. fourteen-cylinder water-cooled radial driving a two-bladed Integrale propeller direct.

Data

Span top	54ft
Span bottom	44ft 6in
Chord	6ft 9in
Gap inboard	6ft
Stagger	1ft 6in
Dihedral	3 deg
Area	600 sq. ft
Length	36ft 6in
Height	10ft
Hull	20 ft long 4ft 4in beam 3ft 6in deep
Weight	2,300 lb.
Weight allup	3,120 lb.
Speed range	40-70 mph
Endurance	4 1/2hr (180 mile radius)
Rate of climb	500ft per min

SOPWITH mockup torpedo machine (Type TT)

This machine, which was ordered on 21 November 1913, in advance of the Special Tractor Seaplane No.170, resulted from the earliest Admiralty interest in torpedo dropping from aircraft.

The machine was to be powered by a Green engine supplied by the Admiralty and consisted of a tractor fuselage, mounted on widely spaced, twin main floats, which had two steps and were spring mounted at the rear attachments. A cylindrical float supported the tail. The machine, with its shortspan wings, was intended to be used for taxiing only.

The fuselage, which accommodated two crew in tandem, was uncovered from aft of the rear cockpit. The torpedo and its suspension gear hung below the fuselage, between the float mounting struts. The tail unit had a balanced rudder, which extended down to act as a water rudder, together with a conventional semicircular tailplane and divided elevators. The wing trailing edges were hinged along most of their length to act as ailerons. Radiators were fitted on either side of the fuselage alongside the front cockpit.

The total cost of £632-15-0 was absorbed into Contract CP02007/14X2800 dated 18 February 1914 for No.170. The

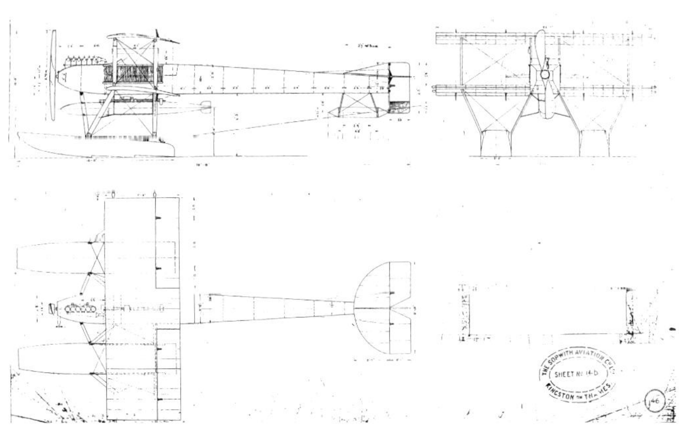

Sopwith Type TT was built early in 1914 to develop torpedo dropping. It was designed for taxiing only.

The RNAS conducted trials with the Sopwith Type TT at Calshot.

machine was completed on 25 March 1914 and was reported to be taxiing at Calshot on 28 April 1914, in the hands of Lt. R.P. Ross, and it continued in use until dismantled at the end of June.

Power: 100hp Green six-cylinder inline water-cooled driving a 8ft 9in diameter propeller

Data (From Sopwith drawing No.146)

Span	22ft
Chord	7ft 6in
Gap	5ft 3in
Area	330 sq. ft
Area tailplane	38 sq. ft
Span tailplane	9ft
Area elevators	16 sq. ft
Span elevators	9ft
Area fin	2 1/2 sq. ft
Area rudder	10 sq. ft
Length	39ft 8in
Height	12ft 9in
Main floats	16ft long 3ft beam 2ft deep (spaced 10ft between centers)

SOPWITH '*Sociable*' two-seater biplane (Type Ds)

Work on this machine was initiated on 24 December 1913, confirmed by Contract CP30680/14X457 soon after. Among the details specified were that the accommodation should be 'roomy' and the seats specially upholstered in leather, requirements by the First Lord of the Admiralty, Winston Churchill,

Sopwith Type Ds, also known as the *Sociable* for its side-by-side seating, was built to the instructions of Winston Churchill in 1914.

for his own comfort. The machine became serial No.149 with the Navy and was also referred to as the '*Churchill*' or '*Tweenie*'. It cost £1,528.

The machine was fitted with dual controls and was basically a scaled-up version of the *Tabloid* with two bay wings, all fitted with inversely tapered ailerons.

Pixton flew the '*Sociable*' for the first time at Brooklands on 17 February 1914, and it was delivered to Hendon a few days later, where it was taken over by Lt. Spenser Grey, who piloted the machine on several occasions with Churchill aboard, until it crashed at Eastchurch on 25 April 1914. Sopwiths repaired the machine, which was not ready until the 27 August 1914. Thereafter it went to Antwerp, from where Lt. Grey unsuccessfully attempted to bomb airship sheds at Cologne, and where the machine was left undergoing repair before the German advance. It was deleted from Admiralty lists on 14 October 1914.

Power: 100hp Gnôme Monosoupape nine-cylinder air-cooled rotary.

Data (From Sopwith drawing No.222)

Span	36ft
Chord	6ft 11/2in
Gap	4ft 6in
Area	336 sq. ft
Area tailplane	19.35 sq. ft
Area elevators	19.90 sq. ft
Length	24ft 3in
Height	9ft
Weight	960 lb.
Weight allup	1,640 lb.

SOPWITH '*Tabloid*' tractor Scout biplane (Type SS)

Hawker's two-seater 'Stunt Biplane' was the basis of the single-seater 'Tabloid' Scout biplane, nine of which were ordered for the RFC, on a War Office Contract A/2368/MA3 at a cost of £1,075 each. The entry in the Order Book on 9 January 1914 followed an entry of 9 December 1913, for one of these machines, which Sopwith had authorized for stock but subsequently canceled. Instructions for a further three machines for the War Office were issued to the works on 14 March 1914, delivery of the whole order being effected between April and August 1914. The Admiralty took over three of these aircraft in September, serial Nos.394 (904), 395 (905) and 604 renumbering them 167-169.

Sopwith Type SS the single-seater *Tabloid* scout derived from Hawker's '*Stunt Bus*' was used by both the RFC and RNAS. In front is Howard Pixton.

Sopwith Type SS.3 was a much modified *Tabloid* scout for the RNAS. The wing stagger was deleted and a gun was mounted on the center section.

One of the machines for the War Office cannot be identified and may have been the machine damaged at Farnborough, for which instructions to repair, and take into stock, were issued on 7 May 1914. This machine was 'ex works' on 25 May 1914 and was costed to an order issued on 8 May 1914 from Societé Transaerea Italiano, for a Tractor Scout less engine. An order was issued on the same day for an identical machine for M. Louis Blériot, which was dispatched by rail on 30 June 1914, and another for W. Lebedeff of Petrograd, Russia, cleared on 30 July 1914. *The Aeroplane* 19 August 1914 (p.180) quoted a report from France: 'The Blériot Co. have built a scout biplane under Sopwith license, and it was recently taken out for test by Pégoud who smashed it on its first flight. It is to be rebuilt again with a considerably altered chassis and several other modifications.'

The '*Tabloid*' SS for the RFC was similar in general design to the two-seater prototype except for the addition of a fin and a reshaped rudder. The twin skid undercarriage with four struts, originally fitted, proved to be inadequate on the early machines and was strengthened with additional struts and stronger wheels. Lateral control was by wing warping. Type SS.2 was covered by drawing No.458, dated 15 June 1914 by R.J. Ashfield, and shows ailerons of seven square feet area on all four wings, and the original vee strut type undercarriage. This version does not seem to have been built.

The Admiralty ordered twelve more '*Tabloid*' aircraft on Contract CP58295/14X19587 dated 3 October 1914, these having been provisioned in advance on 26 September 1914 on verbal instructions, and were referred to as Type SS.3, 'to Mr. Alston's specification'. These aircraft were delivered between December 1914 and March 1915.

The '*Tabloid*' SS.3 for the RNAS was considerably modified from the RFC type. The stagger of the wings was deleted and the gap was increased, enabling a forward firing gun to be mounted on top of the center section, to fire over the propeller arc. The interplane and center section struts were of reduced chord, and apparently made of streamlined steel tube. Ailerons were fitted to all four wings and were connected by slender struts. An undercarriage of vee strut type without skids was fitted, together with a raised tail skid with exposed springing.

A further Admiralty Contract CP60619/14X21893, resulted to an order to the works dated 7 October 1914 to provide a further Scout biplane, which was built using parts made as a batch for the foreign sales. This aircraft was 'ex works' on 6 November 1914 and was delivered to Hendon on 10 November 1914 for the 'Fast Flight'. It became serial No.1213 and was referred to as a '*Tabloid R*'.

Power: 80hp Gnôme seven-cylinder air-cooled rotary driving a 8ft 3in diameter propeller

Data		
	Span	25ft 6in
	Chord	5ft 1 1/2in
	Gap	4ft 6in
	Length	20ft 4in
	Height	8ft 5in
	Area	241 sq. ft
	Area elevators	11.8 sq. ft
	Area fin	1.875 sq. ft
	Area rudder	4.27 sq. ft
	Tailplane span	8ft 3in
	Tailplane chord	4ft 1in
	Tailplane area	11.8 sq. ft
	Weight	730 lb.
	Weight allup	1,120 lb.
	Speed max	93 mph
	Climb to 1,200f.	1 min
	Endurance	3hr

Summary of Production.

Purchaser.	Quantity.	Serial Nos.
War Office A/2368/MA3	9	
Contract amendment	1	362, 378, 381, 386-387, 392, 394 (later 904 & 167), 395 (later 905 & 168), 611, 654
Admiralty CP58295/13X19587		1201-1212
Societé Transaerea Italiano		Not completed
M Louis Blériot		Ex works 30 June 1914
W Lebedeff (Petrograd, Russia)		Ex works 30 July 1914
Admiralty CP60619/14X21893	1	1213

It appears that during 1914 Sopwiths fulfilled orders for a total of twenty-seven aircraft, plus the Australian prototype and the special Schneider seaplane. The latter machine was itself converted back to a landplane, which Hawker crashed at Brooklands on 27 June 1914.

Circuit of Britain of 1914, Sopwith Bat Boat (Type 2) BBS.

SOPWITH Bat Boat (Type 2) Circuit of Britain version

A few changes were necessitated by the fitment of the Sunbeam engine in this machine, which was impressed by the Admiralty as No.879 at the outbreak of war. The main change was to raise the wings on four short struts above the hull to provide clearance for a larger propeller, leaving an open center section across the hull. The engine was mounted on beams running fore and aft between the center section struts. The tail unit was the same as on previous machines but with struts bracing the tailplane to the lower booms. Later a triangular fin and unbalanced rudder were fitted. The wingtip floats originally fitted were cylindrical but these later reverted to floats of box shape.

Power: 225hp Sunbeam twelve-cylinder water-cooled vee driving a two-bladed Integrale propeller later changed to a four-bladed propeller in Naval service.

Data As before except for the following:

Span top	55ft
Span bottom	45ft
Height	11ft
Weight allup	3,180 lb.
Speed range	48-75 mph

SOPWITH special torpedo seaplane

Following the order for the Mock-Up Torpedo Machine (Type TT), but before any trials had been carried out, Contract CP02007/14X2800 was officially placed on 18 February 1914. Delivery on 18 July 1914 was quoted, but a later request to expedite delivery was made and resulted in the machine being sent to Calshot on 1 July 1914. On 6 July 1914 the aircraft, No.170, was complete and taxiing tests took place, but the machine would not lift off. It was 9 July 1914 before the ma-

Sopwith special torpedo seaplane Type C (RNAS No.170) was not able to fly with a torpedo and was abandoned after unsatisfactory trials.

chine became airborne, without the observer and with only a small fuel load for its first flight. It soon became apparent that the machine would not carry the intended load without alteration, and during August the aircraft was taken to the Sopwith works at Woolston where, among other changes, the wings were altered, before it was returned to Calshot at the end of the month.

The original purpose, for which the machine was provided, was soon abandoned. In November experimental bomb dropping gear was fitted, but development was soon discontinued and in January the aircraft was at the Salterns works of the late Hamble River, Luke & Co. being dismantled. The machine had cost £2,870.

The Special Torpedo Seaplane had fourbay wings, with the overhang of the top wings braced by struts. Ailerons were fitted to both top and bottom wings. There was some resemblance to No.151, the *Daily Mail* seaplane, in the arrangement of the floats and chassis but with the addition of balancing floats to the wing tips, and springing of the main floats at the four connections. The tail unit included a fin.

The fuselage carried the crew in tandem, with the pilot in the rear seat. The engine was mounted well back from the nose, and drove the propeller through a long extension shaft. Very tall radiators were mounted on either side of the fuselage, which may have provided inadequate cooling as side panels were not fitted ahead of these.

The general layout was similar to Figs.1 and 2, forming part of the patent No.6938 of 19 March 1914 taken out by Capt. Murray Sueter and Lt. Hyde Thomson for a torpedo carrying seaplane, although the crew positions were reversed. The original Sopwith drawing No.219, signed by R.J. Ashfield on 12 January 1913 (surely 1914) shows that changes to the actual machine were made. The wing stagger was reduced and the position of the radiators lowered as on No.138. The original cylindrical tail float was changed to one of box type.

Power: 200hp Salmson 2M7 (Canton-Unné) fourteen-cylinder two-row water-cooled radial driving a 9ft diameter propeller.

Data		
	Span top	66ft
	Span bottom	58ft
	Chord	8ft
	Gap	6ft 3in
	Area	785 sq. ft
	Area tailplane	36.68 sq. ft
	Area elevators	26.40 sq. ft
	Area rudder	10 sq. ft
	Area fin	5.20sq ft
	Length	36ft
	Height	12ft 10in
	Main floats	16ft 6in long 2ft 10in beam 2ft 4in deep 8ft 6in between centers
	Weight	3,243 lb.
	Weight allup	3,705 lb. with 4hr fuel
	Weight allup	4,324 lb. with 2hr fuel & torpedo

Sopwith Type HS Schneider Cup seaplane by Turk's boathouse at Kingston after conversion to twin floats before dispatch to Monaco.

SOPWITH Schneider Cup seaplane (Type HS hydro-tractor seaplane)

The instructions issued to the works on 2 February 1914 for this machine, recorded that it was for exhibition at the Olympia Aero Show. Whether this was really intended is not clear, and did not happen in any case. The machine was worked on as a matter of urgency and was completed at Kingston on 31 March 1914 for entry in the Schneider Trophy Contest to be held at Monaco on 20 April 1914, in which it flew as race No.3.

The machine was taken immediately to the works of Hamble River, Luke & Co., where it was assembled prior to testing by Howard Pixton. The first of April was not a good day, for the machine capsized immediately Pixton opened up the throttle, owing to the central float being set too far aft. The following day the machine was recovered from the water, was then dismantled and returned to Kingston, where, by a great effort, it was reconstructed by late on the 6 April 1914. The following day, at first light, the machine was taken to the Thames at Turk's boathouse for a flotation test, which was satisfactory, but the machine was prevented from flying by Thames Conservancy officials. On 8 April 1914 the men from the works pushed the machine in the early hours to Richmond, where the

The winner of the Schneider Contest at Monaco, April 1914, flown by Howard Pixton.

Port of London Authority was responsible for the river and the machine was flown briefly by Pixton with a misfiring engine.

The aircraft was dispatched from Kingston on 9 April 1914 by rail and sea to Monaco, where it arrived on 16 April 1914. During testing it was decided to change the propeller and to fit a small additional fuel tank, which was done on the day before the race. Pixton was the winner, covering the twenty-eight laps of 280 kilometers at 85.5 mph. He continued for a further two laps, thus establishing a world record for 300 kilometers at a speed of 92 mph. This was the greatest achievement in British aviation to that date.

The Schneider seaplane was a more powerful version of the military '*Tabloid*' scout, mounted on floats. When the machine was taken to Hamble originally it was fitted with a broad main pontoon float and a box like tail float of similar cross-section, mounted on three struts well below the rear fuselage. In the course of repairing the aircraft the pontoon was sawn in half and the insides of the halves were boxed in, forming twin floats, spaced well apart and positioned further forward. Also prior to the contest a new shaped tail float, with water rudder, was fitted and was raised to a position just below the fuselage.

By the 20 May 1914 the machine had been converted to landplane configuration and was at Brooklands, where Pixton flew it the following day in preparation for the Aerial Derby in which Barnwell flew the Schneider unsuccessfully as race No.21. Hawker, back from Australia, flew it as No.14 in a race from London to Manchester. In both cases the aircraft retired due to weather conditions or Hawker's illness. On 27 June 1914, after carrying out aerobatics, Hawker lost control and crashed at Byfleet, wrecking the aircraft without injury to himself.

The Sopwith company received three separate contracts totaling 136 aircraft in 1915 for Schneider type seaplanes, with minor changes to improve the aircraft and adapt it for naval use. An even later development, with a change of engine became known as the Sopwith Baby and was produced in larger numbers by Blackburns and Faireys as well as Sopwiths.

Power: 100hp Gnôme Monosoupape nine-cylinder air-cooled rotary originally driving a 8ft 8in diameter Lang propeller, replaced by a smaller Integrale of coarser pitch.

Data

Span	25ft 8in
Length	22ft 10in
Chord	5ft 2in
Gap	4ft 6in
Area	250 sq. ft
Height	10ft
Weight	1,220 lb.
Weight allup	1,700 lb.

SOPWITH tractor seaplanes Admiralty Types 137 and 138

A Contract CP30775/14X2387 initiated instructions to the works on 11 February 1914 to proceed with the manufacture of two seaplanes of similar type, but with quite different engines, and these became serial Nos.137 and 138 in service with the Navy. The machines were built at Kingston with final assembly of 137 at Woolston and 138 at Calshot. The two aircraft were recorded 'ex works' on the 15th and 5th of August respectively, and were handed over to the Navy at Calshot a few days later, being accepted on the 21st and 12th of August. The total cost was £2,269 for No.137 and £2,633 for No.138.

In service the machines suffered the usual periods of unserviceability, particularly 137, which was crashed badly on 3 September 1914. No.138 was used for torpedo carrying trials in August and September and made several successful drops. Both aircraft remained in the Solent area and both were deleted on 1 January 1916.

The two machines were of the same basic design, with changes associated with the use of different engines. The wings were of unequal span, with struts bracing the overhanging top portions and ailerons on all four wings. The wings of No.137 were of two bay type, but No.138 required more wing area, the increased span being provided by the insertion of a smaller inner bay.

The Schneider seaplane was fitted with a land undercarriage and was raced by Barnwell as No.21 in the Aerial Derby in May 1914 without success.

Sopwith Admiralty Type 137 seaplane, assembled at Woolston in 1914, had an Austro Daimler engine.

The fuselage was a wooden braced girder with the pilot seated in the rear cockpit. A conventional tail unit of fin, rudder, tailplane and divided elevators was fitted. The machine was mounted on a chassis of struts in the form of an inverted-W with crossbars. The twin single step floats were sprung by leaf springs, at the rear attachments only. A tail float was carried on struts well below the rear fuselage, and was fitted with a water rudder.

A Sopwith drawing No.197, dated 20 January 1914, exists of the Salmson-engined version and drawing No.213, dated 8 January 1914, of the Austro-Daimler version, both signed by Ashfield and these show various features which were later changed on the actual aircraft. These changes included the reduction of stagger to overcome tail heaviness on No.138, and the use of a larger tailplane of rectangular form, rather than one with a semicircular leading edge. This latter change appears also on No.137, as did the incorporation of ailerons on all four wings. The original position of the radiators on No.138 had been higher, with their tops being just below the top wing. On No.137 the original water tank under the top center section was retained, but the flank radiators were replaced by one at the nose just behind the propeller.

Power: No.137 120hp Austro-Daimler six-cylinder inline water-cooled later changed in service to that as fitted in No.138.
No.138 200hp Salmson 2M7 (Canton-Unné) fourteen-cylinder two-row water-cooled radial. Initially fitted with a two-bladed Lang propeller later changed to one with four blades.

Data (From Sopwith drawings Nos.197 and 213)

No.137		
	Span top	48ft
	Span bottom	36ft
	Chord	6ft 9in (7ft 6in over top ailerons)
	Gap	6ft
	Stagger	1ft 3in
	Dihedral	1 1/4 deg.
	Area	517 sq. ft
	Length	33ft 2in
	Height	12ft 3in
	Weight	1,980 lb.
	Weight allup	2,650 lb. with 5hr of fuel
	Area tailplane	36.68 sq. ft
	Area elevators	26.40 sq. ft
	Area fin	5.20 sq. ft
	Area rudder	10 sq. ft
	Top ailerons	28 sq. ft each (inc. in wing area)
	Span tailplane	10ft 4in
	Span elevators	12ft
	Main floats	spaced at 10ft centres 15ft long 2ft 4in beam 2ft 1in deep

Sopwith Admiralty Type 138 seaplane was larger than 137 and was fitted with a more powerful Salmson engine.

No138		
	Span top	56ft
	Span bottom	48ft
	Chord	6ft 9in (7ft 6in over top ailerons)
	Gap	6ft 3in
	Area	658 sq. ft
	Length	34ft 10in
	Height	12ft 8in
	Weight	2,190 lb.
	Weight allup	3,450 lb.
	Area tailplane	36.68 sq. ft
	Area elevators	26.40sq ft
	Area fin	5.20 sq. ft
	Area rudder	10 sq. ft
	Top ailerons	30 sq. ft each (inc. in wing area)
	Span tailplane	10ft 4in
	Span elevators	12ft
	Main floats	spaced at 10ft centers 16ft long 2ft 9in beam 2ft 1in deep

SOPWITH tractor land seaplane 1914 Circuit of Britain Type D.3 Admiralty Type 880 Type D.5 nicknamed '*Spinning Jenny*'

Instructions went out to the works on 9 March 1914 to build a new type two-seater military biplane for stock. This emerged as a landplane at Brooklands on 16 July 1914, and was tested by Victor Mahl, prior to being converted to seaplane configuration, to compete in the 1914 *Daily Mail* Circuit of Britain

Sopwith Type D.3 built in 1914 to compete in the Circuit of Britain. Victor Mahl was pilot, Tom Sopwith just behind the machine.

Sopwith Type D.3 as a seaplane bearing the compromised serial No.896.

Despite its reputation, the '*Spinning Jenny*' carried out patrol work from Yarmouth but later in 1915 was used as a trainer.

Contest. This was abandoned at the outbreak of war and the machine was taken over at Calshot by the Admiralty under Contract CP30567/14X, becoming No.880. It carried out patrols from stations on the East Coast until deleted in May 1915.

A second seaplane version of this type, probably built against the Greek order, was in existence in August 1914, and was taken over under Contract CP50330/14X at Calshot. Although this machine was described elsewhere as a 'Type 880 Folder', it had equal span wings, the top center section was cut away, and it was fitted with end plates as on the 'Circuit' machine. Also it bore the Sopwith name on the side at the same time as its serial No.896. The wings were staggered as on the 'Circuit' machine and folding may not have been possible.

The Admiralty was sufficiently interested to place an order for twenty-four of the landplane version with Sopwiths on Contract CP58282/14X19119 of 29 September 1914, at a total cost of £23,677-4-0, with the 80hp Gnôme engines supplied as free issue. Completion of the batch, serial Nos.1051-1074, took place between November 1914 and June 1915.

In appearance the aircraft was an enlarged version of the earlier Scout type, with two bay wings and elongated fuselage, seating the crew in tandem with the pilot behind. Both crew members had windscreens and raised headrests. The tail unit was conventional, with fin, rudder and tailplane, with curved leading edge, and divided elevator. The vertical surfaces were increased after the first landplane was tested. Ailerons were fitted to all four wings and originally were inversely tapered, but on the production machines, the wings were rigged without stagger, the gap was increased and parallel chord ailerons were fitted.

Sopwith Type D.5 developed from the Circuit biplane had unstaggered wings and was nicknamed '*Spinning Jenny*'.

The land undercarriage for testing of the 'Circuit' machine was of the twin skid type but this was changed to one of 'V' type on production aircraft. The seaplane version had twin pontoon type floats, sprung only at the rear attachments by leaf springs, and a box type tail float fitted with a water rudder.

Several aircraft were used for patrol work along the East Coast and a bomb rack was fitted to some operational aircraft. The type acquired a reputation for unprovoked spinning and, although soon relegated to training duties, was out of service by the end of 1915.

Power: 100hp Gnôme Monosoupape nine-cylinder air-cooled rotary driving a 8ft 6in diameter propeller (D3)
80hp Gnôme nine-cylinder air-cooled rotary (D5)

Data From Sopwith drawings No.383 landplane, and an unnumbered drawing for seaplane.

D3 'Circuit' Aircraft

Span	36ft 6in
Chord	6ft 11/2in to 6ft 7 1/2in at ailerons
Gap	4ft 9in
Stagger	1ft
Area	440 sq. ft
Area fin	2.8 sq. ft
Area tailplane	18.5 sq. ft
Area elevators	22.5 sq. ft
Area rudder	6.5 sq. ft
Area ailerons	12.5 sq. ft each
Length	27ft 6in
Height	8ft 8in
Weight	1,160 lb.
Weight allup	1,800 lb.
Speed	60-80 mph

D3 'Circuit' Aircraft additional data for seaplane version

Length	30ft 9 1/2in
Height	10ft
Weight	1,310 lb.
Weight allup	1,950 lb.
Main floats	12ft long 2ft beam 1ft 6in deep
Tail float	8ft long 1ft deep

D5 '*Spinning Jenny*' other data as for D3 aircraft

Gap	5ft 6in
Stagger	Nil
Speed	69 mph
Climb to 3,000ft	20min
Endurance	3 1/2hr

SOPWITH tractor biplanes (Gordon Bennett racers)

The outbreak of war prevented the holding of the contest for the Gordon Bennett Cup of 1914 for which Sopwith had authorized the construction of two machines on 7 May 1914. Both were completed and purchased by the Admiralty against Contract CP60619/14/X21593 dated 21 October 1914, being identified as Nos.1214-1215, and were delivered to Hendon on 19 October 1914 and 25 November 1914 for the 'Fast Flight'. In naval service, No.1214 went to France for a time, before being finally deleted in March 1916. No.1215 was deleted at Chingford in July 1915.

The two machines were quite unlike one another. No.1214 was similar to a 'Tabloid' and, when in service had hoop shaped skids under the wing tips and was fitted with a forward firing Lewis gun with propeller blade deflectors. Originally to have a 100hp Monosoupape Gnôme, an 80hp Gnôme engine was fitted when supplied to the Admiralty.

No.1215 was intended to be a really fast machine, fitted with a 200hp Gnôme engine, but was downgraded to 80hp, when acquired by the Navy. Although this machine was of typical Sopwith Scout layout, it had a more streamlined appearance. Starting with the completely circular, dome shaped nose cone, the fuselage was faired to circular section, progressively changing to a flatter form adjacent to the tail unit. The nose cone had only a small central opening for cooling air to enter around the spinner and the propeller boss.

The vertical tail surfaces consisted of small triangular shaped fins, above and below the tailplane, which carried a

Sopwith Gordon Bennett racer No.1215 was used in the RNAS at Hendon with a 80hp instead of a 200hp Gnôme.

rudder of almost circular form, flat along the hinge line. The tailplane extended well along the fuselage sides and carried a divided elevator. The front center section struts were canted forward and warping control was fitted to both aircraft.

Power: No.1214 80hp Gnôme seven-cylinder air-cooled rotary fitted instead of 100hp Gnôme Monosoupape B nine-cylinder originally planned.
No.1215 80hp Gnôme seven-cylinder air-cooled rotary fitted instead of 200hp Gnôme Monosoupape M nine-cylinder originally planned.

Data No.1215 Speed with 80hp 105 mph

SOPWITH seaplane Type FTS.200 (Admiralty Type C)

The three aircraft ordered on Contract CP37385/14/X7331, costing £3,122 each, were not a good investment for the Admiralty, for they failed badly in their early trials, and it was the rival Short 184 which entered production.

The naval specification included folding wings, carriage of wireless, a gun and bomb gear. Instructions for the three machines were issued to the works on 15 April 1914, and they

Sopwith Gordon Bennett racer No.1214 when taken into the RNAS was fitted with a forward firing gun with deflectors on the propeller.

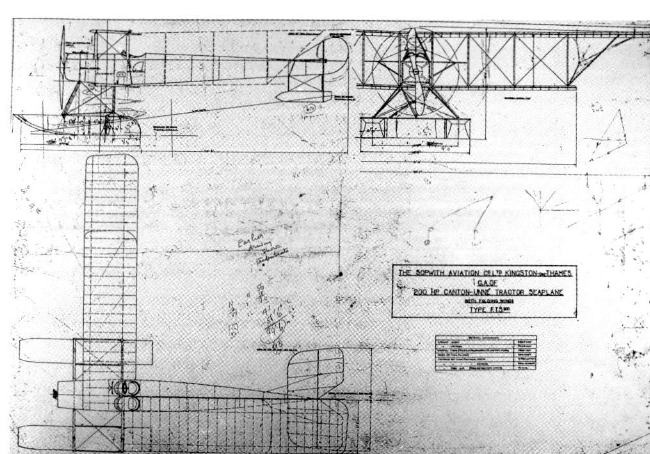

Sopwith Type FTS seaplane Admiralty Type C torpedo carrier. This GA drawing shows side-by-side seating and probably approximates to the actual aircraft of which no photographs have been traced.

were completed in September and October and taken to Calshot in November. After abortive attempts at flight with load aboard the machines were accepted but the engines were removed and development was abandoned. All three were deleted in 1915. No photographs of Nos.157-159 can be traced, but the design bridged the period between the types numbered 137-138 and 170 and the later machine, known as Admiralty Type 860.

An unnumbered Sopwith drawing for a machine of this general specification exists, Type FTS.200, and may represent the type at some stage of its design. It was a side by side two-seater with the pilot in the starboard seat, a note indicated there was an earlier version with tandem seating. The machine was similar in general layout to No.138, being a three bay biplane of unequal span with twin pontoon type main floats, sprung by leaf springs at the rear attachments only and without steps. A box like tail float with water rudder was fitted.

The wings were not staggered and consideration seems to have been given to moving the folding joint further outboard, at the same time folding down the top wing extensions by 'breaking' the bracing struts. The radiator was apparently to be hung on the front cabane struts which were in the form of an inverted vee, to which the radiator shape would conform. The chassis was shown with crossbars, but sketches show a triangulated structure raising the center for the carriage of a bomb or torpedo. For this role the floats would be increased in buoyancy by increasing their length and beam, at the same time bringing their centerlines inboard.

Power: 200hp Salmson 2M7 (Canton-Unné) fourteen-cylinder two-row water-cooled radial driving a 9ft 6in diameter propeller.

Data		
	Span top	58ft 6in
	Span bottom	41ft
	Gap	6ft 3in
	Area	630 sq. ft
	Length	38ft 6in
	Height	13ft
	Chord	6ft 3in
	Area tailplane	39 sq. ft
	Area elevators	28 sq. ft
	Area fin	6 sq. ft
	Area rudder	13 sq. ft
	Main floats	14ft long 2ft 9in beam 2ft 1in deep spaced at 10ft centers
	Later amended	15ft 3in long 3ft beam 2ft 1in deep spaced at 9ft. centres
	Endurance	5hr (90 gal)

SOPWITH tractor biplane (Admiralty Type D)

This single seater aircraft remains something of a mystery for, although ordered by the Admiralty on Contract CP37385/14/X7331 and allocated Serial No.160 it was apparently canceled and there is no evidence of completion. The suggestion has been made that it was retained by the firm for Hawker to use as a means of transport to naval bases. It had been promised for delivery to Eastchurch on 14 September 1914 at a price of £1,473.

The Order Book shows an alteration of engine from 80hp Gnôme to one of 100hp from the Schneider machine.

SOPWITH land scouting pusher biplane. Admiralty Type 806 '*Gunbus*'

A batch of six pusher biplanes, similar to those built for the Greek government but with land chassis, was ordered by the Admiralty on 14 August 1914 on Contract CP02133/14X13857 with Serial Nos.801-806 allocated.

Two types of Sunbeam engine were to be supplied free issue; 110hp. engines were fitted initially to 801-802 and 805, but were soon replaced with the 150hp type, which were fitted to all six aircraft. The redundant engines were returned to the airship shed at Farnborough for disposal. The total cost for six aircraft, including engine changes was £7,893. The first machine flew at Brooklands in early October 1914 and two were delivered to Hendon during that month, the last aircraft did not leave the works until March 1915.

The machines were supplied with a 'telescopic' gun mounting for use by a gunner in the rear cockpit, although later changes to the design of the nacelle were made, in which the positions of the crew were reversed.

The water-cooled engine required a large radiator, which was mounted on the nacelle just behind the rear cockpit. The two wheeled land undercarriage incorporated twin skids and was mounted by vee struts to points on the lower longerons at the main center section spars, with a further strut from the skids to the lower longerons. A rectangular fin was fitted within the booms.

The type was mainly used for training purposes and, although the operational life of the first batch was over by the end of 1915, a further batch of thirty aircraft serial Nos.3833-3862 was ordered in June of that year, from Robey & Co. of Lincoln, of which the last thirteen were not assembled and flown. The service life of the Type 806 was over by the end of 1916.

This was the first order for Sopwith aircraft to be placed with another contractor.

Sopwith Gun Bus Admiralty Type 806 used a Sunbeam engine. Many were built by Robey & Co. of Lincoln.

Power: 110hp Sunbeam eight-cylinder water-cooled vee in Nos.801-802 and 805 initially
150hp Sunbeam Crusader eight-cylinder water-cooled vee driving a 9ft 10in diameter Lang propeller fitted to Nos.801-806 later
150hp Sunbeam Nubian eight-cylinder water-cooled vee in Nos.3833-3862

Data Dimensions as for Type SPAz except as follows:

Length	32ft 6in
Height	11ft 4in
Area	474 sq. ft Quoted in Robey publication.
Area tailplane	28 sq. ft *
Area elevators	28 sq. ft *
Area rudder	13 sq. ft *
Area fin	8 3/4 sq. ft *

*From Sopwith drawing (WO 977) dated 4 May 1915 drawn by R.J. Ashfield passed by H. Smith

SOPWITH tractor seaplane (Admiralty Type 860)

An initial Contract CP02140/14/X13443 for ten 'Torpedo Carrying Seaplanes' appeared in the Order Book on 21 August 1914, with a further entry on 13 October 1914 for another twelve on Contract CP02165/14/X20893. The first batch carried serial Nos.851-860, the second Nos.927-938, of which six machines Nos.927, 932, 934 and 936-938 were not delivered, being either retained at Woolston by the firm for trials, or were not complete when the contract was closed in September 1916. The extent of completion of these aircraft is uncertain, although there is photographic evidence that the last machine, No.938, was completed. Deliveries were made between December 1914 and June 1915. A further batch of 24 aircraft ordered on 7 December 1914, Serial Nos.1280-1299 and 1347-1350 were costed at £53,322, but were canceled well before completion. The Short 184 with the same engine, built to the same requirement, was more successful and entered large scale production. The Sopwith Type 860 had a short service life and was being discarded in the second half of 1915.

This Type 860 at Isle of Grain in 1915 had the enlarged fin and wire braced wing extensions.

Victor Mahl carried out much of the seaplane flying at Woolston, and on 1 November 1914 he took off in 851 with Reg Alston, a senior Sopwith designer as passenger, only to crash soon after. Alston was drowned, but Mahl was thrown clear, only to die himself on 1 April 1915 after an operation.

The fuselage of the Type 860 was of conventional wooden braced girder type, with the pilot in the rear cockpit. The engine was cooled by a front mounted radiator and had a tall exhaust manifold combining the pipes from between the cylinder blocks of the vee-twelve engine On No.852, tall side mounted radiators were fitted, with a cylindrical shaped header tank below the top center section; this latter system became the standard fitment and replaced a box section tank on an earlier machine.

The tail unit was conventional with a divided elevator and a triangular fin with an unbalanced rudder. The fin was later considerably enlarged, No.852 having an ugly rectangular unit, which was later rounded off at the front.

The machine was mounted on twin sprung floats of pontoon type, with a chassis of inverted W-form struts with crossbars, the latter carrying torpedo crutches, a role which was tested but not proceeded with. Wing tip and tail floats, the latter with water rudder, were of pontoon type and hung on four struts well below their mounting points under the wings and rear fuselage. No.855 was initially tested as a landplane at Brooklands.

The wings were first of unequal span, two bay type with Nstrut and kingpost bracing of the overhang on No.851. The

Sopwith seaplane Admiralty Type 860. The prototype with strut braced wing extensions which crashed at Woolston 1 November 1914 killing Reg Alston, senior Sopwith designer.

No.938 at Woolston, with three bay wings and reshaped fin, was the final version of the Admiralty Type 860.

struts were subsequently changed to wire bracing. Pairs of interplane struts at the wing roots, adjacent to the center section, were provided as part of the arrangement for wing folding. A large cutout in the center section was provided to enable a gun to be fitted and to permit an easier exit from the front cockpit in an emergency.

The need for more wing area caused the lower wings to be increased by 100 sq. ft to the same span as the top wing making the aircraft into a three bay biplane. A drawing proposing the new wings and a revised tail unit, provides the following data, although the tail surfaces fitted to the final version of the aircraft were quite different.

Power: 225hp Sunbeam Mohawk twelve-cylinder water-cooled vee driving a two-bladed propeller. No.851 was tested with a four-bladed propeller.

Data

Span	62ft 11in
Chord	6ft 10in
Area	790 sq. ft
Area tailplane	65.4 sq. ft
Area elevators	25.6 sq. ft
Area fin	16.8 sq. ft
Area rudder	12 sq. ft

SOPWITH tractor folder seaplane Type FTGn (Admiralty Type 807)

The first of three batches of seaplanes derived from the 'Circuit' seaplane, was ordered on Contract CP02142/14/X15709 on 29 August 1914; this was for four aircraft and was followed by a further Contract CP012155/X/19188 on 29 September 1914, for a further eight aircraft at a cost of £12,968 with engines supplied free issue. The first batch, serial Nos.807-810, was delivered by the end of the year, the others, Nos.919-926, between December 1914 and March 1915.

The major difference from the '*Spinning Jenny*' landplane was the introduction of folding wings to facilitate shipboard stowage. To accomplish this, the lower wing was reduced in span to allow the interplane struts to clear the tailplane. The upper wing was extended and braced by cables and kingposts to provide additional wing area, and additional struts were fitted to support the wing roots adjacent to the fuselage. A reduction in the width of the inner bays resulted. A small cutout to clear the tail float struts was necessary in the trailing edge of the lower wing at the tip, with the wings folded. A scheme to provide more area by tapering the chord of the top wing does not appear to have taken place, although a small extension of the lower wing seems to have been made on later aircraft.

Sopwith Type FTGn 'folder' seaplane Admiralty Type 807. Only one of this type was in use after mid-1915 as a landplane in Mesopotamia.

A special long nosepiece was fitted to the engine and the top cowling extended forwards to the front bearing. Aircraft taken to East Africa and the Mediterranean had their front cowlings cut away to improve cooling, exposing the front bearing supports. The twin main pontoon floats with crossbars, were mounted on inverted W-struts and were sprung, fore and aft, by leaf springs. The tail float was hung well below the rear fuselage and was fitted with a water rudder.

The overseas operations were generally unsatisfactory in hot conditions and the machines, being lightly built, were easily damaged. With the exception of No.920, which continued to operate as a landplane in Mesopotamia until the end of 1916, all the others were discarded by the middle of 1915.

Power: 100hp Gnôme Monosoupape nine-cylinder air-cooled rotary driving a 9ft 3in diameter two-bladed Chauvière propeller.

Data (From Sopwith drawing No.604)

Span top	43ft 6in
Span bottom	33ft
Span folded	11ft 4in
Chord	5ft 8in
Gap	5ft 9in
Area	405 sq. ft
Area tailplane	24 sq. ft
Area elevators	24.5 sq. ft
Area fin	4 sq. ft
Area rudder	10.25 sq. ft
Area ailerons	19.5 sq. ft
Length	30ft 9in
Height	11ft 2in
Weight	1580 lb
Weight-allup	2,440 lb
Main floats	12ft long 2ft beam 1ft 6in deep
Tail float	5ft 9in long 1ft 9in beam 1ft 3in deep

SOPWITH tractor biplane '*The Sigrist Bus*'

An entry on 7 December 1914, in the Order Book, authorized a 80hp two-seater tractor to Mr. Sigrist's instructions, to be built for stock. The machine was built by the Experimental Dept in the Rink at Kingston incorporating Fred Sigrist's ideas to aid production, as well as advancing performance. Sigrist was basically responsible for the Works, but was sufficiently powerful to influence design, backed by Hawker's experience and with Sopwith's approval.

The Type 807 Folder, itself derived from the '*Circuit*' and '*Spinning Jenny*' biplanes, was the basis for the new aircraft. The fuselage displayed little difference, but it was to the wing structure and tail unit that changes were made.

Sopwith tractor biplane 'The *Sigrist Bus*'.

The top wing was built in halves, meeting on the aircraft centerline and was rigged with dihedral. The mounting was by two sets of W-form struts, the outer arms of which were widely splayed to support the wing well outboard, and enabling the use of a single bay. Ailerons were fitted to all wings, with the control cables running externally across the span of the lower wing, which was slightly shorter than the upper wing. To help to overcome the spinning problem of the earlier type a taller fin and rudder, of greater area, were fitted. The configuration of the top wing support structure was used in the very successful 1 1/2 strutter.

The aircraft was flown at Brooklands for the first time on 5 June 1915 and the following day Hawker flew it to Hendon in thirteen minutes and then proceeded to set a British Altitude Record, by flying to a height in excess of the recorded 20,000ft, although this was later adjusted down to 18,293ft due to the inaccuracy of the barograph.

Power: 80hp Gnôme Monosoupape nine-cylinder air-cooled rotary, Lang propeller.

Data (From unnumbered drawing by R.J. Ashfield countersigned H. Smith 15 December 14)

Span top	31ft 6in
Span bottom	29ft 9 3/4in
Chord	5ft 2in
Area	298 sq. ft inc. ailerons
Area fin	4.4 sq. ft
Area tailplane	25 sq. ft
Area rudder	8 sq. ft
Area ailerons	10.2 sq. ft each
Area elevators	16 sq. ft
Length	25ft 6in
Height	9ft 9in
Dihedral	top 7 deg bottom 2 deg
Weight	920 lb.
Weight allup	1,580 lb.

SOUTH ESSEX Gliding Club (80 Medway Rd., Bow, London, E)

A biplane was reported under construction and due for completion in February 1912.

SOUTH-WEST OF ENGLAND Aeronautical Society (A.J. Fransella, Secretary, 51 St. Leonard's Rd., East Sheen, Surrey)

This club was building a monoplane in early 1910, the heavily cambered wings and triangular section fuselage girder of which were illustrated in *The Aero* under construction, in the works at Down Place, King's St., Hammersmith. The motor was reported as 'well in progress' in May. The society foundered and the machine was taken over by some members of the Aeronautical Society, who, according to the brother of the secretary Bernard Fransella, flew it successfully.

South West of England Aeronautical Society monoplane under construction.

SPENCER Family of Aeronauts and Aviators (Highgate, London, N)

The first member of the family to be airborne was Edward Spencer (1799-1849), who made a balloon ascent on 28 May 1836, from the Surrey Zoological Gardens. His son, Charles Green Spencer (1837-1890), named after his godfather the famous balloonist, was also an aeronaut and important in the development of cycling; he founded the firm of C.G. Spencer & Sons. Originally residing at 14 Ringcroft Rd., Holloway and later at 56a Highbury Grove, he had eleven children. Of these children, the eldest son, Percival (1864-1913), founded Spencer Bros. Ltd., aeronauts, at 45b Aberdeen Park, Highbury with his brother, Arthur Charles (1866), and concentrated on ballooning and parachuting, although Charles Spencer made a glider in 1868, which was shown at an exhibition, organized by the Aeronautical Society at the Crystal Palace. Henry (1877-1937) worked with his brother-in-law, Auguste Gaudron, at Alexandra Palace and with Short Bros. during the war and was also involved in both ballooning and parachuting. He won the Gordon Bennett Balloon Race in 1922, and restarted the firm of Spencer Bros. in 1923. Stanley Edward (1868-1904) built airships and experimented with these at the Crystal Palace from 1902 and flew across London in 1903. Another brother, Sidney (1879), was also involved in ballooning and served in the RNAS and RAF in 1916-1919 and later became a chemist.

The youngest son, Herbert Spencer (1884-1949), like his brothers, became involved with ballooning and parachuting but, in addition, he made and flew his own aircraft and received Aviators Certificate No.124 in August 1911. His first authenticated machine was a collaborative venture built by C.G. Spencer & Sons, but later he operated from 40 Sackville St., London, W, and from Shed No.9 at Brooklands, where he ran a flying school from the spring of 1912 until the end of 1913.

A monoplane was built by C.G. Spencer & Co. in 1905-1906 for Alvarez, (*q.v.*), which was taken aloft by a balloon at Hendon. The general style of the machine can be seen to be rather primitive and unlikely to have been a success. Apart from the 1868 glider, this is the first heavier than air machine made by the Spencers of which there is evidence.

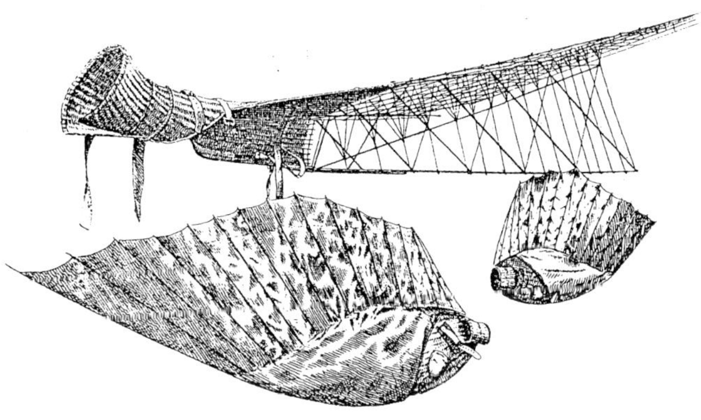

Spencer glider of 1868 was exhibited at the Crystal Palace. Charles Spencer claimed to have flown from a slope in this strange device.

SPENCER glider (Charles Green Spencer)

The glider shown at the Crystal Palace in 1868 was made from umbrella wires and wickerwork. The framework of the wings was covered with silk. It was tested by Charles Spencer, who weighed 140 lb., and claimed a flight of 120ft had been made by him, after taking off from a slope. The pilot had to insert himself into the device, which was a tight fit and uncomfortable. Spencer refrained from giving a demonstration at the exhibition.

Suitable fin propellers were fitted to the body and worked by the pilot through universal joints. British patent No.1178 of 1868 was refused provisional protection.

Data		
	Length of tail	18ft
	Length of wing	7ft
	Width at the end	8ft
	Chord at widest point	4ft
	Depth of keel at the end	4ft
	Area of wing	15 sq. ft
	Area of tail	72 sq. ft

SPENCER-STIRLING monoplane (C.G. Spencer & Sons, 56a Highbury Grove, London, N)

The machine jointly designed by Herbert Spencer and W Stirling was exhibited on the stand of the Berliet Co. at the Olympia Aero Show in March 1910. It was not a success and there were no further reports after the Show; the engine was used in a later aircraft.

Spencer-Stirling monoplane was exhibited at Olympia in March 1910 but was soon abandoned.

The machine was a tractor monoplane with uncovered triangular section wooden fuselage girder with the single longeron at the top. The triangular section was extended into an A-shaped frame at the front, the upper portion housing a fuel tank and two radiators. The lower extensions formed part of the undercarriage, which included twin skids and four wheels on twin axles. A sprung tail wheel was fitted.

The tail unit consisted of a balanced rudder, with the tailplane mounted well in front on the lower longerons. The elevators were pivoted on the outer ends of the tailplane. A warping wing, with parallel chord and rounded tips, was positioned on the top longerons.

The engine, carried in the center of the A-frame, drove twin propellers by chain, on a structure extended outwards from the A-frame. A gearbox on one side only provided contra-rotation of the propellers.

Power: 40-70hp RH four-cylinder inline water-cooled driving two 6ft 6in diameter two-bladed propellers at half engine speed of 600 rpm.

Data		
	Span	34ft
	Chord	6ft
	Area	210 sq. ft
	Area tailplane	19.25 sq. ft
	Area elevators	9.65 sq. ft
	Area rudder	5.80 sq. ft
	Weight	650 lb.

Spencer-Stirling biplane with 40hp RH engine. Herbert Spencer used this machine from November 1910 until it was crashed by another pilot in February 1912.

SPENCER-STIRLING biplane (Herbert Spencer, 40 Sackville St., London, W)

Flight reported Spencer flying straights on this machine on 9 November 1910 and again on 28 November 1910. The machine was of Farman type, with strut braced top wing extensions and single acting ailerons on all wings. It was initially fitted with the RH engine from the monoplane and carried Spencer's name and address on the rudders.

Little more seems to have been achieved through the winter, until more straight flights were reported to have taken place on 27 May 1911. Henri Pecquet tested the machine in June and soon after a Gnôme engine was fitted, which was being ground run on 27 July 1911. From then onwards the machine appears to have been successful, for Spencer passed his aviators tests on it in August, and it continued in use for competition and tuition flying until crashed by Watkins on 25 February 1912.

At the beginning of its life the machine was referred to as a Spencer-Stirling, but Stirling seems to have given up his interest and later the name Spencer only was used.

Power: 40-70hp RH four-cylinder inline water-cooled.
50hp Gnôme seven-cylinder air-cooled rotary.

The Spencer-Stirling biplane was later fitted with a 50hp Gnôme engine.

SPENCER biplane 1912 (Herbert Spencer, Brooklands, Shed 9)

Having lost his earlier biplane, Spencer acquired the old Macfie *Empress* of 1910, and fitted it with a Farman type tail and front elevator, but kept the low wheel base and high engine and pilot's position. The machine was similar to a Sommer pusher type of the time, with single acting ailerons on all wings. It had an undercarriage with short skids and two pairs of wheels on separate axles.

The machine was ready for flight at Brooklands towards the end of March 1912 and made its first flight on the 31 March 1912. Thereafter it was flown successfully by Spencer for competition and instructional work, until the end of the year, including a visit to Dorset in August, where an exhibition was given at Sherborne on 5 August 1912. Spencer continued to use the machine throughout most of the following year.

Power: 50hp Gnôme seven-cylinder air-cooled rotary.

Spencer biplane. Herbert Spencer used part of the Macfie *Empress* biplane to construct this machine which made its first flight at Brooklands in mid-1912.

SPENCER biplane Henry Farman type (Herbert Spencer, Shed 9 Brooklands)

Another biplane built by Herbert Spencer, this was an obvious copy of the Henry Farman F.20 and was used at Brooklands at the Spencer Flying School in 1913/14.

The 50hp Gnôme seven-cylinder rotary engine was mounted at the rear of a long nacelle which had seats for two. It was taken over by the Admiralty in September 1914 as No.200 and was flown at Hendon as a trainer. It was struck off charge there in mid-1915.

Spencer biplane of Henry Farman type was used as a trainer at Brooklands and impressed in September 1914 as RNAS No.200.

STAMPS monoplane glider (N. Stamps, Birmingham)
This glider was completed in March 1914, and was to be followed by a powered monoplane of 'original design'. The glider was largely constructed of bamboo. No further reports.

Data	Span	30ft
	Chord	7ft
	Length	20ft

STANIER monoplane glider (John Stanier, Lansdowne Rd., Crumpsall, Manchester)
This glider was built in 1909-1910 and was exhibited at the Manchester Aero Club Show, which started on 4 March 1910.

Star monoplane. The first version tested at Dunstall Park before being shown at Olympia in March 1910.

STAR monoplane (Star Engineering Co., Frederick St., Wolverhampton)
The proprietor of the company engaged Granville Bradshaw, later the designer of ABC engines, to begin aviation work. The first version of the machine was designed by Bradshaw and tested by him at Dunstall Park in January and February 1910. It was not flown prior to being dispatched for exhibition at the Olympia Aero Show in March.

The fuselage was a braced ash girder of triangular section with the single longeron at the bottom and was fabric covered. The undercarriage consisted of two inverted vee frames connected fore and aft by members which extended forward as twin skids, and carried the wheels, which were originally of aluminium. The machine rested nearly horizontally on a tall tail skid.

The wings, which had skids at the tips, had a slight taper on the trailing edge, did not warp, and were braced by wires to a single post. By the time the machine was exhibited, wheels with tires were fitted and a tail wheel superseded the tail skid. Also an inverted vee pylon replaced the single post, and this was extended down to the level of the wheel centers. The undercarriage had been completely changed by the addition of a frame with cross members, carrying the wheels in forks which pivoted on what was now an A-frame structure. A central skid replaced the twin skids of the original version.

Star monoplane. The second version was tested at Dunstall Park and moved to Brooklands in early 1911.

The system of control was unique. A long fixed fin carried a rhomboidal shaped rudder above the fuselage; a similar rudder below was supported at its lower end by the tail wheel support member, which replaced the tail skid. Fixed tailplanes and elevators of similar shape to the fin and rudder were also fitted. The control surfaces could be moved in the conventional way to control pitch and yaw and in unison to give a twisting motion for lateral control. The system was not a success and was changed later.

After the Aero Show the machine was tested at Dunstall Park, and it was reported to have been flown by Bradshaw for 300 yards at a height of 20-30 feet in late August, but was damaged on 3 September 1910, when avoiding a car.

By November, the original car type engine had been replaced with a lighter engine, designed for aircraft use, and the long tubular radiators alongside the fuselage replaced by a block radiator behind the engine.

When the machine appeared in 1911 and was described by *Flight*, the wings had been reduced in span and had rounded tips and warping for lateral control. The differential control of the tail unit of the tail unit was abandoned, together with the lower rudder, and the top rudder was reshaped with a square trailing edge outline. The original divided elevators were retained, as was the tailplane, but without the differential movement of the elevators. A steel spring tail skid was fitted. The undercarriage reverted to a twin skid type, much strengthened with additional struts to the lower longerons and a cross axle with two wheels. The wing bracing and warp wires were carried to a single post.

The new Star aero engine, which replaced the car type engine, was reported to weigh 100 lb. less. The aircraft flew for the first time with this engine on 10 November 1910 at Dunstall Park. Bradshaw and Joseph Lisle, the son of the proprietor, Edward Lisle, are both believed to have flown the machine. The monoplane was transferred to Brooklands Shed 11 during November, and Bradshaw flew it there for the first time on 2 December 1910, intending to use it to obtain his certificate. It was flown again on 22 December 1910 and 10 January 1911 by J. Valentine and thereafter there were no further reports. The directors of the Star company had become disillusioned with aviation, and Bradshaw parted company with the firm. He remained at Brooklands where, in association with R.L. Charteris, the ABC Engine Co. was formed and began its operations.

The Star engine survived after the monoplane was dismantled, and was adapted for use in an unidentified ornithopter. It remains on exhibition in the RAF Museum at Hendon.

Power: 30hp Star four-cylinder inline water-cooled car type engine (RAC rating 15hp) driving a 6ft 6in diameter Star propeller.
30-40hp Star four-cylinder inline water-cooled driving a 6ft 8in diameter Clarke propeller.

Data	**First Version**	
	Span	42ft
	Chord	8ft to 6ft 6in
	Area	290 sq. ft
	Area tailplane	28 sq. ft
	Area elevators	20 sq. ft
	Area rudders	20 sq. ft
	Length	32ft
	Weight	550 lb.
	Weight allup	750 lb.
	Speed	36 mph
	Price	£450
Data	**Second Version**	
	Span	37ft
	Chord	7ft
	Area	231 sq. ft
	Area tailplane	30 sq. ft
	Area elevators	18 sq. ft
	Area rudder	10 sq. ft
	Length	32ft 6in

STAR biplane

A biplane built on Farman lines was reported complete and ready to go to the flying ground in November 1910. This machine was also abandoned.

STEWART monoplane (Norman S. Stewart, Scout Aero Club)

This very small machine was exhibited at the Stanley Show in November 1910 and was tested at Cuffley, Hertfordshire. in January 1911 with unknown success.

Stewart monoplane shown at the Stanley Show by the Scout Aero Club in November 1910.

The aircraft was of conventional tractor monoplane layout, with wooden fuselage girder, mounted on a high Blériot-style undercarriage and tail wheel. The wings were heavily cambered and rigged at a large incidence angle. Single acting ailerons were hinged on the trailing edge. A fixed tailplane and dorsal and ventral fins were fitted through the girder, with a universal pivoting cruciform tail unit behind for control.

Power: 20hp Alvaston two-cylinder horizontally opposed water-cooled.

Data	Span	16ft
	Area	75 sq. ft
	Length	15ft

STOCKS & COCHRANE biplane (Lts. A.D. Stocks and Cochrane., RN Submarine Depot, Gosport, Hampshire)

The machine was constructed privately by the two officers with volunteer assistance. The aircraft was tested during the Spring of 1909, and on 29 May 1909 the driving chains broke. A trial on 20 November 1909 resulted in severe damage just as the machine lifted.

The biplane was of Wright type with front and rear elevators, rear mounted rudder and ailerons for lateral control. The 40hp motor, of unknown make, drove twin propellers by chain.

Data Span 40ft

Stoddart ornithopter. An optimistic design of 1907-1909.

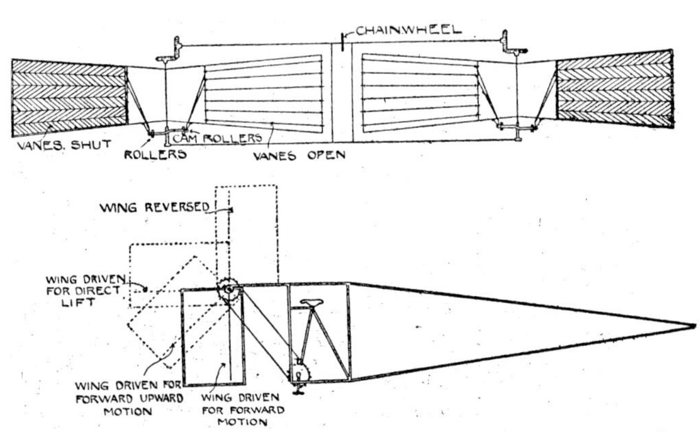

STODDART ornithopter (J. Stoddart, Eltham, Kent)
Nothing is known of this device, which probably dates from 19079, other than what can be deduced from the photograph of the machine undergoing tests, while hanging from the roof of a shed. It will be seen to consist of a tubular structure, incorporating cycle pedal gear, driving through various linkages to open and close the vanes in the 'wings'. The vanes were presumably intended to provide thrust as they moved, and lift when closed. No doubt the machine had still to be completed, or perhaps was just a ground test device, to establish the feasibility of the design.

STONE biplane (Melville Stone, Macclesfield, Cheshire)
This machine was reported to be under construction in 1910.

STONY STRATFORD Aero Club glider (Old Stratford, Buckinghamshire)
This model club was presented with the Wynn Glider (*q.v.*) at the beginning of 1913. This had been made originally as a powered machine, but was later tried as a glider. The first test on 15 February 1913 showed the need to fit wheels and a seat. The machine was a monoplane of 'tail-first' type with a span of 32ft and weighed 50 lb.

STOREY aeroplanes (Isaac Henry Storey, Loughrigg Brow, Ambleside, Cumbria)
Whilst not an actual aircraft constructor himself, Isaac Henry Storey patented a number of original designs and collaborated with two other pioneers John Kitchen and Cedric Lee.

Storey was a local magistrate and wealthy industrialist (linoleum manufacturers) and had an inventive turn of mind. His patent 28203 of 1909 covered an elegant egg shaped streamlined nacelle incorporating cockpit, fuel tank and engine for attachment to wings and/or fuselage. His remaining patents 9635 of 1909 and 885 and 29531 of 1910 covered circular or annular winged aircraft which he believed would prove much more stable in flight than conventional machines with straight wings.

Storey's ideas were developed by Kitchen, Lee and Tilghman Richards and flown in various prototypes. (q.v. Kitchen and Lee).

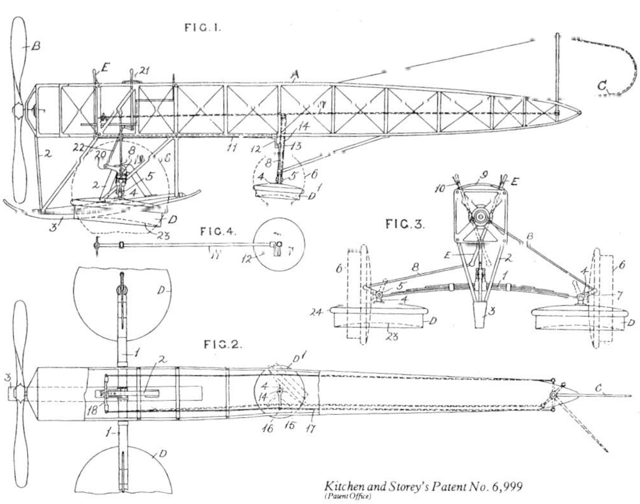

Storey aeroplanes. Storey patented various designs in 1909-1910.

STRINGFELLOW model aircraft (John Stringfellow, Chard, Somerset)
John Stringfellow and William Henson worked closely together on various aeronautical projects between the years 1835 and 1847. Henson emigrated to the U.S.A. in 1848 and it seems likely that it was Stringfellow who was entirely responsible for the first ever engine driven model aeroplane to make a free flight. This was in 1848.

Stringfellow continued to design and fly revolutionary model aeroplanes, including the triplane which was shown and flown at the 1868 Crystal Palace Exhibition. (see also the earlier notes on William Henson).

STRINGFELLOW models (Frederick John Stringfellow, Candle Cottage, Chard)
Following his father's death in 1883, his son Frederick designed and built several model steam engines and various biplanes, triplanes and even a quintuplane. Some of these were displayed at the Aeronautical Exhibition at Alexandra Palace in 1903.

Stringfellow models. Some of the many models built by John Stringfellow and his son between 1835-1900.

SUMMERFIELD glider. (S. Summerfield & Co., 34 Sherrard St, Melton Mowbray, Leicestershire.) (Melton Aeroplane Co., Nottingham Street, Melton Mowbray)

No information has been traced on this biplane glider which was reported built in 1911. Summerfields were mainly builders of model aeroplanes and engines.

Summerfield glider was built at Melton Mowbray in 1911.

SWANN monoplane/biplane (Rev. Sidney Swann, The Vicarage, Crosby Ravensworth, Cumbria)

After attending the Blackpool Meeting of October 1909, the Rev. Swann was inspired to make an attempt to win the *Daily Mail* £1,000 prize, which was eventually won by Moore-Brabazon on 30 October 1909, before Swann's machine was completed.

The aircraft, which was manufactured by the Austin Motor Co. at Longbridge, Birmingham, was delivered on 17 November 1909 to Aintree Racecourse, from where Swann proposed to compete for the *Hartley* £1,000 prize for a flight from Liverpool to Manchester. First tests took place on 19 November 1909, when taxiing only was carried out, but later tests on the 26th and 30th resulted in short hops and extensive damage.

During the following weeks a lower wing was added and the machine was tested in biplane form on 3 January 1910, but engine, and other troubles, convinced Swann that a completely new design was needed. The time limit for the *Hartley* prize had expired but a new *Daily Mail* prize of £10,000 for a flight from London to Manchester was then the incentive to proceed.

In its initial form, the aircraft had a high monoplane wing mounted above a wooden girder structure, which curved forward into twin skids and extended aft as a deep tail boom. A fixed tailplane was mounted on the top of the boom with a movable elevator at the extreme rear. There was no rudder. Widechord single acting ailerons, with curved, scalloped trailing edges were hinged to the rear spar and extended well past the trailing edge. The undercarriage consisted of two wheels on a cross axle and a long tail skid. The pilot sat behind the

Swann monoplane was built by the Austin Motor Co. and tested at Aintree racecourse in 1909.

engine and radiator. The wing was built in three sections and was braced by wires and kingposts.

Power: 35hp ENV type D eight-cylinder water-cooled vee driving a 7ft diameter pusher propeller by belt and shaft.

Data Span 36ft
Length 39ft

Swann biplane. After little success with the monoplane it was extensively redesigned as a biplane but was again a failure.

SWANN biplane 1910

This machine used the original engine and may have incorporated other parts, but it was completely different from the earlier machine. It was basically a two bay biplane with single pusher propeller mounted direct on the crankshaft. There was no tail, the machine being a true 'canard' with long bamboo booms, tapering to a point at the front. This carried a tall rudder within the booms, and a separate biplane elevator pivoted on the front extremity of the booms. The top wing carried single acting wide chord ailerons, with curved and scalloped trailing edges.

Swann became disillusioned by his lack of success and, after removing the engine, burned the remains of the machine.
Power: 35hp ENV type D eight-cylinder water-cooled vee.

SWANSEA monoplane (Ernest and George Sutton, Picton Mews, Swansea, Glamorgan)

It is not clear if this machine was a genuine Blériot, or one constructed on similar lines by the Sutton brothers who were motor engineers. It was being flown by Ernest Sutton, with occasional damage occurring, at Oxwich Bay, at intervals throughout 1911. Ernest Sutton was the first aviator to fly in South Wales.

T

TALBOT-QUICK waterplane (J.J. Talbot and W.B. Quick and manager Croft, Fambridge, Essex)

The machine was constructed in Pemberton Billing's old hangar by local boat builders, supervised by Albert Pink, an engi-

Talbot-Quick waterplane was launched on the River Crouch in August 1914 and was a complete failure.

neer from London, in the first half of 1914. It was launched on the 20 August 1914 on to the River Crouch. In the course of this, one of the launch crew was drowned. Modifications were required and later, while making an attempt to take off, the machine capsized after 300 yards had been covered at speed on the water, fortunately without injury to the pilot. The engine was then removed and the machine abandoned.

Even by 1914 standards, the design was crude and obsolete. The machine was a single-engined pusher flying boat with wings of parallel chord, curved at the tips to meet together, this being a change made after the first test. From the center structure, top and bottom booms extended fore and aft, to taper to points at both extremities, in side elevation. The front boom carried a biplane elevator, interconnected to a similar unit at the rear. Twin rudders were fitted within the rear booms.

The center structure contained the water-cooled marine engine, of unknown make, with radiators on either side and drove a four-bladed pusher propeller. The pilot was seated ahead of the engine on the lower center section. The whole of this structure was mounted above a boat hull with flared stern. The center of gravity was obviously high and the machine was apparently unstable laterally, requiring the addition of pontoon type floats under the wing tips, together with ailerons between the wings after the first attempts.

The machine received little mention at the time and no further details can be traced except for a patent No.6829 which was registered in 1913.

TASKER aeroplane (Tasker Bros., Cycle & Motor Engineers, Bisham Rd., Southport, Lancashire)
This machine was reported being built in December 1912, fitted with a 75hp six-cylinder engine.

TAYLOR flying machine (F. Taylor, London)
Mr. F. Taylor was reported to be experimenting with "a new flying machine" in Gladstone Park, Willesden, North London in April 1909. Nothing further is known.

TAYLOR monoplane (George Crosland Taylor, 'Ravenscar', Helsby, Cheshire)
This monoplane, built in 1909, was reported to have wire gauze covering for the wing and may well have been built by the Crosville Motor Company Ltd., Crane Wharf, Chester, of which Crosland Taylor was a director.

TAYLOR monoplane & biplane (R.W. Taylor & Son, 54 South Clough Lane, King's Lynn, Norfolk)
This concern advertised the following machines in 1909, of which nothing more was heard. Biplane with 30hp English engine £350. Monoplane with 18hp English engine £250.

TEASDALE-BUCKELL vertical lift machine (C. Teasdale-Buckell, RN, HMS *Ariadne*, Portsmouth and 265 The Strand, London, WC)
This attempt to achieve vertical flight was made in 1910, the machine being built and reported to have lifted off on test.

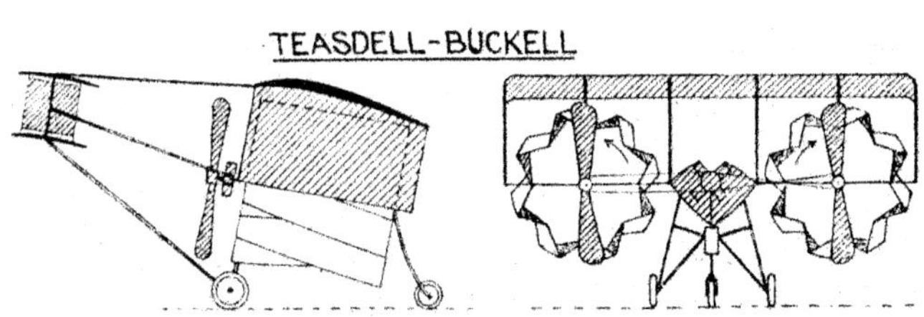

Teasdale-Buckell vertical lift machine of 1910 was not successful in achieving flight.

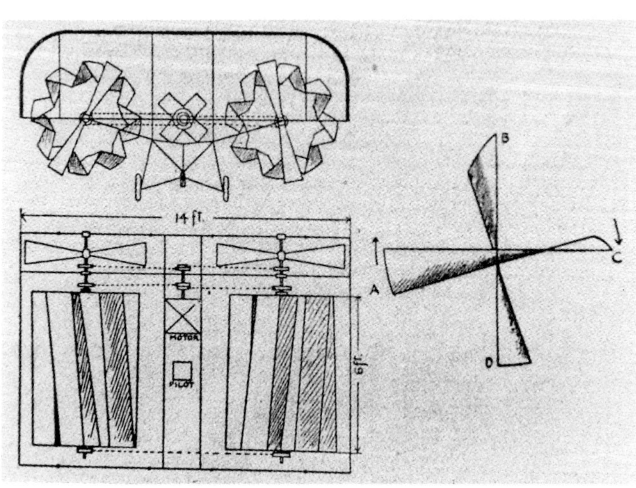

The design was original and covered by patent No.30143/1909. The machine was a high wing monoplane, with the wing tips curving down to meet members supporting the bearings of the twin propeller shafts, thus forming shrouds extending down to their centerlines. Behind, and on the same centerlines, were a pair of 'helically corrugated cylindrical surfaces' or 'helixplanes', also driven by belts or chains from the same central motor. These cylinders were designed to provide lift from the slipstream impinging on the corrugations, at their outboard sides as they rotated, being driven in opposite directions by the crossed drive.

The pilot sat in the center behind the 30hp motor, which was cooled by a fan; an elevator and tailplane were left as an option to be fitted later if tests indicated the need.

Data		
	Span	14ft
	Area	166 sq. ft
	Length	18ft
	Helix planes	5ft diameter
		6ft long
	Corrugations	2ft deep
	Weight	500 lb.
	Speed	70 mph

THAMES BANK WHARF Co. aeroplanes (Westminster, London W, and Halliford, Middlesex)
This company built the two Hornstein biplanes (*q.v.*) in 1909-1910 and when described in an article in *The Motor Boat* magazine in December, reference was made to 'a monoplane and three hydroplanes'. The first was in structure form at the time and was for a Russian owner, Hornstein perhaps? The aerohydroplanes were to be monoplanes fitted with 'catamaran hulls'. Messrs. Livet and Ernest de Loriere were involved in their construction, as they also were for the Hornstein aircraft.

A report in the press in January 1911 referred to 'one complete biplane, two unfinished monoplanes and two unfinished biplanes' at the company premises at Upper Halliford, where flying of the earlier Hornstein aircraft took place.

THIERSCH monoplane (Alexander Thiersch, Plumstead, London SE)
Thiersch bought a wing from Handley Page, which was made at the Barking works in 1910. He probably bought the 20hp Advance engine, originally fitted to the HP type A *Bluebird*,

Thiersch monoplane was partly built by Handley Page and was tested on Erith Marshes in 1910.

for this was apparently the type of engine fitted. Thiersch completed the monoplane to his own design and is reported to have made successful flights near Belvedere on Erith Marshes on 26 May 1910, including one of three quarters of a mile.

The monoplane, a single-seater was of conventional tractor layout, was unusual in the fitment of oval shaped surfaces pivoted through their lengthwise centers, below and ahead of the wing tips. These would appear to serve as both elevators and ailerons.

Power: 20hp Advance four-cylinder air-cooled vee.

THOMPSON & JAMES
No information traced.

THRELFALL glider (Stafford Threlfall, 30 Withy Grove, Manchester)
This glider was at the Manchester Aero Club Show at the White City in March 1910, of which Threlfall was the organizing secretary. This may also have been known as the Threlfall and Leigh Glydoplane.

Tinline biplane model is believed to be a scale version of the machine tested at Eastchurch in 1910 without success.

TINLINE biplane (James D. Madden Tinline, Clifden, Teignmouth, previously of The Grange, Rockbeare, Exeter, Devon in 1897)
Tinline designed a biplane in 1910, which was tested at Eastchurch without success due to engine problems. In March he had exhibited a large model of a Voisinlike biplane with a four and a half horse power two-cylinder motor, possibly a scale version of the full size machine.

TRANMER glider (J.W.F. Tranmer, Scalby Road, Scarborough, Yorkshire)
The Aero of 11 January 1910 advertised this glider for £10. It was built of bamboo and best piano wire with extra strong canvas. The wing span was 21ft and it had a pivoted tail. It had been used only a few times.

TRANMER biplane (J.W.F. Tranmer)
This tractor biplane was built in 1910, but was not flown successfully. It was fitted with a 40hp JAP eight-cylinder vee air-cooled engine and Chauvière propeller.

Travers-Ayers helicopter model took off under its own power.

TRAVERS-AYERS hovering aeroplane and ornithopters. (Travers-Ayers, 19 Dornton Road, Balham, London SW12)
Although the 'Hovering Aeroplane' was only made in model form in 1913, it showed considerable ingenuity and was seen to lift off from water successfully. The ornithopters were tested full size on Tooting Common in 1914.

The 'Hovering Aeroplane' was described as being in the form of a tetrapod with cross bracing. The four upright members came together at the top and were spread apart at the bottom and each carried a float near the lower end. The cross structure formed a biplane wing of some three bays, which were covered at the outboard bays only, this area being in the form of louvers, which would open for vertical flight and close for the horizontal case. Elevators and a rudder were carried on booms.

The rotary engine drove, through a gearbox, a vertical shaft carrying contra-rotating propellers, rotors, at the top and towards the bottom of the structure. A shaft at right angles drove a propeller for horizontal flight. The propellers could be selected by the engagement of suitable clutches.

Two ornithopters were made by Travers-Ayers. The first consisted of a light framework weighing 50 lb., in which a man stood on a platform to manipulate the birdlike wings of some 25ft span. A tail elevator of 5ft. span was fitted.

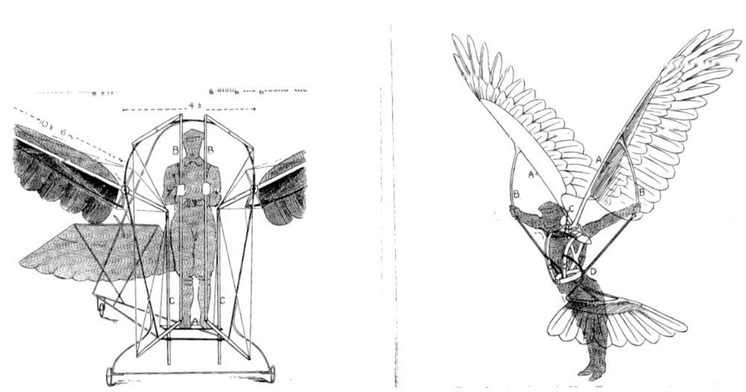

Travers-Ayers ornithopters were tested on Tooting Common in 1914.

The second ornithopter was strapped directly to the man's back. The wings were of 16ft span, constructed with artificial feathers, which could be raised and lowered by the operator pulling on arms, forming part of the wings.

TRYCKLE biplane glider
See Birmingham Aero Club gliders.

TUBBS direct lift aeroplane (Cyril Tubbs, Elcot, Kintbury, Berkshire)
Little is known of this machine, which was offered for sale at £500 in March 1910, other than that it was a self raising aeroplane (without a run). A full-size model had been built, together with the gearing for the 'lifters'.

TUCKER cycloplane (G. Tucker & Co., Wimbledon, Surrey)
No information available.

TURNBULL glider (John Turnbull. Built and flown at Box Hill, near Dorking, Surrey)
In 1910 a young engineer John Turnbull designed and built a small glider in or near the top of Box HIll in Surrey, an ideal location for making trial flights.

The framework was of bamboo, braced with wire and covered with sailcloth. The glider was flown as a hang glider controlled by the pilot shifting the weight of his body and was launched from a four-wheel trolley down the steep slope of the hill.

After some successful glides Turnbull fitted the glider with a one-cylinder Gnome engine; later he decided to fit a more powerful two-cylinder Green (probably a motorcycle engine). Pressure of business and lack of finance forced him to give up his involvement with aviation.

Turnbull was helped in his experiments by Dolly Shephard, the intrepid balloonist and parachutist.

Turnbull glider at Box Hill (courtesy Molly Sedgwick).

TWINING AEROPLANE Co. (E.W. Twining, 29b Grosvenor Rd., Hanwell, Middlesex)
E.W. Twining began as a model maker and progressed to the construction of full size gliders and powered aircraft. In May 1910 his advertisement offered three types of glider. No.1 of 22ft span cost £15; No.2 costing £25 was probably of 24ft span and placed the pilot in the prone position. No.3 was of 30ft span and cost £65.

TWINING type 1 glider 1910
An example of No.1 was sold and taken by the owner, K.R. Murray, to St. Moritz in February 1910, to glide on the snow-covered slopes. This type was a simple two bay boxkite of Chanute type, without camber, with tail booms carrying a biplane tailplane and fin. The operator stood in the center section gap, the machine being supported on curved skids. In Switzerland the owner equipped himself with skis.

Twining type 1 glider of 1910 on the snow at St. Moritz.

TWINING type 2A glider 1910
This was a more advanced machine with three bay wings and cambered surfaces. The operator sat in the structure forward of the wing. The machine stood on two skids, which curved upwards to meet struts running from the top wing, forming a nose boom for the front elevator. There was no tail or rudder, both directional and lateral control being obtained by 'dihedral lateral balancers' ailerons set at 45 degrees to the horizontal, between the outboard interplane struts, these providing or dumping lift as directed.

Twining type 2A glider of 1910 was a more advanced machine with an original lateral control system.

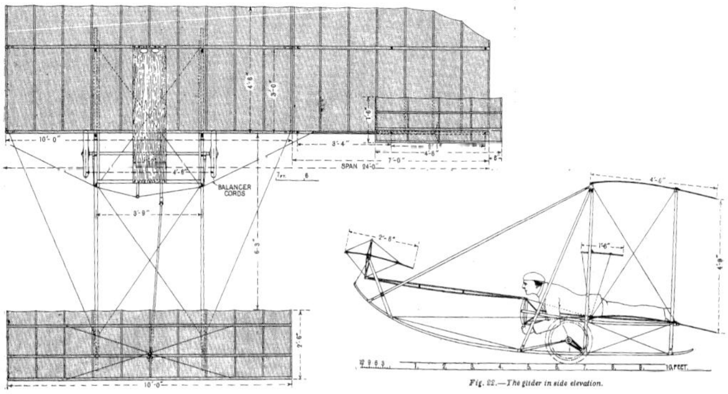

Twining type 1 glider of 1911 had three bay wings and ailerons mounted on the outboard interplane struts.

TWINING type 1 glider 1911

In 1911 Twining wrote articles in *The Aero* describing several versions of his gliders. The first was a variant of the 1910 No.1, the Chanute type, with three bay wings and ailerons mounted on the outboard interplane struts. The twin fins and tailplane were fixed, control being obtained by weight shift by the operator, who was supported on bolsters under his arms. The machine was constructed of ash and spruce.

Data	Span	22ft
	Chord	4ft
	Area	170 sq. ft
	Length	12ft 2in

Twining type 1A glider of 1911. Was an improved version of the previous machine with the addition of a wheeled undercarriage with skids and a seat for the pilot.

TWINING type 1A glider 1911

This was a further variant of the type 1 with changes proposed by Maurice Wright, later an RNAS officer and Fairey Aviation director. It featured a wheel undercarriage with skids, a seat for the pilot and an elevator.

Data	Span	25ft
	Chord	4ft 6in
	Area	225 sq. ft
	Area tailplane	17 1/2 sq. ft
	Area elevator	11 sq. ft
	Length	18ft 2in
	Weight	75 lb.

TWINING type 2 glider 1911

A tailless glider, similar to the early Wright type, was described in August and October. In this the pilot was in a prone position and had a front elevator control. Ailerons were mounted on the outboard interplane struts of the three bay wings.

Data	Span	25ft
	Chord	4ft 6in
	Area	210 sq. ft
	Length	13ft 3in
	Area elevator	25 sq. ft

TWINING No.I biplane

This machine was exhibited at the Aero Show at Olympia in March 1910, without the engine. There were no reports of flight having been achieved. The similar glider, No.2A of smaller span, appeared a few months later, with the same type of 'dihedral lateral balancers' and may have been a reconstruction of the powered aircraft. It was priced at £350.

The biplane had no tail, but had a front elevator as on the glider. Twin front rudders were however fitted and the machine was mounted on four wheels.

Power: 20hp Phoenix two-cylinder horizontally opposed air-cooled rotary driving a 6ft 3in diameter propeller.

Data	Span	28ft
	Chord	4ft 6in
	Gap	5ft 6in to 4ft 6in
	Area	252 sq. ft
	Area elevator	25 sq. ft
	Area rudder	6 sq. ft
	Length	14ft 7in (14ft 10in?)
	Weight	440 lb.
	Weight allup	605 lb.
	Speed	35 mph

Twining No.I biplane was similar to the type 2A glider and was exhibited at Olympia in 1910.

Twining No.II biplane was a larger version of No.I and appeared late in 1910.

TWINING No.II biplane

A larger machine was completed later in 1910, but no details of its use, or whether it actually flew, are recorded. This had the addition of a fixed tailplane on booms, otherwise the layout was similar to No.I, including the 'dihedral lateral balancers'. Twining had patented a universal control column with a wheel for controlling all movements, including the rudder.

Power: 30hp two-cylinder horizontally opposed driving a 7ft diameter propeller.

Data		
	Span	33ft
	Chord	5ft
	Gap	5ft 6in to 5ft
	Area	372 sq. ft
	Weight allup	580 lb.

TWINING No.III biplane

The only information on this machine was that it was to be fitted with a 40hp Phoenix motor and the wing span was 40ft.

TYZACK Glider (R. Tyzack, Scarborough, Yorkshire)

No information is available on this machine. A manlifting kite was also reported to have been made in April 1910.

U

UNITED SERVICES COLLEGE glider (Windsor, Berkshire)

See Hampshire Aero Club.

UNIVERSAL AVIATION Co. (166 Piccadilly, London W, & Shed No.17 Brooklands)

This company which had been known previously as L.D.L. Gibbs & Co., up to May 1911, was refinanced by E.V. Sassoon, later Sir Victor Sassoon, and the name changed to the new title. The company provided tuition and supplied aircraft and aviators for demonstration work, apart from offering for sale various types of aviation equipment, including ABC engines from August 1911. The company engaged in construction work after being established, and offered for sale their *Birdling* monoplane and biplane. The chief pilot was H.J.D. Astley, although other pilots, including Gustav Hamel were engaged at times. R.O. Cary, later to become Sopwith's general manager in 1913, managed the business affairs.

The monoplane was virtually a copy of a Blériot XI, possibly a reconstruction. The biplane, referred to as a *Birdling*-Sommer in the press, was a copy of that type or the sole machine may have been one imported earlier, that Astley was flying in January 1911.

The company used various machines, apart from the Birdlings, including Farman, Lane-Anzani monoplane and the Blériot '*Big Bat*'. Sassoon, under the name of 'Smith' received instruction on this machine but, on 30 November 1911, he crashed it badly. The company did not prosper and by April 1912 was in liquidation, with a Blériot and two Gnôme engines being offered for sale by the receiver.

UAC *BIRDLING* monoplane

The *Birdling* monoplane first appeared at Brooklands as a competitor in the *Daily Mail* Circuit of Britain Race of July 1911. It raced as No.2 flown by Astley, as a substitute for the Macfie Racer, which was not ready. Astley retired at Harrogate but flew back to Brooklands. The aircraft was bought by F.K. McClean soon after the race, and was No.9 in his fleet of air-

UAC *Birdling* monoplane with an ABC engine which was at Brooklands in late 1911.

craft. A second *Birdling* monoplane with modified chassis was referred to by Astley in October 1911. When it emerged at Brooklands it was fitted with a 40hp ABC engine. The undercarriage was a substantial twin skid type. There were no features of the first *Birdling* to identify it from a Blériot.

UAC BIRDLING-SOMMER biplane

It is probable that this machine was imported and was the machine which Astley was flying in January 1911, and used at intervals throughout the year. It may have had a major reconstruction at Brooklands to justify the name, but it is uncertain if the machine was an original, built entirely by the company. No details of the machine are available.

UNIVERSITY COLLEGE SCHOOL triplane glider (University College School, Hampstead, London NW3)

This glider was designed by G.T.R. Hill whilst at school in 1911. It was a triplane with an Antoinette type tail. He later built another glider in 1914 with his brother R.M. Hill. It was reported that a powered machine was to be built.

University College School triplane glider of 1911 under construction.

V

VAUGHAN gliders (Horace W.H. Vaughan, Dinant, Burrell Rd., Sutton, Surrey)

Vaughan was an enthusiast for full size gliders, corresponding and writing frequent articles on the subject in the aeronautical press from the latter part of 1909. He described and illustrated his first glider in *Flight* on 4 December 1909 and followed this with articles in *The Aero* on all aspects of gliding, and later on Aviettes, from January 1911 until February 1913, shortly before the magazine was discontinued.

Vaughan glider No.I of 1909 was of Chanute type.

VAUGHAN glider No.I 1909

This was a Chanute type two bay biplane with fixed boxkite type tail, the pilot standing in the lower center section and controlling the machine by weight shift. During the trials the top, and then the whole of the tail, were removed with disastrous results, convincing Vaughan of the need for proper control surfaces. He described a revised version in January 1911.

Data	1909 Version	1910-1911 Version
Span	22ft	24ft
Chord	5ft	5ft
Gap	5ft	4ft 9in
Area	260 sq. ft	300 sq. ft
Area tail	49 sq. ft	47 1/4 sq. ft
Area fin	25 sq. ft	19 1/4 sq. ft

VAUGHAN glider No.II 1910-1912

After his experiences with No.I, Vaughan decided to build a Wright type glider with control surfaces and span increased to 33ft. This had a biplane elevator in front and a single rudder on tail booms. In the summer of 1912, Vaughan presented it to the Polytechnic Flying Society. It flew very successfully with this society at its gliding camps at Amberley on the slopes of the South Downs.

Vaughan glider No.II of 1910 was of Wright type.

VICKERS AIRCRAFT (Vickers House, Broadway, Westminster, London SW, with works at Erith, Crayford and Bexleyheath. Flying at Joyce Green, Dartford and Brooklands)

This company became interested in aircraft manufacture in 1911, and formed a small drawing office at Vickers House, and a section of the works at Erith for manufacture. In 1912 a flying school was established at Brooklands, which continued to operate up to the outbreak of war. Captain Wood was in charge of the Aviation Department, A.R. Low was the first draughtsman and G.H. Challenger, an engineer from Bristols; they were joined later by Howard Flanders when his company collapsed.

The first move was to acquire a license on 1 January 1911, from Robert Esnault-Pelterie (REP), to build his monoplane types, and to purchase one complete monoplane and a rear fuselage structure from France. The fuselage was incorporated into a machine, which became Vickers No.I monoplane.

Vickers No.1 monoplane of 1911 was a license built REP utilizing a French built fuselage and wings made at Crayford. It was tested at Crayford and Brooklands.

VICKERS No.I monoplane

The Vickers No.I monoplane was built at Erith and tested at Vickers new airfield at Joyce Green, Dartford by Capt. H.F. Wood in July 1911, and was then taken to Brooklands, where it was reflown by Lt. H.E. Watkins. Within a few days it was packed and shipped to Australia, for use by Dr. Mawson's Antarctic Expedition, for which Watkins was to be the pilot. Unfortunately the aircraft crashed at Adelaide and, although the fuselage was taken to the Antarctic to use as a sledge, difficulties arose which prevented its use. The remains of the fuselage survived in the Antarctic for many years.

This first Vickers monoplane was a tandem two-seater, with the crew positioned between the spars of the shoulder mounted wing. The front fuselage was rectangular in section with a triangular section below, and all covered with aluminum sheet. Aft of the pilot's seat the section became diamond shaped, and was constructed of steel tube sweated at the joints into welded steel fittings, and the whole was covered with fabric.

The tapered wing was built on steel tubular spars with wooden ribs and incorporated warping for lateral control. The wing was braced to a tripod pylon, and to the bottom member of the fuselage, with high tensile steel tapes. The warping was actuated through a separate mast by cables and tapes. The tailplane and fin extended well forward to fair into the fuselage. The divided elevator was unbalanced; the rudder had a small balance area forward of the hinge post. The undercarriage was constructed of steel tube with twin wooden skids, mounting two pairs of wheels on separate axles. Shock absorption was by means of rubber cords incorporated into guides on the fuselage sides, to which the landing loads were taken by the main chassis struts, the radius arms being free to pivot from the bottom fuselage members.

Power: 60hp REP five-cylinder air-cooled semi-radial driving a 7ft 11in diameter Regy Frères propeller.

Data		
	Span	47ft 6in
	Chord	8ft 2in to 5ft 2in
	Area	290 sq. ft
	Area tailplane	54 sq. ft
	Area elevators	16 sq. ft
	Area fin	11 sq. ft
	Area rudder	9 sq. ft
	Weight	1,000 lb.
	Speed	56 mph

Vickers No.II monoplane at Brooklands in 1911-1912.

VICKERS Nos.II-V monoplanes

These machines were basically similar to No.I, but with changes progressively introduced by experience.

No.II first appeared at Brooklands on 11 October 1911, flown by E.W.B. Fisher, who crashed it the following day. It

Vickers No.V monoplane at Brooklands. Major changes to the basic design were incorporated in this version, the fuselage being much deeper.

Vickers No.III monoplane at Brooklands.

was repaired and ready for flight again on 11 November 1911, being flown subsequently by R. Kemp and Captain Wood. The Vickers-built version of the REP engine was fitted at some stage.

No.V was the most radically altered, with a much deeper fuselage to provide better crew protection and a new fin and rudder. This aircraft joined the Vickers School at Brooklands in May 1912 and was retired in April 1914.

No.IV did not figure in the weekly reports of flying at Brooklands and it seems not to have been used by the school. This may have been the Vickers 'all steel' monoplane delivered to Larkhill on 29 March 1912 for Capt. Sykes to compete in the Mortimer Singer Competition. On 29 March 1912, after tests, it was decided to change the wings, but on 31 March 1912 it was damaged during a bad landing. This was the last report of this machine.

From No.III onwards the REP method of construction of the joints was changed, the brazing being deleted in favor of welded and bolted joint fittings. The bracing tapes were changed to piano wire. The triangular section top of the fuselage was squared off to give the crew more room. Parallel chord wings were fitted.

Power: 60hp REP or REP Vickers five-cylinder air-cooled semi-radial.

Data (From *Aeronautics* May 1912 for No.IV)

Span	38ft 6in
Chord	6ft
Area	256 sq. ft
Area tailplane	30 sq. ft
Area elevators	17 sq. ft
Area fin	3 3/4 sq. ft
Area rudder	6 sq. ft
Length	28ft 6in
Weight	900 lb.
Weight allup	1,300 lb.

Vickers No.VI monoplane was a major redesign with side-by-side seating.

VICKERS No.VI monoplane

This machine was a major redesign of earlier types for entry as No.3 in the Military Trials Contest of 1912. It was built at Erith and was tested by L.F. MacDonald at Joyce Green and then flown by him to Brooklands on 15 June 1912. To provide more power, a Viale engine of 70hp was fitted, but this proved troublesome and caused the aircraft to miss most of the trials. Later a 70hp Gnôme was fitted with a cowling extending back to the wing, over the top half of the engine.

The main difference was a change to side by side seating in a wide fuselage which was parallel in plan throughout its length. There was a new two-wheeled undercarriage with central skid, sprung by leaf springs. The split axle was connected to the skid and the single radius arms were sprung by rubber cords in the fuselage guides as before. The wing was braced to a pair of inverted vee struts tied together, forming the cabane, and was of reduced span. An aerodynamically balanced rudder of new shape was fitted without a fixed fin.

Power: 70hp Viale seven-cylinder air-cooled semi-radial.
70hp Gnôme seven-cylinder air-cooled rotary.

Data

Span	35ft
Area	220 sq. ft
Speed	63 mph with 70hp Gnôme

Vickers No.VI modified to take a 70hp Gnôme.

VICKERS No.VII monoplane

This aircraft reverted to the tandem cockpit layout with the earlier four wheel, twin skid undercarriage. The fuselage was similar in section to No.VI, but narrower, and a fin was reintroduced with a new shaped unbalanced rudder. A front mounting for the engine was provided, which modified the type of cowling used previously with rotary engines. An unusual three-bladed propeller was fitted. No.VII was flown by Barnwell at Brooklands for the first time on 1 January 1913. It was overturned by a pupil on 4 October 1913 and was probably not repaired for further use.

Vickers No.VII monoplane with 100hp Gnôme was the most powerful of the type.

Power: 100hp Gnôme fourteen-cylinder two-row air-cooled rotary with three-bladed propeller.

Data		
	Span	34ft 6in
	Area	220 sq. ft
	Length	25ft
	Weight	730 lb.
	Weight allup	1,200 lb.
	Speed	70 mph
	Range	350 miles

VICKERS No.VIII monoplane

The last of the early monoplanes was a side by side two-seater, with the undercarriage type of No.VI. The wing span was reduced and the chord increased; a balanced rudder and elevators were fitted. Transparent Cellon windows were fitted into the bottom and sides of the fuselage.

The aircraft had been flown on test at Joyce Green from November 1912, using a 70hp Gnôme, prior to being exhibited at the Aero Show at Olympia in February 1913. It went to the school at Brooklands in April, but was crashed on Weybridge Golf Course, when being flown by Barnwell on 29 April 1913.

Power: 70 or 80hp Gnôme seven-cylinder air-cooled rotary driving a 8ft diameter propeller.

Data		
	Span	34ft 6in (35ft ?)
	Chord	7ft
	Area	228 sq. ft
	Area tailplane	10 sq. ft
	Area elevators	16 sq. ft
	Area rudder	10 sq. ft
	Weight	700 lb.
	Speed range	45-69 mph (70hp)
	Rate of climb	300ft per min full load
	Price	£1,300

Vickers No.VIII monoplane had side-by-side seating and a 70 or 80hp Gnôme engine.

VICKERS tractor biplane. 1912

This obscure type has been reported as a biplane based on the No.VI monoplane, with a wooden structure and powered by a 70hp Gnôme engine. It was completed in December 1912 but crashed on 13 January 1913 in the Thames near Joyce Green, as a result of engine failure, killing L.F. MacDonald and passenger H. England. No details or photographs can be traced.

Vickers No.26 *'Pumpkin'* biplane with 50hp Vickers-Boucier engine.

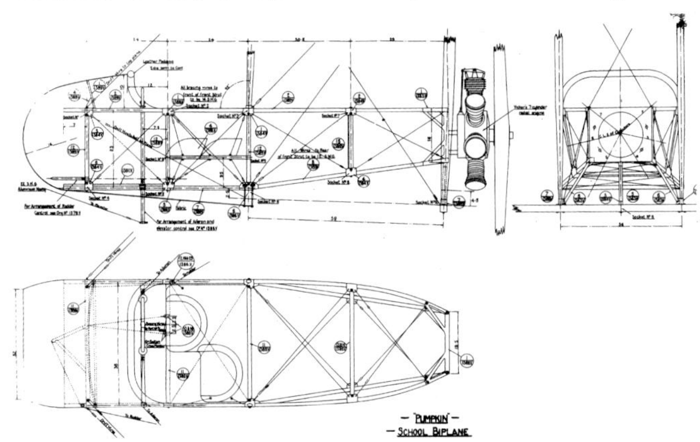

VICKERS No.26 *'Pumpkin'* biplane

At the flying school at Brooklands, Vickers had operated three Farman type pusher biplanes built by Hewlett and Blondeau. Vickers may have made changes to these machines, identified as Nos.19-21, and referred to as Vickers Boxkites (see Hewlett & Blondeau).

In 1913 a version of the Boxkite was built by Vickers with a nacelle seating two crew side by side. The engine was a 50hp Vickers-Boucier seven-cylinder air-cooled radial, later replaced by a 70hp Gnôme. The machine was first tested on 17 September 1913 with Vickers engine at Brooklands by Barnwell and on 17 February 1914 with the Gnôme.

VICKERS Hydravion Type 14 land/seaplane

This machine was designed on Farman lines and was an enlarged version of the familiar Boxkite, fitted with a nacelle to accommodate the pilot and three passengers. The tail booms, with three unequal bays were parallel in plan, and carried the tailplane on top, in line with the top wing. Twin rudders were fitted.

The three bay wings were fitted with top wing extensions braced by wires and kingposts. The seaplane version was the only one built, but it crashed at Dartford during an early test flight in 1912. The duralumin floats were made at the Dartford works.

A version to carry a 37mm cannon, with crew of three, was schemed and known as Type 14B. It would have been powered by two 100hp Gnôme engines in the nacelle, driving twin tractor propellers, mounted outboard in the wings, through shafts and gearing.

Power: 100hp Gnôme nine-cylinder air-cooled rotary.

Data			
	Span	72ft 8in	72ft 6in *
	Area	819 sq. ft	840 sq. ft *
	Length	43ft	43ft 6in *
	Height	12ft 2in	12ft 6in *
	Weight allup	2,400 lb.	3,000 b. *
	Speed range	32-51 1/2 mph	

* Type 14 mod.

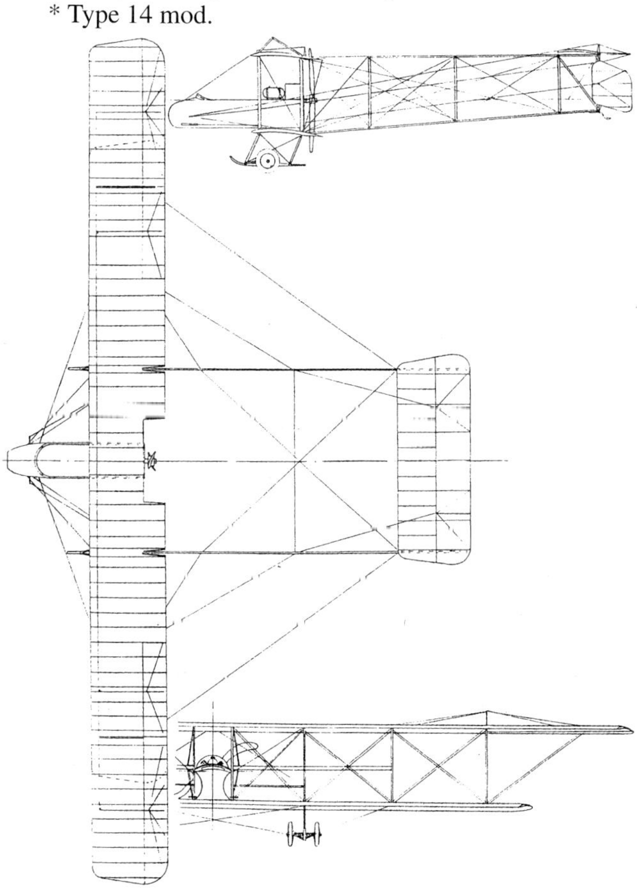

Vickers Hydravion Type 14 land/seaplane was an enlarged version of the Vickers Boxkite.

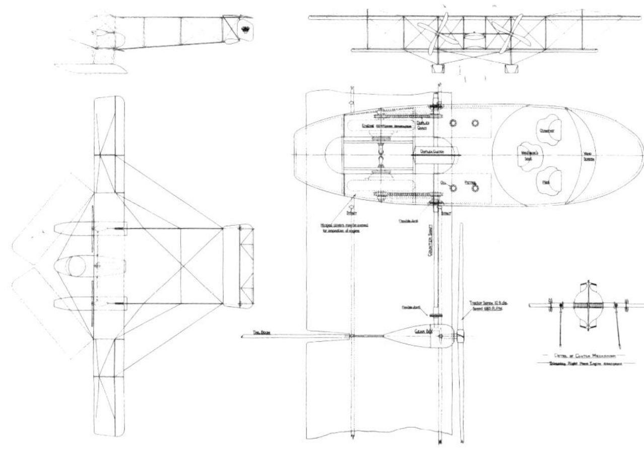

Vickers Type 14B had two 100hp Gnôme engines in the nacelle driving twin tractor propellers mounted outboard on the wing struts.

VICKERS EFB.1 (Experimental fighting biplane No.1) *Destroyer*

This machine, which was the first purpose built armed aircraft, shared the stand at the Aero Show at Olympia in February 1913 with the No.VIII monoplane. Nothing more was reported following the crash of the aircraft at Joyce Green on its first attempted flight. Nevertheless the EFB.1 was the first of a succession of gun carrying biplanes, which later went into production and were used with success in the early war years.

The machine was a pusher biplane with two bay wings with top wing extensions, and lateral control by warping. The fuselage consisted of a steel tube structure, enclosed by a duralumin nacelle containing the crew, with the gunner in front of the pilot and engine installation behind. The tail booms had staggered uprights and were of tubular steel, forming a vee in plan. The tailplane was mounted on the top booms, in line with the top wing and carried a divided elevator. The rudder, pivoting behind the booms, had a balance area forward of the hinge tube, to which the tail skid was attached, and which steered with the rudder.

The undercarriage was a two wheel, central skid type, mounted on inverted vee struts of steel tube. Half axles piv-

Vickers EFB.1 *Destroyer* at Olympia in February 1913. It crashed on its first flight.

oted at the center, and were sprung by rubber cords. The ammunition for the Vickers machine gun was kept in a box at the center of gravity position, and was drawn forward by cables to a position under the seat, when required for use.

Power: 60-80hp Wolseley eight-cylinder vee with air-cooled cylinders and water cooled exhaust valves driving a 8ft diameter propeller.

Data		
	Span top	40ft
	Span bottom	30ft
	Chord	5ft 6in
	Area	385 sq. ft
	Length	27ft 6in
	Height	11ft 11in
	Speed	70 mph (40-65 mph)*
	Rate of climb	450ft per min
	Endurance	4 1/2 hr (6hr)*
	Weight	1,760 lb
	Price	£1,800

*Alternative figures

VICKERS EFB.2 Type 18 and 18A

The second experimental fighting biplane, although of similar layout to the first, was a complete redesign, the most obvious difference being the elimination of stagger of both the wings and the tail booms. The heavy Wolseley engine was replaced by a more powerful rotary and the airframe was reduced in weight. Barnwell tested the Type 18 at Bognor and Brooklands from October 1913, and variations were tested and incorporated in the same machine, identified as Type 18A from 26 November 1913.

Vickers EFB.2 Type 18 was tested at Bognor and Brooklands late in 1913.

The EFB.2 was built with steel tube as the main structural material in the nacelle and tail booms, the nacelle being enclosed in a duralumin shell, with large celluloid windows in the sides. The warping wings had wooden spars and ribs; there was a small overhang of the top wing, which had a cutaway to the rear center portion. The tailplane was of semicircular form with divided elevator and rudder, similar in shape to that of EFB.1. The undercarriage used streamlined steel tubes and twin ash skids with a tubular axle, sprung by rubber cords.

The Vickers gun was carried on a trunnion mounting and protruded through the nose of the nacelle.

Power: 100hp Gnôme Monosoupape nine-cylinder air-cooled rotary.

Data		
	Span	38ft 7in
	Area	380 sq. ft
	Area tailplane	35 sq. ft
	Area elevators	30 sq. ft
	Area rudder	13 sq. ft
	Length	29ft 2in
	Height	9ft 7in
	Weight	1,050 lb.
	Weight allup	1,760 lb.
	Speed	60 mph
	Climb	200ft per min
	Range	150 miles

VICKERS EFB.3 Type 18B

Further development of the type took place in the next aircraft, which arrived at Brooklands in January 1914. These were less fundamental, but resulted in improved performance. The warping wings had been changed to wings of equal span, fitted with ailerons top and bottom. The windows in the nacelle had also been deleted, together with the top wing cutout.

EFB.3 was exhibited at the Aero Show at Olympia in March 1914. By mid-year a triangular fin had been fitted. A dual-control version SB.1 (School Biplane) with a 100hp Anzani radial was proposed early in 1914, but was not built. Length of this version would have been 29ft 2in. The gunner's cockpit was replaced by a pupil's position fitted with dual controls and the nacelle was slightly raised above the lower wing on short struts.

Vickers EFB.3 Type 18B had a metal nacelle and was flown at Brooklands in January1914.

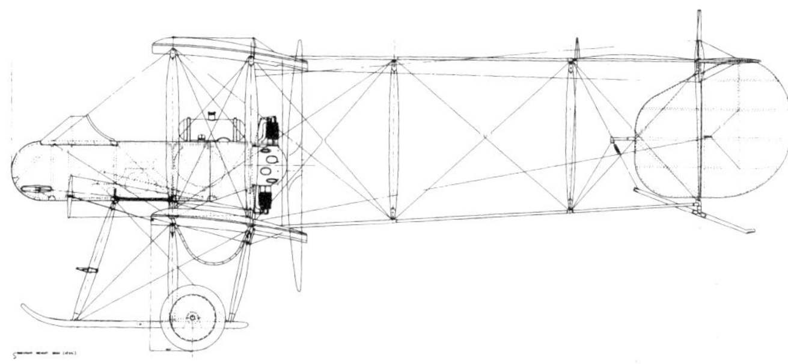

Vickers SB1. A school biplane design based on EFB.3 with pupil's position in the front cockpit.

Power: 100hp Gnôme Monosoupape nine-cylinder air-cooled rotary.

Data		
	Span	37ft
	Area	385 sq. ft
	Length	27ft 6in
	Weight	960 lb.
	Weight allup	1,760 lb.
	Speed range	45-70 mph
	Endurance	4 1/2 hr
	Armament	Vickers gun with 300 rounds

VICKERS EFB.4 and 5 and FB.5-FB.7 (Type 18 Variants)

Both the Admiralty and the War Office became interested in the Type 18, and a prototype was ordered by the Admiralty on Contract CP 03330/12/53661 dated 19 December 1912. This aircraft was the subject of much development by Vickers, and was presumably the aircraft, serial No.32 delivered in 1914, providing the identity, Admiralty Type 32, for naval aircraft ordered later.

The War Office apparently ordered one of each Type 18A and 18B on Contract No.A2033 in 1913. These may have been retained by the firm, or possibly, transferred to a further Contract No.A2321 dated 18 December 1913 for six Type 18B aircraft. These were development aircraft and may have been identified as Nos.FB2-FB.7, before the Type FB5 was finally selected for production. No.FB4, the FB5 prototype, became No.664 and FB6 became No.704 in service, the latter with a top wing of increased span, braced by cables and kingposts appearing at Brooklands in the early part of July 1914.

Vickers FB.5 Gunbus was the version produced for the RFC and RNAS and used in some numbers in the early months of World War I.

Production orders were to follow for both the RFC and the RNAS, by which time the FB.5 was colloquially known as the Gunbus.

Production of the FB.5 was carried out at Crayford, and the first delivery to the RFC was in September 1914. Many changes were made to the type, which continued in production into 1917, with the final FB.9 version. The aircraft was made under license also; in France and Denmark in wartime.

The use of the armament was of prime importance, and the gun was eventually positioned on a pillar mounting, above the nacelle, involving several changes to the front of the na-

Vickers FB.6 had a top wing of increased span; the sole example was built in July 1914.

Vickers EFB.4 was the prototype of the FB.5.

celle before being finalized. The original semicircular tailplane was replaced by one of rectangular shape and finally the rudder was increased in area and acquired a curved trailing edge.

It is difficult to be certain of quantities built, partly owing to transfers of aircraft between the RFC and the RNAS and the renumbering that took place. However, it does appear that forty-eight production aircraft were on contract to Vickers to the end of 1914.

Variants of the basic type which were proposed included the EFB.4, the GFB.1, 2 and 3, the latter three, including twin-engined versions, were for the German government, none of which were built.

Power: 100hp Gnôme Monosoupape nine-cylinder air-cooled rotary.

Data for FB.5

Span	36ft 6in
Area	382 sq. ft
Length	27ft 2in
Height	11ft
Weight	1,220 lb.
Weight allup	2,050 lb.
Max speed	70 mph. at 5,000ft
Climb	16 min to 5,000ft
Ceiling	9,000ft
Range	250 miles
Armament	Lewis gun.

Prototype and Pre-Production Aircraft

	Qty	Contract No	Date	Type	Serial Nos.
Admiralty	1	CP 03330/12 /53661	19 January 1912	18	
War Office	1	A.2033	1913	18A	
War Office	1	A.2033	1913	18B	
War Office	6	A.2321	8 December 1913	18B	649, 664 (FB.4) 686,704 (FB.6) to RNAS as 862

Production Contracts in 1914

	Qty	Contract No	Date	Serial Nos.
Admiralty	4	CP 02132/14/X	14 August 1914	861/4
War Office	8	A.2978	14 August 1914	2340/7
War Office	12	A.2614	19 September 1914	1616/1627
War Office	12	A.2705	2 November 1914	1628/1639
War Office	12	A.2806	26 November 1914	1640/1651

VICKERS tractor Scout biplane

This machine appeared at the Aero Show at Olympia in March 1914 alongside the EFB.3. The type was not developed since the firm was deeply involved with work on the Gunbus.

The aircraft was a conventional two-seater with single bay staggered wings of equal span with a cutaway in the top center section. The decking behind the pilot sloped steeply down to the semicircular tailplane, positioned well forward on the top longerons. Divided elevators and an aerodynamically balanced rudder were fitted. The undercarriage consisted of two pairs of vee struts of streamlined steel tube with a rubber cord sprung cross axle.

Power: 100hp Gnôme Monosoupape nine-cylinder air-cooled rotary with Levasseur propeller.

Data

Span	25ft
Area	270 sq. ft
Length	20ft 7in
Weight	600 lb.
Weight allup	1,200 lb.
Speed range	45-100 mph
Endurance	2 hr

The Vickers archives contain drawings of the following machines which may have been built but of which no photographic evidence has been found.

Vickers Scout biplane shown at Olympia in March 1914. Only one was built.

VICKERS biplane glider

Possibly an apprentice exercise of simple conventional design.

Data

Span	25ft
Chord	5ft
Length	18ft 9in

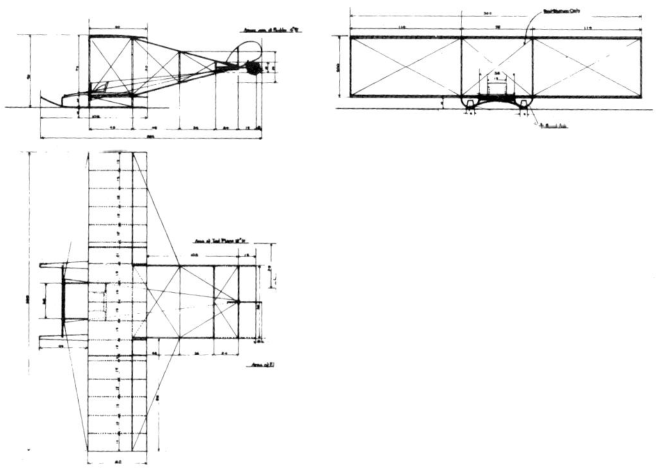

Vickers biplane glider. Possibly an apprentice exercise of simple conventional design.

VICKERS Blériot-type monoplane No.22 Project

In May 1913 Vickers bought a Blériot XI monoplane from Gustav Hamel for experimental purposes. It was fitted with a 50hp Gnôme engine.

The company produced a design for a single-seat monoplane of similar size to the Blériot, but with improved performance. A 70hp Gnôme was specified, a redesigned undercarriage was fitted and the wings were arranged to fold.

Data	Span	29ft 2in
	Wing area	171 sq. ft
	Weight empty	550 lb.
	Length	25ft 4in
	Max speed	63 mph

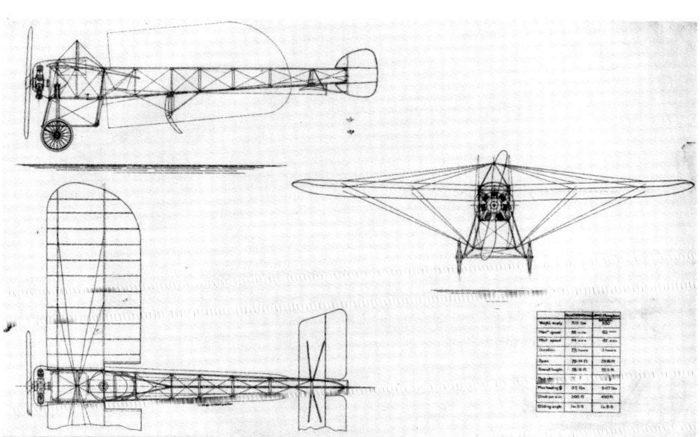

Vickers Bleriót-type monoplane Type 22. An improved Bleriót with revised undercarriage and folding wings.

VICKERS German Fighting biplanes GFB.1/3

Three advanced designs for fighting biplanes were prepared for the German Government in early 1914. Known as the GFB.1, GFB.2 and GFB.3 they had some features in common with the Types 14 and 14B *Hydravions,* but had a conventional fuselage. The engine or engines were installed inside the fuselage driving two tractor airscrews, mounted outboard on the wing interplane struts, through shafts and bevel gearing. Clutches could cut out one or both engines. The engine of the GFB1 was a 200hp Gnôme. The GFB2 would have had two 100hp Gnôme Monosoupape and the GFB3 two 140/165hp Austro-Daimler.

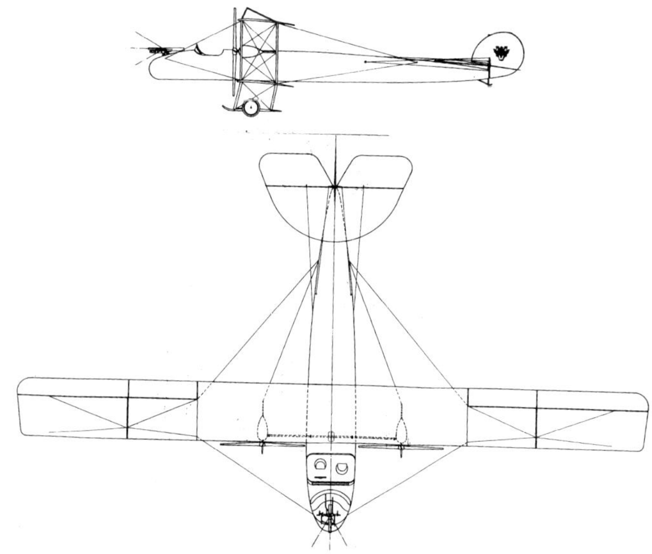

Vickers German fighting biplane, probably the GFB.1 of 1914, was not completed.

A crew of three would have been carried, with pilot and observer behind the engines and a gunner in the nose with a 37mm one-pounder cannon. Equipment specified included wireless and fifty rounds of ammunition.

It is possible that some work was carried out on one of the machines but was canceled at the outbreak of war in August 1914.

Data	Span (upper)	72ft (folded 59ft)
	Gross weight	4000 lb
	Max speed	72 mph
	Endurance	5 hours

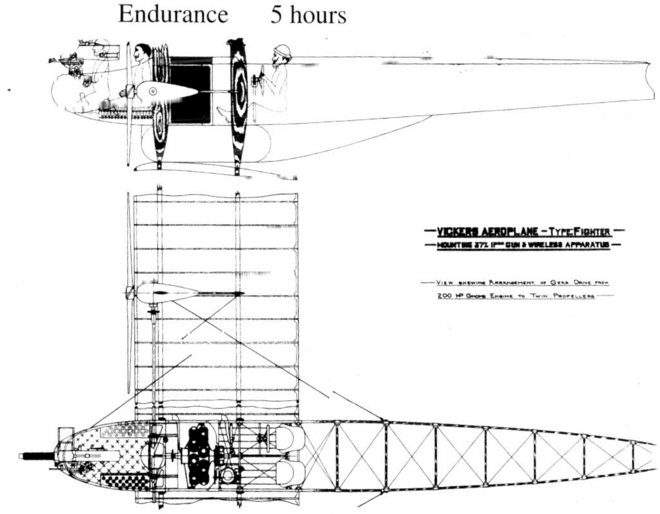

Vickers GFB.1 showing drive from 200hp Gnôme to the wing mounted propellers.

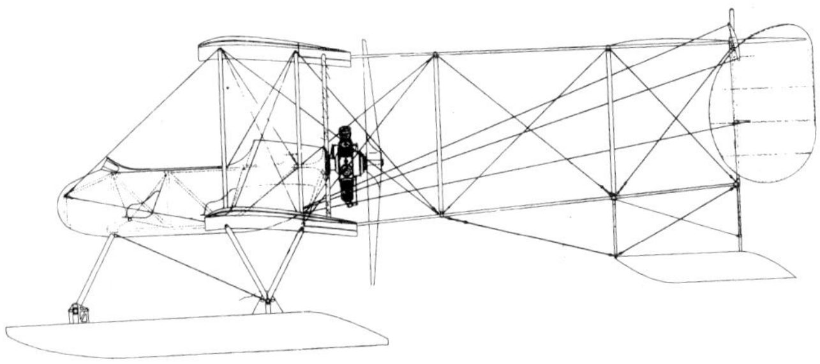

Vickers seaplane Type 43. A float seaplane based on the EFB.5 as a potential Gunbus for the RNAS. It was not built.

VICKERS seaplane Type 43

A float seaplane, a biplane based on the EFB.5, was designed in 1914 as a potential 'Gunbus' for the RNAS. An earlier version had been based on the EFB.3. A Vickers machine-gun was mounted in the nose and 300 rounds of ammunition would have been carried

Data		
	Span	52ft 10in
	Chord	5ft 6in
	Wing area	468 sq. ft
	Length	30ft 2in
	Gross weight	2,020 lb.
	Max speed	60 mph

VICTOR AND GOBLE aeroplane (J. Victor and T. Goble, Felpham Bognor, Sussex)

An aeroplane was built in 1910 but no details have been traced.

W

W.F. and E. biplane (Wright Forge & Engineering Co., Tipton, Staffordshire)

This was the company to which Howard Wright had been apprenticed at the time it was owned by his father. In 1914 the firm built a tractor two-seater, single bay training biplane designed by W. Westwood. The cockpits were in tandem with the instructor in the rear seat. The machine was tested by Rene Desoutter at Dunstall Park, Wolverhampton. There was no production of the type.

Power: 40/45hp Anzani six-cylinder air-cooled radial.

Data		
	Span	26ft
	Chord	5ft
	Gap	5ft
	Area	245 sq. ft
	Length	22ft
	Area tailplane	15 sq. ft
	Speed range	50-65 mph

W.F. & E. biplane. Built at Tipton in 1914 and tested by Rene Desoutter at Dunstall Park.

WALKER glider (Cameron Walker, Bristol)

A biplane glider was reported in the press on 9 March 1910, with the following details.

Data		
	Span	18ft
	Area	190 sq. ft
	Length	15ft
	Height	7ft
	Weight	170 lb.

WALLBRO monoplane (Horace S. and Percy V. Wallis, 12 St. Barnabas Rd., Cambridge)

The brothers built this tractor monoplane in 1909-1910, testing it in a field at Fulbourn, Cambridgeshire from late May onwards, after the machine had been exhibited in public at Whitsun on 16-17 May. The aircraft left the ground on a number of occasions, including a mishap on 4 July 1910, when the machine turned over without causing injury to P.V. Wallis who was at the controls. There is no record of sustained flight and the project ended in October, when the machine was wrecked in its shed during a storm.

Wallbro monoplane built at Cambridge and flown with some success in July-October 1914.

In general layout the aircraft resembled a Blériot XI, but the type of construction was original. The primary structure was of 1"o/d. x 20swg steel tube from Accles and Pollock, and employed brazed joints using Chater Lea fittings, similar to motor cycle frame construction. This method was used for the construction of the fuselage, and tubing was also used for the wing and tailplane spars and for the rudder hinge post. The balanced elevators pivoted at the ends of the tailplane on a tubular spar passing through the main tailplane spar. Wooden ribs were used for the wings and wood was used for the framework of the control surfaces. The lateral control was by wing warping.

The undercarriage incorporated twin skids of ash on a separate chassis, sprung by tension springs; the skids, on coming into contact with the ground would tilt the chassis, bringing the wheels into contact. Steel tubing of 16 and 18swg was used for the undercarriage. A large sprung tail wheel was fitted. The fuselage was covered with fabric to the rear of the pilot's seat.

A replica of the Wallbro monoplane was built in 1974-1976 by the son of H.S. Wallis, Wing Cdr. K.H. Wallis and his cousin, G.V. Wallis. This used brazed instead of Chater Lea fittings, fitted to the tubes with Loctite adhesive, and secured by the eyebolts for the bracing wires. The engine employed was a 72hp McCulloch four-cylinder horizontally opposed, air-cooled, two-stroke target-drone engine. The replica aircraft was flown successfully at Swanton Morley, Norfolk.

Power: 25hp JAP four-cylinder air-cooled vee driving a 6ft 6in diameter Avro adjustable pitch propeller with metal blades.

Data		
	Span	30ft
	Area	180 sq. ft
	Area tailplane	12 sq. ft
	Area elevators	18 sq. ft
	Area rudder	8 sq. ft
	Length	25ft
	Weight	330 lb. with fuel & oil

WALTON-EDWARDS *Colossoplane* (Walton & Edwards Aeroplane Co., Shed 16, Brooklands)

Edwards had been involved with an earlier large machine at Brooklands, the Rhomboidal, and, presumably with a new financial partner, embarked on the *Colossoplane*, also known as the *Elephantoplane*. The machine emerged at Brooklands in August and was flown for the first time on 25 September 1911. Later a number of straight flights were carried out at low altitude, and a turn was reported to have been made on 6 December 1911, but no activity was reported after the middle of December. The machine was flown by Harold Blackburn and was notable for the noise of its engine and for being the largest aircraft at Brooklands at the time.

Walton-Edwards *Colossoplane*. This huge biplane made mainly straight flights at Brooklands towards the end of 1911.

The *Colossoplane* was a mixture of tractor and pusher design in that it had a nacelle, mounted high in the gap on struts, and tail booms, but the engine was mounted in the nose and drove a tractor propeller. To adjust the balance of the aircraft the engine and crew could be moved fore and aft. The three bay wings had a wide gap and were of thin section, with considerable reflex curvature on the lower surface. Most unusual was the system of warping of the leading edges of the wings for lateral control, struts connecting top to bottom surfaces to actuate the control.

The tail booms carried a biplane tail with elevator on the top only. One triangular rudder was fitted within the tail, and two similar rudders midway along the tail booms. The undercarriage consisted of a central skid and two wheels, the track being narrow and out of keeping with the proportions of the machine. A tail skid was fitted.

Power: 100hp Clerget 4W four-cylinder inline water-cooled.

WARREN glider (Frank Warren., Birmingham)

A monoplane glider with a span of twenty feet was reported built in October 1911.

WATKINS monoplane (C. Horace Watkins, Cardiff, Wales)

This machine, constructed in 1908-1909, was designed and built entirely by Watkins, including the engine. The aircraft flew successfully on the occasions when it was possible to avoid the encroachment by people at the flying ground, which was often only early or late in the day. The original aircraft was flown as late as 1918 and is still in existence, being exhibited for many years at RAF, St. Athans.

The fuselage of the monoplane was a conventional braced wooden girder, uncovered except around the pilot's seat. The wings were braced from four pylons, two below and two above the fuselage, each of four struts. The undercarriage consisted of two pairs of vee struts with a cross axle sprung by rubber cords. This could be located in three positions, the rearmost was the flight position, the other two were for testing the engine and for taxiing.

The wing tapered towards the tips, had a scalloped trailing edge, and was fitted with ailerons. A balanced rudder, but no fin was provided, and this carried the maker's initials. The rectangular oil tank was of aerofoil section and was mounted above the engine, in the slipstream. Owing to the need to avoid the crowds, Watkins mainly flew at dawn or dusk and to assist his judgment of height for landing, he designed a height indicator. This consisted of two weighted cords, twelve feet and

Watkins monoplane built in 1908-1909 flew successfully for a number of years and still survives in a museum in Wales.

six feet long hung from springs in the cockpit which operated electrical switches when in contact with the ground, causing dashboard lights to operate.

Power: 40hp Watkins three-cylinder air-cooled fantype semi-radial.

Data		
	Span	32ft
	Length	21ft 6in
	Weight allup	390 lb.
	Height	8ft
	Speed range	25-70 mph
	Range	180 miles

WATSON biplane glider (Preston A. Watson., Dundee, Scotland)

This first machine was a primitive Wright type glider, built in 1903, and fitted in 1906 with a small French engine. This was not a success and the aircraft was discarded. Many years later, members of the family claimed that flight had been achieved, which, if substantiated, would have been among the earliest in Britain.

WATSON biplane No.1

Watson built this machine in 1908-1909, incorporating a novel type of control in the form of a 'rocking wing'. This was intended to control the machine in both roll and yaw.

The aircraft was built about a tubular structure with A-frames, to which the main wings were attached low down, with the rocking wing pivoting at the apex of these. To the bottom

Watson No.1 biplane with rocking wing for lateral control. Built at Dundee in 1908-1909.

of the frames were attached twin skids and wheels. Tail booms of bamboo, tapering to a point in elevation, carried a boxkite elevator, the end panels being fixed. The upper boom tapered to a point where it met the rear A-frame. The pilot sat on the lower center section and controlled the rocking wing with a hanging stick.

The first attempts at flight were probably made as a glider, but later a Dutheil et Chalmers two-cylinder horizontally opposed engine was fitted. In 1910 the machine was reinstated as a glider and was passed to the Dundee Aero Club, who experimented with it in the summer of 1911.

WATSON biplane No.2

An improved version built on the same general lines as No.1 appeared in 1910. The revised center structure was mounted on a four wheeled undercarriage, and bracing struts were added below the lower wings. The engine was a 30hp Humber three-cylinder, air-cooled, fan type semi-radial driving a tractor propeller. The boxkite tailplane, which was first fitted, was later changed for a monoplane with elevator, and a triangular portion of each of the tail booms was filled in to provide vertical surface area.

In this latter form the machine was known to have been flying at Errol, Perthshire in 1912.

Watson No.2 biplane of 1910.

Watts ornithopter No.2 of 1911.

WATSON biplane No.3

This version, built during 1913 had a structure entirely of steel tubes, encased with streamlined fairings. The tail booms were straight lengths of tube, well braced by wires. A wickerwork nacelle was located on the center section, through the base of which the pilot could raise his legs through slots. The upper wing was still controlled by a hanging lever, to provide both lateral and directional control, the small vertical tail surface with a protective skid at the bottom being fixed. A 45hp Anzani six-cylinder, air-cooled radial provided the power.

By 1914, Watson had sufficient confidence to take the machine to Buc in France to compete in the 'Concours de la Securité en Aeroplane', where it was flown by Summerfield without success, as the pilot was classified as a novice and excluded.

The principle of the 'rocking wing' had no future and, with the advent of war, Watson joined the RNAS, but was killed while serving as a Flight Sub-Lieutenant in 1915.

Watson No.3 biplane. The final rocking wing machine was built in 1913-1914.

WATTS ornithopter (Alfred H. Watts, Coventry, Warwickshire)

Watts made several versions of his manpowered, flapping-wing machine after studying bird flight from 1900. His No.1, built in 1906, was a failure. No.2, which followed in 1911, was covered by patent, and was reported to have performed a bouncing type of lifting motion in the owner's garden. This machine had a bamboo framework and was propelled by horizontal levers in a similar way to the oars of a rowing boat.

No.3 was preceded by further model tests and did not materialize until 1920. This machine of bat-like design was reported to have flown 150 yards following a running takeoff from the brow of a bridge near Stoke, Coventry in March 1920.

L.G. Watts, grandson of A.H. Watts, an instructor at the Coventry Gliding Club, also experimented with this type of machine in 1956.

WEAVER ornithopters and helicopters (Morton & Weaver, later renamed Coventry Victor Motor Co., Cox St., Coventry, Warwickshire)

William Arthur Weaver was the designer of several machines which incorporated flapping wing devices, or rotors, with which he hoped to achieve vertical flight. He began a long correspondence in *Flight* with a Mr. C.J. Reynolds, in which details of some of his machines were given.

Ornithoplane No.1 of 1904-1905 had flapping wings powered by a 9hp two-cylinder De Dion engine and was tested on the golf course of Hampton-in-Arden where it was destroyed in a gale. A modified version in 1906 had fixed wings and is reported to have reached a height of 50ft at the Hampton-in-Arden golf course. It is probable that it was fitted with a 20hp engine made by Weaver and Morton.

Weaver Ornithoplane No.2 was built in Coventry and flew at Hampton-in-Arden in 1910.

Ornithoplane No.2 was converted from parts of the No.1 machine and was re-engined in 1907-1908 with a special lightweight water-cooled engine of about 35/40hp built by The Alpha Engineering Company, Coventry. The radiator was made by a Mr. Randle. It was a two-seater monoplane of conventional layout, with a fuselage of triangular section, using bamboo and aluminum in its construction. The machine was mounted on a twin skid, two wheeled undercarriage, the front uprights of which, in the form of an A-frame, carried the radiator for the water-cooled engine and, at the top, the bearing for the propeller shaft. The propeller was chain driven, and was claimed to have variable pitch in flight, which seems to have been achieved with power provided by a small auxiliary propeller revolving behind the main one. The wings were braced to the lower longerons by lift struts and had steep camber at the leading edge, and considerable reflex curvature to the lower surface. The wing warping and tailplanes were actuated by servo mechanism driven from the engine flywheel. Unusually, the pilot's seat and control wheel were mounted on the starboard bottom longeron. The tail unit consisted of a rectangular rudder and large elevators, which flapped to provide lift. A tail wheel, below the rudder post, supported the machine on the ground.

The machine made various short hops at Hampton-in-Arden between 1906 and 1908 and these culminated in a flight of a quarter of a mile on 17 May 1910.

Power: 35/40hp Alpha four-cylinder water-cooled driving a 8ft diameter metal propeller with variable pitch aluminum blades.

Data		
	Span	42ft
	Length	28ft
	Weight	under 15 cwt.

Ornithoplane No.3. A sketch of this single-seater monoplane in *Flight* of 20 May 1911 was of a twin rotor helicopter type with a fuselage of wood and aluminum. A 100hp water-cooled engine buried in the fuselage was to be used to drive the rotors through bevel gear boxes. A separate 50hp rotary engine at the nose, drove a tractor propeller and a pusher propeller at the tail by means of a long shaft. This machine may not have proceeded beyond the proposal stage.

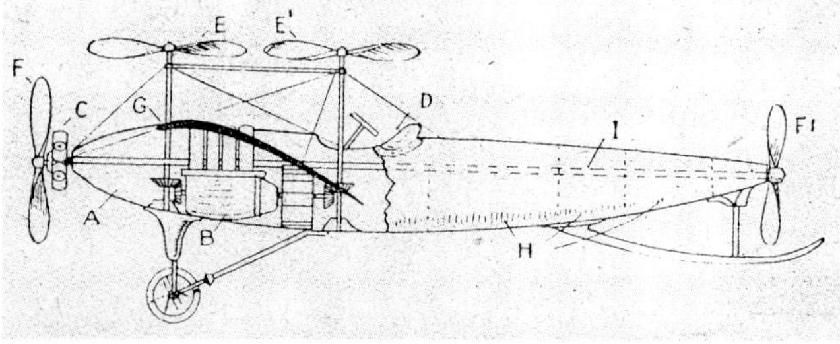

Weaver Ornithoplane No.3. An ambitious design with two engines which was not built.

Ornithoplane No.4. A reference to a further machine may have been to No.2 with a more powerful 40/50hp engine.

WEBB-PEET monoplane (Webb-Peet & Co., Westgate Ironworks, Gloucester)

The aircraft was designed by two brothers, Bateman David Scott and Wilfred Edward Scott, who in 1907 had designed the Scott ornithopter. There is no evidence that the ornithopter was built, but the monoplane certainly was, under an arrangement with the Webb-Peet Co. The aircraft was not a success and did not fly, being eventually offered for sale as part of the company's effects at an auction on 20 May 1913, when the firm was liquidated.

The machine was a tandem wing monoplane with a triangular section braced wooden girder fuselage, in the center of which the engine was positioned. This drove twin wooden propellers by chain on outriggers ahead of the rear wing. One report stated that one chain was crossed, but an earlier report claimed that an endless chain was to be used and the rotary engine would run in reverse to the propellers, to counter the torque.

The fuselage was mounted on a chassis with twin skids and with two wheels on a cross axle at the front, and with a larger third wheel behind. The pilot and passenger sat ahead of the engine in tandem, behind a raised decking, in the enclosed portion of the fuselage, which extended to the nose.

The front wing was of shorter span than the rear and served as the elevator, the operation of which simultaneously lifted the front wheels, thus reducing drag at takeoff. The rear wing incorporated warping for lateral control, the cables for this, and the wing bracing wires, were taken to a pylon on the cutaway center section; the lift wires were taken to the lower member of the chassis. Both wings were of complicated 'gullwing' shape with dihedral over the central portion, curving down to

Webb-Peet monoplane was built at Gloucester in 1910-1911 but failed to fly.

anhedral at the tips. In one report the rear wing was said to be hinged at the center of pressure, but it is difficult to imagine this as being workable in the air, bearing in mind the complication of bracing wires and warping control, plus the fact that the control was not irreversible. Twin rectangular shaped rudders were fitted at the rear on either side of the fuselage for directional control.

The description in the technical press was at variance with patent No.10 253/1910, applied for by W.E. Scott and W.W. Peet on 27 April 1910 and accepted in April 1911. In this both wings were pivoted, and the front wing also carried ailerons. There was no mention of warping.

The Webb-Peet rotary engine was on the stand of the Weston-Hurlin Co. at the Aero Show at Olympia in March 1911. It was described as having no cooling fins, but relied on the effect of rotation and water injected into the-cylinders for cooling purposes.

Power: 25/30hp Webb-Peet five-cylinder air-cooled rotary with additional cooling by water injection. (Patent application No.6215/1911)

Data	Span rear	40ft
	Span front	25ft
	Chord rear	8ft
	Chord front	5ft
	Area	410 sq. ft

WEISS glider '*Olive*' (José Weiss, Houghton House, Amberley, Sussex)

José Weiss, a Frenchman living in England, was an artist and engineer who experimented with models from 1902 to 1907, basing his work on the shape of bird's wings. In the course of his experimental work, Weiss arrived at a swept wing shape, with curved leading edge, and thick section inboard, tapering outwards to a thin flexible section, which he patented under cover of patent No.17150/1908. In 1908-1909 he made a full size glider, named *'Olive'* after one of his five daughters, which was tested on the Sussex Downs at Amberley in 1909, being flown by Gordon England, Graham Wood, Gerald Leake and others.

The glider was a small tailless single seater, with lift struts and a central skid and retractable wheels. The wing incidence changed from positive inboard, to negative outboard. The wing was made in halves for ease of transport, the spars projecting into steel tubes in the nacelle, to which they were bolted. Control was by 'righting planes', namely small flaps on each inboard trailing edge, serving as elevators and ailerons. The machine was crudely made largely of bamboo, referred to as Tonkin Cane.

Weiss formed the Weiss Aeroplane and Launcher Syndicate on 10 June 1908, to exploit his patent. Handley Page, with whom he had collaborated on earlier research, took a financial interest in the syndicate. The two men had become acquainted through the Aeronautical Society and Handley Page went on to build several of his early machines with Weiss type wings.

Data	Span	26ft
	Area	108 sq. ft

WEISS man-powered monoplane

After his first full size glider, the *'Olive'*, Weiss made a pedal powered monoplane, possibly using the wing of the glider, to which it showed some similarity. However, it was mounted on four wheels and was propelled by a tractor propeller at the nose; this was driven by a long drive shaft terminating in a bevel gearbox driven by chain from the pedals.

The wheels were not driven and the machine was to be started from the top of a ramp, fifteen feet high with a forty-five feet base, at a location on Bury Hill, Sussex. It is doubtful if sustained flight was achieved. The machine may then have been powered and identified as *'Elsie'*, which was itself subsequently converted back to the glider *'Olive'*.

Weiss man-powered monoplane in incomplete form.

Weiss glider '*Olive*' was built and tested at Amberley in 1908-1909.

Weiss man-powered monoplane on the ramp near Bury Hill, Sussex.

Weiss pusher monoplane '*Madge*' shown at Olympia in March 1909 but never flew.

Weiss tractor monoplane No.1 '*Elsie*' was a reconstruction of the 1909 glider.

WEISS Monoplane. '*Madge*'

A pusher monoplane was built by Weiss, and was exhibited on the Handley Page stand at the Olympia Aero Show of March 1909. The stand bore a notice that a second machine was being built by Handley Page for the Weiss Aeroplane and Launcher Syndicate. The machine was priced at £500 with a guarantee of flight, although it never did fly.

The machine was a single seater with Weiss type wing and was crudely constructed mainly of bamboo. The short nacelle housed the pilot in front of the engine and tapered to a point at the rear. There were no tail surfaces. The undercarriage consisted of a pair of curved skids for landing, with a separate four wheeled chassis, from which the machine was to detach itself at takeoff. It had been tested at Fambridge late in 1908 using rails and a pylon, as had the Wright brothers, but without success.

The engine drove twin pusher propellers by chain and gearing at the trailing edges of the wing, at a reduction ratio of 3:1. Behind the propellers were 'righting planes', which could be independently operated by foot pedals, thus serving as both ailerons and elevators.

Power: 12hp Anzani three-cylinder air-cooled fan type semi-radial driving two 6ft diameter pusher propellers.

Data		
	Span	34ft
	Area	150 sq. ft (226 sq. ft also reported)
	Weight	360 lb.
	Weight allup	500 lb.

WEISS tractor monoplane No.1 '*Elsie*'

This machine was constructed from the 1909 glider, and was being tested on the sands at Littlehampton in tailless form on 16 April 1910, when damage occurred. By August it was at Brooklands being tested by Gerald Leake, who accomplished a few straight flights. On 21 September 1910 it was rammed by a Howard Wright, damaging the engine and propeller, which finished operations.

In its initial form the machine was tailless, although an extension of the glider nacelle had been added, to carry a tail skid or tail wheel. The main undercarriage was an ungainly affair of struts, with two main wheels and a pair of smaller wheels on the central skid. Later the narrow track of the main wheels was increased in width, and the wing tip skids removed; a central beam with a smaller, more robust nose wheel, was tried, also a tail wheel at one stage. The final version at Brooklands featured widely spaced skids and nose wheels, and a further extension of the nacelle to move the tail skid further aft.

The engine installed was a 25hp Warren-Simpson and later a similar Anzani. The control system consisted of the pedal operated trailing edge flaps of the earlier monoplane, but instability caused the fitting of a cruciform tail, and the covering of the extension in September, just before the demise of the machine. After this it was converted back to '*Olive*', fitted with a vertical tail and presented to the Polytechnic Gliding Club in 1912.

Weiss '*Madge*' at Fambridge. Gerald Leake in the cockpit.

Weiss monoplane No.1 '*Elsie*' at Brooklands in 1910 with tail added.

Power: 25hp Warren-Simpson three-cylinder air-cooled fan type semi-radial.
25hp Anzani three-cylinder air-cooled fan type semi-radial.

Weiss tractor monoplane No.2 '*Sylvia*' was tested at Brooklands in 1910-1911.

WEISS tractor monoplane No.2 '*Sylvia*'

Weiss still relied on the design of his wing to provide lateral stability, but braced the tips with kingposts and wires, however he stated that there was a possibility that warping control would be tried. A conventional tail unit was introduced consisting of fin, rudder, tailplane and divided elevator.

It was a larger and more powerful machine that Gordon England tested at Brooklands on the 19 December 1910, after some initial tests at Littlehampton. On 22 December 1910 Gordon England came to grief in the sewage farm, but the machine was soon repaired, and the trials continued in the spring.

The general construction, as with earlier Weiss machines, was mainly of bamboo with the joints laced with twine. A neater undercarriage with twin skids carried smaller bumper wheels at the tips than on No.1. A hoop shaped tail skid was fitted. The engine was mounted on a structure of steel tube and plate, mounted high in the fuselage, with the radiator lying almost horizontally below the bottom longerons.

Power: 35hp ENV type D eight-cylinder water-cooled vee driving a 7ft diameter propeller.

Data		
	Span	34ft
	Max. chord	7ft
	Area	164 sq. ft
	Area tailplane	19 sq. ft
	Area fin	6 sq. ft
	Area elevators	8 sq. ft
	Area rudder	6 sq. ft
	Length	23ft
	Weight allup	750 lb.

WEISS glider '*Joker*'

Weiss continued his interest in gliding, even while the powered aircraft were being built and tested. In 1911 a glider named '*Joker*', a nickname for one of his sons, was tested at Amberley.

Weiss glider '*Joker*' of 1911-1912 was tested at Amberley and later given to the Polytechnic Gliding and Flying Society.

The wing took the same form as used in earlier Weiss machines, but the fuselage was long with the pilot's seat at the front, ahead of the uncovered center portion, which extended to behind the wing. The rear fuselage was covered and tapered to a point at the tail and carried a vertical tail surface. The fuselage was triangular in section with the single bottom longeron serving as a skid.

In 1912 Weiss presented the glider to the Polytechnic Gliding and Flying Society, who held gliding camps in the summer months at Amberley, where it was flown mainly by B. Graham Wood, who was responsible for the control of the activities. Despite all the effort that Weiss put into the design and building, it would appear that he never carried out the actual testing of either his gliders or powered machines, for he was not a young man and had considerable family responsibilities, but he concentrated much time and effort to model work.

WELFORD monoplane (Robert Welford., Mansion House, North Hylton, Sunderland, County Durham)

Welford patented improvements to flying machines under cover of patent No.8086/1909, which was finally accepted on 24 March 1910. Two main features were covered; the first being alternative means of assisting takeoff, using a spring drum on the undercarriage, or elastic cord tensioned by a winch in the machine. The second feature was the use of large shield shaped, pivoting surfaces attached at the tips of the short span wing, which itself could be adjusted for incidence. A rudder was coupled into the controls and there was a fixed tailplane.

It is not clear whether the machine built by Welford and his two sons, George and Charles Welford, conformed precisely to the patent, but it was probably on similar lines. A contemporary report described the machine in the following manner:

"The machine is a monoplane somewhat similar to the Antoinette and Blériot types, but differs distinctly from both in certain respects, especially in the landing carriage, fuselage and tail. The mainplane is about 29ft by 6ft, and the elevating surface is 6ft by 3ft, whilst the weight carrying tail is also 6ft by 3ft. At each end of the mainplane there are continuations for the purpose of adding stability. The back part of the monoplane has a triangular surface, which distinctly differs from any other machine, offering greater surface to the wind and thereby assisting in the turning, and by steadying the machine, preventing it from bucking to any extent."

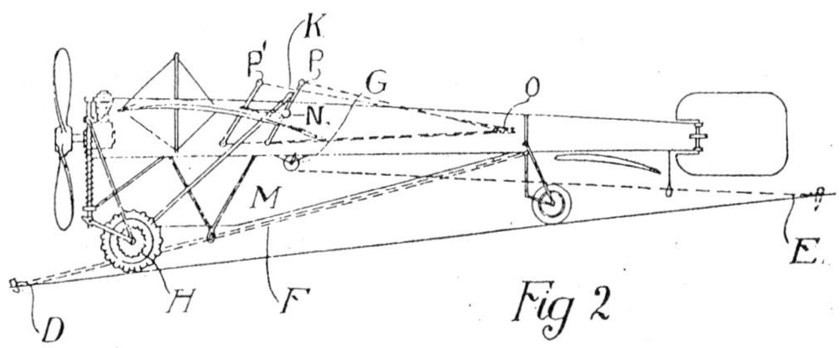

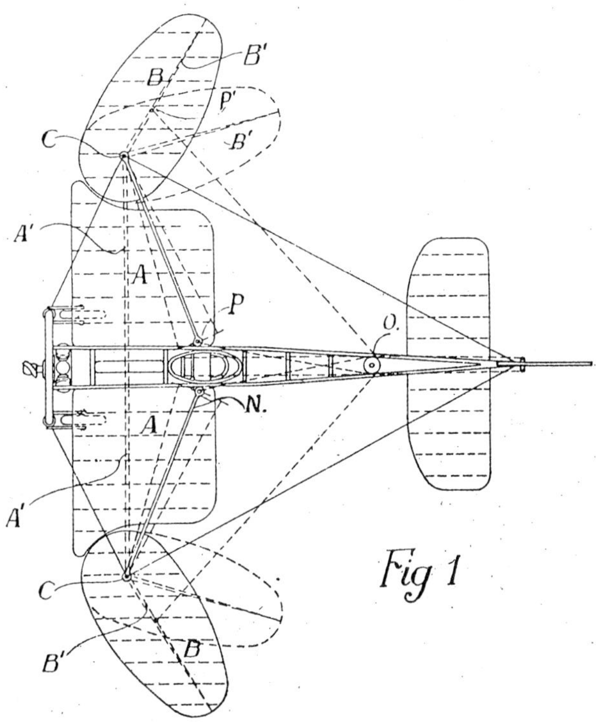

Welford monoplane. The machine due to be tested at Boldon in 1909 was probably based on this design.

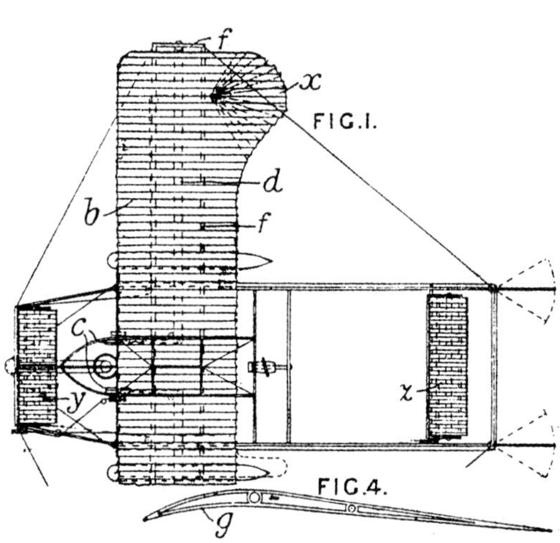

Welford pusher monoplane patented in 1911 does not seem to have been built.

The construction of the fuselage was of bamboo, the wings and other surfaces of ash and silver spruce covered with varnished calico. Two main wheels and skids and a tail wheel were fitted.

The machine was due to be tested on Boldon Flats in the summer of 1910, but there are no further reports and no illustrations can be traced. A later patent No.29058/1911 was also taken out for an entirely different design. This was a pusher monoplane with front and rear elevators on twin booms. The special feature was the ability to draw the wing ribs and fabric inboard towards the center of the machine for storage purposes. The frames or spars on which the ribs were to slide, could also be telescoped. The idea was to prevent wind damage, particularly to a seaplane on the water. The machine does not appear to have been built.

Power: Four-cylinder air-cooled of unknown make driving a Chauvière propeller.

Data		
	Span	29ft
	Chord	6ft
	Area tailplane	18 sq. ft
	Weight	530 lb.
	Area elevator	18 sq. ft

WELLS Reo biplane (Wells Aviation Co. Ltd., 10a Elystan St., Chelsea, London SW)

This small concern was established in 1914 and went on to build this conventional single-seater tractor biplane, which appeared at Hendon in April 1915, reported to be ready for testing in a few days. The aircraft may also have been flown at an airfield at Cobnor, near Portsmouth, which were taken on by the firm in 1915. The low powered machine was not suitable for military use and proved to be purely a design exercise, which the firm was unable to build on. This was Wells' only original design but, in wartime, the firm went on to produce 100 Sopwith 1 1/2 strutters and 50 DH.9 aircraft, until liquidation in 1917, when it was absorbed into the Waring and Gillow organization.

The machine had two bay wings, all fitted with inversely tapered ailerons. The fin and tailplane extended well forward along the fuselage, and were fitted with semicircular divided elevators and a balanced rudder. The vee undercarriage was

Wells' *Reo* biplane was built in 1914-1915 and tested at Hendon.

unusual; the struts of each ending in a fork into which the wheels were fitted. The rear legs of each vee were connected together but not attached directly to the fuselage, but through leaf springs below the longerons. Flat radiators were fitted to either side of the fuselage just forward of the cockpit.

Power: 35hp Green four-cylinder inline water-cooled.

WENHAM glider (F.H. Wenham)

F.H. Wenham was a respected engineer and founder member of the Aeronautical Society, who lectured at the opening meeting on 27 June 1866, on work carried out with his glider of 1858-1859.

This was a machine with five superimposed wings, in which the pilot lay in a prone position. His experiments with this impressed on Wenham the importance of wings of high aspect ratio to achieve the maximum possible lift.

The glider was in the nature of a test rig and not a great success, nevertheless Wenham's findings had considerable influence on later experimenters.

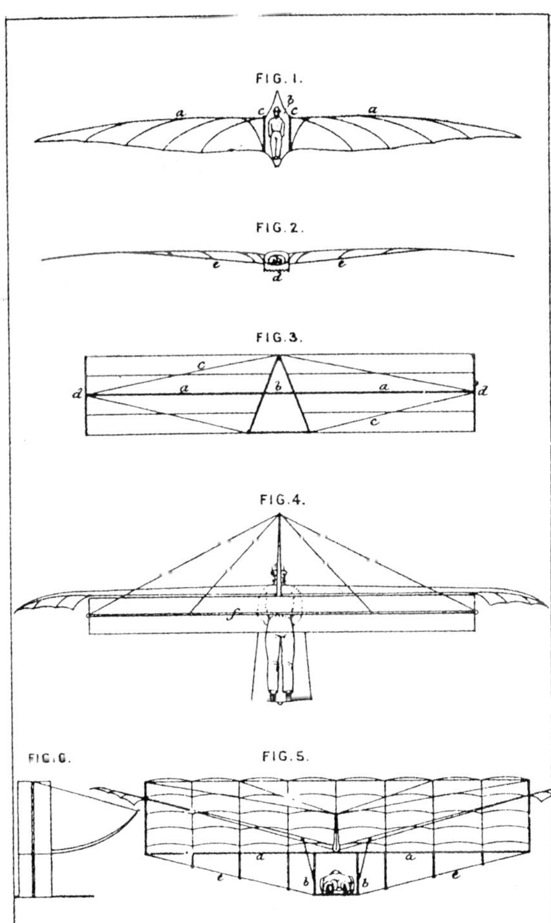

Wenham multiplane glider was tested in 1858-1859 with no great success.

WEST monoplane (H. West, Wolverhampton, Staffordshire)

Little is known of this machine, probably a glider of 250 sq.ft, reported briefly in July 1911.

WESTLAKE biplane (A. Westlake, 26 Great Ormond St., Bloomsbury, London WC)

Westlake was an automobile engineer, who had experimented from 1904 in the aeronautical field, and was reported in March 1911 to have built a biplane of Cody-Wright type with the same type of engine as used in his later monoplane. No details have been traced and this may have been the Fitch biplane (q.v.).

Power: 18hp Westlake four-cylinder horizontally opposed air-cooled. An unidentified 35/40hp engine was also tried out.

WESTLAKE monoplane (A. Westlake, East Anglian Aviation Co., Clacton-on-Sea, Essex. Head Office, 26 Shaftesbury Avenue, London W)

Westlake eventually designed and built a small monoplane in 1913. It was a single-seater tractor and made a number of straight flights, but was under-powered and had little potential.

The fuselage was built as a girder with four longerons, to which a triangular lower section was added below the front portion, this triangular shape being continued aft of the cockpit by bringing the lower longerons together at the tail. In contrast to this the upper longerons were swept outwards halfway along the rear fuselage to form the tailplane boundaries. A balanced rudder was fitted above the center of the tailplane, on a post inclined forward, tending to lift the tail when in use. The shape of the tailplane extended to the elevator forming a large fan shaped surface with a balance area forward of the hinge line.

A substantial A-frame formed the front of the cabane and extended down to the skids of the landing chassis, which was braced by lighter struts and wires, and had a rubber sprung cross axle. Inverted vee-shaped struts formed the rear of the cabane (although *Flight* described and drew another heavy A-frame), which carried the fuel tank and provided anchorages for the bracing and control wires.

Westlake monoplane was completed in 1913 and made a number of straight flights.

The wings had curved tips, with ailerons whose hinge lines were set at an angle across the wing, and these partly overhung the trailing edges. The ailerons could be operated downwards in unison, to improve lift, or independently. Westlake proposed to revert to wing warping for his next machine, of which nothing more was heard. The engine, which was of Westlake's own construction, used de Dion cylinders on a new crankcase. The monoplane and a spare Lascelles engine were auctioned on 20 October 1913.

Power: 18hp Westlake four-cylinder horizontally opposed air-cooled.

Data		
	Span	34ft
	Chord	5ft 6in
	Area	175 sq. ft
	Area elevator	18 sq. ft
	Area rudder	5 sq. ft
	Length	23ft 6in

WESTON HURLIN biplane or bimonoplane (Weston-Hurlin & Co., 191 Cambridge Rd., and 2 Edbrooke St., Paddington, London W. Office at PF Building, 13 Milford St., Strand, London WC)

This company, which stemmed from the Hurlin Coachbuilding Co., became involved with the supply of materials and parts for models and later, in 1911, for full size aircraft. They advertised as an agency for the sale and exchange of aircraft, engines and accessories. In 1911 they announced their intention of constructing complete aircraft, starting with a monoplane and a biplane of original design. This probably referred to the Harper monoplane (*q.v.*) and the Weston-Hurlin biplane, also referred to as a bimonoplane, both of which were seemingly not completed, and the company faded away soon after. H. Hurlin, formerly the secretary of the Paddington Aero Club, was the designer and the machine was said to be the result of two years of experiment.

The biplane was first described in March 1911, and the associated drawing showed an open braced girder fuselage with longerons curving up to meet the top members at the front, on which a forward elevator was mounted. The tail unit was comprised of elongated triangular dorsal and ventral fins, with an unbalanced rudder. The tailplane was fixed and similar in shape to the two fins, but with a scalloped trailing edge. The front portions of the bottom longerons were terminated halfway along the fuselage to form skids. A central skid formed the front of the single lower longeron, which continued to the tail; this skid carried a pair of wheels on a cross axle, sprung by five rubber wheels in a rubber band, which took the initial landing loads. There was a further skid to support the tail.

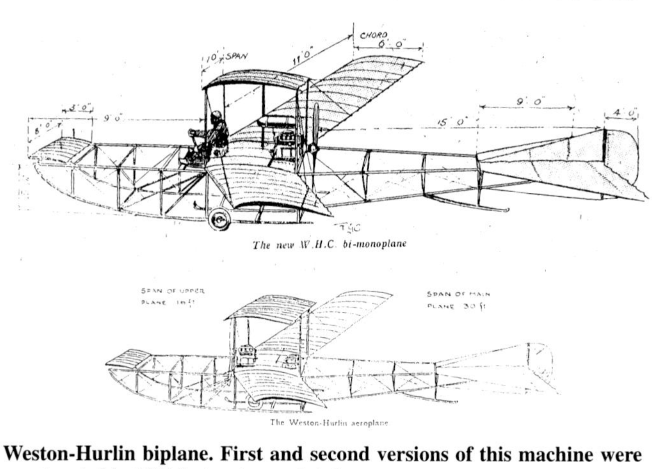

Weston-Hurlin biplane. First and second versions of this machine were constructed in 1911 but not completed.

The lower wing was parallel in chord and was swept back at a considerable angle, perhaps thirty degrees, and was mounted above the top longerons. The upper wing was straight and of short span and was mounted well above the center section of the lower wing. The pilot was seated on the front of the latter, and the engine was behind driving a pusher propeller, which revolved inside the top longerons, and which were widely spaced to clear the tips. The engine shown was a four-cylinder inline, probably a Green, with fuel supplied by gravity from a tank above.

However, by September 1911, a new drawing was published which showed that various changes had been made. A major change was to reverse the engine and pilot's positions and to use a Webb-Peet rotary, instead of the inline engine. Certain dimensions now differed from those published previously.

Data		
	Span bottom	30ft
	Span top	10ft later 16ft
	Chord	6ft
	Length	38-40 ft approx.

WHITE monoplane (White, 6 Woodville Rd., Thornton Heath, Surrey)

A two-seater Blériot type monoplane with folding wings, fitted with a 12-15hp engine, was reported to be under construction in October 1909.

J. SAMUEL WHITE & Co. Ltd. (East Cowes, Isle of Wight)

This old established shipbuilding company appointed Howard T. Wright as manager and chief designer of their Aviation Dept. in November 1912, and he immediately began the design of their first type. This was a seaplane and although some landplanes were built, the company concentrated principally on seagoing aircraft, mainly for the Admiralty under the Wight name.

WIGHT No.1 seaplane

This aircraft was built with some urgency, in order to appear at the Aero Show at Olympia opening on 14 February 1913. There was insufficient time and the machine was incomplete when displayed. After the Show it was returned to Cowes and was completed in the following May.

On the thirteenth of May Howard Wright attempted his first flight as a seaplane pilot, but crashed on takeoff. The remains were rebuilt with some minor changes.

The rebuilt machine was ready on 10 July 1913 and F.P. Raynham was the pilot, but was also involved in an accident causing less damage. The machine was again rebuilt, but this time greater changes were made, resulting in a change of identity to No.2.

Wight No.1 was a two-seat pusher biplane, with the nacelle mounted in the gap and carrying its tail unit on booms, which tapered to a point at the tail in elevation, but were parallel in plan. The machine was built with a broad center section, the outboard ends of which were connected by pairs of interplane struts, which continued below to the floats. These struts were only part of the complex of struts, connected to the center section and nacelle, which comprised the chassis for the long narrow floats with three steps.

The two bay wings, outboard of the center section, were slightly swept back and the top wing had greater span by a small amount. The ailerons were mounted on the interplane struts at midgap. The wings had an aerofoil section incorporating 'double camber', covered under patent No.24822/1912, which aimed to improve lift, but not without a penalty, for drag was increased. The tailplane fitted inside the vee of the booms and overhung slightly each side.

The elevator was divided by a tall rudder of symmetrical shape, hinged to the fin, which also extended symmetrically above and below the tailplane.

The nacelle was an elongated enclosure of shallow depth, with the pilot in front, protected by a raised decking; the passenger behind was unprotected, and both crew were rather exposed by the low cockpit sides. The engine at the rear was completely exposed, as was the forty gallon fuel and fourteen gallon oil tanks just behind the crew.

Neither the first nor rebuilt second version was acquired by the Admiralty.

Data:		
	Span top	44ft
	Span bottom	40ft
	Chord top	6ft 6in
	Chord bottom	5ft 6in
	Area	500 sq. ft
	Length	30ft
	Height	10ft
	Floats	21ft long by 2ft 6in beam
	Weight	1,350 lb.
	Weight allup	2,000 lb.
	Speed	70 mph
	Price	£2,500

Wight No.1 seaplane with Curtiss-type ailerons mounted midgap.

WIGHT No.2 Navyplane

After the crash of 10 July 1913, the changes made to the design of the earlier machine were more drastic. The work was completed by the 21 August 1913, when the machine, now known as the Navyplane, was launched and flown successfully two days later by Gordon England.

The changes introduced were mainly to the tail booms and tail unit. The top and bottom members of the booms were opened up in elevation, with an extra bay introduced, and slightly towed in in plan. A larger rectangular tailplane was mounted on top with a one piece elevator and twin rudders of a new shape below. There was no fin.

Although the floats were unchanged initially, considerable changes were made to the chassis. Struts were introduced between the floats, and these were anchored at the center to vee struts under the nacelle and center section. Long struts were added from the top center section to near the tips of each float, when new floats were fitted in September. Soon after, a land undercarriage was tried and the machine was tested with this in fields near Cowes at Northwood, until the machine was dam-

Wight No.2 Navyplane. This improved version of the No.1 seaplane first flew in August 1913 but was not bought by the RNAS until the outbreak of war.

Wight No.2 Navyplane. The second version with additional struts to front of floats

aged early in 1914. The machine was then put aside until refurbished as a seaplane at the outbreak of war, when it was acquired by the Admiralty on Contract No.CP53665/14, becoming Serial No.884. It was fitted with dual controls and used as a trainer at Calshot and Warsash until June 1915.

Data		
	Span top	44ft
	Span bottom	40ft
	Chord top	6ft 6in
	Chord bottom	5ft 6in
	Floats	21ft long by 2ft 6in beam
	Area	500 sq. ft
	Length	31ft 6in
	Height	11ft 6in
	Weight allup	2,400 lb.
	Speed range	35-70 mph
	Climb	400 ft per min
	Endurance	2 3/4 hr

WIGHT Navyplane 1914 type

This version was exhibited in a near complete condition at the Aero Show at Olympia in March 1914. It was an enlarged version of the 1913 Navyplane, with considerably increased span and a more powerful engine, but was built to the same basic layout.

The wings now consisted of four bays outboard of the center section and were without sweep back. The rudder shape was more rounded, and had a balance area below the bottom tail booms. Tall radiators, of square cross section, were positioned on either side of the nacelle ahead of the engine, which was provided with a compressed air starter. Howard Wright had developed his 'double camber' section also for propellers, and a large two-bladed propeller of this type was fitted. Accommodation was provided for three crew and a radio installation.

A total of seven of these machines was built, six for the RNAS and one for the German government. The first two, Serial Nos.128-129, were ordered on Contract CP57922/13 and these flew on 8 April 1914 and 2 May 1914 respectively. No.128 crashed on 4 June 1914, killing two naval officers; No.129 was wrecked on 29 January 1915. No.155 ordered on CP40819/14 flew on 5 May 1914 and was lost on 11 April 1915.

Wight Navyplane 1914 type. This enlarged version was first seen at Olympia in March 1914. Seven were built including four for the German Navy.

The fourth machine was launched at Cowes on 16 May 1914 and was flown at Kiel in Germany by Gordon England on 24 June 1914, but crashed with fatal results for the crew on the following day.

The fifth to seventh machines, also ordered by the German government were taken over by the Admiralty under Contract No.CP50287/14 and were all flown for the first time in July and August 1914 becoming serial Nos.893-895.

Power: 200hp Salmson (Canton-Unné) fourteen-cylinder two-row water-cooled radial.

Data		
	Span top	63ft
	Span bottom	59ft
	Chord top	6ft 6in
	Chord bottom	5ft 6in
	Area	735 sq. ft
	Weight (Br)	2,600 lb.
	Weight allup (Br)	3,500 lb.
	Weight (Ger)	2,901 lb.
	Weight allup (Ger)	4,175 lb.
	Speed range	35-69 mph (80 mph for Nos.893-895)
	Endurance	6hr
	Climb	400 ft per min

WIGHT type A.I, Improved Navyplane

Seven of these 'Improved Navyplanes' were ordered by the Admiralty on Contract CP37385/14/X7333 in May 1914. The aircraft, serial Nos.171-177, were delivered at various times from October and into the early part of 1915. Four of the aircraft saw undistinguished service in the eastern Mediterranean. The operations from the seaplane carrier Ark Royal were some of the earliest, and conditions at sea and aboard ship caused the aircraft to deteriorate rapidly. The Wight seaplanes never did live up to their original promise and the Navy were soon disillusioned with them.

This type was fitted with folding wings, with a system patented by Howard Wright under patent No.16924/14, to cater for storage aboard ship. The folding was actuated by a rack and worm wheel at the leading edge, adjacent to the fold joint. As a safety measure the engine could not be started until the

Wight type A.I Improved Navyplane. Seven of these (Nos.171-177) were ordered in 1914 for the RNAS.

wing locking bolts were home. This feature and the general strengthening and additions that had taken place added to the weight and affected the performance. In service the folding gear was removed and the wings were folded manually.

The wings were increased in span, the top wing overhang being supported by struts.

Wing tip floats were fitted, but again these were removed in service. The ailerons were connected top to bottom by struts. A further tailplane was fitted below the top booms and rectangular fins partly filled the last bays. Cross bracing and cables were added to stiffen the tail booms. Water rudders were added to the floats. The radiators were moved forward and raised to protrude above the level of the top wing. Radio equipment, a gun and bombs could be carried.

Power: 200hp Salmson (Canton-Unné) fourteen-cylinder two-row water-cooled radial driving a 9ft 4in diameter Wight 'double camber' propeller.

Data

Span top	71ft
Span bottom	60ft
Chord top	6ft 6in
Chord bottom	5ft 6in
Gap	5ft 7in
Area	780 sq. ft
Length	34ft 2in
Length folded	41ft 9in
Floats	27ft long
Height	12ft
Weight	3,467 lb.
Weight allup	4,500 lb.
Speed range	35-80 mph
Endurance	5hr
Climb	21min to 3,000 ft

WIGHT type A.II Improved Navyplane, Admiralty Type 177

Further changes to the Navyplane resulted in the A.II version, known in service as the Admiralty Type 177, serial number of the last of the previous batch from which it was derived.

More power was provided by the Sunbeam engine, to match the increasing weight, and anticipating the carriage of a torpedo, for which provision was made. The improvement in engine reliability aimed for was not achieved and, with general airframe problems, the type was not well liked in the Navy. The torpedo experiment was abortive as the machine would not lift off the water.

Only four type A.II Navyplanes were ordered on Contract CP02143/14/X16079 on 2 September 1914, identified as serial Nos.831-834, the first being delivered to Calshot on 2 March 1915.

The next Wight-type produced was a tractor machine.

Power: 225hp Sunbeam Mohawk twelve-cylinder water-cooled vee driving a 9ft 6in diameter Lang four-bladed propeller.

Data Dimensions as for type A.I Improved Navyplane.

Weight	3,467 lb.
Weight allup	4,500 lb.
Speed	86 mph
Climb	20 min to 3,000ft

WIGHT tractor seaplane, Admiralty Type 840

The production of this machine began in the Autumn of 1914, with the first of three batches on Contracts CP01243/14 for serial Nos.835-840 and CP68765/14 for Nos.1300-19 and 1351-4, totaling thirty aircraft built at Cowes by the parent company. The first machine was completed in April 1915, and production continued at Whites until the end of the year. Other contracts were placed in 1915 with Beardmores for thirty-two aircraft and Portholme for twenty-two aircraft, the manufacture of which continued until late in 1916, although not all of these were finally assembled.

The Type 840 was a conventional tractor biplane, with four bay folding wings and top wing overhang, braced by wires and kingposts. Wing tip floats with skateboards were fitted. The wings were of 'double camber' section and carried single acting ailerons on the top wing only. The fuselage was of tapering rectangular section with a curved top decking, with the crew located well behind the wings. The engine exhausted through a funnel like pipe above the top wing, and was cooled by tall square section radiators on each side.

The triangular fin was braced by a strut to the tailplane tip each side. A triangulated group of struts under the tail end was an anchorage for the tailplane lift struts, but after the prototype had flown, a further triangular fin was inserted in this area. The tailplane was rectangular and carried divided elevators. The chassis was a complex affair of inverted vee struts with curved crossbars to permit the carriage of a torpedo. The long

Wight type A.II Improved Navyplane. Four of these were built with various modifications including a 225hp Sunbeam engine.

Wight seaplane Admiralty type 840 was produced in quantity from late 1914 onwards.

floats were similar to those of earlier Wight machines with three steps and water rudders. On later machines the requirement to carry a torpedo was deleted and the chassis was increased in height with four single struts to each float and straight cross-bars. At least one machine was flown as a landplane with the tailplane repositioned and with a modified fin and rudder.

Power: 225hp Sunbeam Mohawk twelve-cylinder water-cooled vee driving a 11ft or 11ft 6in diameter two-bladed Lang propeller or a 10ft diameter four-bladed propeller.

Data		
	Span top	61ft
	Span bottom	46ft
	Span folded	12ft 2in
	Chord	5ft 6in
	Area	568 sq. ft
	Length	41ft
	Height	14ft 4in
	Weight	3,408 lb.
	Weight allup	4,453 lb.
	Max speed	75 mph
	Endurance	7hr
	Climb	3,000 ft in 16 min

WIGHT twin-fuselage landplane

This machine was ordered by the French government in August 1914 and was the largest machine built in Britain to that time. Completion at Cowes took place on 24 July 1915 and the machine was then taken by sea to Eastchurch for flight trials by Gordon England, the first taking place on 6 August 1915. In the following month a French pilot crashed the machine on to its nose and the contract was canceled. The machine was returned to Cowes in pieces and was not rebuilt.

The aircraft was designed with a central nacelle seating the crew in tandem with the pilot behind. On either side was a fuselage, each with a tractor engine and deep enough to carry bombs stowed vertically. The bottom, parallel chord, tailplane joined the two fuselages at the lower longerons and a similar tailplane was attached above to an open structure. The horn balance of the rudders swung above the top tailplane, to which a one piece elevator was hinged. As a result of spinning problems during the flight trials, long fins were fitted above the fuselages and top tailplane, displacing the rudder balance areas, and a bottom elevator was added.

The folding wings consisted of five bays outboard of the fuselages, with top overhang braced by kingposts and cables. There were ailerons at all four wing tips, connected by struts top to bottom. The radial engines mounted in the nose of each fuselage were cooled by four tall, square section radiators fitted alongside the fuselages, between the trailing edges of the wings. Separate undercarriages, under each fuselage, consisted of pairs of vee struts with twin skids and two pairs of wheels at each mounting; small bumper wheels were fitted to the tip of each skid, which were supported by struts to the lower longerons.

Power: Two 200hp Salmson (Canton-Unné) fourteen-cylinder two-row water-cooled radials driving 10ft diameter propellers.

Data		
	Span top	84ft
	Span bottom	71ft
	Span folded	28ft
	Chord top	7ft
	Chord bottom	6ft 6in
	Area	1,048 sq. ft
	Weight allup	10,000 lb.

WIGHT twin fuselage seaplane, Admiralty Type 187

The Admiralty ordered a seaplane version similar to the previous machine on Contract No.CP02108/14. This aircraft, serial No.187 was virtually complete, less engines, in September 1915, but was not delivered to Felixstowe until July 1916. It was found to be unsuitable for carrying the large Mk IX torpedo for which it was intended and was deleted in October. Two further machines ordered in 1915, Nos.1450-1451, were delivered in April and July 1916 and were soon deleted as unsuitable.

The first machine showed a few changes from the Twin-land-plane. Most noticeable were the engine cowlings and ex-

Wight twin-fuselage landplane was ordered by the French government in August 1914 but crashed on test at Eastchurch.

Wight twin-fuselage seaplane Admiralty type 187. A modified version of the Twin landplane. Three were built but found unsuitable as torpedo carriers.

tended noses, with front bearing supports for the new extension shafts, introduced for the Salmson engines. The radiators were moved forward, just ahead of the wing leading edge. The central crew nacelle was discarded and the two crew were given separate cockpits in each fuselage. Long main floats, under each fuselage, supported the aircraft, and small cylindrical floats with skateboards protected the wing tips.

The two later machines were fitted with redesigned tail units and the top engine cowlings were deleted.

Power: Two 200hp Salmson (Canton-Unné) fourteen-cylinder two-row water-cooled radials driving two-bladed propellers by extension shaft.

Data

Span top	84ft
Span bottom	71ft
Span folded	28ft
Chord top	6ft 6in
Chord bottom	6ft
Area	972 sq. ft
Length	43ft 9in
Main floats	23ft long
Max speed	80 mph
Endurance	5 hr

WHITE & THOMPSON Ltd. (Dr. John Douglas Campbell White and Norman Arthur Thompson, Aeronautic Works, Middleton-on-Sea, Bognor, Sussex)

Thompson was an established engineer who, in 1907-8, began to be interested in aeronautics through reading the books of F.W. Lanchester, the well known writer on aerodynamics and consultant to the Daimler company. In 1909, Thompson made contact with Dr. White, an old college friend, and together the pair approached Lanchester to discuss their plans for developing their aviation interests, and persuaded him to act as their consultant from March of 1909. Dr. White agreed to provide finance for an experimental machine, and to administer the new concern that the pair established at Middleton-on-Sea, near Bognor, in early 1910. Thompson managed the practical aspects of a works and manufacturing facilities, adjacent to the sands, which were to be used for flight testing. A private company, with the two men as directors, was finally established in March 1912.

Meanwhile Lanchester was designing a machine, work on which was commenced at the Daimler works in Coventry pending the availability of the facilities at Middleton. This machine was identified as the Thompson-Lanchester No.1 biplane, and was probably Lanchester's only design for a complete aircraft.

W&T: THOMPSON-LANCHESTER No.1 biplane '*The Grey Angel*'

The machine was intended to satisfy a military role, but the initial design by Lanchester, of a single-seater, single-engined aircraft evolved into a twin engined two-seater pusher, when the official views on military requirements became known. The fuselage was built at the Daimler Motor Co. works and, in the summer of 1910, with the aircraft only partly built, work was transferred to Middleton, where a small works had been set up adjacent to the sands and the machine was completed there. The sands were in poor condition when the time for the trials, carried out by Thompson, arrived, and these were abandoned when the machine overturned. Modifications were made in 1911 before further trials by Capt. Nicholson RN, which were again unsuccessful and the machine was scrapped.

The fuselage was of streamlined shape at the front, and was mounted on a faired in four wheeled undercarriage. This employed a hydraulic, pneumatic design patented by Lanchester No.18384/1909, which proved unsatisfactory and was replaced

White & Thompson Thompson-Lanchester No.1 biplane '*The Grey Angel*' was partly built by Daimler and was wrecked at Middleton in 1911.

with sprung wheels on outriggers. A large sail like rudder and divided elevator were fitted to the nose. The crew were seated side by side in front of the wing, from which point the fuselage flattened out to provide clearance for the two propellers. A fixed biplane boxkite tail, containing four vertical panels, comprising the tail unit was made of wood with 26 swg aluminum skinning. The fuselage was built of ash with sheet steel skinning, in the same manner as contemporary car bodies.

The wings were of unequal span, parallel in chord, and with squarecut tips and were braced, Warren girder style, with three pairs of struts plus diagonals each side. The construction was all metal with 23 swg aluminum alloy skins, which no doubt gave rise to the name *'Grey Angel'*, by which the machine was known.

To maintain continuous power in the event of a failure, the two engines were coupled by a crossed belt and initially drove four-bladed, adjustable pitch metal propellers, with blades, which tapered to the roots. These were replaced by three-bladed propellers, in which the blades tapered in the reverse direction, the broad blades having pointed tips.

Power: Two 50hp Gnôme seven-cylinder air-cooled rotaries, driving 5ft 2in diameter four-bladed propellers by crossed belt. Three-bladed propellers used later.

Data

Span top	25ft
Span bottom	20ft
Chord	2ft 6in
Area	105 sq. ft
Length	14ft
Weight	1,200 lb.
Speed	75 mph

W&T: NORMAN THOMPSON No.1 biplane

Following the failure of the first machine, Thompson proceeded to design a second landplane which was built during 1913. This was a side by side pusher biplane with the engine in the center of the nacelle, driving the propeller through an extension shaft. The propeller was just behind the crew, whose cockpit was at the extreme rear of the short nacelle. The tail unit was carried on slender steel tube tail booms, and consisted of top and bottom fins and one piece rudder, with a fixed tailplane and divided elevator. The nacelle, which was covered with steel sheet, was mounted above the lower center section on a chassis of steel struts, passing through, or clear of, the lower center section to the nacelle. At the connections to the nacelle there were rubber cord sprung mountings covered by a system under patent No.23744/12. The machine rested on two small diameter wheels with balloon tires, and on the rear end of the central skid.

The engine was a 100hp ABC water-cooled type mounted in the center of the nacelle with the coolant radiator on the nose, and the fuel tank between the engine and the crew, supplying fuel by gravity. The pusher propeller was driven by a long drive shaft passing through the cockpit.

The two bay wings tapered slightly and were rigged with slight dihedral, and carried tapered and curved single acting ailerons hinged behind the trailing edges. These were made of sheet fiber board and were initially fitted to the top wings only, but later were added to the bottom wings as well. The wings were constructed with metal spars and ribs with fabric covering. A series of wires stretched from the root to the tip of each wing, on each surface and the fabric was threaded through these wires. The method of wing construction was covered under patent Nos.21604/1912 and 1909/1913.

W&T Norman Thompson No.1 biplane was flown in 1913 but was not developed.

The machine was flown by E.R. Whitehouse and J.C. Porte, both RN officers and R.L. Charteris, managing director of the ABC company, also by Norman Thompson. Although reported to be quite successful, the machine was not rebuilt after Porte overturned it in 1914. The firm had by this time acquired the rights to the Curtiss flying boat from Capt. E.C. Bass, which became their main interest from then on. Later the interpretation of this agreement became the subject of a legal wrangle with the Admiralty.

Power: 100hp ABC eight-cylinder water-cooled vee driving a 8ft 9in diameter three-bladed metal, adjustable-pitch Norman Thompson patented propeller.

Data (Dimensions estimated)

Span	30ft
Chord at root.	4ft 5in
Gap	5ft 5in
Length	24ft
Height	12ft
Length of nacelle	9ft

W&T: CURTISS type flying boats

Initially acting as an agency for the Curtiss-made boats, White and Thompson imported a few (possibly two machines), for sale and for setting up a flying training school at Middleton. The experience with the demonstration machine revealed shortcomings in design and construction which the firm decided could be improved upon. The emphasis thus became more on building improved English boats, rather than importing from America.

Changes were also incorporated in the demonstration machine, particularly the replacement of the seat back yoke for lateral control with a more conventional system. This machine

W&T Curtiss type flying boat. An improved version of the Curtiss F was flown at Middleton in March 1913.

was sold to Ernest Bass and taken to the south of France in the spring of 1914, but was crashed badly and returned to Middleton for rebuilding. This work included replacing the Curtiss type wings, fitted with midgap ailerons, with improved units of a new section and with ailerons in the top wings. A 100hp Anzani engine, instead of the Curtiss OX, was fitted and the machine, then referred to as the Bass-Curtiss Airboat, showed a marked improvement in performance.

White and Thompson constructed their first machine incorporating changes, and this was flown on the 27 March 1914 by J.C. Porte at Middleton. The hull had provision for four people ahead of the lower wing, in the wide forward portion, which tapered aft to a vertical knife edge. The underside was flat tapering to a vee bottom at the single step, behind which the hull bottom was flat. The top of the hull was raised in front of the crew, for protection against wind and spray. The structure was an ash framework covered with two layers of mahogany planking, with the addition of a waterproofed canvas interlayer in the planing bottom. Bulkheads were arranged to provide watertight compartments. Dual controls were fitted.

The wings were divided into two bays, outboard of the center section struts, and the top wing had overhang braced by a strut, cables and kingposts. A large cutout for clearing the propeller was provided in the top wing. Ailerons, whose area extended mostly behind the wing trailing edge, were fitted in the top wing only.

Flatsided metal wing tip floats were fitted flush to the underside of the wing and had a ski-shaped springboard to cushion the shock of alighting on water. A tail skid served to protect the hull when the boat was being moved on land, and the main hull was fitted with an external keel for similar protection.

W&T Bass-Curtiss Airboat was a major reconstruction of a Curtiss F fitted with an Anzani engine in June 1914.

The engine was positioned on a mounting of struts high in the gap, and was also braced forward by a strong diagonal member. The main fuel tank was under the rear seat, the fuel being raised by an engine driven pump.

The tail unit consisted of a long triangular fin and partly balanced rudder, which was reinforced and covered with copper sheet to form a water rudder. The triangular tailplane was set at a negative angle to offset the high thrust line and was raised above the hull on a steel tube mounting and carried a divided elevator.

Power: 100hp Curtiss OX5 eight-cylinder water-cooled vee driving a 8ft diameter two-bladed propeller.

Data		
	Span top	41ft
	Span bottom	30ft
	Chord top	5ft 3in
	Chord bottom	5ft
	Area	378 sq. ft
	Area tailplane	27 sq ft
	Area elevators	24 sq. ft
	Area fin	10 sq. ft
	Area rudder	13 sq. ft
	Gap	5ft 6in
	Length	27ft 4in
	Max speed	60 mph

Curtiss type F 'The Bass-Curtiss Airboat' (W&T reconstruction)

Power: 100hp Anzani six-cylinder air-cooled radial.

Data		
	Span	41ft 8in
	Chord	5ft 6in
	Gap	5ft 6in
	Length	26ft

WHITE & THOMPSON flying boat '*Seaplane No.1*'

This was a twin engined flying boat based on Curtiss design, and was developed by the firm to compete in the *Daily Mail* Seaplane Circuit of Britain, which was canceled by the outbreak of war in 1914. The fin carried the company name and 'Seaplane No.1'; the rudder bore No.9, the entry number in the contest, although in fact the machine would not have been ready in time.

The Curtiss H.1 type named '*America*', was a machine being prepared to fly the Atlantic with J.C. Porte as one of the pilots. This was bought by the Admiralty, together with two other similar machines, Curtiss type H.2, which became known as '*Small Americas*'. The White and Thompson machine was sometimes referred to by the same name, after it was taken over by the Admiralty as serial No.883. It remained with the company until late in 1915 in use as a test and trial installation machine, until developments at Felixstowe, for which Porte

W&T flying boat 'Seaplane No.1' was to have flown as No.9 in the 1914 Circuit of Britain but was taken over by the RNAS as No.883 instead.

was largely responsible, rendered it obsolete. Later the H.4 version of the '*Small America*' was bought by the Admiralty in some quantity from Curtiss, who also established a Canadian branch for their production, and from the Aircraft Manufacturing Co., Airco.

The White and Thompson machine was basically a scaled-up version of the single-engined Curtiss in appearance, but it embodied some new features. The hull was of rectangular section at the front, changing, behind the step, to circular towards the tail and it accommodated two crew side-by-side with dual controls, just ahead of the wing. A ninety gallon fuel tank was positioned behind the pilot's seat. The hull was built by Williams & Co. boat builders of Littlehampton. Gordon England, when testing the machine in September and October 1914, was unable to get off the water, until flared planing fins were added to the sides of the hull.

The wings were of RAF 6 section, and had squarecut tips and large tip floats to which springboards were fitted, otherwise the general configuration was similar to Curtiss. The tailplane was of rectangular shape, still on top of the fin, with a revised arrangement of struts. The tail surfaces were all larger in area. The Curtiss OX engines were made by the Austin Motor Co., who were licensees from Curtiss, and drove three-bladed, variable pitch metal pusher propellers. The engines were mounted in the center section, by an arrangement of inverted vee struts, and this was wider, and supported by bracing struts to the lower part of the hull.

Power: Two 90hp Curtiss OX eight-cylinder water-cooled vee.

Data		
	Span top	52ft
	Span bottom	40ft
	Chord	5ft.6in.
	Gap	5ft 6in
	Area	500 sq. ft
	Area tailplane	59 sq. ft
	Area elevators	38 sq. ft
	Area fin	12 sq. ft
	Area rudder	18 sq. ft
	Length	32ft 3in
	Height	10ft 6in
	Weight	2,000 lb.
	Weight allup	3,000 lb.
	ax. speed	70 mph
	Endurance	6 hr

WHITE & THOMPSON flying boat 'Seaplane No.2'

This single-engined machine built in 1914 was also entered for the Seaplane Circuit of Britain which was abandoned. It carried entry No.6 on the rudder and 'Seaplane No.2' on the fin. As with No.1, it was taken over by the Admiralty under Contract No.CP53624/14 for £2,400, becoming serial No.882. The aircraft first flew on 1 August 1914 and, after service at various stations, was wrecked on 8 June 1915.

Although basically of Curtiss design, a number of modifications were made. The wings, of RAF 6 section with squarecut tips, were transferred from the Bass-Curtiss Airboat. The top wing carried a new rounded fin on the centerline, serving also as a king post for bracing wires; the outboard kingposts and bracing wires were deleted. Conventional ailerons were fitted to the top wing only. The hull, built by S.E. Saunders & Co. of Cowes, was covered with two skins of mahogany with an interlayer, and was sewn by the 'Consuta' method. At the front the decking was rounded, the greater width at the top tapering down to a vee bottom, which continued to the step. Aft of this, the hull was of round section with a flat top. No tail skid was fitted and detachable wheels were provided for ground handling. There was a general scaling up of dimensions and more power was provided.

Power: 120hp Austro-Daimler built by Beardmore driving a four-bladed propeller.

Data		
	Span top	45ft
	Span bottom	34ft
	Chord	5ft 6in
	Gap	5ft 9in
	Area	400 sq. ft
	Area tailplane	30 sq. ft
	Area elevators	27 sq. ft
	Area fin	12 sq. ft
	Area rudder	16 sq. ft
	Length	27ft 6in
	Weight	1,600 lb.
	Weight allup	2,400 lb.
	Max speed	70 mph
	Endurance	6 hr

W&T flying boat 'Seaplane No.2' was to have been No.7 in the 1914 Circuit of Britain but was impressed into the RNAS as No.882.

W&T flying boat No.3 (NT2). One of a batch of six built in 1914-1915 for the RNAS.

WHITE & THOMPSON flying boat No.3 (NT.2 and NT.2A)

This machine was a version of No.2 with improvements; six aircraft were acquired by the Admiralty on Contract No.CP60401/14, the machines bearing serial Nos.1195-1200. Delivery of these commenced in January 1915 and were followed by two more, Nos.3807-3808, (NT.2A) ordered later that year on CP45183/15 at £1,600 each.

Changes made from W&T No.2 included enclosure of the lower center section, which was then faired into the hull. The forty-five gallon fuel tank was slung above the floor, instead of resting on it, which had earlier resulted in damage. The wing tips were rounded and the single fin on the top wing was replaced by two of lower height above the inner interplane struts, later fitted above the outer struts on Nos.3807-3808. The struts bracing the top wing extensions were replaced by kingposts and wires. The main fin area was increased and an additional small fin was fitted above the tailplane, replacing the rudder balance area.

The Austro-Daimler engine drove a four-bladed propeller and was fitted with a starting handle in front of the radiator, accessible from the cockpit. A Hispano-Suiza engine was fitted to the last machine No.3808, anticipating its use in the wartime NT.2B. This last machine was also provided with dual controls. Experience with the first machine led to the fitting of two struts from the hull to the engine mounting and an improved type of hull using copper riveting; these modifications were introduced in the NT.2A and fitted retrospectively to earlier machines.

A windscreen and machine gun were fitted to some of the machines, which were used for oversea patrol operations, one of which resulted in No.1199 being forced to alight in neutral territory, where it was acquired by the Dutch Navy, remaining in service until 1917.

Power: 120hp Austro-Daimler six-cylinder inline water-cooled built by Beardmore driving a four-bladed propeller.
150hp Hispano-Suiza eight-cylinder water-cooled vee driving a four-bladed propeller (NT.2A No.3808 only)

Data		
	Span top	45ft
	Span bottom	34ft
	Chord	5ft 6in
	Length	27ft 6in
	Height	10ft 6in
	Weight	1,850 lb.
	Weight allup	2,803 lb.
	NT.2A	
	Climb to 500ft	2min
	Climb to 1,000ft	4min
	Climb to 9,000ft	20min

WHITE & THOMPSON biplane NT.3 '*Bognor Bloater*'

The company engaged F.P.H. Beadle, previously a partner in Perry Beadle & Co., soon after Perry was killed in France on 16 August 1914 and their company was liquidated. His first type as chief designer flew in March 1915, and was the first of a batch of twelve ordered by the Admiralty. These were serial Nos.1171-1182, known in the RNAS as *'Bognor Bloaters'*, all delivered by September 1915, although the last two were not assembled by the firm.

The NT.3 was in the same category as the BE.2C, which the Admiralty required for training purposes, and was fitted with similar wings of unequal span and without stagger. The aircraft was a two-seater of conventional tractor layout, but incorporated a rather original feature in the form of a monocoque fuselage, the wooden skins of which were copper

W&T flying boat No.3 (NT2). Modified with additional struts from the nose of the hull to the engine mounting. The NT2A was similar.

W&T biplane NT3, nicknamed *'Bognor Bloater'*, was a BE2c derivative used by the RNAS.

sewn by the Saunders 'Consuta' process at the works of Williams and Co. of Littlehampton. The undercarriage was of the two wheel, twin skid type.

The wings were of unequal span, the top overhang being unbraced, and consisted of two bays with pairs of full height interplane struts adjacent to the fuselage. There were ailerons on all wings, connected top to bottom by struts, whose wide chord caused these to extend beyond the general line of the wing.

Dr. White, a medical man who had largely financed and managed the company, withdrew his support in June 1915 and joined the RAMC. Thereafter the company was reconstituted in September as the Norman Thompson Flight Co.

Power: 70hp. Renault eight-cylinder air-cooled vee driving a 10ft 4in diameter four-bladed propeller.

Data	Span top	37ft
	Span bottom	32ft 6in
	Chord	6ft
	Tailplane span	12ft
	Tailplane chord	7ft 6in
	Length	28ft 3in
	Height	12ft.

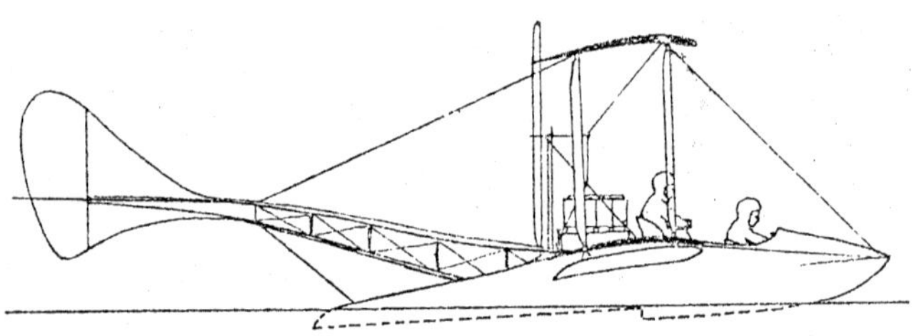

Wigram flying boat designed by Arthur Wigram of Sydney to take part in the Seaplane Circuit of Britain race.

WIGRAM flying boat (Arthur Wigram, Wigram Flying Boats Ltd., E. Cowes, Isle of Wight)

Reported in February 1913, designed by Mr. Wigram of Sydney, Australia, the machine was a development of one he proposed in 1910. It was constructed for him by S.E. Saunders Ltd. of Cowes, but was never completed.

The description and sketch in *The Aeroplane* showed a single step hull housing the crew of two in tandem. The tail unit was carried by a curved girder, shown open, although covered by plywood skinning on the actual hull, projecting upwards and backwards. The tail surfaces were curved in outline, and consisted of a tailplane with divided elevator, and a fin with an additional ventral area and an unbalanced rudder.

The engine was placed in the hull behind the pilot, who was seated at the leading edge of the lower wing, and this drove a pusher propeller by chain. No detail of the wing structure is available, but it is known that wing tip balancing floats would be fitted.

Wigram flying boat under construction at SE Saunders' works. It was not completed.

The machine was not completed, presumably due to lack of funds and did not compete in the Seaplane Circuit of Britain, for which it had been entered.

Power: 100hp Green six-cylinder inline water-cooled driving by chain a pusher propeller.

WILDING cycle-aeroplane. (Wilding, Standish, Wigan, Lancashire)

These 'Aviette' type machines were reported and advertised in the Press, and were made by a German workman, living in Lancashire. Two types were publicized in 1910-1911.

WILLETTS flying bicycle (Charles and Harold Willetts, Colonial Works, Cradley Heath, Staffordshire)

This was just another type of 'Aviette' which went the way of all the others.

WILLIAMS monoplane (William Ellis Williams, Bangor University, North Wales. Flight trials at Llanddona, Anglesey)

Ellis Williams began theoretical work on aeronautics at the University in conjunction with Professor G.H. Bryan, and early in 1904, published certain results in a paper entitled *The Longitudinal Stability of Aeroplane Gliders.* After further study the design and construction of a monoplane commenced in 1909.

Williams monoplane. The first version in 1910 was not a success.

The machine, which was expected to weigh 700 lb., was made of ash and bamboo. The fuselage was of triangular section, with the single longeron at the top and had twin skids attached to the two lower members. The undercarriage consisted of a single sprung wheel adjacent to each skid, together with a long tail skid. The wing was of Eiffel section, eight inches deep, with flat lower and contoured upper surfaces, and with sharp leading and trailing edges. A large triangular tailplane provided additional lift area and there was a rectangular rudder mounted above. Wing tip ailerons were provided for lateral control; two vertical fins were fitted above the wings to aid stability.

The expense of providing an engine was to delay the project throughout, but with financial assistance from a wealthy benefactor in October 1909, Williams acquired a secondhand 35hp six-cylinder air-cooled engine, of unknown make, for £100 from a Mr. P. Pillion.

The machine was transported by boat to the remote Red Wharf Bay at Llanddona, where a hangar had been built, and the first tests were carried out there along the beach in the late summer of 1910. These proved to be unsuccessful and flight was not achieved, the attempts being discontinued with the onset of winter.

In the period before further trials could commence, changes were made to the machine and a replacement engine, a 35hp JAP, was hired from A.V. Roe for the months of August and September 1911, during which time flight was achieved, before the engine was wrecked and the trials were discontinued.

It is probable that the changes made for the 1911 season were the major modifications to save weight shown in the surviving illustrations. The triangular fuselage was discarded and replaced by a single tail boom and the twin skids were replaced by a single member. The streamlined fuel and oil tank fitted ahead of the pilot was probably also supplied by A.V. Roe.

For the summer of 1912 Williams hired a 40hp Clément-Bayard from Weston Hurlin Ltd. for three months. This engine, which replaced a hired REP, did not arrive until the 20th of July, and little progress was made with the trials. The Weston Hurlin company was in liquidation by September and Williams was able to buy the engine for £25.

More successful results were achieved in 1913, and it is recorded that on 3 September 1913 a flight took place at a height of seven feet, at a speed of 37 mph., in which an experiment to obtain pressure distribution data over the wing was carried out, for correlation with wind tunnel data

Williams monoplane. The second version was flown in 1912-1913 at Llanddona, Anglesey.

No further flights were carried out in 1914 and the machine was presumably broken up, for the hangar was taken over in August by a Mr. Albert Roberts.

The various reports are not specific as to the type of engine used and in cases the details conflict with records of early engine types. The photograph of the front view of the monoplane shows a fantype engine apparently with three-cylinders such as an Anzani. The 1908-1909 fantype Anzani was rated at 25hp and weighed 143 lb., and may have been the engine used in 1910. There were several sizes of fantype Anzanis made between 1907 and 1910 with ratings up to 50hp.

Power: 1910 35hp six-cylinder weighing 180 lb.
35hp JAP air-cooled presumed to be the eight-cylinder type.
REP air-cooled, either 25hp five-cylinder weighing 121 lb or 35hp seven-cylinder weighing 132 lb.
40hp Clément-Bayard vee type. This type is not recognizable from contemporary records. It may have been a four-cylinder inline type.

Data		
	Span	32ft
	Length	37ft
	Area	200 sq. ft
	Tailplane	45 sq. ft

WILKES biplane

This machine was reported built in 190910. Fitted with a 50hp Gladiator engine, it had a wing area of 533 sq. ft.

WILSON helicopter (Edgar E. Wilson, 77 Glamorgan St., Pimlico, London, SW 1)

In 1906 Wilson constructed a helicopter in a building on top of the Watkin Tower, Wembley Park. The machine, named *The Clipper of the Clouds,* had four elevating and propulsion screws of 10ft 6in diameter, extending from the sides; 'stepped', bird-like surfaces were fixed above. The machine was based on a girder frame and was 30ft long. A 6hp engine was fitted. Finance for a larger motor was not forthcoming and the machine was abandoned.

WILSON ornithopters

Edgar Wilson made many experiments in early flight initially concerned with ornithopters and direct lift devices.

He built an ornithopter with Pettigrew type wings of 22ft span and a tail five feet long and three feet across. Weighing around 250 lb with pilot and a 6hp engine it was tested whilst suspended from a specially constructed 60ft high staging. Further trials were made when it was launched from the top of the 160ft Watkin Tower at Wembley Park and glided successfully to the ground.

A second ornithopter had a pair of wings, each 8ft long and 6ft wide and a fanshaped tail. Again a 6hp engine was used.

Wilson ornithopter had wings of the Pettigrew type.

The third ornithopter had 8 wings measuring 18ft from tip to tip, flapping alternatively and was to be fitted with a 50hp engine.

Wilson then turned to the construction of a giant direct lift machine. This was 70ft long and 10ft wide and had thirty screws 5ft in diameter to provide lift and four screws of 7ft diameter for propulsion. The power was to be provided by a 100hp engine. It was being built on the top of the Watkin Tower in early 1906 but finance for a motor was not forthcoming.

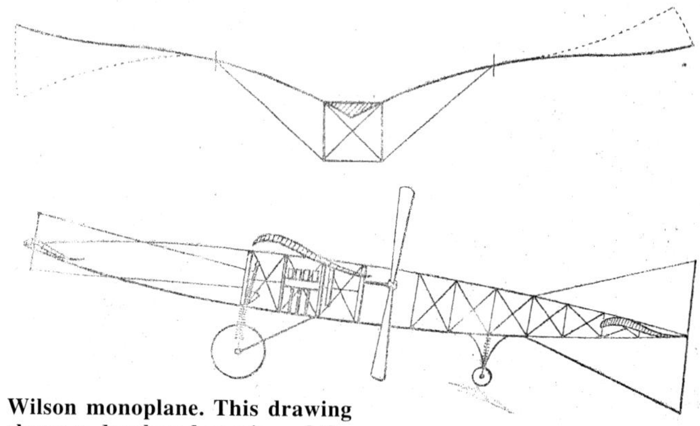

Wilson monoplane. This drawing shows a developed version of the machine built in 1909.

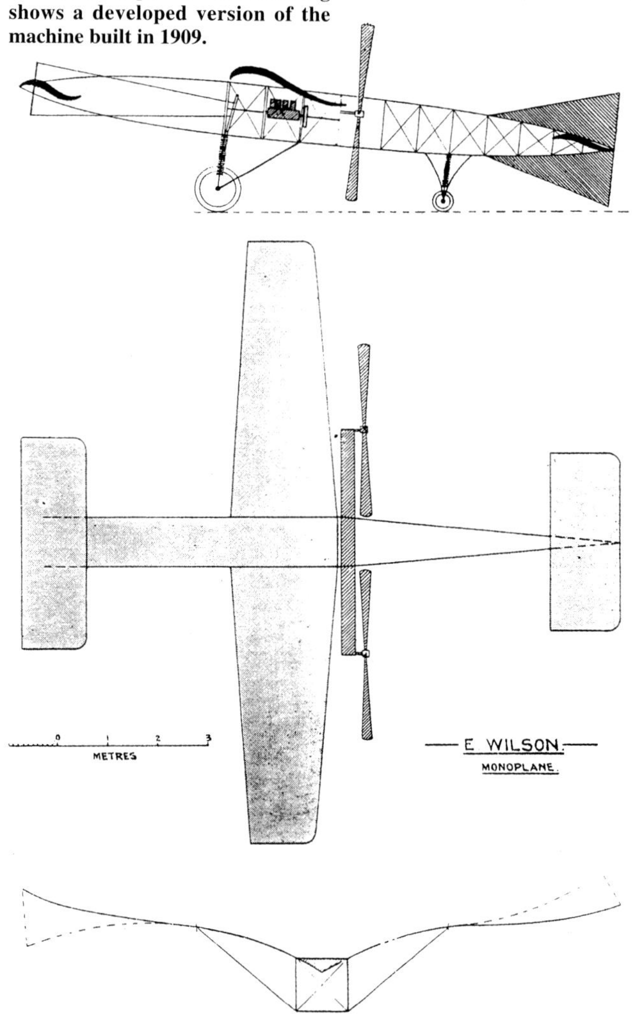

Wilson ornithopter was launched from the top of a 60ft staging and glided to the ground.

WILSON monoplanes and biplanes

Although Edgar Wilson was mainly interested in achieving flight by vertical lift, he built a monoplane in 1909 at Wembley Park, which was shown at the Imperial Exhibition held at Shepherds Bush in June, in a dismantled condition.

A scheme for a later monoplane was published, with a letter in *The Aero* in June 1909. This was to be a canard type with twin pusher propellers and tapered warping wing, with bird-like upward curves. The engine was to be a 25hp Smith made by David J. Smith & Co. of London.

Wilson also claimed at the time to be working on an improved Wright type biplane, with front boxkite elevator and boxkite tail, to compete for the prize for the London to Manchester flight. A pair of gyroscopes for control of lateral stability were to be employed. Nothing further was heard of these projects, or of his proposals for a helicopter and an ornithopter. Data for the biplane is given below.

Power: 16hp engine driving two 6ft 6in propellers.

Data		
	Span	32ft
	Chord	6ft 6in
	Gap	6ft
	Area	528 sq. ft
	Length	32ft
	Front elevator	10ft by 3ft
	Tailplane	8ft by 5ft
	Weight	600 lb.

WILSON-GIBSON monoplane (Motor Repair Co., Heath Rd., Twickenham, Middlesex)

This single-seater machine was built in 1910 by W.B. Wilson, his brother and W.E. Gibson of the Motor Repair Co., using a structure of camphor wood, made by a Mr. Hebblewhite of Gould Rd., Twickenham. It was illustrated in skeleton form in May 1910, and complete in February 1911 at the London Aviation Ground at Acton, where it was to be used by the Allan Knight Co. as a practice machine. It is believed to have been among a number of machines destroyed by fire on 14 May 1911.

Wilson-Gibson monoplane built in 1910 and destroyed by fire at Acton in May 1911.

The machine resembled an Antoinette, with a parallel chord wing, braced by wires and kingposts to a central pylon and fitted with ailerons hinged on the trailing edge. The aerofoil section was deeply cambered and tapered in thickness from 5in to 3in along the span. The fuselage was of triangular section, and was fabric covered over most of its length.

The undercarriage had two pairs of skids and two pairs of wheels on separate axles, and a large fork mounted tailwheel. The fin and tailplane extended well forward along the fuselage and carried a triangular rudder and divided triangular elevators. The pilot's control was by wheel and column, as distinct from the Antoinette type of wheel control. Two radiators made by the firm of Lamplough were mounted one on each side of the fuselage.

Power: 60hp NEC six-cylinder inline water-cooled two-stroke.

Data		
	Span	46ft 6in
	Chord	8ft
	Area	356 sq. ft
	Area tailplane & elevators	90 sq. ft
	Length	40ft
	Area rudder	15 sq. ft
	Weight allup	950 lb.

WILSON-JONES glider (Owen Wilson-Jones, Surbiton, Surrey)

No information available.

Windham glider of Chanute type was built by apprentices in 1909.

WINDHAM glider (Capt. Walter G. Windham, Windham Detachable Motor Body Co., 22 St. John's Hill, Clapham Junction, London, SW)

The glider was built in the early part of 1909 by apprentices at Windham's motor body works; Messrs. A.M. Grose, T. de Guerin, H. Cutter, H. Turrell and A.N. Feary, were those involved with the construction and testing, which took place at Wembley Park in the summer, when a few flights were made before damage occurred and the tests were terminated.

The glider was of Chanute type with biplane wings and boxkite tail, carried by booms tapering in elevation. The operator supported himself in the cutaway center section of the bottom wing, relying on weight shift for control. The materials used for construction were mainly poplar with some deal and bamboo.

Data	Span	18ft 6in
	Chord	5ft 6in
	Area	180 sq. ft
	Weight	60 lb.

WINDHAM biplane No.1

Although not of British construction, the Windham biplane, which was exhibited at the Aero Show at Olympia in March 1909, incorporated a number of his design requirements and was further modified during later trials. Despite a report in *The Autocar* that the machine was being built by Voisin Frères, it was actually made by de Pischoff and Koechlin, who, before delivery, guaranteed that it would fly 300-400 meters. The machine was tested, probably at Wembley Park, but there were no reports that it was a success and proposals to build further examples in Britain did not proceed.

The two-seater biplane had parallel chord wings, spaced by four pairs of interplane struts, the front outboard struts carrying large pivoting, balanced 'ailerons' operated by sideways movement of the seat back. There was a fixed boxkite tail on booms tapering to a point in elevation, with a central rudder between. The front booms were shorter and carried a biplane elevator. The machine was supported on two pairs of wheels in tandem. The engine, mounted on the lower center section was cooled by a circular radiator in front of it, and drove twin pusher propellers by chains; the propeller shafts and their drives being contained by a structure of steel tubes.

Later Windham made changes to improve the stability of the aircraft by lengthening the nose and tail booms. Also wheels were attached to the wing tips.

Power: 20hp Dutheil-Chalmers two-cylinder horizontally opposed water-cooled or 40/45hp Dutheil-Chalmers four-cylinder horizontally opposed water-cooled.

Data	Span	35ft
	Area	495 sq. ft
	Weight	530 lb.

Windham biplane No.1 was originally built by de Pischoff and Koechlin in France but was much modified later by Windham. It was shown at Olympia in March 1909.

WINDHAM tandem monoplane No.1

This unusual looking machine appeared at Wembley Park in August 1909, being first reported as under construction in June, thus succeeding the de Pischoff machine, which may already

Windham monoplane No.1. This tandem monoplane was at Wembley Park in August 1909 but was not capable of flight.

have been abandoned. Although the design was described at the time as 'ingenious' the machine was not capable of flight. The machine consisted of single top and bottom longerons of bamboo, spaced by vertical struts and braced by wires. Extending from the top longeron were single spars of bamboo for the front and rear wings, which were set at a pronounced dihedral angle. The wings were merely diamond shaped panels of fabric, laced to wire leading and trailing edges. Set below the wings, were long triangular shaped panels provided as fins. A small square elevator and a rudder were fitted at the extreme rear. The undercarriage consisted of two pairs of wheels, mounted separately, below each wing spar.

The engine was mounted at the front driving a tractor propeller and was cooled by a circular radiator, as used on the earlier biplane, but mounted end-on to the airflow, above the front wing spar. The pilot apparently sat on the lower longeron.

Power: 35/45hp Dutheil-Chalmers four-cylinder horizontally opposed water-cooled.

Data	Span	24ft
	Length	50ft
	Weight less engine	125 lb.

WINDHAM monoplane No.2

Windham's second monoplane appeared at the Doncaster Meeting on the first day, Friday 15 October 1909. It caused a sensation when a light gust of wind caught the machine, while Windham was seated in it on the ground, and the back of the fuselage broke. The machine was repaired, but was involved in a collision with a car on the third day, after which it was no longer in evidence and was abandoned.

The Aero described the machine as being 'on distinctly Blériot lines, and reproduces that machine with more or less accuracy except in a few details'! The important detail was the fuselage girder, which was of light construction with weak longerons at the top. A biplane type tail, with two elevators, was fitted. The engine was of vee type, reported to be either a JAP or a 25hp Advance and drove one of the unique Beedle propellers, with large curved metal blades. Elsewhere it was referred to as a Green or a Dutheil-Chalmers, but these were of a different configuration and water-cooled. The latter may have been confused with the use of that make of engine in the first monoplane, for which Windham had the agency.

Windham monoplane No.2. At the Doncaster meeting in October 1909 the machine collapsed on the ground.

Data	Span	30ft
	Chord	6ft
	Length	25ft

WINDHAM biplane No.2

In 1910 Windham built a biplane for Grahame-White, who intended to use it for a flight from London to Paris, which he never carried out. This machine was exhibited at Gamages in the summer of 1910, and is already described as the Grahame-White Biplane Type 1 (*q.v.*). It was made before Grahame-White had established his own manufacturing facilities.

WINDSOR MODEL AIRCRAFT & GLIDING CLUB

This club was started as a model flying group in April 1912, and progressed to full sized gliders and finally commenced the building of a powered aircraft. The secretary of the club was Sydney Camm, later to become famous as the chief designer of many very successful Hawker aircraft, and who contributed frequent reports to *Flight* on the club activities. His younger brother, F.J. Camm was also a member, and became well-known as a writer on radio, model engineering and model aircraft, in magazines which he edited.

Although some work was done on the powered machine as late as April 1915, most of the members had enlisted in the forces or were engaged on work of national importance and the club faded away.

WINDSOR glider WMA and GC No.1

The glider, which was of Chanute type, relied on weight shift for control and was ready for trials by September 1912. By the end of the year Sydney Camm reported that five members had

Windsor glider WMA & GC No.1 was flown with and without wing extensions in 1912-1913.

flown with only one mishap. It was tested originally as an equal span biplane, but was soon fitted with top wing extensions. The machine stayed in use until the end of March 1913, when it was wrecked during a gale.

Data	Span	32ft
	Chord	4ft
	Tailplane	10ft by 2ft

Windsor glider WMA & GC No.2 was built in 1913.

WINDSOR glider WMA and GC No.2

Experience with No.1 indicated the need for a more advanced type with controls, and work on this commenced in April 1913. The machine was complete and exhibited at the Royal Counties Show in June and, in August, was tested in towed flight. It would seem that, with the coming of winter and the need to travel to a suitably sloping site, little was accomplished with No.2; by November the members were considering a powered machine.

The glider was an equal span biplane with double acting ailerons on the top wing only. The outer wing panels each of 9ft 6in span, and comprising two bays, were detachable for transport. The chassis had twin skids and the two 24in diameter wheels were sprung by rubber cord. The chassis contained a bucket seat and was braced by long struts to the top wing spar joints.

Data	Span	25ft
	Chord	4ft 6in
	Gap	4ft 6in
	Area	225 sq. ft
	Length	15ft

WINDSOR biplane WMA and GC No.3

The powered machine had been under consideration since August 1913, when negotiations to purchase a 25hp Anzani engine were started. This engine was not obtained, but design and construction of the machine proceeded, so that by June 1914, the fuselage was nearly completed, and by September, the wings were ready for covering.

The engine eventually acquired was the Edwards two-stroke from the Bland *Mayfly*. War had been declared by this

Windsor biplane WMA & GC No.3 was started in 1913 but was not completed when war was declared. The Edwards engine still exists in the Brooklands Museum.

time and although the engine was fitted, work on the machine was abandoned. The engine survived in the possession of the Camm family and, after being exhibited in a small museum in Windsor, was handed over to the Brooklands Museum, where it is on display and has recently been reconditioned.

A photograph of the front of the machine and undercarriage gives some indication of the design, which perhaps inspired Sydney Camm to design the Hawker *Cygnet* much later.

Power: 15/20hp Edwards two-cylinder, horizontally opposed two-stroke made by H.W. Cowley & Co.

Data	Area	320 sq. ft
	Est. weight	less than 300 lb.
	Speed	35 mph

WOKINGHAM Flyer (A.M. 'Jesse' Farbrother, 44 Westcott Rd., Wokingham, Berkshire)

This unusual machine was described in a local paper and in other journals and, although lacking any information direct, *The Aero* published their own description. Patents for certain features of the machine were applied for under Nos.121634/1909. It was known locally as the *'Whale'* and was built to compete for the prize of £1,000 offered by the *Daily Mail* for the flight across the Channel, eventually won by Blériot.

The machine was a large cigar shaped structure, with the engine in the center driving the propeller, known as a 'Rotoscope', by a long drive shaft. The 'Rotoscope' was described as being spiral and cone shaped pointing outwards.

The accommodation provided for an engineer, several passengers, self balancing hammocks and a lavatory. Electric power for lighting was generated by the main engine in the center section, upon which the end compartments were to slide to and fro. The car was built of 1/8th. inch thick three-ply wood and the sixteen 'portholes' on each side were to accommodate 'poles', namely spars for the wings which were to be covered by stout canvas. The machine was telescopic in all directions as shown by the dimensions.

Wokingham Flyer. This is the fuselage of this large machine on its way to Windsor for engine installation in January 1910. The Flyer was not completed.

The machine was completed towards the end of October 1909 and was taken to St. Leonards Rd., Windsor for the engine to be fitted in January 1910. Farbrother, who was a journeyman carpenter, had ruined himself financially and the machine was not completed. It was returned to Wokingham where it was broken up.

Power: 80hp of unknown make weighing 616 lb. Was referred to as a 'Blériot' engine.

Data		
	Length	60ft extending to 140ft
	Width	14ft extending to 20ft
	Height	16ft extending to 31ft

Wokingham monoplane. Little is known about this machine of probably 1909-1910 vintage.

WOKINGHAM monoplane.

The Flyer was not the only aircraft built by Farbrother in Wokingham. The photograph reproduced here was sent to *The Wokingham Times* many years later, as a result of publicity given to Farbrother's Flyer. This is the machine, with ribs like those of an umbrella, reported to have risen about a foot off the ground on its first and only test. It was constructed of white pine in the summer of 1910 and Farbrother intended to fly from London to Edinburgh.

Power: The engine was a 35hp six-cylinder.

Data		
	Span	16ft
	Length	18ft
	Chord	9ft
	Weight	280 lb.

Ref *The Wokingham Times* 12 July 1910 (p.12)

WOLFE & ASHBURNER monoplane (Samuel Wolfe and Joseph Ashburner, Burnley, Lancashire)

On the morning of 21 May 1910, the partners took their machine to Rosegrove near Burnley for trials, but a weak undercarriage caused these to be abandoned. In August the machine was put on view in Burnley with a view to raising funds for a more powerful motor and a better propeller. A provisional patent No.12866/1910 was taken out to cover the 'Collapsible and Variable Planes', which adjusted the span of the wing and tailplane for storage and transport. No progress was made towards obtaining funds and consequently work on the machine was abandoned; no doubt this was all for the best.

The machine is best described by the illustration. It had a 3hp engine and a span of 21ft, but no other details are available.

Wolfe & Ashburner monoplane made at Burnley in 1910.

WOLLASTON helicopter/parachute (Herbert and Ernest Wollaston, Miller's Flat, Elmstead, Kent)

This device was also referred to as a Brolly-Screw, and was the invention of the brothers in the years 1885-1888 and covered by patent No.1885/1908. It is best described by quoting from the letter to *Flight* by a writer signing himself 'Pioneer of Orpington', who was the pilot for all the tests, and weighed 104-7 lb. at the time.

'The machine was a soarer only, not a flyer, being always held captive. It was never started from the ground, but from a bent iron rod at the end of a 22ft long ash pole. This pole was balanced near the half length, on a forked pivot, so that it could be swung clear after release of the brolly. The pole itself was on a tripod 12ft in height, starting as follows: from one end of pole to the other, through guides, ran a cord to a snatch link on bent iron at end of pole. The snatch link closed by spring into a ring, attached to screw spindle cap on top of brolly. About 15ft of strong rope was fastened to other end of pole, some length of snatch link cord hanging free. The rope was ready to pull as soon as the pilot was seated and strapped himself secure. At a signal, a seesaw motion was begun, steadily up and down, by pulling and giving on the rope until the screw had obtained its greatest momentum, then when the brolly end of the pole was at its greatest height, the snatch link cord was pulled, the pole being swung clear instantly. After release the brolly slowly descends a few feet, but the pilot, working quickly, the brolly soon gathers purchase and rises. The tether ropes were attached to the footrest by swivel.

After having reached a height thought advisable to ascend, two grip catches were let go by the pilot, which engaged with tether ropes, those holding such go wider out, and so stop spinning motion; 60-80ft was about the usual, although of course very much higher could have been attained. Over 100 ascents were made, and on June 16, 1887, Mr. Shadbolt, balloonist was a very interested assistant with brolly and gliders, tripod and pole. The fitments were made by Dean, ladder maker of Redhill'.

WONG *Tong Mei. Dragonfly* biplane (Tsoe K. Wong, Shoreham Aerodrome, Sussex)

This conventional single-seater tractor biplane was built at Shoreham in the early part of 1913 and was tested there in the summer. There were no further reports of the machine, which it was intended should be produced later in the Far East, and it is not known if the machine was actually flown. When described in *Flight* the engine and radiator had not been installed and the aircraft was not covered.

The fuselage was a spruce girder structure with aluminum panels ahead of the cockpit, the rest was to be fabric covered. The two bay wings were staggered and of equal span. They had additional full depth interplane struts adjacent to the fuselage for ease of assembly and dismantling. The leading edge member was the front spar, to which the front interplane struts were attached. The trailing edge outboard of the center interplane struts was inversely tapered and flexible. The tail unit consisted of a tailplane mounted on the top longerons and a balanced rudder, shaped to clear the one piece elevator. The machine rested on a chassis with twin skids and two pairs of rubber cord sprung wheels on separate axles. A steerable tailskid was fitted to the extension of the rudder post.

Wong Tong Mei *'Dragonfly'* biplane built at Shoreham in 1913.

Power: 40hp ABC four-cylinder inline water-cooled driving 8ft 6in diameter propeller

Data		
	Span	30ft
	Chord	5ft tapering to 5ft 6in at tips
	Gap	5ft 4in
	Area	304 sq. ft
	Area tailplane	20 sq. ft
	Area elevator	14 sq. ft
	Area rudder	5 sq. ft
	Length	30ft.
	Weight	504 lb. airframe less engine & fabric
	Endurance	3 1/2 hr

WONG biplane (T.K. Wong Ltd., 17 Ironmonger Lane, London EC)

In December 1913, Wong established a company with himself, W.F. Skene and E.H. Lawford as directors. Their first type, constructed in 1914, was a two-seater development of the *Tong Mei.* The machine was completed, and after initial tests, was dispatched to the Federated Malay States. It crashed, during a demonstration, on the racecourse at Kuala Lumpur, Selangor during 1915 and was probably not rebuilt.

Very few changes from the earlier machine are apparent, apart from the air-cooled radial engine and the revision for two

Wong biplane was a two-seater built at Shoreham in 1913.

cockpits. The wings were not staggered and were attached to the fuselage top longerons by vertical struts; ailerons were incorporated in the top wing only. The front spars, unlike those of the Tong Mei, were set back from the leading edge and the interplane struts attached to these. Windows were fitted in the fuselage sides, ahead of the front cockpit and the chassis incorporated diagonal bracing struts.

Power: 100hp Anzani ten-cylinder air-cooled radial.

WOOD triplane glider ornithopter (B. Graham Wood., 17 Glenthorne Rd., New Southgate, London N)

Wood was a lecturer at the North London Polytechnic and chairman of the Polytechnic Gliding Club, founded in January 1912. He took his pilot's certificate at the Hewlett and Blondeau School at Brooklands.

In addition to arranging club activities, he constructed a triplane glider with 'beating wings', which was ready for testing in October 1912. The glider was very lightly constructed with the three wings reducing in span, with the shortest at the top. The wings had considerable stagger and dihedral and the single pairs of interplane struts were canted outwards. The wings were made in halves and pivoted at the center. They were connected by rods and cables to actuate the downward movement, by means of a foot bar. A semicircular fin on a light boom, was braced by cables. The intention was to develop the machine as a glider before operating the flapping mechanism. It was not recorded if the machine was tried as an ornithopter.

Wood triplane glider ornithopter of 1912.

WOOD Aeroscope Polyplane (Wood, c/o H. Arnold, Tonbridge St., Leeds, Yorkshire)

No details of this machine, which was offered for sale in July 1911, are available.

WOOD & MAAS biplane glider (B. Graham Wood and J.C. Maas)

Wood was involved with a partner in the construction of a biplane glider in 1910, which was tested on the South Downs in the summer.

The glider was an equal span two bay biplane, with a long front boom mounting a nose elevator and small triangular fin. A large rectangular rudder was pivoted direct to the trailing

Wood & Maas biplane glider was tested on the South Downs in 1910.

edges of the wings. Lateral control was by unusual type ailerons, hinged between the outboard interplane struts. The machine was mounted on a twin skid chassis, but was lightly constructed and almost immediately sustained damage and was soon wrecked.

Data	Span	42ft
	Chord	5ft
	Area	408 sq. ft

WOODS biplane glider (Frederick Woods, Fleetwood, Lancashire)

This glider was built by Woods and was tested at Knott End in the summer of 1912. It was well constructed and designed on the lines of a Wright, but with single acting ailerons and a monoplane front elevator. Twin tail booms were fitted to mount the tall rectangular rudder. The wings were of three bays, with marked camber, and were double surfaced.

Woods biplane glider of Wright type was built at Fleetwood and tested in 1912.

Suitable ground and flying conditions were difficult to find and little more was heard of Woods' glider. In later life he moved to Chelmsford, where he served for a period as Mayor.

WORSWICK monoplane (Alfred Worswick, The Woodlands, Gathurst, Wigan, Lancashire)

Worswick designed and constructed his first monoplane during 1909, after several years of study of aeronautics. The machine was constructed in sheds, which he also built for the purpose. Despite all the thought and effort expended, the machine apparently never flew due to excessive weight, mainly because of the heavy engine.

The fuselage was of oval section with a flat top, changing to a flat vertical section at the rear forming an integral fin. The internal structure included oval and horseshoe shaped frames and longitudinal members; the forward section was covered with aluminum sheet, the rest with fabric. The pilot was said to be protected and entirely covered in, being provided with a 'gelatin' screen. If required as a military machine, the aluminum could be replaced with bulletproof metal. The two halves of the tapered wing, were attached at the top longerons on either side of the fuselage and were braced to triangular overhead pylons, without dihedral.

The elevator was made in halves and was mounted through the integral fin; each half could be operated independently for roll control. Each half elevator was triangular in shape and about the same length as the fin. The whole cruciform section of the rear fuselage was moveable for control in pitch. The pilot sat between the wings and operated the irreversible controls, by means of three concentric steering wheels. The engine, which was behind the pilot, drove twin pusher propellers behind the wings, through shafts and universal joints.

Certain features of the machine were covered by patent No.8591/1908 and Nos.1656 and 5525 of 1909.

Power: 20hp eight-cylinder vee of unknown type.

Data		
	Span	33ft
	Chord	9ft tapering to 5ft 3in at tips
	Area	240 sq. ft
	Length	14ft
	Area elevators	40 sq. ft
	Area rudder	10 sq. ft
	Weight	600 lb.
	Weight allup	1,000 lb.

Worswick monoplane was built during 1909 at Gathurst near Wigan but was not flown.

WORSWICK motorplane type C No.1 (Motorplanes Ltd., Gathurst, Lancashire)

Alfred Worswick formed a company in 1909, of which he was managing director, to exploit his designs. This monoplane was a development of the earlier machine and it appeared in the early part of 1910.

The aluminum skinning had been discarded and a lighter, lower powered motor reduced the weight. The wings of this version were joined on the centerline, with seven degrees of dihedral, the whole being mounted on vertical and splayed struts above the fuselage. The wing was flat on the lower surface, and heavily cambered on top forming a deep section, and was designed as a cantilever. The trailing edge for 2ft 6in of the chord was flexible for lateral control, presumably superseding the differential tail surfaces used previously. The aircraft included ash, pine and mahogany in its construction with a steel tube undercarriage which incorporated spiral springs.

The whole of the business, equipment and stock was auctioned on 24 May 1911. This included one monoplane chassis with wings and 10-12hp Humber engine and also a monoplane chassis assumed to be that of the original machine. Worswick died in 1917.

Power: 10-12hp Humber driving two two-bladed 4ft diameter pusher propellers.

Data		
	Span	31ft
	Chord	9ft tapering to 5ft 3in at tips
	Area	190 sq. ft
	Area elevator	40 sq. ft
	Area rudder	22sq ft
	Weight	450 lb.
	Weight allup	850 lb.

Worswick motor plane type C No.1 appeared in early 1910 but never flew.

WRIGHT glider (Maurice E. Wright, F.G.T. Dawson and V. Nicholl, Cambridge and Eastchurch)
These three friends were Cambridge undergraduates, who built a glider and tested it at Eastchurch in the late summer of 1913. It was also flown by Eng. Lt. Briggs RN. Later all three joined the RNAS and became directors of the Fairey Aviation Co.

WRIGHT & WILTON glider (F.J. Wright and G.N. Wilton, Doncaster, Yorkshire)
These members of the Conisbrough Gliding Club (*q.v.*) constructed a monoplane glider of the Demoiselle type, in the latter part of 1911. The machine was in skeleton form in December but it is not known if it was completed.

Wright & Wilton glider was built by two members of the Conisborough club in 1911.

WYLLIE helicopter. (see also Ottino-Wyllie)
Nothing is known of this unusual machine built in 1911. The constructor is also reported to have built a multi-plane in 1910. Neither machine achieved any success.

WYNN monoplane (Charles and William Lea Wynn, Buckinghamshire)
These brothers built a monoplane in the early part of 1909, incorporating a Peugeot racing motorcycle in its construction.

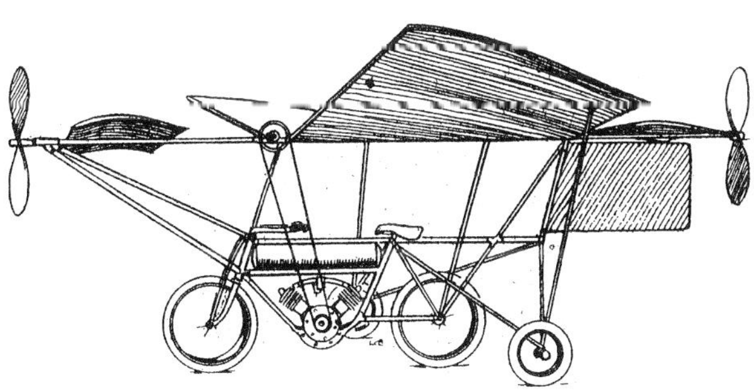

Wynn monoplane of 1909 lifted off the ground briefly before being wrecked.

A chain drive from the engine was taken to a shaft above the machine, where a bevel gearbox turned the drive through a right angle. Tractor and pusher propellers were mounted on the shaft, which was carried in bearings in a tubular member extending fore and aft beyond the length of the motorcycle. The wing, front elevator, tailplane and rudder were also mounted on the fore and aft member. Additional tubes carried outrigger wheels for stability on the ground.

In the summer of 1909 the machine lifted briefly from their small ground, except for one of the outer wheels, and suffered damage as a result.

The engine was later removed and the monoplane was flown as a glider. It was eventually presented to the Stoney Stratford Aero Club (q.v.).

Power: Peugeot vee twin-cylinder air-cooled of 97m/m bore & 120m/m stroke.

Data Span 26ft
Chord 8ft

Y

YARMOUTH aeroplane factory
The reference to this in *Flight* of 15 May 1909 (p.282) is almost certainly in connection with the Sanders Aeroplane Co. (*q.v.*).

YATES monoplane (Victor Yates, Cavendish Place Eastbourne, Sussex)
This monoplane was built by Yates in April 1911 prior to his becoming a pupil of F.B. Fowler of the Eastbourne Aviation Co. Yates aviator's certificate No.306, was issued in September 1912.

The machine was a two-seater tractor of conventional type built of steel tubing, with Chater-Lea lugs. No wood was used in the construction. The fuselage was a triangular section open girder mounted on a wide chassis with nose wheel, in addition to a tail wheel. Two radiators were mounted in the form of a vee below the engine. Control was by double rudder, ailerons and elevator. The wing ribs were sewn into pockets in the canvas covering so that broken ribs could be replaced without having to strip the whole of the wing fabric. The machine was tried out at Mr. Gribble's farm at Wilmington, but the machine was apparently incapable of flight and Yates eventually flew a Blériot monoplane

Power: 35hp Alvaston two-cylinder horizontally opposed water-cooled.

Data Span 34ft
Chord 7ft 6in
Length 25ft

Yates monoplane was built at Wilmington, Sussex in 1912 but was incapable of flight.

X

XL'ALL monoplane (XL'All Specialities Ltd., 67 Moseley Street, Birmingham)

This Blériot pattern monoplane was offered for sale for £250 in *The Aero* in July 1910.

The 25hp Riley water-cooled engine was set low in the fuselage and drove the propeller by a chain. The undercarriage had two wheels and ash skids

This may have been the same firm as Birmingham Aero Company (q.v.) although they are listed in the 1910 Kelly's Directory as "cycle makers".

Data	Span	32ft
	Chord	6ft 6in
	Wing area	c.200 sq. ft

Addendum

GROSSART monoplane (Robert Dykes Grossart, Milton near Moffat, Scotland)

In June 1910 the Dumfries & Galloway Standard reported that Robert Gossart was building a monoplane which he intended to take to the Lanark Aviation Meeting in August of that year.

Grossart joined the Royal Flying Corps and was killed on 9 February 1917 when his BE2c crashed at Beverley, Yorkshire.

Power 30 hp engine of constructor's own design driving a 6 ft diameter propeller situated behind the mainplane.

Data	Span	30 ft
	Length	28 ft

Airships

AIRSHIPS Ltd. (Merton, Surrey)

Naval Airship. SS.2. ex-Willows No.VI

E.T. Willows, having been unsuccessful in obtaining government orders for his airships, became Chief Engineer in 1914 of Airships Ltd., the company started in the spring of that year by G. Holt Thomas. Willows designed his sixth airship to meet the Admiralty need for small airships for submarine scouting purposes, in competition with prototypes made by Armstrong Whitworth Ltd. and Short Bros. It was not accepted for several reasons, among which the mechanism and controls were regarded as too complex, making the ship longer and heavier than the Navy had envisaged, nor was it a simple airship to build and maintain.

Armstrong Whitworths secured contracts for a number of their type, although Airships Ltd. received orders for large numbers of a smaller ship based on the SS1, the Navy's own development, consisting of a BE2c fuselage slung below a spare envelope from Willows No.VI (Naval Airship No.2). SS2 was not used operationally and was struck off charge in May 1915. Due to shortages, an unsuitable fabric was used for the envelope and this was prone to leakage.

The envelope was fitted with top and bottom fins with rudders and tailplanes with elevators, which were in addition to the Willows belt driven swiveling propellers, as used on his earlier ships, the drive for which required three gearboxes.

Power: 90hp Curtiss vee eight-cylinder water-cooled, driving two tractor swiveling propellers.

Data	Capacity	60,000 cu ft
	Length	140ft
	Diameter	30ft

ARMY AIRCRAFT FACTORY Airships. (Farnborough, Hampshire.)

Originally known as the Army Balloon Factory until late 1909 and became the Royal Aircraft Factory from April 1912.

NULLI SECUNDUS I & II. (Army Dirigible No.1)

The first version of *Nulli Secundus* had its origins in an envelope of gold beaters skin, work on which was started in 1902, but which on completion, was found to be too heavy. This work was carried out when the Army Balloon Factory was at Aldershot, before it moved to Farnborough in 1905, where Col. J.E. Capper RE. became responsible for the establishment with S.F. Cody largely controlling the constructional side.

This heavy envelope was then used as a mould for a new envelope, which was completed in 1906. Below the envelope a triangular section keel of tubular steel and joints made of aluminum fittings, was suspended by fabric bands and carried below it a boat shaped car at its center. The car carried the crew of two or three and the engine. A rudder and elevator were attached to the keel together with wings, which were soon removed.

The first flight took place on 10 September 1907 and on 5 October 1907 the airship flew over London, circling St. Paul's Cathedral, but was forced to land at Crystal Palace, due to adverse winds. It was deflated in the night to prevent its loss in a gale.

The second version flew on 24 July 1908 in a much modified form. The airship was now a semi-rigid with a fabric cov-

***Nulli Secundus* I (Army Dirigible No.1) first flew on 10 September 1907.**

ered keel, of reduced height, directly below the envelope. The car was raised and changes were made to the control surfaces. It was all to no avail, for after only a few flights, including one of four hours, with several incidents with damage occurring, *Nulli Secundus II* was deemed uncontrollable and was deflated at the end of August 1908.

Nulli Secundus was the name of a royal racehorse. The second version has also been referred to as *Alpha*, conforming to the later method of designation by the Factory.

***Nulli Secundus* II was Army Dirigible No.1 extensively modified. First flew on 24 July 1908.**

Power: 40/50hp Antoinette vee eight-cylinder water-cooled driving two 9ft diameter two-bladed pusher propellers by belts.

Data

	NS I	**NS II**
Capacity	85,000 cu ft	56,000 cu ft
Length	111 ft	122 ft
Diameter	30 ft	26 ft
Speed	16mph	
Useful lift	800 lb	
Gross lift	3740 lb	
Speed		22mph

BABY Airship. (Army Dirigible No.3)

Work on this small experimental non-rigid was begun in late 1908 and she was flying early in 1909 fitted with two small Buchet engines coupled to drive a single propeller. This power plant was unsatisfactory and the craft was unstable. The car and control surfaces were soon reconstructed and the engines changed for a single REP. These changes, which were made after six flights, showed no great improvement. After five more flights, the last on 23 December 1909 work commenced on a major reconstruction to improve the craft, which then became known as *Beta*. (*q.v.*).

The envelope of gold beaters skin was streamlined, but rather short, and carried one vertical and two horizontal inflated surfaces. These latter were unsatisfactory and were replaced with rigid fins. The car hung well below the envelope and carried a crew of two or three with the engine amidships. It was fabric covered with a fin and rudder and elevator at the rear. The use of the car with swiveling propellers, from the

***Baby* Airship (Army Dirigible No.3). A small experimental non-rigid. First flew early in 1909.**

Beedle airship, was abandoned after trials proved that the thrust was inadequate.

Power: Two 12hp Buchet two-cylinder vee-twin air-cooled driving a single propeller.
20hp REP five-cylinder air-cooled fantype.

Data

Capacity	21,000 cu ft
Length	84ft
Diameter	24ft 8in
Speed	20mph

Beta Airship I. _Baby_ was renamed _Beta_ after major reconstruction and first flew in May 1910 with a Green engine.

BETA Airship I and II. (Army Dirigible No.3 and Naval Airship No.17)

Baby was renamed Beta after major reconstruction and first flew in this form in May 1910. The envelope had been split and lengthened by twenty feet. The control surfaces now consisted of a large fin and rudder below the tail of the envelope and fixed horizontal surfaces with elevators. The car was adapted to take a water-cooled Green engine and could carry a crew of two or three.

Considerable use was made of *Beta* by the Air Battalion during 1911 for army maneuvers and for experimenting with wireless. Flights of up to four hours duration were made and, early in the following year, the RAF. carried out trials with a mooring mast. The envelope, which had deteriorated, was replaced in August 1912 with one of larger capacity and a new Clerget engine was fitted. The craft then became *Beta II*. In this form she served with the Army throughout 1913 until handed over to the Navy on 1 January 1914 together with the other Army airships. *Beta II*, with the Navy, was used mainly for training and was identified as Naval Airship No.17 until struck off charge in May 1916.

Beta Airship II (Army Dirigible No.3 and Naval Airship No.17). This had a new envelope of larger capacity and a Clerget engine.

Power: 35hp Green four-cylinder inline water-cooled driving twin pusher propellers by chain. (*Beta I*).
50hp Clerget four-cylinder inline water-cooled driving twin pusher propellers by chain. (*Beta II*).

Data

	Beta I.	***Beta II.***
Capacity	33,000 cu ft	42,000 cu ft
Length	104ft	116ft
Diameter	25ft	26ft
Lift	800lb	
Speed	26mph	

GAMMA Airship I and II. (Army Dirigible No.2 and Naval Airship No.18)

Planning for a second military airship was begun in 1907, but funding was not available until 1909, consequently it was not until 12 February 1910 that the first flight was made. The craft flew satisfactorily until April. when a trial of the ripping panel caused unexpected damage.

In its original form the envelope was rubber proofed and was a product of the Astra Co. of Paris. It incorporated bulbous inflatable horizontal surfaces at the tail. The car was a long open girder structure with the Power: amidships. The twin propellers were mounted outboard and were driven by shafts and gearing and could be swiveled for vertical control. Originally the amount of movement was limited but was later increased to give a full 360deg. rotation. *Gamma* was the first military airship to use swiveling propellers for control. At first there were small triangular shaped horizontal surfaces on either side of the car at both front and rear and a rudder, which were all discarded later. A fixed fin and rudder, below the envelope, provided for directional control.

Gamma Airship I (Army Dirigible No.2). Made its first flight in February 1910 and had bulbous inflatable horizontal tail surfaces.

Gamma **Airship II (Army Dirigible No.2 and Naval Airship No.18). This had a new larger envelope and a biplane tail with vertical surfaces. It had two 45hp Iris engines with swiveling propellers.**

Gamma **II at mooring mast.**

During the course of repair, following the ripping panel test, further changes were made. These included replacing the single engine by two Iris engines and removable of the horizontal inflatable surfaces, which were replaced by rigid ones.

By April 1912 *Gamma* had again been reconstructed and then became *Gamma II.* The envelope had been replaced by one of larger capacity made by E.T. Willows & Co. and the car had been enclosed and fitted with a biplane tail incorporating vertical surfaces, this tail being later removed.

Gamma II continued in service with the Army, carrying out radio trials and on maneuvers and, although suffering damage at times, gave generally good service until the end of 1913. The ship was then handed over to the Navy becoming Naval Airship No.18, and was used for training purposes until July 1914. *Gamma II* could carry a crew of four or five and the swiveling propellers allowed the airship to operate without a mooring party under certain conditions.

Power: *Gamma I.* 80hp Green eight-cylinder water-cooled vee driving twin two-bladed metal propellers. The engine was mounted across the car with extension shafts on either side and bevel gearboxes to enable the propellers to be rotated from the vertical to the horizontal position.

Gamma I & II. Two 45hp Iris four-cylinder horizontally opposed water-cooled each driving a four-bladed wooden propeller with the same swiveling feature.

Data

	Gamma	*Gamma I.*	*Gamma II.*
Capacity	75,000 cu ft	80,000 cu ft	101,000 cu ft
Length	154ft	162ft	169ft
Diameter	25ft	25ft	34ft
Speed	30mph	32mph	30mph
Weight			4,300 lb
Lift			2,500 lb
Endurance			8hr

DELTA Airship. (Army Dirigible No.4 and Naval Airship No.19)

Delta was the fourth military airship to be built and it was larger and faster than its predecessors. The craft was originally built as a semi-rigid with two girders built into the underside of the envelope. Soon after the first flight in September 1912 these were removed and the ship then became a non-rigid.

Delta was used by the Army on maneuvers and for wireless work, but was never so successful or handy to operate as *Gamma. Delta* was taken over by the Navy in January 1914 and was at the Spithead Review in July, but the envelope failed in August and the ship was not flown again.

Like the other early airships, changes were made to the control surfaces, apparently being finalized with a cruciform tail on the envelope. The car was quite short but could carry five or six crew, with the engines mounted low down and the swiveling propeller shafts and gearboxes mounted well above and driven by separate chains.

Power: Two 110hp White & Poppe

Data

Capacity	173,000 cu ft
Length	198ft
Diameter	39ft 6in
Weight	8,932 lb
Lift	3,635 lb
Speed	44 mph
Range	400 miles

Right: ***Delta*** **Airship (Army Dirigible No.4 and Naval Airship No.19) Made its first flight in September 1912 and was larger and faster than its predecessors.**

ETA Airship. (Army Dirigible No.5 and Naval Airship No.20)

The last Army airship to be built had a very short career. After completion by the Factory on 18 August 1913 she was passed to the RFC. on 20 September 1913 and proved to be completely successful on maneuvers and radio trials. *Eta* became Naval Airship No.20 on 1 January 1914, but was wrecked on 19 November 1914 near Redhill after breaking away from moorings on the way to Flanders and was not rebuilt. In 1916 the Navy built an entirely new ship named *Eta II* for experimental purposes.

Eta gave her name to the special patches developed for the attachment of the rigging to the envelope. These were entirely successful and were used extensively on future non-rigids and kite balloons.

The car had a long tapering tail and housed a crew of five or six with the engines located amidships, driving separate swiveling propellers mounted well above the car, by chain drive.

Power: Two 80hp Chenu four-cylinder inline water-cooled.
Also reported fitted with:
Two 80hp Salmson (Canton-Unné) seven-cylinder water-cooled radials.

Data		
	Capacity	118,000 cu ft
	Length	188ft
	Diameter	33ft 6in
	Weight	6,100 lb
	Lift	2,700 lb
	Speed	42 mph
	Range	500 miles

***Eta* Airship (Army Dirigible No.5 and Naval Airship No.20) Was completed in August 1913 and went to the Navy on 1 January 1914.**

EPSILON Airship.

The design and construction of two new airships was started at the Royal Aircraft Factory in 1913, but work was discontinued with the transfer of responsibility for operating airships from the Army to the Navy. This concluded airship design and construction at Farnborough.

BARTON Airship No.1 (Dr. F.A. Barton and F.L. Rawson, Alexandra Palace, London, N.)

Dr. Barton experimented with airships from 1883 and built a successful 30ft flying model in 1900. He patented one design No.24848 in 1899 and the British War Office showed interest.

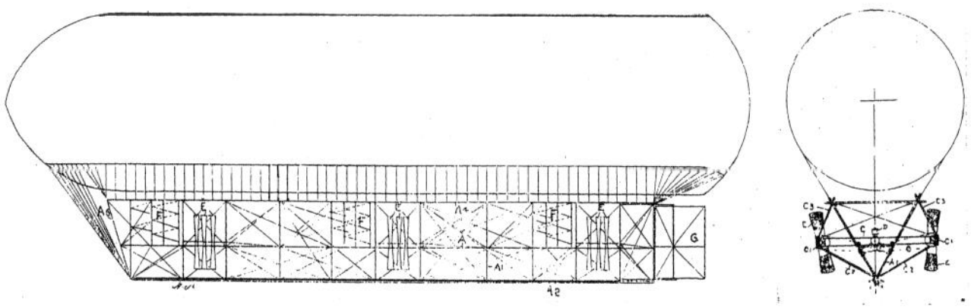

Barton Airship. The design for the original version.

However, the negotiations were protracted and a contract for an improved type was not concluded until 1902.

His designs included the use of three tiers of horizontal planes mounted on the airship's car which could be used to control the pitch of the airship without the necessity of dropping ballast or venting gas.

The balloon was made in the large Banqueting Hall of the Alexandra Palace and the framework of the car was assembled in a large, specially erected shed in the grounds of the Palace, whose Trustees had agreed to pay for this on condition that the public were allowed to view the work in progress. The Trust would keep the admission charge.

A deep triangular section keel, made of bamboo and steel tube, extended practically to the full length of the envelope and was suspended by steel cables from the reinforced lower portion. The envelope of Jap silk incorporated ballonets, was filled with hydrogen gas and was parallel sided with a pointed reinforced nose and a spherical rear end. The power was provided by three 45hp Buchet engines mounted for and aft and amidships on the keel. The engines were controlled independently and each drove three outboard mounted propeller clusters on each side of the keel through gearing and belt drives. Each cluster consisted of three separate two-bladed propellers. Clutches were fitted to engage or disengage the drive. Water in tanks at each end of the keel could be transferred to correct the balance. Elevation was controlled by the tiers of planes mounted on the keel and directional control was by a large rear mounted rudder.

The work took longer to complete than had been anticipated and the War Office canceled the contract.

Power: Three 50hp Buchet eight-cylinder air-cooled vee.

Data			
	Capacity	235,000 cu ft	* (156,000 cu ft)
	Length	176 ft	* (180ft)
	Diameter	43ft	* RAeS Journal 1929 (p.933)
	Gross lift	8,172 lb	
	Speed	15 1/2 mph	

BARTON Airship No.2

Barton continued to work on his airship and patented various modifications to the original design in patent Nos.18920 of 1903 and 12503 of 1904.

Initially the horizontal planes were increased in number to three banks of ten planes (five on each side of the three engines) each of which measured 15ft x 3ft, giving a total area

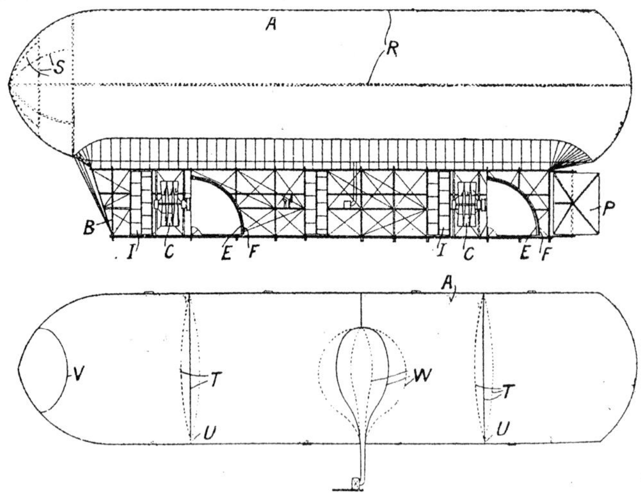

Barton Airship. The final design.

of 1350 sq ft. The envelope was slightly reduced in length, but increased to 43ft in diameter, giving about 235,000 cu ft capacity. The car and keel was made wholly of bamboo.

Various problems, including an explosion of one of the hydrogen generators which badly injured Dr. Barton and damage by a gale which tore off one of the valves from the top of the envelope all delayed the completion. By now Dr. Barton and his colleagues decided that the airship needed to be lightened to increase the available lift; the damaged envelope was increased in length to 180ft but reduced in diameter to 40ft which decreased the capacity to 200,000 cu ft. The tiers of movable planes were removed and replaced by four small winglets, two on each side of the car. The engines were reduced to two 50hp Buchets driving four propellers, two on each side of the car.

The first flight was made with six crew aboard in July 1905. The crew for the flight included Rawson as helmsman, Henry Spencer controlled the motors and Auguste Gaudron was aeronaut in charge. The gas bag was leaking badly, some of the guy ropes were rotten and the ship was virtually uncontrollable and was wrecked on landing in Essex. By now Dr. Barton was heavily involved with the Barton-Rawson hydroaeroplane at St. Helens, Isle of Wight and no further development of the Airship took place.

Barton Airship. In flight at Alexandra Palace in July 1905.

Power. Two 50hp Buchet eight-cylinder air-cooled vee driving 7ft diameter two-bladed propellers

Data		**1903**	**1905**
	Capacity	235,000 cu ft	200,000 cu ft
	Length	176 ft	180 ft
	Diameter	43ft	40ft
	Car length	120ft	127ft

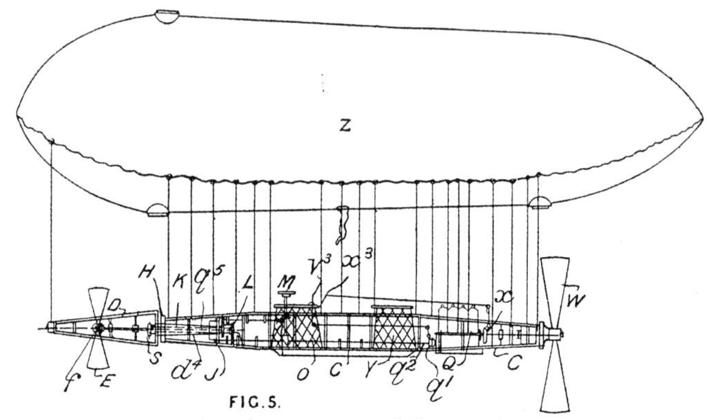

Beedle Airship. The drawing from patent 15139 of 1903.

BEEDLE Airship. (Capt. William Beedle, Walham Green, London SW, and Alexandra Palace, London N and Cardiff, S Wales)

This airship was built in 1901-1903 incorporating a steering system covered by Patent No.15139/1903 taken out by Beedle. The airship was brought out for test at Alexandra Palace on 3 November 1903, too late to compete for the second Deutsch Prize of 1902. Beedle had also previously made a 30ft long model for test purposes, which was reported to have been flown on 3 November 1903, but exploded soon after on 18 November 1903.

The steering device consisted of a frame which was capable of moving through an angle of 180 degrees, mounted on a pivot at the front of the car. A propeller of 8ft. diameter was mounted in the frame, which could be controlled by the pilot to direct the thrust of the propeller in the desired direction, so altering the line of flight. At the rear end of the car a 11ft 6in diameter propeller driven at 300 rpm. provided the propulsion.

Late in 1904, Beedle moved to Cardiff, by which time the capacity of the envelope had been increased from 26,000 to 28,000 cubic feet and the car had been reduced in weight from 860 to 780 lb. The envelope remained the same length but increased to 23ft diameter. The ship was still unable to fly, but

Beedle Airship. The "hull" complete with engine and fore and aft propellers.

Beedle Airship on test at Alexandra Palace in November 1903.

Bell's Locomotive Balloon. This shows the first ascent from Vauxhall Gardens on 22 July 1850.

the steering system had potential and was the cause of a collaboration with E.T. Willows, which lasted until 1908.

Towards the end of this period, the War Office purchased the car of an unflown airship which the two men had been constructing, with the object of providing an example for use on the airship *Baby*. In the event the Beedle car was not used and the first military airship with swiveling propellers was *Gamma*, which flew later.

Power: 12/28hp Blake four-cylinder air-cooled.

Data		
	Capacity	26,000 later 28,000 cu ft
	Length	93ft
	Diameter	23ft
	Weight	1,260 lb
	Length of car	56ft
	Weight of car	860 later 780 lb

BELL'S Locomotive balloon

This airship was built in 1850 by a Dr. Hugh Bell and the envelope was made of the finest white silk with a covering network of stout amber silk bands two inches in width.

The car suspended from the envelope was built like a boat in case of an enforced descent into the sea. On either side of the car was a large fanlike screw propeller and at the rear was a hinged rotating rudder. All of these were operated by the aeronauts.

The first ascent from Vauxhall Gardens was on 22 July 1850 but the hand-turned propellers were not powerful enough to counter the effect of the wind and the airship was found to be uncontrollable.

Data		
	Length	c.50ft
	Diameter	22ft
	Capacity	c.15,000 cu ft
	Gross lift	500 to 600lb
	Propeller diameter	6 ft

BENNETT airships (F.W. Bennett, Penge, London SE)

Bennett was a grocer with an obsession with the idea of powered flight. In about 1895 he invented what he termed an 'aerial bicycle'. This was a framework, with bicycle pedals and rotating handles, which would be attached to a balloon to convert it into a dirigible.

A large fan airscrew projected over the front of the balloon car and was driven by the bicycle pedals. Two horizontal fans, rotated by handles, would have driven the balloon up or down.

A trial was made at the Crystal Palace when the device was attached to the basket of one of Spencer's balloons. The balloon rose to a height of about 60 feet, but was held captive by four men with ropes. Furious pedaling made the balloon move forwards, but very slowly, and it would have made no headway against even a light breeze.

The Bennett device was exhibited at the National Cycle Show but no more was heard of it.

The success of the Spencer Mellin airship in 1902 inspired Bennett to invent another dirigible. Lack of funds prevented the construction of a full-size machine but a model was built with ten electric motors, five on each side of a triangular frame and each driving two propellers. The frame was suspended beneath a gasbag, above which was a canopy which could be opened by the pilot and would provide additional lift or act as a parachute if the balloon accidentally deflated. Again nothing further was heard of this machine

BOYD Airship (Dr. M.B.B. Boyd; Dutheil-Chalmers & Boyd Ltd., Tothill St. London SW1)

This airship is something of a mystery, for its constructor carried out tests at night in order to preserve secrecy. This inspired the *Daily News* to name it the 'Scareship' from the reaction of members of the public who witnessed the nocturnal activities. Dr. Boyd claimed that it was his machine that was seen by people at Northampton on 15 May 1909. It was also seen by others at Caephilly mountain and at Cardiff Docks on 18 May 1909. He also claimed that on the same day the ship had crossed to Belfast, between nine o'clock at night and four o'clock the following morning, being less than four hours actually over the water. No flights in daylight appear to have been made and no photographic or conclusive evidence of its existence were ever found. Patents were due to be taken out, but were not registered. £20,000 had been spent to the middle of 1909, but a syndicate was to be formed to provide a fund of £250,000.

Dr. Boyd described his airship as having 'no car suspended from the envelope, which was oval in shape and divided into three separate bags, the "works" being placed between them, and the motors having a closed in compartment to themselves at the end'. Wings extended each side, like an aeroplane, and there were three pairs of landing wheels. Two acetylene lamps were carried on night flights and a crew of three men, which would be increased to eight on a reconstructed ship. It was housed in a hangar in a park at an unnamed location within an hour car ride of London.

To add to the considerable consternation, at about the same time, a small unmanned advertising airship, illuminated by lights, and towed by a car, was also being flown at night. This was apparently a Sizaire-Mors ship, owned by Jarrett & Letts Ltd., motor agents, and was wrecked on Chalk Hill Down, Dunstable in the early morning of 25 May 1909. It was found by Mr. L. White, who was rewarded with £5 by the company.

The Boyd airship was never mentioned again in the aeronautical press, and lacking any photographs or evidence, it remains a mystery.

Power: 300hp motors driving four propellers to be increased to 500hp driving eight.

Data	Length	120ft being increased to 200ft
	Fuel	600 galls to be increased to 1,200 galls
	Crew	3 being increased to 8

BOYMAN Airship (Richard Boyman)

Boyman registered a patent No.3262 in 1866 for a most ambitious airship design; built of sheet steel, it would have had four boilers installed which were gasfired from the envelope and the steam generated together with consumed gas, was exhausted through venturis to four swiveling nozzles which both propelled the airship in either direction and gave additional lift for take-off (c.f. the modern Harrier jump-jet)

Data Length 1320ft Diameter 200ft Gross weight 600 tons

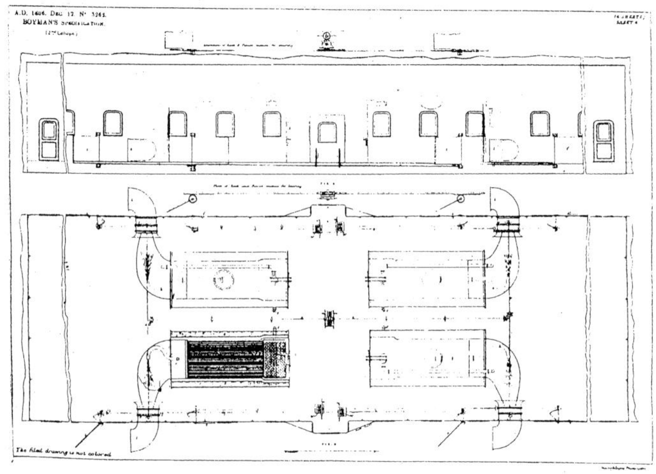

Boyman Airship. The patent drawing of 1866 shows the layout of this novel design.

CAYLEY Airships (Sir George Cayley, Brompton Hill, Yorkshire)

Sir George Cayley did not venture into the field of lighter-than-air aircraft until 1816, many years after he had commenced his experiments with helicopters, gliders, ornithopters, parachutes and aero engines.

The first airship in 1816 was a large side-by-side twin-hull design employing hot air to provide lift. It was 300 feet long and of semi-rigid construction.

The second design was to have had a single envelope filled with hydrogen and was 432 feet long powered by tiers of "flappers".

The third and most practicable design was of an elegant elongated dirigible filled with hydrogen and propelled by twin airscrews on outriggers and also with tiers of "flappers".

Unfortunately none of these interesting projects was constructed.

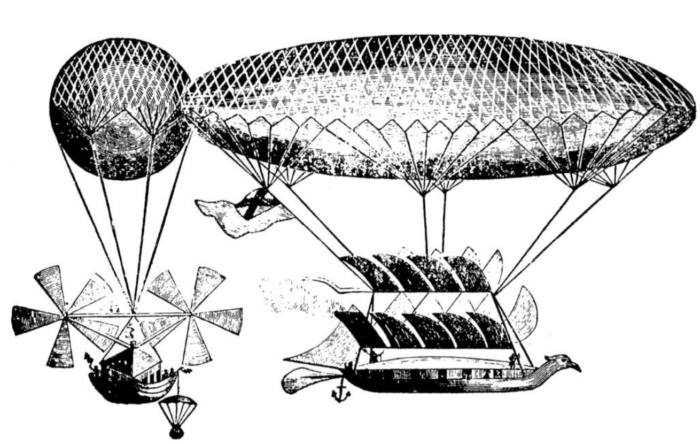

Cayley Airship. His third design published in 1817.

COSHAM Airship (built (?) at Cosham, Portsmouth)
Jane's All the World's Airships for 1909 makes brief mention of a dirigible at Cosham, Portsmouth which burned in about 1902. Nothing else has been uncovered regarding this obscure machine.

FOX Airship (A.H. Fox)
This small airship was built by A.H. Fox in 1903. The car was suspended below the cigar shaped envelope and had a propeller mounted on each side.
The airship does not seem to have been developed.

Friese-Green Airship. The patent drawing shows the twin envelopes.

FRIESE-GREEN Airship (W. Friese-Green, Brighton)
A novel type of airship was patented in 1908 by William Friese-Green, one of the early pioneers of cinematography.

Two gasbags were linked side-by-side with the car slung underneath and so designed that, if one bag should collapse, the other bag would provide sufficient lift to avoid a sudden descent. The airship was to have had a gyroscopic balancing device powered by one of the four motors; another drove the propeller and two others were kept in reserve.

The airship could be made to rise or descend by a movable weight traveling on an endless band from end to end of the machine. Two planes were also to be used to alter the angle of the ship.

A small version of the airship, intended to carry one man, was under construction in January 1909 but no further reports have been found.

GAUDRON Airship (Auguste E. Gaudron, Gaudron Ltd., 45 Outram Road, London N.)
The *Daily Sketch* of 7 March 1919 published a photograph of an airship flying over London. The caption stated that it had been made by Gaudron in 1896 and had flown from Alexandra Palace to Epping Forest where it had crashed.

Gaudron Airship at Alexandra Palace in 1896.

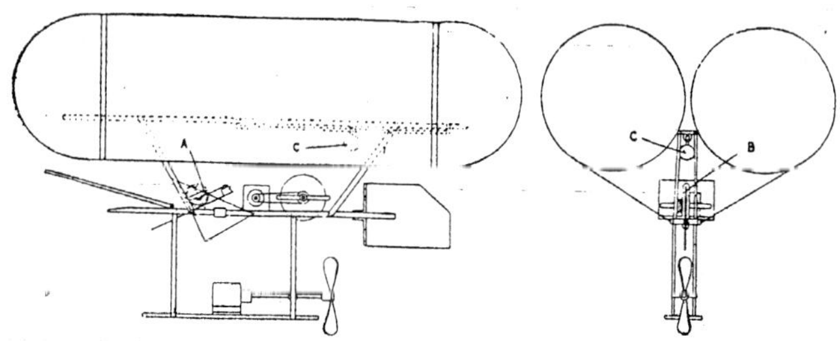

Gaudron were major manufacturers of balloons and worked closely with Spencers. Auguste had married the sister of the three Spencer brothers.

Nothing further has been found regarding this early airship.

GRAHAM'S Dirigible (George Graham, London)
This was built in 1852 but no details have been found. George Graham and his wife Margaret were professional aeronauts between 1823 and 1853.

GREEN's Voyage balloon (Charles Green, London)
Green designed a dirigible balloon in 1840 in which he intended to cross the Atlantic the following year from America to the European Continent. No doubt he intended to make use of favorable winds, but the balloon was also fitted with a moveable tail and a propeller which was probably worked by foot treadles This ambitious flight did not take place.

LENNOX *Eagle* Airship (Colonel Comte de Lennox, Vauxhall Gardens, London SW)
The airship illustrated was apparently built inside a wooden enclosure in Victoria Road, Kensington and then moved to Vauxhall Gardens for exhibition throughout September 1835, when the public was charged an entry fee of one shilling per head (5p). The capacity of the envelope was 2,700 cu ft.

No other information or evidence of flight has been found. In the event nothing happened and the whole scheme collapsed.

The RAeS showed some fragments of the *Eagle* on their stand at Olympia in July 1929.

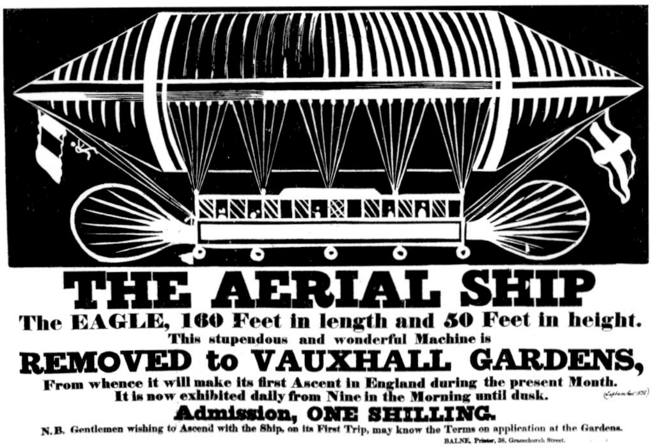

Lennox *Eagle* Airship was exhibited at Vauxhall in September 1835.

LESTER-BEST Airship (E.J. Lester, W.G. Best & E.A. Whitby Lester-Best Airship Ltd.)

This large model airship, with a novel control system, was exhibited at the Stanley Show in January 1910.

The inventors claimed that the control paddles enabled the machine to lift without discharging ballast; to descend without venting gas and to steer without using rudders. The paddle boxes deflected the sir from the part of the wheel which they were covering and were moved to alter the direction of the airship up, down or left or right (NB c.f. the control system on modern Harrier jump-jet aircraft).

The model was powered by an electric motor and tethered to its power lead.

Lester-Best Airship. The scale model exhibited at the Stanley Show in January 1910.

An improved model was tested in March 1910 and was shown in the hall of the Borough Polytechnic on 20 April 1910.

Data	Length Envelope	14ft
	Length Car	6ft

LOVELACE Airship (Capt. T.T. Lovelace, Hopton, Suffolk)

Reported on 14 August 1908 with a capacity of 10,000 cubic feet. No other details are known other than that it was destroyed by fire in August 1908.

LUNTLEY Dirigible (Luntley)

This was built between 1847 and 1851 and was a cross between an ornithopter and an airship. It was 63ft long and had flapping wings powered by a motor burning gas fired by an electric current.

Monck Mason Airship. The scale model of 1843 was powered by clockwork.

MONCK MASON Airship (Thomas Monck Mason)

The first elongated dirigible balloon in England was a model powered by clockwork and built by the Irishman Monck Mason in 1843. It utilized a helical airscrew and the full size version would have had a car suspended from a rigid ovalshaped framework, which seated the captain and the passengers; steering was by a large paddle shaped rudder. The full size version does not seem to have materialized. The model was 44ft long and when tested at the Royal Adelaide Gallery in the Strand it achieved 6 mph.

NAPIER Airship (D. Napier & Son, Vine St. Works, Lambeth, London SW & The Motor Power Co. Ltd., 14 Regent St., London SW)

At the end of 1901, the makers of Napier cars proposed to build airships, offering a craft with Spencer envelope for £950 on six months delivery.

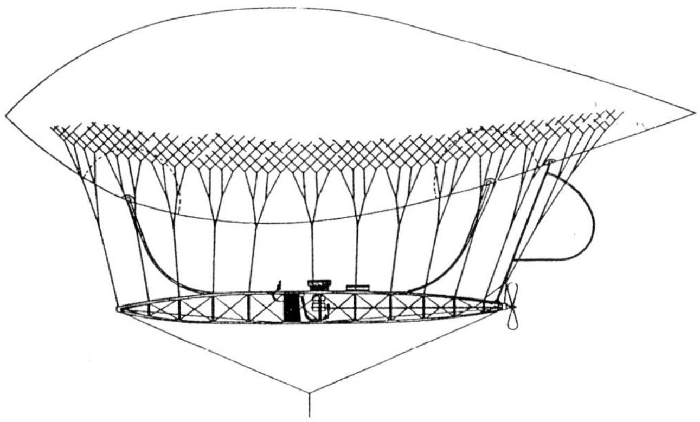

Napier Airship of 1901 had a Spencer envelope. It does not seem to have been completed.

The airship was said to have been designed on Santos-Dumont principles and carried a long open girder suspended at numerous points, below the envelope. A large rudder hung between the envelope and the car for directional control. The engine, mounted amidships, drove a pusher propeller at the rear of the girder through a long drive shaft.

Power: 14hp Napier four-cylinder inline water-cooled.

Data		
	Capacity	25,000 cu ft
	Length	75ft
	Diameter	25ft
	Length of girder	50ft

NEALE II Airship (J.V. Neale, Neale's Aeroplane Works, Baker St., Weybridge, Surrey)

In November 1910 Neale advertised in *Flight* magazine for readers to subscribe £45,000 to enable him to build a dirigible as 'an Aerial Cruiser primarily for the purposes of war and to carry quick firing guns for protection against air craft.' It was to be fitted with searchlights, long-range wireless equipment and telephotographic apparatus.

It would appear that the funds were not forthcoming and the Neale II was not built.

Power: Engines developing a total of 3,000hp

Data		
	Capacity	1,139,340 cu ft
	Length	600 ft
	Width	120 ft
	Height	40 ft
	Weight	74,286 lb
	Lift	18 ton
	Est. speed	100 mph maximum 20 mph cruising
	Range	22,000 miles at max speed
		110.940 miles at cruising speed

PANUZZI Airship "*Spezia*" (designer Frederick Walker, Oxford)

"*Flying*" of June 1902 announced that Mr. Panuzzi had instructed Mr. Frederick Walker of Oxford to design his air yacht "*Spezia*".

The frame of rolled aluminum bronze was similar to that of an inverted ship, the car being part of the rigid structure and the thirty-two gas-bags were contained within a double waterproof envelope. The bows were protected by a metal shield.

The 55hp motor was a petrol turbine connected to a dynamo which operated two electric motors, each driving an eleven foot diameter propeller.

The longitudinal balance was maintained by a long pivoted tube containing 56lb of mercury, tilted by a small motor controlled by a gravity balance.

The airship was expected to be ready for a trial trip early in August 1902, but no reports have been found relating to this and it seems likely that it was never completed – perhaps it was never even started.

Data		
	Length	230ft
	Load	6000 lb
	Lift	21.200 lb

PARTRIDGE Airship (J.M. Partridge)

This design in 1848 was especially interesting in that the envelope, which included a ballonet, consisted of a light rigid framework covered with fabric. The complete envelope was covered with a light wire netting from which the car was suspended. The car was fitted with three propellers at the stern. The center propeller faced rearwards for propulsion and similar propellers on either side at right angles provided side thrust for maneuvering. In addition two large triangular sails were positioned on each side, which may have been for control purposes or for providing extra lift. Power was provided from a compressed air supply.

It was reported that the airship made several short trips but control was inadequate.

PAULY & EGG *Dolphin* Airship (John Pauly & Durs Egg, Knightsbridge, London W)

These partners were Swiss citizens resident in England, who began construction in about 1816, of a non-rigid airship with a fish shaped envelope, made of gold beaters skin, which incorporated a ballonet and was provided with a tail. The boat shaped

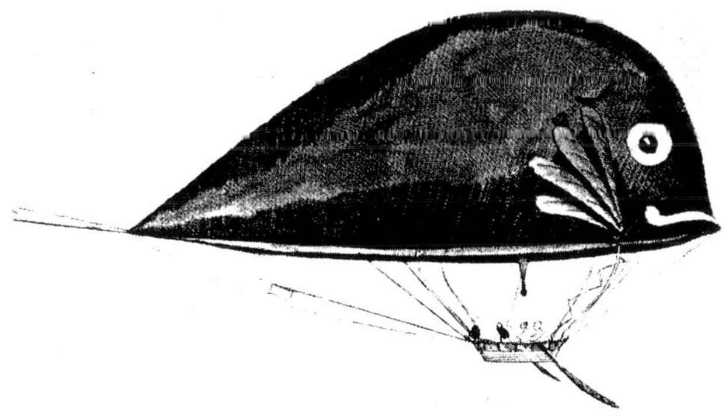

Pauly & Egg *Dolphin* Airship. This elegant fish-shaped dirigible was not completed.

car featured oars for propulsion driven by a kind of atmostpheric steam-engine, unveiled by a Mr. Collier, and a sliding weight, to adjust the attitude of the craft, was slung between the rear of the car and the tail of the envelope. It was not completed but the ballonet was converted by P.T. Barnum into a small balloon in which the famous dwarf 'Tom Thumb' made an ascent from Surrey Gardens.

PENNINGTON Airship (Edmund Joel Pennington, London & Coventry)

Pennington was an American from Indiana who was forced to leave the U.S.A. following fraudulent activities to finance his ventures, mainly in the automobile field. In Britain his vehicles were built by the Humber company and the British Motor Syndicate in Coventry, but with little success. In addition to taking out many patents for automobile designs, in 1895 he prepared a scheme for an airship and registered it in patent No.25050/1895.

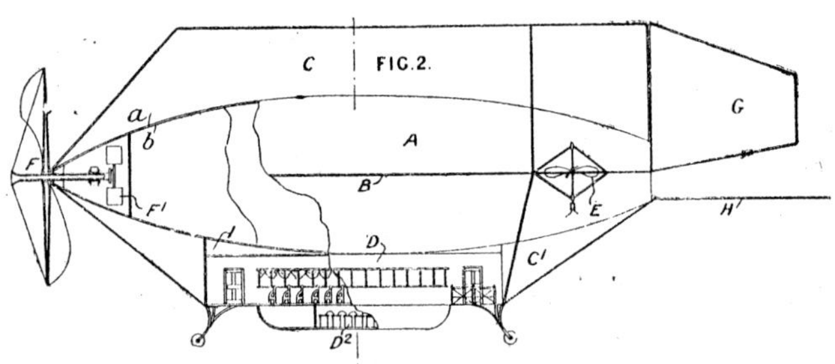

Pennington Airship was patented in 1895 but finance for its construction was not forthcoming.

The machine was to be largely constructed of aluminum and featured a fin stretching along the full length of the main body, to which a rudder was attached. Planes were attached at the center of the body on either side and a large single elevator at the rear. A large car for passengers was hung below the body and also carried electric batteries to drive lifting fans on either side at the rear of the body and to power the controls of the elevator and rudder.

Propulsion was by a motor in the nose, fueled by hydrogen, driving a large diameter propeller. Apart from the storage of hydrogen in compartments in the main body, additional hydrogen was carried inside doors, berths, furniture, cushions and even in hollow propeller blades.

The airship was a further example of Pennington's outlandish ideas aimed at obtaining financial support and it is doubtful if construction was started. It was reported that he again became involved aeronautics later in life, but this was after his return to the U.S.A. where he died in 1911.

SPENCER Airships (C.G. Spencer & Sons, 56a Highbury Grove, London N)

Charles Green Spencer (1837-1890) was a balloonist throughout much of his life and founded the firm of C.G. Spencer & Sons. in Holloway to carry on the manufacture of balloons. He had a family of eleven children who were mostly involved in the business, particularly Percival (1864-1913), Arthur Charles (1866) and Stanley (1868-1904) who carried on after their father's death. The third son, Stanley was the first Englishman to have built and flown a dirigible, which was operated from the grounds of the Crystal Palace.

Stanley Spencer attempted, in 1903, to fly round the dome of St. Paul's back to the Crystal Palace, but adverse winds and lack of power caused the airship to drift across London and to eventually land at Barnet. He received a visit from Count Zeppelin at Highbury before his early death from malaria in 1904.

After Stanley died, Henry (1877-1937) assisted by Auguste Gaudron, was working at Alexandra Palace on the Barton airship, where he also continued the work of the company. In addition a separate company, Spencer Bros. Aeronauts, covered the operating of balloons and parachuting from 45b Aberdeen Park, Highbury before the war. After wartime employment with Short Bros. he revived the company in 1922 at Bouverie Rd., Stoke Newington.

The Spencers made a number of airships between 1902 and 1914, but no authentic record of these remains and information from various sources is confused. The following record is the best that can be ascertained from the surviving references. It is known that the Spencers supplied a 20,000 cu ft envelope to Willows for his No.II. The frame of a Santos-Dumont gondola was also reported to be in Spencer's works in about 1914, which may have been that of his No.6, which he flew round the Eiffel Tower.

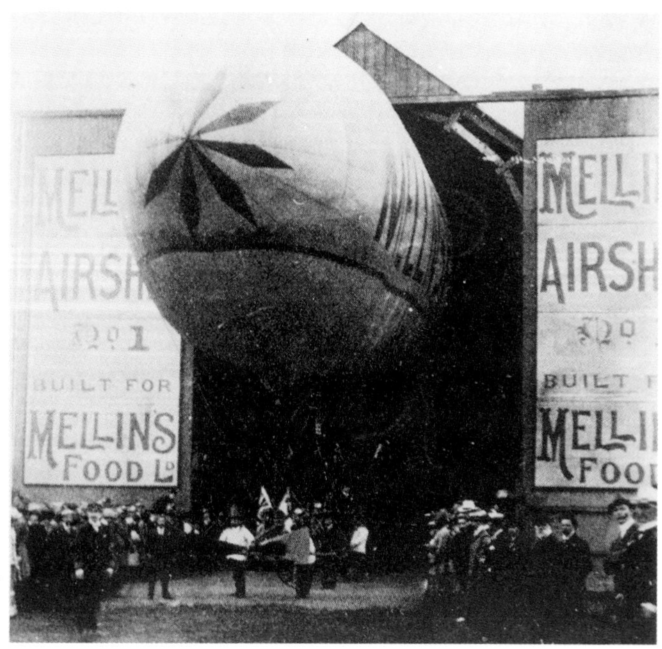

Spencer Airship No.1. This single-seater machine was flown by Stanley Spencer in 1902.

SPENCER Airship No.1. 1902

This was a small airship built for experimental work and for carrying one person only. It was operated by Stanley Spencer from the Crystal Palace in 1902. It was reported to be fitted with a 30hp Simms engine, but this is more likely to have been a less powerful engine, possibly a 3 1/2hp unit.

Spencer No.1 has been referred to as the Spencer-Moreing airship by Lord Ventry and was also the *Mellins Food* advertising airship. It may have been sold later to Moreing of the Australian Electrical Co. and used by Roberts and Healey at Dagenham in 1909-1910 (*q.v.*).

Data	Length	75ft
	Diameter	20ft
	Capacity	20,000 cu ft
	Car length	45ft

SPENCER Airship No.2

Spencer's second airship was test flown at Ranelagh in July 1903 and proved somewhat of a disappointment, being improperly balanced so that the nose dropped and the front mounted propeller struck the ground and was broken. The engine was therefore removed and the airship made a flight as a free balloon and landed near Chipping Ongar in Essex.

Spencer Airship No.2. Was test flown at Ranelagh in July 1903 but was not successful.

The car had three longitudinal members of bamboo braced with diagonal wires. The engine was subsequently moved further aft to improve the balance.

Power. Simms 20/24hp water-cooled driving 12ft diameter two-bladed propeller

Data	Length	93ft
	Capacity	24,000 cu ft
	Length of car	50ft

SPENCER Airship No.3. 1903 '*Evening News*'

This was a larger ship than No.1 and, on 17 September 1903, Stanley Spencer attempted to fly it from the Crystal Palace around St. Paul's and back. Adverse winds prevented this and the airship drifted across London, eventually landing at Barnet. A second attempt was made on 1 October 1903 but was less successful than the previous flight. Spencer soon got lost as the day was misty and he could not make out any landmarks. Leakage of gas caused the airship to descend and it eventually landed in Peckham Rye Park in South London.

The *Evening News* sponsored these flights and the paper's name appeared on both sides of the envelope.

The airship was 93ft long and 24ft in diameter and a 20/24hp Simms engine was reported to be fitted which drove a 12ft diameter tractor propeller at the front of the 50ft long car built of bamboo and aluminium. A large sail-like rudder was hung between the envelope and the rear of the gondola.

Spencer Airship No.3.

SPENCER Airship. 1908-1909 '*Urban*'

This small craft was tested from the grounds of the Crystal Palace in the spring of 1908, before being taken to the Wandsworth Gasworks for a flight round St. Paul's in May 1908. However the wind carried it back to Croydon, where it landed near Factory Lane after two hours in the air.

It was probably Percival in charge of the flight, accompanied by a well-known explorer and photographer named Cherry Kearton. The same airship seems to have been used by Percival in 1909, bearing the slogan 'Votes for Women' on both sides, although the rudder appears to have been changed for one of triangular shape. The airship carried a passenger, a suffragette named Miss Matters, over the opening of Parliament ceremony, who used a megaphone to spread the word.

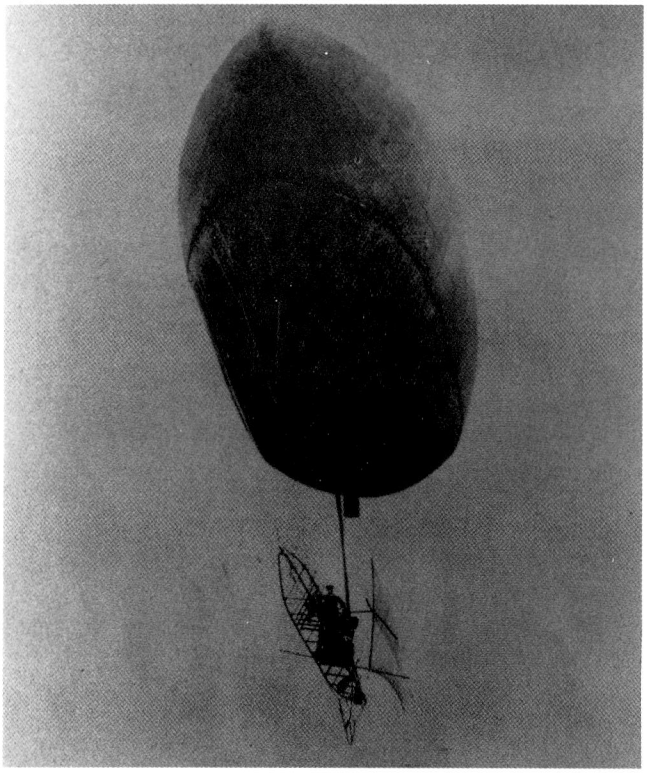

Spencer Airship 1908-1909 *Urban*.

The envelope was of low aspect ratio and had both a pointed nose and tail. The gondola was attached by many suspension cords to a band around the lower periphery of the envelope. It was an open girder of inverted vee section, tapering to points at the nose and tail and was approximately two thirds as long as the envelope. The girder was made of bamboo members, strapped or lashed at the joints and made rigid with wire cross-bracing.

The engine was mounted just forward of the center of the gondola and drove a tractor propeller by a long shaft. A passenger could be accommodated in a balloon basket, but the pilot needed to stand on the longerons and move to reach the engine or to adjust the balance. A huge sail-like rudder was fitted for directional control.

Power: 7hp Simms vee-twin air-cooled driving a two-bladed tractor propeller.

Data		
	Capacity	40,000 cu ft
	Length	84ft
	Diameter	32ft
	Gross lift	2,720 lb
	Speed	1215 mph

SPENCER Airship 1908 *'Gamage 1'*

This medium size airship was built to the order of Gamages Ltd., the large London departmental store.

It made its trial flight on 9 December 1908 and, after departure from Wandsworth, it passed over Central London and landed at Bentley, near Brentford, Essex a flight of some thirty-five miles.

The crew consisted of L. Talbot (aeronautical advisor to Gamages). Henry and Herbert Spencer.

Gamage 1 was probably similar in layout to the 1908/9 *Urban* airship and had a large sail-like rudder.

Power. 6hp Simms vee twin air-cooled.

Data		
	Capacity	40,000 cu ft
	Length	80ft
	Max diameter	30ft

Spencer Airship 1909-1910 "*Carbic*" leaving Wandsworth Gasworks.

SPENCER Airship. 1910 *'Carbic'*

This airship was similar to the *'Gamage 1'* and may even have been the same machine. It was operated from the Wandsworth Gasworks from March 1910 for advertising work. 'Carbic' which it is known was displayed on the airship, was the name of a new acetylene lighting generator marketed by C.C. Wakefield Ltd., purveyors of 'Castrol' lubricating oil.

Motor trouble and strong winds prevented one attempt to fly round St. Paul's. After reaching a height of 7,500 feet the

Spencer Airship at the 1911 Olympia Aero Show.

airship drifted over London to land at Broadway Common, Broxbourne, Hertfordshire.

Data	Length	80ft
	Diameter	30ft

SPENCER Airship. 1909-1910 (C.A. Moreing, A.J. Roberts and F.J. Healey, The Australian Electrical Co.)

This small nonrigid airship, referred to as the Moreing airship, was at the Dagenham experimental ground of the Aeronautical Society on 5 August 1909 where it was being used to demonstrate the possibility of the remote control of flying machines, boats and torpedoes by Hertzian (radio) waves.

Moreing, an Australian had retired to England; Roberts and Healey, employees of the Australian Electrical Company, were fellow Australians who came to England to carry out the experimental work with the airship and other radio controlled devices. The airship was still at Dagenham in a purpose-built shed uninflated in June 1910.

The envelope was similar to that of an earlier Spencer craft, with a long triangular section girder slung below; this was of upright vee section and carried a triplane tail surface, which presumably was moveable under radio control.

The airship was flown unmanned and controlled by radio and was used in conjunction with the torpedo experiments being carried out in the nearby Dagenham Breach.

Power: 100-120hp ENV of unknown type. There is doubt about the power reported for this engine, which is more likely to have been a 60hp type F.

Data	Capacity	20,000 cu ft
	Length	105ft
	Diameter	18ft

SPENCER Airship. 1913. '*Bovril*'

This nonrigid was commissioned for advertising '*Bovril*' in July 1913 and carried out a number of successful flights, including one from Hendon to Sutton in Cambridgeshire, with a stop at Sunderland Farm near Biggleswade overnight, during which a height of 4,700 feet was reached. In August and September, flights were made over London at heights between 100 and 1000 feet, in one of which a mock battle with a biplane took place. The airship pilot was R.F. Dagnall who later formed RFD Ltd. to make balloons, parachutes, dinghies etc. The advertising campaign started on 5 July 1913 bearing the slogans 'Give him Bovril' on one side and 'Give her Bovril' on the other, the flights operating from near the Old Welsh Harp at Hendon.

The envelope was filled with hydrogen gas. There was a rudder, but control in yaw was mainly by differential operation of the propellers. There were two horizontal fixed surfaces above the top of the gondola, but attitude was probably varied by weight shift of crew and ballast along the 56ft long structure.

The Bovril booklet stated that the ship was of 40,000 cu ft and was 90ft long. Claxton refers to the Spencer Type A with 35-40hp Green engine, and capacity of 26,000 cu ft, 88ft long and 24ft diameter. A change to the envelope could account for the differences and an assumption has been made that the references are to the same basic craft. Another reference gives the capacity as 40,000 cu ft and the length as 96ft.

Spencer Airship 1913 *Bovril.*

SPENCER Airship. 1914 (A.J. Roberts, Sydney, Australia)

In 1914 it was reported that a Spencer airship was inflated at the Royal Agricultural Society's ground at Sydney. This was probably the 1909-1910 airship with which Roberts and Healey carried out the radio controlled experiments at Dagenham, although it used a much smaller two-cylinder engine. The car was described as of triangular section with triple elevator and the airship could lift two men and ballast.

SPENCER Airship. Type B. 1914

Claxton also refers to a much larger airship of 100,000 cu ft capacity, 150ft long and 35ft diameter. Power was provided by two 50-60hp engines and accommodation for ten crew and passengers was provided in the car, which was made in sections of steel and aluminum for easy transport. There is no evidence that this airship was actually built.

VICKERS Naval Rigid Airship No.1 '*Mayfly*' (Vickers, Sons & Maxim Ltd., Barrow-in-Furness, Cumberland)

The Admiralty was authorized to build the first British rigid airship by the approval of a sum of £35,000 in the Naval Bud-

Vickers Airship No.1 *Mayfly* at Cavendish Dock, Barrow-in-Furness in 1911.

get for 1909-1910. Vickers were the design authority and constructors and Capt. Murray Sueter RN was appointed to oversee the work, assisted by Lt. Boothby RN and Comdr. Oliver Schwann RN.

The structure of the airship was to be built of duralumin, an aluminum alloy of recent development by Vickers. This massive undertaking involved also the construction of a large shed at Cavendish Dock, Barrow to house the craft. This alone was a considerable expense, borne by the company, with little hope of recovery against the contract, even if the ship was a success, which it proved not to be.

Construction began in April 1910 and the ship was ready to be floated out on 22 May 1911 for the initial tests. These revealed that the ship had inadequate lift and it was returned to the shed on 25 May 1911 for modification. The gas capacity was increased, for which purpose the ship was cut in half and lengthened; some strengthening took place, which did not include the keel, which was considered by some to be under strength. The airship was again judged ready to be floated out of the shed on 24 September 1911, but it broke it's back against the shed and finished up in two halves, which gradually settled down on the water. These were recovered, but the damage was such that the craft was not rebuilt. The cause of the accident is not certain and the Admiralty criticized the design, but inexperience of the handling party and changing wind directions no doubt contributed.

The structure consisted of twelve longitudinal members and fifteen main frames with subframes in the nose and tail. The hydrogen gas was contained in seventeen gasbags made by Short Bros. who also made the gas control valves and the envelope. The engines were contained in two separate boat-like cars spaced well apart below the keel, with an external interconnecting walkway of triangular section.

Long horizontal and vertical fins were mounted towards the tail with quadruplane rudders above and below and triplane elevators either side of the tapering tail. A small triplane rudder was fixed below the ship just behind the rear car and a small elevator under the bows. Access to the top of the ship could be gained by a rope ladder inside the envelope, which was silver gray on top and yellow below to minimize the effect of temperature changes.

The Admiralty took no responsibility for the design and Vickers, with no experience in aeronautical or light engineering construction, learned the hard way that, although shaped like a submarine, construction of airships was an entirely new matter and there was much to be learned. The airship had cost £41,000 before it was modified and the total project was assessed at £100,000.

Power: Two 180hp Wolesley eight-cylinder water-cooled vee. The engine in the front car drove two four-bladed pusher propellers mounted on outriggers. The engine in the rear car drove a single two-bladed pusher propeller. The power of the engines has also been quoted as 160 and 200hp Wolesley exhibited at the 1909 Aero Shows in London and Paris showing an engine of 180hp 'suitable for dirigibles'.

Data		
	Capacity	706,330 cu ft (663,518 cu ft also quoted)
	Length	512 ft
	Diameter	48 ft
	Length of car	20ft
	Weight	43,879 lb
	Gross lift	Twenty-two tons including a crew of twenty-two
	Est. speed	40-42 mph

WILLOWS Airships (Ernest Thompson Willows, 2 Dumfries Place, Cardiff, South Wales.
E.T. Willows Ltd. 2432 Villa Rd., Handsworth, Birmingham. May 1911 to March 1928.
Willows Aircraft Co. Ltd., 2 Dumfries Place, Cardiff. March 1913-1919)

In 1904, with the support of his father, Ernest Willows at the age of 18, planned the building of an airship. He made contact with Capt. William Beedle, who had moved to Cardiff during that year, bringing with him his unsuccessful airship. Some form of cooperation was established, which lasted until about 1908, when it seems that Willows had the initiative to develop new and more successful ideas and so later went to work on his own. Over the years his efforts resulted in the construction of five airships before the war, the government taking only little interest in these. His sixth design was prepared when he was with Airships Ltd. of Merton, where he became Chief Engineer in 1915, and this was completed in March 1915. Also at Airships Ltd. as Works Manager from 1914 was R.F. Dagnall who had worked for Willows from March 1910 to the end of 1912, when he moved to Spencers and was later to form his own company, RFD Ltd. to make balloons, parachutes and life saving equipment. In 1916 Willows transferred to the AID for inspection work in connection with kite balloons and later, in mid-1918, revived the Willows Aircraft Co. at Cardiff for the manufacture of kite balloons under government contract. After the war, with little demand for balloons and airships, he developed a small shipping concern, which foundered after the wreck of his vessel in October 1921, in which he lost most of his assets and almost perished himself.

In 1923 he again became involved with ballooning, carrying out advertising and passenger flights at various places throughout the country. The business was not a financial success and Willows was heavily in debt when, on 3 August 1926 he lost his life, together with four passengers, in an accident to the Spencer-made balloon '*Marana*' at Kempston, Beds.

Willows work on kite balloons and balloon aprons was the subject of a claim to the Royal Commission for Awards to Inventors, but he was rewarded with only a very small fraction of his claim. Willows was the leading pioneer of British airship building, who was responsible for many practical developments, such as his swiveling propeller system, which received scant recognition from his country, before his affairs and his life ended in disaster and tragedy.

In May 1911 his company E.T. Willows Ltd. had been formed in Birmingham, where he had moved his operation with local support. At the end of 1912 it was bankrupt, although it remained in being nominally until March 1928. In March 1913, with his father's support, he headed a new company, Willows Aircraft Co. Ltd. registered in Cardiff. Operations were based on grounds adjacent to the Old Welsh Harp at Hendon, initially providing training on spherical balloons; later Willows began the construction of his fifth airship. He also worked for Grahame-White part of the time to assist with his finances.

Willows was a great protagonist of swiveling propellers, originally covered by Patent No.14439/1909 but these found little favor with the Admiralty, who considered the advantages in operations insufficient to justify the higher manufacturing costs for the small scouting airships used extensively in naval operations during the war.

WILLOWS Airship. No.I and IA

This was the first practical British airship and was constructed at Cardiff, where it made it's first flight at Splott Farm, East Moors lasting one hour twenty-five minutes on 5 September 1905.

The Japanese silk envelope was made by a local tailor, G.A. Secombe & Co. and it was filled with hydrogen gas, as were later airships made by Willows. In the first version of the craft it adopted a curved back shape under the load of the original car. The car itself was an open structure of light steel tubes, tapering down to a boss at each end. At the rear end a large two-bladed pusher propeller provided forward thrust, while at the front end a pair of two-bladed propellers were fitted to provide control, serving as both rudder and elevator. The motor was mounted amidships and drove both sets of propellers through long shafts, with clutches arranged to enable any propeller to be engaged or disconnected. The axis of the forward propellers could be rotated to either the horizontal or vertical position for control in pitch and yaw.

Willows Airship No.I. The first ascent at Cardiff by E.T. Willows on 18 August 1905.

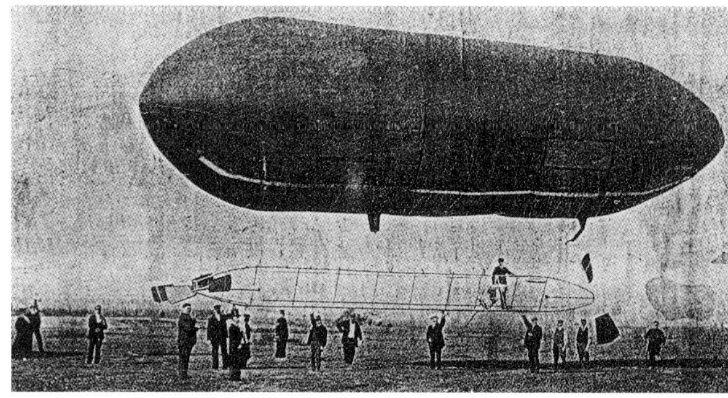

Willows Airship No.Ia, as modified with extended framework. The double swiveling propellers are clearly seen.

A number of successful flights with a crew of two were made which demonstrated the effectiveness of the swiveling propellers. The shape of the envelope improved with the fitting of a lengthened car, No.IA, which approached the length of the envelope itself and permitted the use of more suspension ropes to spread the load more evenly.

The Willows airship was reported by the national press to be exhibited at the Aero Show at Olympia in March 1909, although its presence remained unrecorded by *Flight,* which gave publicity to the American Wellman airship. Willows was by this time embarking on the construction of his second craft and No.IA was not seen again.

Power: 7hp Peugeot vee twin-cylinder air-cooled driving the 10ft diameter pusher propeller and the steering propellers.

Data	**No.I.**	**No.IA.**
Capacity	12,000 cu ft	13,000 cu ft
Length	72ft	
Diameter	17ft	18ft
Weight of envelope	200 lb	
Weight of car	400 lb	450 lb
Length of car	30ft	45ft

WILLOWS Airship. No.II

Willows' second airship was constructed in his shed at East Moors, Cardiff in 1909 and it took to the air for the first time on 26 November 1909. Problems were experienced with the drive, nevertheless the tests were generally satisfactory. Two other flights were made before the end of the year. In March 1910 the airship was exhibited at the Olympia Aero Show and, after further improvements, Willows again flew the airship on 4 June 1910 at Cardiff. Further local flights to Cardiff center followed. In July he took the craft by rail to Cheltenham and

Willows Airship No.II. Original version with two part bamboo and steel keel in late 1909.

Willows Airship No.II modified with one piece keel at Cardiff early in 1910.

on 9 July 1910, not only demonstrated as planned, but flew back to Cardiff, a distance of seventy miles, arriving back just before midnight.

He then decided on a much more ambitious flight to London to the Crystal Palace. This took place overnight on 6-7 August 1910, but he ran out of fuel and drifted beyond his goal to Lee near Eltham, arriving at the Palace the following evening. While based at the Crystal Palace on 10 September 1910, he flew round St. Paul's and over the Houses of Parliament. Willows had now become involved with W.J. Potter of the Aerial Manufacturing Co. of Great Britain and Ireland who proposed schemes to raise funds and became a contractor for the envelope for his next airship.

The envelope for No.II was made of varnished cotton by Spencer Bros.; below this was suspended a long boom of bamboo and steel tubes of triangular section, which was increased in depth at the center to accommodate the power and crew. A fixed fin between the envelope and the boom terminated as a rudder and tailplane of cruciform shape. A ballonet was incorporated in the lower portion of the center of the envelope.

The power was provided by a JAP air-cooled engine driving, through a belt and shafts and bevel gearboxes, the outboard mounted tractor propellers, which could be rotated to provide vertical lift as well as forward thrust. It worked very satisfactorily and was a marked improvement on the Beedle system. The original propellers, made by Willows, were crude paddles made of tube and sheet metal, but these were replaced by wooden ones made by Handley Page.

Power: 35hp JAP eight-cylinder air-cooled vee driving two tractor propellers on outboard swiveling mountings at half engine speed.

Data		
	Capacity	21,000 cu ft
	Length	86 ft
	Diameter	22 ft
	Length of boom	58 ft
	Weight	1,021 lb
	Lift	1,170 lb

WILLOWS Airship. No.III *City of Cardiff*

The manufacture of the third Willows airship, which was a nonrigid, was carried out in the autumn of 1910 at the Crystal Palace. The envelope was ordered from the Aerial Manufacturing Company of Great Britain and Ireland (W.J. Potter) using rubberized fabric supplied by the North British Rubber Co. Ltd. The final assembly of the airship took place at Wormwood Scrubbs and it was ready for first flight to take place on 29 October 1910 carrying Frank W. Goodden, later to become a well respected test pilot at Farnborough, as passenger.

It was planned to demonstrate the capability of the airship by a flight from London to Paris, which started at Wormwood Scrubbs on 4 November 1910 with the same crew aboard, making the Channel crossing at night. Mechanical problems curtailed the flight at Corbehem, near Douai after ten and a half hours, some 100 miles from Paris, where they landed causing some damage. This was repaired at the Breguet works, but the flight could not be resumed owing to severe winds, so the ship was deflated and transferred to the Clement Bayard hangar at La Motte for further repair work. The final attempt to reach Paris was on 28 December 1910 and this was success-

Willows Airship No.III *City of Cardiff* taking off from Wormwood Scrubs, London en route for France.

fully concluded after dark, when they landed at Issy-les-Moulineaux. After carrying out a number of local flights, over a period of several days, the *City of Cardiff* was deflated and returned to England in early January.

Willows III had a larger envelope than its predecessor and incorporated a divided ballonet for adjustment of trim with changing fuel loads. The car was a modification of that used on No.II, accommodating a crew of two and using the same engine and propeller mechanism. Willows and four passengers were carried on at least one flight in September 1911.

After returning from France, the airship operated at Watford in March 1911 and moved to Wolverhampton later that month, then on to Birmingham in the autumn. Meanwhile, Potter began legal proceedings against Willows for costs relating to the envelope, which Willows now regarded as unsatisfactory due to excessive leakage. This led to other court actions involving the supplier of the fabric connected with the loss of costly hydrogen. Willows was unable to justify his claim and this resulted in a considerable financial loss and led to his bankruptcy by the end of 1912. The JAP engine is still preserved at Old Warden.

Power: 35hp JAP eight-cylinder air-cooled vee driving two 6ft 4in diameter Beney tractor propellers in outboard swiveling mountings at half engine speed.

Data		
	Capacity	32,000 cu ft
	Length	120 ft
	Diameter	23 ft

WILLOWS Airship. No.IV (Naval Airship No.2)

Despite his financial losses, Willows was able to proceed with the building of his fourth ship at Birmingham and it was ready to fly by the end of May 1912. The Admiralty purchased the airship for £1050 on the recommendation of an inspecting officer, it was then delivered to the Naval Wing of the RFC at Farnborough. It was operated by No.1 Squadron and was identified as Naval Airship No.2. It was purchased essentially as a training and experimental airship with little thought of any military potential.

Willows Airship No.IV (Naval Airship No.2) as originally built. On trial at Farnborough.

This airship, smaller than No.III, was of improved shape and the car, in particular, was efficiently streamlined. The envelope was made of oiled cotton, a new development, below which hung a long bamboo boom, at the center of which the engine and propeller mechanism were mounted. A triangular fin and rudder were attached at the rear end of the boom; there were fixed surfaces on either side of the envelope. The cruciform surfaces were later moved from the boom further aft to a position just below the envelope. In later service a fin was added to the top of the envelope and the lower fin increased in area. The car was rigidly attached to the boom and accommodated the two man crew in tandem and was constructed of steel tubing with aluminum sheet skinning. The swiveling propeller

Willows Airship No.IV (Naval Airship No.2) with modified fin and rudder at rear of envelope; seen at Farnborough.

Willows Airship No.IV (Naval Airship No.2). Final version.

mechanism driven by a wide leather belt was retained. Various changes were made later when the airship was in service including increased capacity envelopes and a different engine.

A spare Willows envelope of 35,000 cu ft capacity was fitted with a BE2c fuselage by RNAS Kingsnorth early in 1915. This was identified as SS1 and became the basis for a large number of maritime patrol airships with various types of car, which were used with considerable success in wartime. These submarine scout airships were made by contractors and assembled at various RNAS bases.

Power: 20-24hp JAP four-cylinder air-cooled vee driving two four-bladed 6ft diameter Rapid propellers.
35hp Anzani three-cylinder air-cooled fantype semi-radial fitted later.

Data	**First Version**	**January 1914**	**May 1914**
Capacity	20,000 cu ft includes	24,000 cu ft	35,000 cu ft made
	3000 cu ft in ballonets of gold beaters skin		
Length	110 ft	100 ft	
Diameter	18ft 6in	20ft	
	Range		75 miles
	Speed		25 mph

WILLOWS Airship No.V

Construction of No.V took place at Hendon in the second part of 1913. This was Willows' largest airship with an envelope of rubberized fabric incorporating two ballonets. Fixed fins were mounted above and below the envelope, the lower carrying the rudder, together with horizontal tail surfaces. The car was slung directly onto the envelope and was fully enclosed with a tapered nose and long tapered tail. The engine, mounted ahead

Willows Airship No.V.

of the crew, drove the outboard propellers by the latest patented system of flexible belt drive, which avoided the use of gearing, and provided separate control of the angle of each propeller. Upward and downward thrust could thus be selected as desired.

The airship was eventually flown in December 1913 and after several local flights from Hendon, Willows flew with a passenger around St. Paul's Cathedral and back in one hour and twenty five minutes. Up to four passengers could be carried.

A gale damaged the airship in the hangar on 16 March 1914 and it was deflated. This was the end of airship building on his own account by Willows.

Power: 60hp ENV type F eight-cylinder water-cooled vee driving two 8ft diameter two-bladed tractor propellers.

Data		
	Capacity	50,000 cu ft
	Length	130 ft
	Diameter	26 ft
	Length of car	42ft
	Speed	38 mph

Models

In no other engineering industry can the amateur model makers have had so much influence on the progressive design of the full size machine as had the aeromodellers from 1900 onwards.

Almost every city and town in Britain had its model aeroplane building and flying club or society and many of these organizations progressed to the design, construction and flying of full size kites, gliders and powered aircraft.

Many of the famous names in the aircraft industry such as A.V. Roe, Handley Page, Sydney Camm, W.O. Manning and F.W. Lanchester started their careers as model aircraft builders and their designs were often scaled up and became full size machines. Sometimes these pioneers also started their careers as bicycle makers (e.g. the Wright Brothers), or simple carpenters (e.g. Sydney Camm) and often progressed to motorcycle and automobile engineers on their 'way to the stars'.

Handley Page certainly acknowledged the importance of model building and gave a lecture entitled 'The Dependence of Aviation on Experimental Model Work' at the Caxton Hall on 9 January 1914.

Model aeroplane flying was a most popular sport prior to World War I and valuable prizes were offered to club members who achieved the longest or highest flights. Rubber power was universally popular, but petrol engines, steam engines, compressed air or Carbonic Acid gas engines were also used. Model gliders were also very popular. Some of the models were of considerable size; 10ft wing span, or even more, was not uncommon.

As will be seen from the photos on the page opposite, some of the designs were quite farfetched, but others were obviously well thought out and extremely practical.

Chubb helicopter.

Crouchley double biplane.

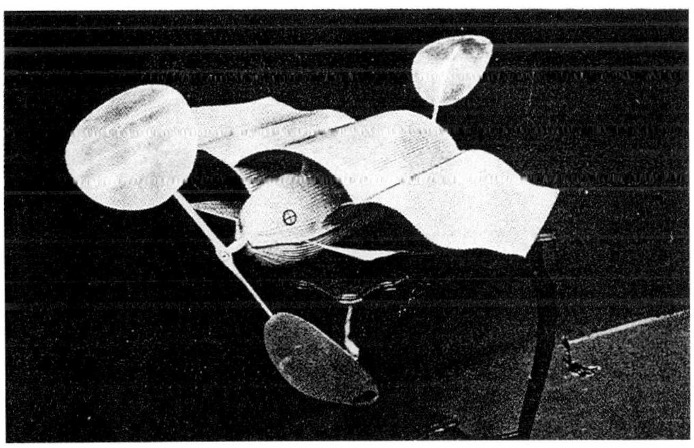

Cochrane aluminium monoplane.

Deixler monoplane.

Fell biplane.

WO Manning models of biplane and monoplane.

George Goodall's large monoplane (8ft span) with petrol motor. Wings made by Handley Page.

Jose Weiss "*Albatross*" monoplane.

WO Manning twin-propeller monoplane.

WO Manning large monoplane with CO2 motor.

AV Roe Langley-type monoplane.

AV Roe biplane.

Unidentified British Aircraft

The author would be delighted to hear from any readers who might be able to identify any of these machines.

Glider biplane with long skids.

Very early ornithopter with push-rods operating flapping wings. Date appears to be 1860-1880.

Much modified Bleriot monoplane probably used as a photographer's "prop" – Josie Roth, photographer, Onchan Road, Isle of Man.

Cody-type biplane in unknown location.

Demoiselle-type monoplane with large radiator in front of wing. Possibly a collection of old aircraft parts being used as a photographer's "prop."

Unknown monoplane of Blériot type possibly photographed in Sussex.

Bibliography

Books

Andrews, Charles F. *Vickers Aircraft since 1908.* Putnam 1969

Andrews, C.F. & Morgan, Eric B. *Supermarine Aircraft since 1914.* Putnam 1981

Aspin, Christopher *Dizzy Heights Helmshore Local History Society* 1988

Axted *The Hon. C.S. Rolls 1877-1910*

Barnes, C.H. *Bristol Aircraft since 1910* Putnam 1964

Barnes, C.H. *Shorts Aircraft since 1900* Putnam 1967

Berriman, A.E. *Aviation* Methuen 1913

Brew, A. *History of Black Country Aviation* Alan Sutton 1993

Broomfield, G.A. *Pioneer of the Air* Gale & Polden 1953

Brown, Alan *They Flew from the Forest* Alan Brown c.1998

Brett, R. Dallas *The History of British Aviation* John Hamilton 1933

Brewer, R.W.A. *The Art of Aviation* Crosby-Lockwood 1910

Bruce, Gordon *Charlie Rolls - Pioneer Aviator* Rolls-Royce Heritage 1990

Bruce, J.M. *British Aeroplanes 1914-1918* Putnam 1957

Bruce, J.M. *The Aeroplanes of the R.F.C. (Military Wing)* Putnam 1982

Clayton, D.C. *Handley Page Album* Ian Allan 1970

Cole, K.A. *Aviation in Essex* Southend Branch Royal Aeronautical Society 1967

Collyer, D.G. *Flying-Pioneer Years in Kent* North Kent Books 1982

Connon, P. *In the Shadow of the Eagle's Wing* P. Connon 1982

Cullingham, G. *Patrick Y Alexander* Cross Manufacturing 1984

Cullingham, G. *F.J. Camm - The Practical Man* G. Cullingham 1996

Dodsworth, Ted *Wings over Yorkshire* Hutton Press 1986

Duval, G.R. *British Flying Boats & Amphibians* Putnam 1967

Gibbs-Smith, C.H. *The Adventure of the Aeroplane 1799-1909* Faber 1965

Gibbs-Smith, C.H. *Sir George Cayley* H.M.S.O. 1962

Gillies & Wood *Aviation in Scotland* 1966

Goodall, M.H. *The Norman Thompson File* Air Britain 1995

Goodall, M.H. *The Wight Aircraft* Gentry Books 1973

Gould-Lee, A *The Flying Cathedral* Methuen 1965

Grahame-White *Company Catalogue* 1913

Green, G., *Bristol Aerospace since 1910* G. Green 1985

Hannah, J. *Avro* Key Publishing 1983

Harlin, E.A. & Jenks, G.A. Avro Aircraft Album Ian Allan 1984

Hare, P. *Royal Aircraft Factory* Putnam 1990

Hughes, J. *A Steep Turn to the Stars* Benevenagh Books 1991

Jackson, A.J. *Avro Aircraft since 1908* Putnam 1965

Jackson, A.J. *Blackburn Aircraft since 1909* Putnam 1968

Jackson, A.J. *De Havilland Aircraft since 1909* Putnam 1962

Jane, F.T. *All the World's Airships* Sampson Low Marston 1909

Jane, F.T. *All the World's Aircraft* Sampson Low Marston 1910/14

Jarrett, P. *Another Icarus* Smithsonian Institute 1987

King, H.F. *Aeromarine Origins* Putnam 1966

King, H.F. *Sopwith Aircraft 1912-1920* Putnam 1980

Lewis, P. *British Aircraft 1809-1914* Putnam 1962

Maxim, Sir Hiram *Artificial & Natural Flight* Whittaker1908

McKinty, A. *The Father of British Airships* Kimber 1972

Mowthorpe, C. *Battlebags* Alan Sutton 1995

Negus, G. & Staddon, T. *Aviation in Birmingham* Midland Counties Publications 1984

New, P.T. *The Solent Sky* *P.T.* New 1976

Oberhalzer, H. *Pioneers of Early Aviation in South Africa* National Museum Bloemfontein 1974

Oughton, J.D. *Bristol Aircraft Album* Ian Allan 1973

Penrose, H. *British Aviation- the Pioneer Years* Putnam 1967

Penrose, H. *An Ancient Air* Airlife 1988

Preston, J.M. *A Short History* 1978

Ransom, S. & Fairclough, R. *English Electric Aircraft & their Predecessors* Putnam 1987

Redman, R.N. *Yorkshire's Early Flying Days* Dalesman 1981

Robertson, B. *Sopwith-the Man and his Aircraft* Air Review 1970

Robinson, B.R. *Aviation in Manchester* Manchester Branch Royal Aeronautical Society 1977

Rolt, L.T.C., *The Aeronauts* Sutton 1985

Russell, D.A. The Book of Bristol Aircraft Harborough 1948

Sangar, R. *The Martinsyde File* Air Britain 1999

Shepherd, Dolly *When the 'Chute went up* Skyline 1996

Sloan, T *Early Aviation in North Wales* 1962
Sturtivant, R. & Page, G.R.N. *Aircraft - Serials & Units 1911-1919* Air Britain 1992
Tagg, A.E. *Power for the Pioneers* Crossprint 1990
Tagg, A.E. & Wheeler, R.L. *From Sea to Air* Crossprint 1989
Talbot-Weiss *The Centenary of Jose Weiss* 1959
Talbot-Weiss *Soaring Flight* 1923
Tapper, O. *Armstrong-Whitworth Aircraft since 1913* Putnam 1973
Taylot, J.W.R. *Aircraft Annual - 1966, 1969, 1970, 1974* Ian Allan
Turner, C.C. The Old Flying Days Sampson Low Marston 1920's
Ventry, Lord & Kolesnik, E.M. *Pocket Book 7 - Airship Development* Macdonald 1976
Walker, P.B. *Early Aviation at Farnborough Vol.I* Macdonald 1971
Walker, P.B. *Early Aviation at Farnborough Vol.II* Maconald1974
Wallace, G. *Claude Grahame-White* Putnam 1960
Warwick, A.R. *The Phoenix Suburb* Blue Boar Press 1982
Williams, P. *The James Brothers* Pembrokeshire Aviation Group 1992

Newspapers

Birmingham Despatch 1894
Birmingham Post 1914
Blackburn Times 1910
Blackburn Weekly Telegraph 1910
Bolton Evening News 1938
Bridgnorth Journal 1966
Burnley Express 1909/10
Burnley Gazette 1909
Cambridge Chronicle 1910, 1912
Cambridge Express 1906
Cambridge Independent News 1910
Cambridge Daily News 1910
Coventry Evening Telegraph 1959
Daily Graphic 1909
Daily Mail 1910
Daily Sketch 1910
Daily News 1909
Dartmouth & Brixham Chronicle 1894
Dartmouth & South Hams Chronicle 1894
East Coast Illustrated News 1909
Eastern Daily Press 1908/9
Eastern Morning News 1910
Eastbourne Gazette 1911
Essex Advertiser 1909, 1911
Evening Standard 1909
Halifax Evening Courier 1909
Hampshire Telegraph 1910
Harrow Gazette 1893
Keighley News 1910
Kentish Gazette & Canterbury Press 1910
Lowestoft Journal 1908/9
Lowestoft Weekly Press 1908/9
Liverpool Daily Post & Mercury 1910
Lloyd's Weekly News 1910
Manchester Evening News 1909
Mansfield & North Notts. Observer 1910
Newcastle Evening Sentinel 1961
Newcastle Weekly Chronicle 1909/10
Northern Mail 1909
Reading Observer 1909
Rossendale Free Press 1910
South Wales Graphic 1906
Staffordshire Sentinel 1910, 1931
The Times 1893, 1908, 1910
West Sussex Gazette & Advertiser 1954, 1969
Western Evening Herald 1956
Wigan Observer 1909/10, 1917

Periodicals

Aero 1909-12
Aerocraft 1908-10
Aeronaute 1894
Aeronautical Journal 1897-1914, 1968, 1970
Aeronautics 1907-14
Aerospace
Aeroplane 1911-14
Aeroplane Monthly 1976, 1984, 1987
Air Britain Digest 1984
Auto 1900-14
Autocar 1900-14
Automotor Journal 1900-10
Ballooning & Aeronautics 1907
Brooklands Society Gazette 1983-4
Car 1900-10
Car Illustrated 1900-10
Cross & Cockade International Journal 1982
Cycling 1910
Engineer 1902
Engineering 1885, 1893, 1908
English Mechanic & World of Science 1909
Flight 1909-14
Flying 19013
Flypast 1986
The Golden Penny 1898
Illustrated London News
Illustrated Weekly Mercury 1913
Mechanics Magazine 1852
The Mirror 1840
The Motor 1910
The Motor Boat 1909
Motor Car Journal 1911
Motor Cycling
Popular Flying 1933
Popular Science Review 1868-9
Schweizer Luftfahrt 1941-2
Scientific American
The Sketch 1894

Index

Y